D0984160

Handbook of
Solid Waste Disposal
Materials and Energy Recovery

Handbook of
Solid Waste Disposal
Materials and Energy Recovery

Joseph L. Pavoni

Vice-President
TenEch Environmental Consultants, Inc.
Louisville, Kentucky

John E. Heer, Jr.

Professor and Chairman, Civil Engineering
University of Louisville

D. Joseph Hagerty

Associate Professor of Civil Engineering
University of Louisville

VAN NOSTRAND REINHOLD ENVIRONMENTAL ENGINEERING SERIES

VAN NOSTRAND REINHOLD COMPANY

NEW YORK   CINCINNATI   ATLANTA   DALLAS   SAN FRANCISCO
LONDON   TORONTO   MELBOURNE

Van Nostrand Reinhold Company Regional Offices:
New York Cincinnati Chicago Millbrae Dallas

Van Nostrand Reinhold Company International Offices:
London   Toronto   Melbourne

Manufactured in the United States of America

Published by Van Nostrand Reinhold Company
450 West 33rd Street, New York, N.Y. 10001

Published simultaneously in Canada by Van Nostrand Reinhold Ltd.

15 14 13 12 11 10 9 8 7 6 5 4 3 2

Library of Congress Cataloging in Publication Data

Pavoni, Joseph L.
    Handbook of solid waste disposal.

    (Van Nostrand Reinhold environmental engineering
series)
    Includes bibliographies and index.
    1. Refuse and refuse disposal.   2. Recycling (Waste,
etc.) 3. Refuse as fuel. I. Heer, John E., joint
author. II. Hagerty, D. J., joint author. III. Title.
TD791.P3   628'.44   74-26777
ISBN 0-442-23027-3

This book is sincerely dedicated to Mr. & Mrs. Barry Bingham, Sr. and Mr. Cyrus MacKinnon in honor of their untiring efforts to improve the quality of life of their fellow citizens.

Van Nostrand Reinhold Environmental Engineering Series

THE VAN NOSTRAND REINHOLD ENVIRONMENTAL ENGINEERING SERIES is dedicated to the presentation of current and vital information relative to the engineering aspects of controlling man's physical environment. Systems and subsystems available to exercise control of both the indoor and outdoor environment continue to become more sophisticated and to involve a number of engineering disciplines. The aim of the series is to provide books which, though often concerned with the life cycle—design, installation, and operation and maintenance—of a specific system or subsystem, are complementary when viewed in their relationship to the total environment.

Books in the Van Nostrand Reinhold Environmental Engineering Series include ones concerned with the engineering of mechanical systems designed (1) to control the environment within structures, including those in which manufacturing processes are carried out, (2) to control the exterior environment through control of waste products expelled by inhabitants of structures and from manufacturing processes. The series will include books on heating, air conditioning and ventilation, control of air and water pollution, control of the acoustic environment, sanitary engineering and waste disposal, illumination, and piping systems for transporting media of all kinds.

Van Nostrand Reinhold Environmental Engineering Series

ADVANCED WASTEWATER TREATMENT, by Russell L. Culp and Gordon L. Culp

ARCHITECTURAL INTERIOR SYSTEMS—Lighting, Air Conditioning, Acoustics, John E. Flynn and Arthur W. Segil

SOLID WASTE MANAGEMENT, by D. Joseph Hagerty, Joseph L. Pavoni and John E. Heer, Jr.

THERMAL INSULATION, by John F. Malloy

AIR POLLUTION AND INDUSTRY, edited by Richard D. Ross

INDUSTRIAL WASTE DISPOSAL, edited by Richard D. Ross

MICROBIAL CONTAMINATION CONTROL FACILITIES, by Robert S. Runkle and G. Briggs Phillips

SOUND, NOISE, AND VIBRATION CONTROL, by Lyle F. Yerges

NEW CONCEPTS IN WATER PURIFICATION, by Gordon L. Culp and Russell L. Culp

HANDBOOK OF SOLID WASTE DISPOSAL: MATERIALS AND ENERGY RECOVERY, by Joseph L. Pavoni, John E. Heer, Jr., and D. Joseph Hagerty

Foreword

When Heracles cleaned out the Augean stables by diverting a river from its channel and flushing it through the stalls, the myth does not tell us what he did with the ordure. One hopes that he sent it coursing out over the fields of King Augeas, where, following the orderly rhythms of nature, it became a part of what Homer calls the "all-nourshing earth."

Today, the whole world, over-populated and heavily industrialized, is becoming an Augean stable. The composition of that waste, however, would have overwhelmed the ingenuity even of Heracles. For what could the hero have done with throw-away bottles and cans, bleach containers, spent facial tissues, old refrigerators, and cannibalized automobiles, in addition to human and feedlot wastes?

The sludge made from the garbage the scows of New York City have deposited by the billions of tons outside Ambrose Light has crept back toward the Long Island beaches. Our lakes and rivers are polluted with the detritus of our throw-away economy. Almost too late we are awakening to the nightmare possibility of our teeming world strangling in its own excrement and garbage.

I believe Professors Pavoni, Heer, and Hagerty have given us in their book a vision, rising out of an observed, already functioning technology, of an escape route out of this nightmare. Surveying the whole problem of waste disposal at home and abroad, they have gathered together for the first time all the information that exists about the various futile or promising methods which men employ to put waste out of sight and out of mind.

And to that innate sense of satisfaction which all of us harbor in the idea and practice of frugality, in the art of making something out of nothing, their solution is immensely welcome and heartening. Out of our mountains of rubbish and ordure we can make useful energy. In this exercise, we kill two huge, menacing birds with one stone. What could be more exhilarating?

Here is a meticulously researched, sober and optimistic answer to one of the world's most daunting problems. With the publication of this book, only one thing remains to do: remember the admonition of St. James (in a somewhat different context): "Be ye doers of the Word, and not hearers only."

MARY BINGHAM

Preface

No book comes to fruition through the efforts of the authors alone, and this manuscript is no exception. Many individuals have guided our efforts and lent us support from the conceptual development of this text through its various stages of editorial revision and production and, although we cannot acknowledge everyone individually, we extend to all our sincere appreciation. At the same time, several persons have given so generously of their time and talent that we would be remiss if we did not recognize them specifically.

Thanks must go to our colleague, John J. Reinhardt, of the Wisconsin Department of Natural Resources for contributing pictures of various solid waste disposal and resource recovery systems throughout the world, and for lending invaluable advice concerning the organization of our study of European solid waste management sites.

We extend deep appreciation to the following European solid waste management professionals who so graciously gave of their time and effort to acquaint us with the solid waste disposal techniques currently being utilized in their countries: Mr. P. K. Patrick of the Greater London Council in London, England; Mr. Pieter Houter of Vuilafvoer Maatschappig in Amsterdam, Holland; Mr. Jacques Sigwalt of Gondard, S.A., in Paris, France; Mr. Hans-Joachim Wallek of Zentraldeponie Emscherbruch in Essen, Germany; and Sig. Squatriti of the Municipality of Rome in Rome, Italy.

The authors wish to acknowledge the assistance of our students, Stephen M. Evans, Roger D. Setters, and Michael S. Evans, in data compilation, and the efforts expended by William H. Sloan and William C. Dues of the University of Louisville in developing illustrations for the text. Grateful thanks are also extended to Sharon Loehr, LeAnne Whitney, Angela Reed, and Judith Pavoni for secretarial assistance, proofreading, and preparation of the index.

JOSEPH L. PAVONI
JOHN E. HEER, JR.
D. JOSEPH HAGERTY

Contents

Introduction

In an age of mushrooming technology and scientific innovation, it is ironic that one of man's oldest problems is becoming increasingly acute. The collection and disposal of modern waste products is a monumental task. Moreover, technological change has produced affluent, throwaway societies in many of the industrialized countries of the world. It is becoming apparent that one of the major impacts of technological development is a significant increase in the generation of solid wastes.

Even a quick perusal of life in the United States today indicates that solid waste management is a national problem of considerable magnitude. The severity of this problem caused attention to be focused on the improvement of solid waste management techniques in the early 1960s. This interest spurred the passage of the Solid Waste Disposal Act of 1965 under which research funds became available for the first time from the federal government for basic investigations of solid waste generation and management. Consequently, data regarding waste generation and disposal were gathered during the next several years throughout the United States.

Fig. 1-1   The ever-increasing mountain of solid wastes—a disposal problem requiring an urgent solution (*courtesy* John J. Reinhardt).

In 1970, the Resources Recovery Act was passed by Congress; as a result, the focus of attention was shifted from simple disposal techniques to the development of technology necessary to recover and utilize materials being discarded as waste. During the first two years of this decade many new materials recovery systems were developed in response to this new approach. With the advent of the "energy crisis" in 1973, the focus of attention was again shifted to concentrate on energy recovery systems. Because of technical innovations developed as a result of these actions, and because of conflicting and ambiguous claims made for many of the new so-called "resource recovery" systems, it became apparent that a comprehensive national study of the state-of-the-art of solid waste disposal and resource recovery should be undertaken. This text is the result of a two year state-of-the-art study conducted by the authors at the University of Louisville.

To assess the state-of-the-art of refuse disposal and resource recovery, this particular study was structured to include a survey of scientific and technical literature, a comprehensive series of site visitations, and a number of consultative sessions with recognized experts in the solid waste management field both in the United States and abroad. This three-phase investigation was necessary because of the uncertain state of knowledge concerning many of the innovative disposal and resource recovery methods. Nurtured by the national wave of popular enthusiasm for resource recovery, many schemes and methods have been developed to capitalize on apparently available public funding and support. Almost without exception, these methods have not been objectively and comprehensively evaluated by independent investigators. In many cases, manufacturers' promotional literature is the only source of information available on these new recycling systems. Thus, a major purpose of this study was to establish the feasibility, limitations, and advantages of recently developed methods for materials and energy recovery.

The study on which this text is based was limited to the consideration of disposal techniques and recovery methods. Collection methods were not evaluated for the following reasons:

1. Collection methods as yet have only a secondary influence on resource recovery.
2. Few innovations have been made, to date, in conventional collection techniques.
3. Evaluations of existing collection methods have been accomplished in previous studies by other investigators.

The major emphasis in the evaluation of the systems described in this volume was placed on practicable systems. More time and effort were expended in the study of techniques currently operational or in pilot-plant status than in evaluation of the myriad of systems existent only in designers' after-dinner speeches or preliminary drawings.

The first portion of this book is limited to consideration of systems developed in the United States or widely used in this country. The last portion of the volume is devoted to an examination of European methods and practices.

The information in the following chapters constitutes a state-of-the-art assessment designed to be of primary utility to professionals in the field of solid wastes management and students at an advanced level. For purposes of clarity the presentation has been divided into five main topical areas: conventional disposal operations, innovations in disposal techniques, materials recovery methods, energy recovery systems, and European practice.

1-1. CONVENTIONAL DISPOSAL OPERATIONS

Refuse disposal has been conventionally carried out according to one of three basic methods: sanitary landfill, incineration, or composting. Sanitary landfilling, permanently placing refuse under maximum density in the earth with daily cover, is the predominant method wherever sufficient land is available at low cost near the sources of waste generation.

Where land is not available at economical prices within reasonable distances from the centers of refuse generation, central incineration presents distinct advantages. In fact, because of the small land requirements and apparent weight (and volume) reduction possible with incineration, municipal incineration has been the usual method of refuse disposal in most large cities for several decades.

Despite the ever more stringent air quality standards which are being applied to incinerator effluents and the consequent increasing control costs, incineration still must be considered feasible and economically attractive in many situations. Composting, the aerobic degradation of waste, is not practiced to any significant degree in the United States because of the relative availability of arable land and the abundance of inexpensive fertilizers in North America. With impetus for resource recovery coming from all strata of society, composting may be attempted with more frequency in the future. Certainly aerobic decomposition of degradable waste constituents could be a valuable adjunct to other methods of refuse management.

These conventional disposal techniques are analyzed in-depth in the text since they constitute the bulk of solid waste management methods employed today.

1-2. INNOVATIONS IN DISPOSAL TECHNIQUES

The conventional methods of solid waste disposal have been modified and altered in innumerable ways in individual installations, but only two major innovations appear to offer significant promise. In the first of these innovations, refuse incineration has been modified through combustion controls such as the use of preheated air, the addition of auxiliary fuel, and the modification of combustion chambers to achieve a more complete volume reduction of wastes at much higher operating temperatures. This method has only been implemented on a pilot plant scale to date (1974).

A second major innovation in solid waste disposal, landfill with leachate recirculation, has evolved from the development of leachate management techniques to improve the sanitary landfilling process. The degradation of wastes is accelerated in the landfill during the leachate recirculation process by adding water to the deposited refuse, collecting the generated leachate, adding any necessary conditioners to maximize degradation, and pumping the modified leachate back into the fill. This innovative method also has only been tested in a pilot plant installation to date (1974).

These innovative disposal methods are fully examined in the text to ascertain both the advantages and disadvantages inherent in the operation of these modified disposal techniques.

1-3. MATERIAL RECOVERY OPERATIONS

Materials recovery connotes for many persons in the general public the collection of waste paper "to save trees" and other similar operations which may be simplistic in concept if not impracticable or uneconomical in operation. At the present time, tremendous amounts of materials are recycled on-site or in-house.

For example, industrial scrap is often returned directly to the manufacturing process rather than the waste pile and agricultural residues may be returned to the soil or used for food, raw material, etc. Although vast quantities of industrial process scrap are recycled in-house each year, a large percentage of other goods are wasted and, consequently, there is considerable room for improvement in the recovery of certain types of manufacturing wastes.

Recovery of the materials in mixed municipal wastes appears to be an obvious solution to growing shortages of resources. However, recovery of a potentially valuable material from the solid waste stream is much more difficult than the recovery of homogeneous, uncontaminated manufacturing scrap. Mixing of the constituents in municipal refuse renders some portions of the wastes practically unuseable and, consequently, salvaging these wastes requires expenditures of power and labor in subsequent sorting and separation operations.

Several integrated materials recovery systems have been developed during the last several years to recover materials from mixed municipal refuse including the U.S. Bureau of Mines systems, the wet-pulping method, a combined composting-pyrolysis process, and a new proposed network of materials recovery systems located throughout the United States. These integrated materials recovery systems are described and evaluated.

1-4.   ENERGY RECOVERY SYSTEMS

Energy recovery systems are designed to achieve resource conservation through the retrieval of the energy contained in the combustible fraction of the solid waste stream. Much of the combustible portion of solid wastes consists of replaceable or renewable energy-rich materials. Utilization of this energy source in turn decreases the rate of depletion of irreplaceable resources such as coal or oil. This concept is neither unique nor novel since heat recovery has been practiced in European incinerators for more than 30 years. Present developments in the United States include: the installation of the water-tube walls in new heat-recovery incinerators and in the secondary combustion chambers of existing units, the removal and recovery of noncombustibles followed by the burning of the combustible portion of refuse in existing power station boilers, the development of special energy-recovery units designed to operate specifically on refuse, the formulation of pyrolysis systems which are capable of converting refuse to oils, gas, and carbonaceous char, and an anaerobic fermentation process developed to produce natural gas (methane) from solid waste.

With the increasing depletion of currently available energy sources, these new energy recovery systems will probably move to the forefront of solid waste management practice during the next decade. Consequently, each of these systems is analyzed completely and predictions are made concerning their potential for adoption.

1-5. EUROPEAN PRACTICE

The United States' solid waste management practice is compared to European techniques in the final section of the text. This discussion is based on a survey of solid waste disposal and resource recovery facilities in six European countries by the authors in late 1973. Economic, operational, and technical data are presented regarding the following European operations:

1. Power production through waste heat utilization.
2. Composting with front-end and back-end separation.
3. Landfilling to provide recreational areas.
4. Transfer station operations.
5. Animal food production from solid wastes.
6. Paper pulp recovery from solid wastes.

These European solid waste management practices are evaluated for their overall applicability in the United States and recommendations are presented which focus upon the improvement of solid waste disposal and resource recovery system implementation in this country.

1-6. SUMMARY

Refuse disposal and resource recovery methods are undergoing a period of alteration and modification, with rapid development of many small-scale operations for which expansive claims are being made. A comprehensive and critical examination of the state-of-the-art of refuse disposal and resource recovery methodology is urgently needed. The information presented in the following chapters is intended to fulfill, to some extent, this need.

2

Composting

Composting is the biochemical degradation of the organic fraction of solid waste material, having as its end product a humus-like substance that is used, primarily, for soil conditioning. Composting has been practiced for many years in the Orient by placing vegetable matter and animal manures in piles and allowing decomposition to proceed. Attempts to systematize the composting process began in the 1920s when Sir Albert Howard developed the Indore process in India and Becarri patented his process in Italy. The Indore process involved the anaerobic degradation of leaves, garbage, and animal manures for 6 months in soil pits (Ref. 2-40). This process was later modified to include turning the compost during degradation to encourage aerobic digestion to occur. The Becarri process was quite similar, except that it utilized concrete cells as digestion chambers (Ref. 2-15).

There has been a considerable amount of information accumulated since the 1920s concerning the practice of refuse composting in Europe. However, European practice and experience are not directly applicable to the refuse disposal situation in the United States because there is a significant difference in refuse composition and the demand for a soil conditioner. For example, there is more paper in American refuse. In addition, there is an abundance of fertile land and synthetic fertilizers, and most farmers have a problem trying to dispose of tons of manure (a natural soil conditioner). Consequently, there is little demand for compost in this country. It should be noted that to operate economically, any compost plant must be able to sell its finished product. The lack of adequate markets, coupled with the fact that the composting process is quite expensive when compared with other solid waste disposal methods, has resulted in failure for most composting efforts in the United States.

2-1. MICROBIOLOGY OF THE COMPOSTING PROCESS

Composting entails the degradation of organic compounds by naturally occurring microbes. The key to commercial composting is to provide an environment in which the microorganisms necessary for decomposition can perform most efficiently, thereby reducing the time required for stabilization to occur. There are a variety of microorganisms involved in, and necessary to, the composting process. During the course of composting, both qualitative and quantitative changes occur in the active microflora. Some species are more active in the initial stages of composting, changing the environment so that other populations may succeed them and continue another step in the process. One of the most important process parameters used to determine the type of microflora present is temperature.

During initial composting development, the mesophilic flora (organisms able to grow in the 77 to 113°F range) predominate and are responsible for most of the metabolic activity that occurs. This increased microbial activity elevates the temperature of the compost, with the subsequent replacement of mesophilic populations by thermophilic flora (which metabolize optimally at temperatures above 113°F). This change in temperature is controlled to a great extent by the amount of oxygen available; that is, the amount of aeration provided. Therefore, an optimum environment must include efficient ventilation for the aerobic microorganisms to survive. Because of the above-mentioned phenomenon, the temperature of the composting mass is normally an indication of the amount of biochemical activity taking place. A drop in temperature usually means that the material needs to be aerated or moistened, or that decomposition is almost completed (see Fig. 2-1). In windrow type composters, temperatures of 170°F have been recorded in the center of the mass, with 140 to 160°F temperatures maintained throughout the mass for up to 3 weeks (Ref. 2-54). In compost digesters where there is forced aeration but no agitation, temperatures of 180°F have been reported. In general, temperatures of 140 to 160°F should be maintained for adequate digestion (Ref. 2-35).

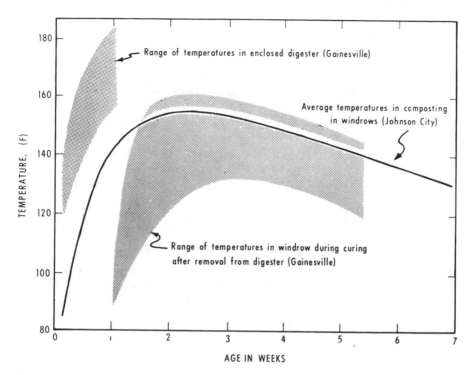

Fig. 2-1 Typical temperature profiles for composting municipal refuse. *Source:* Breidenbach, Andrew W., "Composting of Municipal Solid Wastes in the United States," U.S. Environmental Protection Agency Publ. No. SW-47r, Govt. Printing Office, 1971.

Another important process parameter is the nutrient content available to the microorganisms for growth, which is generally measured by the carbon to nitrogen and carbon to phosphorus ratios within the refuse. After separation and grinding, refuse may contain only 0.5 to 0.9 percent nitrogen. A low nitrogen content requires that the microorganisms metabolizing during the decomposition process recycle this nitrogen through many generations (gradually building up the percentage of nitrogen) as carbonaceous materials decompose through aerobic decomposition to carbon dioxide and water. This process of regeneration is extremely slow and will greatly increase the time required for digestion. Generally, in windrow composting a digestion period of 9 to 12 days will be required if the initial C to N ratio is 20; 10 to 16 days if the C to N ratio is 30 to 50; and 21 days if the C to N ratio is 78 (Ref. 2-18). Composting will generally not take place in an environment in which the C to N ratio is greater than 80. Phosphorus is the other micronutrient in the composting process. The carbon to phosphorus ratio should be 100 to 1 to assure proper microbial growth and digestion (Ref.

2-35). Solid waste streams usually contain a sufficient amount of phosphorus for microbial growth.

To achieve the optimal decomposition rate, the water content of the compost is also important and should be between 50 and 60 percent by wet weight. If the moisture content is in excess of 60 percent, the compost becomes more compact, thereby reducing the amount of air present. This can lead to the development of anaerobic conditions. However, if a sufficient amount of water is not available, the temperature of the entire compost mass is lowered (even though localized high temperatures may be maintained in the center), thereby extending the degradation time.

A final parameter which is important in evaluating the microbial environment is the pH of the refuse. As in the case of temperature, the pH of the compost varies with time during the composting process and is a good indicator of the extent of decomposition within the compost mass (see Fig. 2-2). The initial pH of solid waste is between 5.0 and 7.0 for refuse which is about 3 days old. In the first 2 or 3 days of composting the pH drops to 5.0 or less and then begins to rise to about 8.5 for the remainder of the aerobic process. If the digestion is allowed to become anaerobic, the pH will drop to about 4.5 (Ref. 2-24).

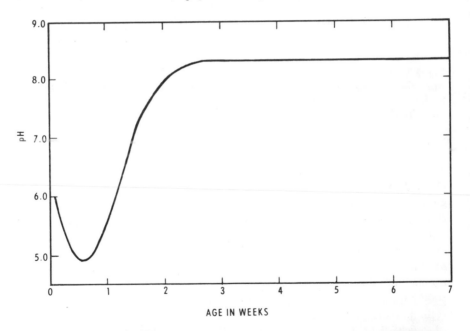

AGE IN WEEKS

Fig. 2-2 Typical pH profile for composting municipal refuse. *Source:* Breidenbach, Andrew W., "Composting of Municipal Solid Wastes in the United States," U.S. Environmental Protection Agency Publ. No. SW-47r, Govt. Printing Office, 1971.

An important facet in the composting process is the determination of the point at which digestion of the solid waste has been completed. Generally, satisfactory stabilization is attained when the compost has the characteristics of humus, has no unpleasant odor, high temperatures are not maintained even though aerobic conditions and desirable moisture content exists, and the carbon to nitrogen ratio is such that the humus can be applied to the soil (if the C to N ratio is too high, the compost will remove nitrogen from the soil). To date, most methods used determine chemically when the digestion period has been completed by measuring the reduction in total carbon, cellulose, and lipids and the increase in percent ash. These methods are adequate for determining the completeness of composting on the assumption that a representative sample has been taken. All have, however, obvious disadvantages such as complicated sample preparation, considerable time, and expensive equipment.

A practical alternative to these methods is the starch–iodine method suggested by Richard D. Lossin of the Bureau of Solid Waste Management (Ref. 2-48). The maximum rate of starch degradation in compost occurs when optimal compost stabilization parameters (temperature, moisture content, pH, etc.) are maintained. Therefore, the amount of starch found in compost decreases with increasing compost age under proper operating procedures. This phenomenon, coupled with the fact that starches form characteristic color complexes when combined with molecular iodine, forms the basis for the starch–iodine method. In this method, a sample of the compost is mixed with perchloric acid and then the mixture is filtered through open-texture filter paper. Two drops of iodine are then added to the filtrate. If the filtrate becomes yellow with very little precipitate then the compost is finished. Poor or unfinished compost will give a dark blue color and a heavy blue precipitate. This method is quick, simple, and inexpensive and yields reasonably dependable results.

2-2. HISTORY OF COMPOSTING IN THE UNITED STATES

Composting was introduced to the United States as a solid waste disposal process with the development of the Frazer-Eweson process in 1949. Since that time, at least 16 different types of composting processes have been practiced in this country (see Table 2-1). During the 1950s, basic research studies on composting were conducted at the University of California, Michigan State University, and by the United States Public Health Service. In 1956, the World Health Organization published a comprehensive monograph entitled, *Composting and Sanitary Disposal and Reclamation of Organic Wastes.* The most important accomplishments of these initial studies included:

1. The fundamentals of aerobic composting and the necessary operating parameters were established for municipal refuse.

TABLE 2-1   Municipal Solid Waste Composting Plants in the
United States (1969)

Location	Company	Process	Capacity (Ton/ Day)	Type Waste	Began Operating
Altoona, Pennsylvania	Altoona FAM, Inc.	Fairfield-Hardy	45	Garbage, paper	1951
Boulder, Colorado	Harry Gorby	Windrow	100	Mixed refuse	1965
Gainesville, Florida	Gainesville Municipal Waste Conversion Authority	MetroWaste conversion	150	Mixed refuse, digested sludge	1968
Houston, Texas	Metropolitan Waste Conversion Corp.	MetroWaste conversion	360	Mixed refuse, raw sludge	1966
Houston, Texas	United Compost Services, Inc.	Snell	300	Mixed refuse	1966
Johnson City, Tennessee	Joint USPHS–TVA	Windrow	52	Mixed refuse, raw sludge	1967
Largo, Florida	Peninsular Organics, Inc.	MetroWaste conversion	50	Mixed refuse, digested sludge	1963
Norman, Oklahoma	International Disposal Corp.	Naturizer	35	Mixed refuse	1959
Mobile, Alabama	City of Mobile	Windrow	300	Mixed refuse, digested sludge	1966
New York, New York	Ecology, Inc.	Varro	150	Mixed refuse	1971
Phoenix, Arizona	Arizona Biochemical Co.	Dano	300	Mixed refuse	1963
Sacramento Co., California	Dano of America, Inc.	Dano	40	Mixed refuse	1956
San Fernando, California	International Disposal Corp.	Naturizer	70	Mixed refuse	1963
San Juan, Puerto Rico	Fairfield Engineering Co.	Fairfield-Hardy	150	Mixed refuse	1969
Springfield, Massachusetts	Springfield Organic Fertilizer Co.	Frazer-Eweson	20	Garbage	1954 1961
St. Petersburg, Florida	Westinghouse Corp.	Naturizer	105	Mixed refuse	1966
Williamston, Michigan	City of Williamston	Riker	4	Garbage, raw sludge, corn cobs	1955
Wilmington, Ohio	Good Riddance, Inc.	Windrow	20	Mixed refuse	1963

Source: Breidenbach, Andrew W., "Composting of Municipal Solid Wastes in the United States," U.S. Environmental Protection Agency Publ. No. SW-47r, Govt. Printing Office, 1971.

2. Inoculums were shown to be unnecessary and those offered for sale to be worthless.
3. Simple turning was shown to be an effective way to maintain needed aeration.
4. Aerobic composting was shown to be a semi-aerobic process requiring far less aeration than the term implied.
5. A composting mass permitted to become anaerobic was found to resume aerobic decomposition if aerated again by turning.
6. The process was found to continue to completion without further aeration after a relatively short period of composting under aerobic conditions.
7. Fly control and destruction of disease vectors were found to occur in the process.
8. Finished compost was found to be a low-grade fertilizer, more valuable for its soil-conditioning and moisture-retaining properties than for its content of nitrogen and phosphorus.
9. The cost of various composting systems was postulated, but it was evident that the process was not destined to be a source of income to cities.
10. Serious doubts were cast upon the marketability of any large amount of compost.

Wide interest in composting in the United States was brought about by several factors. First, the wartime growth of small towns into fair-sized cities overwhelmed the primitive system of open dumps, overran and obliterated the hog farm, and introduced the air pollution factor into consideration because of the need to burn increasing amounts of refuse. Next, an increase in the standard of living resulted in a significant increase in the volume of urban-generated solid waste, thereby compounding the problem of disposal. Finally, opportunistic entrepreneurs, conservationists, and food faddists combined forces to promote the use of composting as a major method of solid waste disposal within the United States.

Early attempts at commercial composting offered little to encourage those who supported the process. Operational experience showed that it was difficult to separate the noncompostable refuse economically; that initial shredding was expensive and destructive to the shredding equipment; that construction and operational costs for composting were high; and that a market capable of absorbing large quantities of compost did not exist in the United States. It was also discovered, however, that one of the major drawbacks in windrow composting, excessive digestion time, could be overcome through the use of mechanical digesters.

In an effort to learn more about the feasibility of composting, the United States Public Health Service, the Tennessee Valley Authority, and the municipality of Johnson City, Tennessee entered into an agreement to undertake a joint research and demonstration project in solid waste and wastewater sludge composting in February, 1966. A similar project was initiated under a USPHS grant to the Gainesville Municipal Waste Conversion Authority in Gainesville, Florida in 1968.

The Johnson City plant utilized the windrow process and was developed to explore health hazards associated with compost, critical parameters in the composting process, the economics of composting, and the effects of the application of compost to agricultural lands. The Gainesville study had similar aims; however, it employed the MetroWaste conversion process and had a capacity of 150 tons per day. A detailed analysis of these operations can be found in section 2-9 which deals with case histories. Because of an inability to generate adequate financing after the termination of the USPHS grants, both of these projects have been forced to cease operation.

Past experience in composting in the United States has shown conclusively that composting is technically feasible and highly desirable as a major method of solid waste disposal. However, since a market for compost does not currently exist in this country, the composting process cannot compete economically with other solid waste disposal methods.

2-3. COMPOSTING METHODOLOGY

There are five basic steps involved in all composting practices: 1) preparation, 2) digestion, 3) curing, 4) finishing, and 5) storage or disposal. Differences between various composting processes may occur in the method of digestion or in the amount of preparation and finishing required. In choosing the type of process to be used and the amount of sophistication required, a number of criteria must be considered. The following discussion will describe a "typical" composting operation and suggest how variable operational criteria could affect the design and operation of a composting plant.

2-3.1 Preparation

The composting preparation process may consist of several steps, depending upon the sophistication of the plant and the amount of resource recovery practiced. Typical preparation steps may include the sorting of salvable material, the removal of noncombustibles, shredding, pulping, grinding, and the addition of wastewater sludge. Most plants also utilize receiving equipment which provide a steady flow of solid waste through the operation. Consistency of flow is accomplished by the use of storage hoppers and a regulated conveyor system.

After the solid waste leaves the receiving area, the bulky items which could damage the grinders are removed by hand. The separation of other noncompostable material and salvables such as glass, metals, rags, plastics, rubber, and paper may be accomplished before or after comminution by either hand or mechanical methods where feasible. (See Chapter 6 which describes the separation of salvable materials from solid wastes.) Those noncompostables which are not salvaged must be ultimately disposed of in a sanitary landfill.

After initial separation, most composting processes require that the solid waste material be reduced in particle size to facilitate handling, digestion, and the mixing of the end product to provide a more homogeneous compost. The three major methods of comminution which have been utilized in composting processes, listed in the order of popularity, include: the hammermill, the rasper, and the wet pulper.

Hammermills consist of high speed swing hammers connected symmetrically to a horizontal shaft and cutter bars which have grate openings through which the refuse is poured. Hammermills have a high power requirement (25 hp/ton/hr) and operate at 1200 to 3500 rpm. Therefore, they are expensive to operate and produce a significant amount of noise and vibration.

Specialized shredders, called raspers, may also be used to reduce the solid waste particle size. In these systems, solid waste material is introduced into a large vertical cylinder that surrounds a vertical shaft on which heavy arms are mounted. These arms rotate over a perforated floor, shredding the refuse and

Fig. 2-3   Rasper utilized in composting plant in Arnhem, Holland.

allowing the smaller particles to drop through the perforations. Raspers operate more slowly than hammermills and require less power; however, their initial cost and space requirements are greater.

Wet pulpers can also be used to comminute refuse. In pulping operations, the slurry produced must be dewatered to a water content of 40 to 50 percent before digestion. Wet pulpers have met with limited application in composting systems in this country.

In a discussion of comminution methods in composting, it is beneficial to note that the preparation of the solid waste constitutes a large percentage of the power requirements for the process. Some estimates place the power requirement of shredding as high as 90 percent of the total process power requirement. Also, various grinders may experience difficulty reducing materials such as cardboard boxes, metals, tough plastics, and other bulky items. Therefore, predigestion grinding may result in frequent clogging and breakdown with concurrent maintenance and possible shutdown. For these reasons, the initial grinding process may be very expensive. However, this grinding expense could be reduced through improved efficiency in the digestion process which would result in a more rapid decomposition of materials in bulk, thereby eliminating the need for initial grinding. It is interesting to note that composting operations in Holland do not include initial grinding and still produce a high quality end product.

Since refuse characteristics are variable from one load to the next, a final step in the preparation phase of composting may be to adjust the moisture and nitrogen content of the solid waste to be composted. Optimal composting process moisture content ranges from 45 to 55 percent of wet weight, while the optimal carbon to nitrogen ratio should be below 30. The majority of solid domestic waste in the United States cannot meet these criteria, especially for nitrogen content, and, therefore, would require some adjustment before composting. Moisture and nutrient adjustments can be accomplished most efficiently through the addition of raw wastewater sludge. The addition of sludge also provides for a sanitary disposal of the sludge itself, resulting in substantial savings in the cost of wastewater treatment, since sludge disposal costs frequently approach 50 percent of the total cost of wastewater treatment. The combined processing of the refuse and wastewater sludge in a composting operation should also result in a significant savings in the construction and operational costs of the combined plant over the cost of separate treatment and disposal facilities. The greatest savings obviously could be realized in a community where a new wastewater treatment facility is being planned and where construction of composting and wastewater treatment plants could be accomplished on adjoining sites. The conventional sludge digestion and drying operations could be eliminated in the wastewater treatment system with the raw sludge simply thickened and pumped to the composting plant. The thickening of the sludge to approximately 88 percent moisture content can be obtained rather inexpensively by gravity filtration

Fig. 2-4 The combined processing of refuse and wastewater sludge is accomplished in this composting plant in Schaffhausen, Switzerland (*courtesy* Klaranlage Verbandes, Schaffhausen).

through cloth. For a slight increase in the net cost of composting, a considerable savings in the wastewater treatment cost can be realized through the processing of the sludge with the refuse. In such a process the composted material will be increased in volume by only 6–10 percent, while the addition of sludge will accelerate the composting operation and produce a better final product in terms of nutrient contents. Therefore, in addition to the savings in the wastewater treatment plant operation, a greater market value of the final end product will result.

2-3.2 Digestion

Digestion techniques are the most unique feature of the various composting processes and may vary from the backyard compost heap to the highly controlled mechanical digester. Basically, composting systems fall into two categories: windrow composting in open windrows or mechanical composting in enclosed digesters.

In windrow composting, the refuse is ground and moistened, and then conveyed to an outdoor decomposition area where it is placed in windrows. The windrows are turned once or twice each week during the composting period of approximately 5 weeks to provide for aeration and mixing. This turning can be accomplished with either a front-end loader or a clamshell bucket on a crane. Machines with a shoveling or screw arrangement are also used in turning operations. These turning machines are designed to pick up the solid waste material from a belt and place it on the ground. Another type of turning machine, with a rotating drum on which teeth are mounted, straddles the windrow and turns the solid waste in place.

Windrow composting has been practiced in many locations throughout the United States, including the joint USPHS–TVA project at Johnson City, Tennessee (see Fig. 2-5). However, windrow composting has many disadvantages which limit its application including; 1) the operation is affected by the local climate since the composting refuse is exposed to the atmosphere, 2) odors from the exposed compost may be difficult to contain and control, 3) the retention time is excessive (30–90 days), and 4) the amount of land required is excessive (a city of 20 thousand would require approximately 55–60 acres of land.) Therefore, this type of operation is only suitable for small cities having an adequate land area and a suitable climate. In an effort to improve the applicability of composting to metropolitan areas, mechanical digesters have become the mainstay of composting systems in the United States.

Because digestion in a mechanical system takes place inside an enclosed digester, mechanical systems have a definite advantage over windrow systems. Enclosed digesters present none of the problems listed above for windrow systems and, therefore, appear to be compatible with the operation of any large metropolitan area. Some typical digesters include the Fairfield system, the Metro-Waste system, the Naturizer system (International Disposal Corporation), and the Varro system.

A composting plant using the Fairfield system (see Fig. 2-6) has been in operation for many years in the city of Altoona, Pennsylvania. This plant has composted Altoona's municipal refuse continuously since 1960 at an average rate of approximately 25 tons per day. This plant has sold every pound of compost produced during its history of operation.

The input to the plant is separated municipal refuse, usually high in moisture

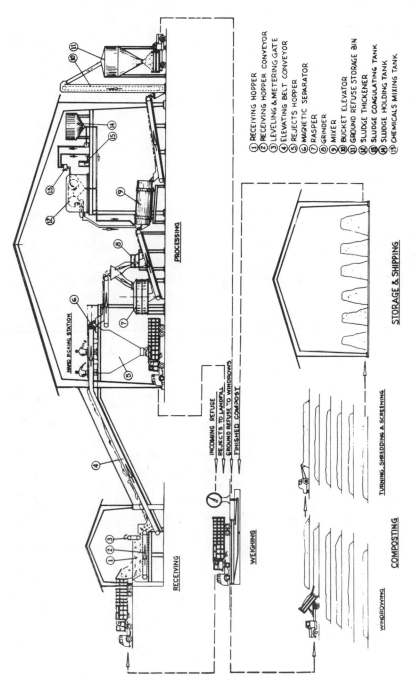

1 RECEIVING HOPPER
2 RECEIVING HOPPER CONVEYOR
3 LEVELING & METERING GATE
4 ELEVATING BELT CONVEYOR
5 REJECTS HOPPER
6 MAGNETIC SEPARATOR
7 RASPER
8 GRINDER
9 MIXER
10 BUCKET ELEVATOR
11 GROUND REFUSE STORAGE BIN
12 SLUDGE THICKENER
13 SLUDGE COAGULATING TANK
14 SLUDGE HOLDING TANK
15 CHEMICALS MIXING TANK

RECEIVING

HAND PICKING STATION

PROCESSING

WEIGHING

INCOMING REFUSE
REJECTS TO LANDFILL
GROUND REFUSE TO WINDROWS
FINISHED COMPOST

STORAGE & SHIPPING

WINDROWING

COMPOSTING

TURNING, SHREDDING & SCREENING

Fig. 2-5 USPHS–TVA compost plant, Johnson City, Tennessee. *Source:* Wiley, J. S., Gartrell, F. E. and Smith, H., "Concept and Design of the Joint USPHS–TVA Composting Project, Johnson City, Tennessee," *Compost Science,* Vol. 7(2), Autumn, 1966, pp. 11–14.

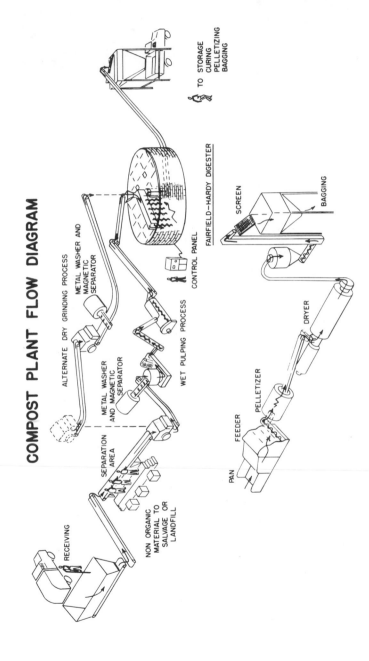

Fig. 2-6 Fairfield-Hardy compost system. *Source:* Drobny, N. L., Hull, H. E. and Testin, R. F., "Recovery and Utilization of Municipal Solid Waste," USPHS Publ. No. 1908, Govt. Printing Office, 1971, pp. 55–65.

content. Because of Altoona's refuse separation policy (separation of rubbish from garbage), the inorganic content of the plant's input stream is relatively low. Trucks dump the solid waste directly into a concrete chute which feeds a hammermill (see Fig. 2-7). The shredded refuse is then conveyed to a ferrous metal separator and the remaining fraction undergoes secondary sizing and water addition. Water is added in an amount sufficient to maintain an optimum solid waste moisture content of 52 to 54 percent. The solid waste moisture content in this facility is never allowed to be less than 48 percent or more than 55 percent. An important point to note concerning the Altoona operation is that no nitrogen or sludge is currently added during the composting process.

The solid waste is next conveyed to the digester which is 37 ft in diameter and 10 ft deep and usually contains 3 to 8 ft of solid waste. The solid waste enters the digester on its perimeter and is continuously mixed during which time it is slowly moved to the center discharge point. Oxygen and temperature probes are located throughout the digester in each of its three digestion zones which read out temperature and oxygen profiles of the digester every 6 minutes. This information in turn controls the air feed to the digester to maintain the temperature-oxygen parameters at desired levels. The oxygen content in the digester is maintained between 12 and 16 percent. The digester's solid waste temperature at the inlet is maintained near 150°F, whereas the temperature at the outlet following digestion is controlled between 160 and 165°F. The pH of solid waste entering the digester averages about 5.5, however, the digested solid waste pH is naturally

Fig. 2-7 Loading of refuse into concrete feed chute at composting plant in Altoona, Pennsylvania.

elevated to approximately 5.9. The average detention time in the digester is 4 to 5 days which is a sufficient detention time for complete pathogen kill. The detention time may be lowered if the plant output is to be utilized as poultry or swine feed. The lower detention period is necessary to yield a high organic content compost which is a good feed supplement.

The digested compost is then conveyed from the digester to a vibrating screen which removes plastics. The remaining solid waste stream is then dried to reduce the moisture content to between 18 and 22 percent and stored. The solid waste then passes through density separators (which remove remaining glass), screens, and is then packaged for shipment.

This process flow is adjusted depending upon the anticipated use of the final end product. For example, if compost is to be used solely for land application, the process ends with digester output.

The output from the Altoona plant is either sold to various parties including fertilizer plants, mushroom plants, and research projects, or stockpiled for shipment to projects which require compost having advanced degradation properties. Stockpiled compost is currently being shipped to the Atlas Peat and Soil Company in West Palm Beach, Florida for use as a soil mixer on golf courses and fruit and vegetable farms.

Of the 25 tons of solid waste processed per day at the plant, only about 32 to 34 percent ends up as compost. The other 66 to 68 percent of the solid waste

Fig. 2-8 Stockpiled compost awaiting shipment at composting plant in Altoona, Pennsylvania.

stream consists of:

1. Ferrous metals which are reclaimed and sold.
2. Glass which is currently used on roads in the area. (The economics of glass recovery for reuse is so poor that no plans are being considered for glass recovery for reuse as glass.)
3. Rags, plastics, and nonferrous metals which are not salvaged.
4. Miscellaneous screenings from the various plant processes.

All the nonsalvaged material is sent to an existing dump across the street from the plant. The Altoona plant is operated by 5 men and a supervisor. Solid waste is delivered to the plant each morning by city sanitation trucks, and the entire input to the composting plant is usually processed by 3:00 pm. The plant operates 5 days per week.

Fairfield-Hardy has proposed (1973) a contract with Altoona to operate a new 150 ton per day composting plant with a minimum contract period of 10 to 20 years. The new plant would accept mixed refuse and process or reclaim approximately 83 to 85 percent of all input solid waste. The new plant would be operated by a total of 19 men and was expected to be operational in November of 1974.

A typical MetroWaste system is the 150 ton per day plant built in Gainesville, Florida with a USPHS grant in 1968 (see Fig. 2-9). In this system, ground and separated solid waste is moistened with either wastewater sludge or a nitrogen-rich solution. This moistened solid waste material is then fed into batch digesters consisting of horizontal tanks with perforated bottoms. These tanks are 20 ft wide, 10 ft deep, 200 to 400 ft long and equipped to impart one or two turnings to the solid waste during the digestion period. Air is forced through the perforations in the tank bottom to provide adequate aeration. The ground solid waste is held in the digester from 4 to 6 days, depending upon the operating conditions at the plant. For convenience, a center belt serves both tanks for feed and take-off; therefore, one agitator can be used for both tanks with periodic shifting of the agitator from one tank to the other. The advantage of this system is that the capacity of the system can be increased by simply adding to the digestion tank while retaining and using the same agitator.

Another system is the Naturizer process, adapted for use in the International Disposal System (see Fig. 2-10). In this system, the separated and shredded refuse is fed into a mixer called a pulverator and a moistening agent (an ammonium nitrate solution) is added. After being moistened, the solid waste material enters a flail mill grinder which shreds the refuse but does not remove or shred plastic items or rags. These items, therefore, enter the composting process virtually intact. The heart of the system is a plug-flow digester which is contained in a vertical building. Within the building are five 9 ft-wide steel conveyor belts arranged to pass material from one belt to the other. Each belt is an insulated

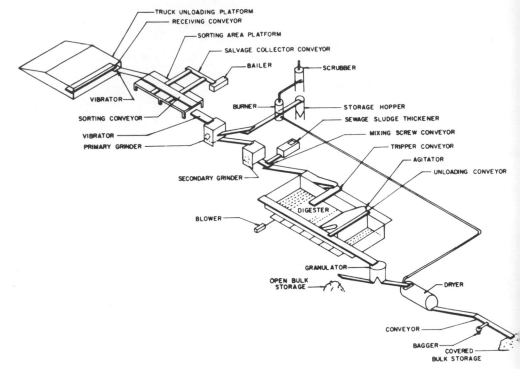

Fig. 2-9   Gainesville Municipal Waste Conversion Authority, Inc. compost plant, Gainesville, Florida. *Source:* Harding, C. I., "Recycling and Utilization," in *Proceedings;* the Surgeon General's Conference on Solid Waste Management for Metropolitan Washington, July 19–20, 1967, USPHS Publ. No. 1729, Govt. Printing Office, 1967, pp. 105–119.

cell and air passes upward through the digester to provide aeration. After 2 days of processing, the material is reground and reinserted into the compost conveyor system. At the termination of the 5-day detention time, the composted material is removed from the digester and passed through a $^3/_4$ in. opening screen which separates nonshredded noncompostable materials from the compost. This process has the disadvantage of requiring the construction of an entirely new digester if the system capacity is to be increased.

The Varro system represents the newest and most sophisticated of the mechanical composting systems (see Fig. 2-11). In this process, separated and shredded solid waste is pulped and placed in storage tanks. This stored pulp is then conveyed to the top of the digester which consists of an eight-step enclosed box in which stationary layers of compost, about 12 in. deep, are moved along the belts by chain-driven rakes. Throughout the process, the temperature, aeration, acidity, and moisture content are monitored, and if any of these factors vary from

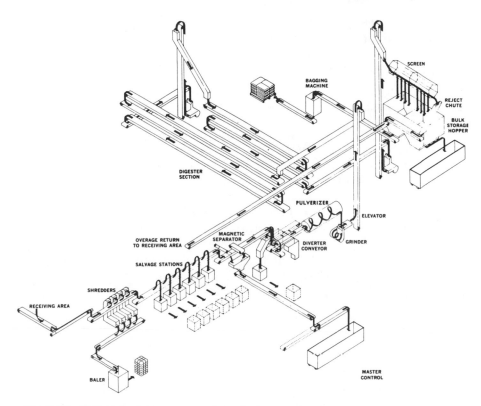

Fig. 2-10 IDC-Naturizer compost plant, St. Petersburg, Florida. *Source:* Drobny, N. L., Hull, H. E. and Testin, R. F., "Recovery and Utilization of Municipal Solid Waste," USPHS Publ. No. 1908, Govt. Printing Office, 1971, pp. 55–65.

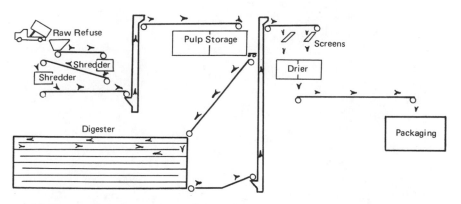

Fig. 2-11 Varro process flow diagram. *Source:* "Completely Controlled Compost Plant Opens," *Engineering News Record*, April 1, 1971, copyright, McGraw-Hill, Inc., all rights reserved.

the optimum they can be automatically adjusted without stopping the process. Because of this process control, the Varro system produces a very uniform final product which may, in the near future, be used to produce wallboard, building block, or other construction materials according to Stephen Varro, its developer (Ref. 2-60).

2-3.3 Curing

The amount of curing necessary at the initial stabilization depends upon the proposed use of the composted material. Compost can be applied with little curing in fields which will not be planted for some months; however, complete stabilization must have occurred before planting to assure that the compost does not rob the soil of nitrogen needed for digestion. If the compost is to be pelletized and/or bagged, it must be completely stable before being finished.

In windrow composting, a period of about 2 weeks is required for proper curing and drying. However, in mechanical composting processes, various curing periods are utilized, ranging from about 3 weeks to no curing time whatsoever.

2-3.4 Finishing

The amount of finishing required is also dependent upon the final marketing of the compost. If the compost is to be sold in bulk or landfilled, it may be disposed of without finishing.

The compost from most processes is not uniform in size and may contain bits of plastic, glass, or other objectionable debris. For this reason, the compost is usually reground and rescreened. The moisture content must also be adjusted so

Fig. 2-12 Finished compost at a composting plant in Meaux, France.

that the compost does not contain more than 30 percent moisture content by wet weight. In some climates, drying may be accomplished by spreading the compost outdoors; however, mechanical dryers are necessary in most applications.

When the compost is to be sold in the luxury garden market, it is usually desirable to pelletize the material to produce a uniform appearance and facilitate handling.

2-3.5   Storage or Disposal

In the final analysis, regardless of the efficiency of the composting process, the success or failure of the operation depends upon the method of disposal. Even where a good compost market exists, provisions still must be made for storage.

Storage is necessary because the use of composting is seasonal, with greatest demand during the spring and fall. Therefore, a composting plant should have a 6-months storage area. For a 300 ton per day plant, this would require about 13 acres of storage area. Many composting operations combine their curing period with the storage time period; however, this requires shallower piles and, therefore, an increase in storage area.

The price at which compost can be sold depends on the benefits to be obtained from its use and what customers are willing to pay for such benefits. Conditioning or improving the product by screening, pelletizing, bagging, and providing well-planned promotion and distribution may result in a greater gross return. Compost may also be sold in bulk, finished or unfinished, as well as fortified with chemical fertilizers.

Fig. 2-13 Windrows storing compost at a composting plant in Arnhem, Holland.

To date, the market for compost has never been developed to any significant extent in the United States. The use of artificial fertilizers and high-yield agricultural techniques have made it unnecessary to attempt to reclaim poorer land for agricultural purposes, and the benefit of the application of compost, when compared to its cost, is still under debate. Therefore, it appears that unless the use of compost is encouraged in this country, the market for compost will never be great enough to support composting operations on a large scale basis.

2-4. LIMITING METHOD FACTORS

There are five basic parameters which may tend to limit the application of composting as a major method of solid waste disposal in a particular community. These parameters include: 1) composition of the refuse, 2) land availability, 3) size of the community, 4) need for secondary disposal, and 5) existence of a viable compost market.

Since composting is a microbial system which depends upon the existence of certain microbes and nutrients in the matter being composted, it seems reasonable to suspect that the composition of the refuse would play an important role in the success of a composting operation. However, it has been shown that the necessary microbes exist in sufficient quantity for composting in all domestic refuse and the necessary nutrients, when not present in the collected refuse, can be supplied through the use of wastewater sludge or other appropriate additives.

A typical sample of domestic solid waste in the United States contains (by wet weight) 15.0 to 50.0 percent moisture, 15.0 to 30.0 percent carbon, and 0.20 to 1.0 percent nitrogen. These figures suggest that for proper composting, a typical American city would need to add wastewater sludge to its municipal waste in order to increase the water content and decrease the carbon to nitrogen ratio to the levels recommended in the foregoing sections. The addition of wastewater sludge will not only improve the composting process but will also lower the cost of sludge disposal—in some instances, by as much as 10 percent.

In some attempts at composting it has been discovered that the solid waste was actually too wet for composting and required drying. The only practical way to dry wet bulk refuse is to allow it to dry in windrows. During this drying period, anaerobic digestion is usually initiated, producing extreme odors, noxious gases, and supporting disease vectors. At least one composting plant has been closed in the United States because its influent solid waste stream required drying.

The physical composition of the refuse is also important to composting operations. Increasingly, composting plants are depending upon the sale of recycled materials as a major source of income since these materials are for the most part nonbiodegradable items and are removed from the refuse at some point in the composting process. However, even after separation, recycling requires extensive sorting and cleaning. Therefore, the salvaged goods must exist in sufficient quantities, and a sufficient market for the recycled item must be available in order to make separation a profitable operation.

It has been shown that the physical and chemical composition of solid waste at times varies greatly from one community to another. However, with reference to the composition of the refuse, composting can be considered as having universal application even though some modification of the process may be necessary.

Land which is suitable for composting plants may be difficult to acquire in large metropolitan areas. Windrow plants may require up to 60 acres, and the appearance of windrowed refuse will most likely be objectionable to those living in the area. Mechanical systems, however, can be operated in an enclosed area occupying about 8 to 10 acres and need not create objections if properly managed. In one instance, a mechanical digester was built in downtown Brooklyn, New York, in an abandoned warehouse.

The size of the community can play an important part in the operation of a composting plant. The average U.S. city produces 9 to 11 lb of domestic and industrial solid waste per capita per day of which approximately 70–80 percent is compostable. The largest composting plant constructed to date in this country has had a capacity of 300 tons per day which could serve a population of approximately 90,000.

The scale of the composting plant is governed either by the volume of wastes available for processing or by the market which will absorb the product. Questions as to the cost and optimum size of composting plants have largely been predicated on the assumption that compost is to be marketed, and hence that the system should be scaled to yield the most favorable economic picture. If the sale of compost is not a realistic objective, the whole question becomes academic. The cost of composting then depends on a scale determined by considerations other than economic optimization. Using this new criterion, it is possible to consider no existing limit to the size of a composting plant other than the volume of wastes available for processing. However, the manufacturers of existing compost systems already know the optimum scale for their respective plants, and at least one has indicated 200 tons per day as the "breakpoint." The basis for this limitation is not known, but apparently is related to market size, off-shelf equipment availability, and physical constraints.

Even if composting is chosen as a primary disposal method, a secondary disposal process (usually landfill) must be provided for the disposal of noncompostable, nonrecycled wastes and unsold compost. As previously noted, nearly 20 to 30 percent, by wet weight, of municipal refuse consists of noncompostable matter which must be landfilled. This means that for every 200 tons to be composted, at least 50 tons of solid waste will have to be landfilled. Furthermore, the market for compost is quite unstable, and the finished compost will have to be landfilled as a final disposal method if it cannot be sold or given away. This requirement can greatly increase the land area needed and the cost of the system when considered as a part of the process. This item should be considered by communities viewing composting as a disposal method where land for a landfill is not readily available.

Without a doubt, the absence of a viable compost market is the greatest hindrance to the successful application of composting systems in the United States. In Europe, where a consistent market for compost does exist, composting has been practiced for many years with great success (see Chapter 8).

The traditional markets for compost have been truck farms relatively close to urban areas, horticulture, and highways and city parks. Compost has also found some use as filler for fertilizers. The potential for development of these traditional markets is somewhat limited. The sale of compost has not developed at any compost plant according to plan, even when distribution problems were thoroughly studied and when there was advance assurance that the compost could be sold. For this reason, many seemingly promising composting plants have had to cease operations because they could not sell, or even give away in some cases, the compost they produced. It should be noted that these attempts were all in the area of small operations of the 50–150 tons per day size. Therefore, anticipated sales should be thoroughly studied and tested before plans are made to construct a composting plant, and alternate outlets for the compost should be provided in any case.

The present development of composting as a waste disposal system has been primarily directed toward semirural areas where a rather small-population city is surrounded by an area with high potential sales for the final compost end product. Little likelihood exists for greater development of composting as a primary method for handling refuse in large urban areas.

2-5. METHOD BY-PRODUCTS

The by-products from any process can be divided into two categories. The first category includes those by-products which are the result of improper management and operation of the process; in the composting process these include odor, disease vectors, noise from grinders, and the survival of pathogens. The second category consists of those by-products which are always produced by the method and are the natural result of the operation of the system; in the composting process these include noncompostable materials and the compost itself.

Composting plants may affect the surrounding environment and the neighborhoods in which they are situated because they are potential sources of odors and noise and may provide breeding places for flies and rodents. If the refuse must be retained for long periods in the storage area before composting, anaerobic digestion will begin to occur, thereby producing offensive odors, noxious gases, and providing a breeding place for flies and rodents. The process itself, especially if mechanical digestion is used, is designed to operate as an aerobic process free from odor and other detrimental characteristics. Therefore, odors and disease vectors need not be associated with a composting plant. However, most persons associate composting with odor (the backyard compost heap is generally anaerobic and quite odorous) and the slightest hint of an odor can bring threats of action against plant operation from irate citizens. For this reason, close monitoring of compost systems for odor, flies, and rodents is highly recommended so that the necessary action can be taken before the situation becomes critical.

A certain amount of noise will be emitted from any composting plant, from the operation of refuse trucks, loading equipment, and other machinery. For this reason the plant should be located in an industrial area. However, excessive noise may be hazardous to the workers as well as annoying to the neighbors. Since hammermills can generate intolerable noise levels, they should be isolated from the building by dampening materials. Materials falling into a metal-sided reject hopper from a separating station may also cause excessive noise. Lining the hoppers with wood or some other insulating material can reduce this noise problem. Internal and external noise levels should be kept below 80 decibels.

Through the years there has been a great deal of debate concerning the health hazards associated with the handling and use of compost because of the possibility of pathogen survival within the composted refuse. The survival of pathogens in compost is dependent mostly on one factor—the temperature to which the treated material is subjected or the heat produced during decomposition. Currently, the consensus of opinion is that the likelihood of pathogen survival becomes remote when the duration of exposure to the critical temperature level (140°F) is sustained over an extended period of time.

Elevated temperature alone cannot entirely account for the reduction in pathogen number observed in composting refuse. An appreciable amount of kill undoubtedly is due to the mutually antagonistic effects found in cultures consisting of the wide variety of types of microorganisms such as those that exist in composting material. In research performed by Knoll, he concluded that in the compost process, biological "self-purification" compensates for temperature inadequacies as far as pathogen kill is concerned (Ref. 2-67). In other words, pathogens cannot compete well with nonpathogenic bacteria.

Although the public health aspects of composting seem favorable with respect to pathogenic bacteria, some doubt exists as to whether this also applies to pathogenic fungi. At some stage in the composting process, fungi become quite conspicuous. In windrows, the greater part of the fungal population is in the outer layer of the compost pile. This layer is aerobic with a temperature range of 115 to 130°F. In mechanical digesters, the occurrence of a fungal stage is observed. However, peak development of the fungal population is not reached until the composting material is cured in windrows due to the brief retention time in the digester. Research has shown that a wide variety of fungi readily survive the composting process, although the extent of survival is a function of the temperature and aeration, based on the characteristics of the fungal group (Ref. 2-67). This fungal survival suggests that truly pathogenic forms may survive, and hence the need for intensive research in this direction is needed. At this time there are no reports of any worker or other handler of compost having been infected by fungi present in compost.

From 20 to 30 percent of municipal waste is noncompostable. For this reason, the compost plant generates a large amount of waste in the form of glass, plastic, rubber, metals, and other noncompostables which are separated from the refuse at some stage of the composting process. The majority of these materials can be recycled, and paper, which slows the composting process, is also recoverable. Therefore, these waste by-products can become an important source of income, or at least, pay for their own disposal, if salvage of these goods is practiced. The other alternative is to landfill these wastes, which can greatly increase the cost of composting. Without exception, some degree of recovery has been practiced in every contemporary composting plant constructed in this country, and should be considered as a necessary part of any composting system being considered as a means for solid waste disposal today.

The primary by-product of any composting operation is, of course, compost. The major argument for composting, other than its claim as the "natural method" of waste disposal, is that the compost can be sold for a profit, therefore removing the cost burden of solid waste disposal from the taxpayer. However, experience has shown that this is seldom the case. In fact, the major cause of failure of composting processes in this country has been an obvious lack of a market for the compost.

Compost does not meet the legal nutrient requirements for designation as a fertilizer. A typical compost is 1 percent nitrogen, 0.25 percent phosphorus, and 0.25 percent potassium. Therefore, compost is generally sold simply as a soil conditioner to improve workability and soil structure related to compaction and erosion, and increase water holding capacity. For this reason, it has been difficult to place an economic value on the use of compost. However, it is generally accepted that compost does, in fact, add to the value of the soil and a good compost should optimize this value.

A good compost, one that could compete favorably with synthetic fertilizers and peat, should possess the following characteristics: directly increase the productivity of the soil, maintain the soil in good condition, increase the fertility of the soil if used regularly over a long period of time, possess a high organic and nutrient content, contain a correct amount of lime (too much lime can be injurious to the soil), be uniform and dustless, and be free of impurities. It should also be noted that compost which is effective in one area will not necessarily be successful in another region; therefore, a study should be made to determine the type of compost which would be most successful in the market area surrounding the proposed plant.

If the compost cannot be sold, or even given away, then it must be landfilled. Moreover, composting can be considered as a size reduction technique, and some authors have suggested composting as an initial step in a solid waste disposal system where the final step would be landfilling of the compost. The biochemical stability, smaller volume, and compressibility of compost results in savings of land area, less groundwater pollution, and in accelerated use of the filled area, which offset processing costs. However, the cost of composting has been set at $10 to $12 per ton, which suggests that the cost of land would have to be extremely high in order to make this suggestion feasible (Ref. 2-15). In any case, the final disposal method for the compost should be determined by pertinent economic situations on the local level rather than the national situation.

2-6. METHOD REQUIREMENTS

The physical requirements for a composting plant include three major elements: land, power, and labor. These requirements differ greatly from one system to another with windrow systems having the higher land and labor requirements, whereas mechanical digesters require more power and a smaller but highly skilled labor force.

There are two basic land requirements for a composting plant. They are the area required for the plant location and the land needed for secondary disposal. The land area needed for the plant site depends upon a number of variables including plant capacity and system type. In general, it can be estimated that the requirement for a 200 ton per day windrow plant would be on the order of 55 to 60 acres, while a mechanical system of the same capacity would need approximately 8 to 10 acres (Ref. 2-15). Approximately 30 acres and 5 acres, respectively, would be needed for 100 ton per day windrow and mechanical plants. Moreover, it has been shown that the capital cost for a plant, which is governed by construction costs and land requirements, is a linear function of plant capacity up to 300 tons per day (see Fig. 2-14).

An example may be used to show the method to be used in computing the land requirement for secondary disposal; assume a plant capacity of 200 tons per day, a noncompostable percentage of 25 percent, and a reduction ratio of 2:1. In simple figures, the plant would produce a total of 50 tons per day of noncompostable waste and 75 tons per day of compost. If one-half of the noncompostable waste is recycled (which is a very conservative estimate), a total of 100 tons per day of refuse would have to be landfilled if the compost was not disposed of in another manner. Even if an existing landfill site is used for secondary

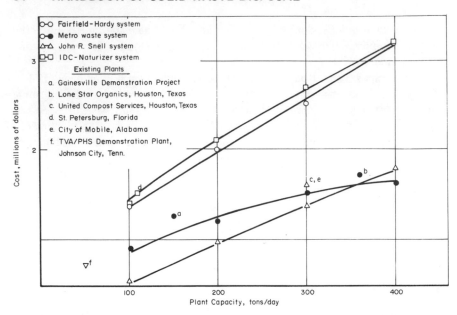

Fig. 2-14 Capital costs of composting systems. *Source:* Drobny, N. L., Hull, H. E. and Testin, R. F., "Recovery and Utilization of Municipal Solid Waste," USPHS Publ. No. 1908, Govt. Printing Office, 1971, pp. 55–65.

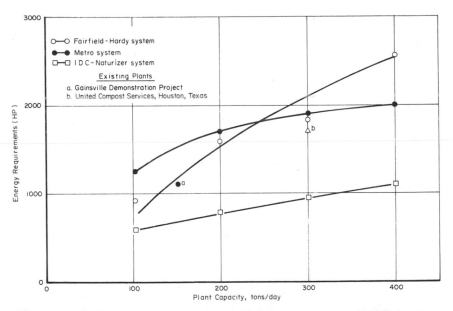

Fig. 2-15 Energy requirements for composting systems. *Source:* Drobny, N. L., Hull, H. E. and Testin, R. F., "Recovery and Utilization of Municipal Solid Waste," USPHS Publ. No. 1908, Govt. Printing Office, 1971, pp. 55–65.

disposal, this land requirement should be included in an analysis of the composting system's operational costs.

Power requirements for composting are generally not excessive, and as would be expected, are basically a function of degree of mechanization and capacity of the plant. Estimates made by Harding (Ref. 2-36) compare the power requirements for three major mechanical digesters: the MetroWaste Conversion system; the Fairfield-Hardy system; and the Naturizer system. Power requirements were given in ranges from a capacity of 100 tons per day to 400 tons per day and were 1250 to 2000 hp, 930 to 2560 hp, and 600 to 1100 hp, respectively (see Fig. 2-15). A windrow system would require less than the smallest of these values.

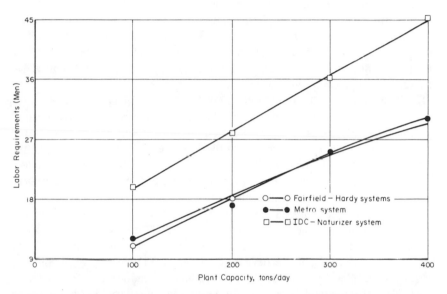

Fig. 2-16 Estimated labor requirements for composting systems. *Source:* Drobny, N. L., Hull, H. E. and Testin, R. F., "Recovery and Utilization of Municipal Solid Waste," USPHS Publ. No. 1908, Govt. Printing Office, 1971, pp. 55–65.

Labor requirements are directly related to the capacity of the plant. Values given by Harding (Ref. 2-36) for the MetroWaste, Fairfield-Hardy, and Naturizer systems were: 12 to 30 men, 11 to 30 men, and 20 to 45 men, respectively (see Fig. 2-16). A minimum labor force for a modern, 200 ton per day plant would consist of a feed operator, 8 sorters, 5 machine operators, 3 general laborers, 1 mechanic, 1 secretary, 1 laboratory technician, 1 foreman, and 1 manager. These figures are necessarily dependent upon the system in question; however, they should be considered as general guidelines when computing the operating costs of a proposed composting installation.

2-7. EQUIPMENT REQUIREMENTS FOR A COMPOSTING OPERATION

A list of equipment necessary to the operation of a composting plant would include: 1) scales, 2) tipping area and receiving bin, 3) various systems of conveyors for refuse transportation within the plant, 4) sorting equipment for salvage or bulk item removal, 5) grinders and shredders, and 6) finishing equipment: final grinding, screening, pelletizing, bagging, etc.

2-7.1 Scales

A necessity for any solid waste disposal facility is to provide for the measurement of the quantity of solid waste received. An accurate method for weighing the solid waste as it arrives at the plant is necessary if meaningful records of plant operation are to be maintained. These records make it possible to evaluate plant efficiency, provide for future needs, and analyze disposal economics. This measurement is accomplished through the use of truck scales.

Small operations (50 to 100 tons) may be able to operate efficiently using wood platforms, manually operated mechanical scales, and handwritten records. However, if the plant is to serve a large community it will be necessary to provide for automatic systems employing load cells, electronic relays, and printed output. The electronic relay scales allow for greater flexibility in platform location in relation to the scale house and are faster and more accurate than beam scales; however, they are considerably more expensive.

The electronic scale used at the Gainesville, Florida plant consisted of four major components: 1) the scale—a Toledo Model 2781 motor truck scale, 50-ton capacity, 60 ft by 10 ft platform, 2) an electronic dial and printweight with nine banks of selective numbering devices, 3) a card reader to receive pre-punched plastic truck identification cards and produce a readout of four digits for truck identification and five digits for tare weight into printweight, and 4) an adding machine, equipped to provide in one operation the gross weight and net weight (minus tare), the ability to maintain accumulated net subtotals, and to total and clear at the will of the operator. The adding machine can operate together with the card reader and the printweight while Mode Selector Switch is in "Agency" position. The system prints a cumulative tape for weighing while the Mode Selector Switch is in "Agency" position, and can also simultaneously print a separate weight ticket for each load. When the Mode Selector Switch is in "Other" position, the cumulative tape does not print, but any number of separate weight tickets showing truck number, gross, and tare weights can be printed if desired. This system can also be operated manually if the automatic system fails to function properly. Such an extensive system is imperative if the composting plant is to be operated at maximum efficiency.

The scale should have sufficient capacity to weigh the largest vehicle anticipated to service the composting plant on a routine basis. The platform should be of adequate size to weigh both axles simultaneously. For the majortiy of collection trucks, a 10 ft by 34 ft platform would be adequate with a scale capacity of 30 tons. The scale described above would be able to weigh any refuse handling equipment in operation today, and should be considered a proper-sized scale for the modern composting plant.

2-7.2   Receiving Area and Hopper

After the trucks are weighed, they dump the refuse into a storage hopper from the tipping area. The tipping area is the flat area adjacent to the storage pit or storage hoppers where trucks maneuver into position for dumping. The area should be large enough to allow for safe and easy maneuvering and dumping. The minimum recommended width of the tipping area is 70 ft; it should be larger when room is available. The tipping area should be enclosed with a minimum ceiling of 24 ft when possible to effect dust control, odor confinement, noise reduction, and general public relations.

An adequate storage area must be provided to compensate for equipment failures which might render the plant inoperable for short periods and for periods when the rate of receipt of solid waste exceeds the composting rate. The storage area is usually designed to contain about $1\frac{1}{2}$ times the 24-hr capacity of the plant. In designing the storage area, the generally accepted average unit weight of domestic refuse of about 0.175 tons/cu yd is used. Therefore, a storage area for a 300-ton capacity plant should provide for 450 tons of storage or $450 \div 0.175 = 2570$ cu yd.

The flow of refuse from the receiving area should be controlled. In many plants, especially those of smaller capacity, the storage hopper is combined with the charging hopper which discharges to an oscillating belt or leveling gate to achieve flow control. Arching or bridging often occurs in the receiving hopper and may be more acute if a leveling gate is used. The operation often proceeds more smoothly if one or both of the hopper's long sides are nearly vertical.

2-7.3   Conveyors

Endless moving belts are widely used to carry solid waste material from station to station. When hand-picking is practiced as a method of separation, the bed of refuse should not be more than 6 in. deep, with belt width and speed also suited to hand-picking. If the refuse is transported outside the building, adequate cover must be provided; i.e., high and wide enough so that the refuse will not catch on it. Sideboards or skirts should also be provided to keep the solid waste from falling from the belt. In freezing weather it may be necessary to heat the belts where they come in contact with the end pulleys.

An oscillating conveyor will break and loosen packed refuse and move the refuse in a smooth, continuous, uniform flow. In effect, it meters the amount of refuse which it discharges. This is accomplished by the upward and forward oscillating motion of a metal trough mounted on inclined reactor legs. A constant stroke eccentric drive provides a powerful surge-proof conveying action.

Other belts used in the plant may be standard conveyors chosen with respect to the amount and rate at which they must move the refuse through the plant. The number of belts used depends greatly upon the area required by the plant and the complexity of the operation. It should also be remembered that ground refuse moves more easily than raw refuse, but narrow openings, restrictions, or chutes must be avoided because ground refuse clogs easily. Bucket elevators and screw feeds may also be used to move refuse through the plant.

2-7.4 Sorting Equipment

Most plants remove as many noncompostables as possible before the refuse reaches size-reducing equipment. Even when this is not done, some hand-picking of bulky items is necessary to protect the equipment. When salvaging is practiced, the material removed is usually classified.

Generally, ferrous metals are removed by magnetic separators, and although the state-of-the-art of solid waste separation is rapidly advancing, the other noncompostable/salvable materials are currently removed by hand sorting. At Johnson City where no salvaging was practiced, 2 pickers handled up to 69 tons of refuse in 6 to 8 hrs; also, in Gainesville, where salvaging was practiced, 6 pickers were used to process 125 tons per day. For detailed information on types of separation equipment which are presently available commercially, refer to Chapter 6.

2-7.5 Grinders and Shredders

With respect to size-reduction equipment, composting operations have represented what is perhaps the most extensive application of such equipment in the field of solid waste disposal. A size-reduction operation is required at one point or another in virtually all of the many types of composting processes. The amount of size reduction practiced during composting operations can vary, depending on the type of composting process and the intended use of the resulting product.

As far as the size-reduction process is concerned, all composting plants in the United States appear to differ from each other (even processes of the same type) possibly because most plants are demonstration plants or pilot projects. Rarely do two plants use the same size-reduction equipment in the same manner and on the same type of solid waste. Size-reduction equipment for composting will be discussed in six broad categories: hammermills, flail mills, vertical-axis rasps, drum rasps, roll crushers, and pulpers.

Hammermills represent the broadest category and cover all types of high-speed crushers, grinders, chippers, or shredders that employ pivoted or fixed hammers or cutters. A mill is composed of high-speed swing hammers connected symmetrically on a horizontal shaft and cutter bars that have grate openings through which the refuse is forced. The refuse is reduced in size through the application of high tensile and shearing forces. Because of the tough and abrasive nature of refuse constituents, high-speed hammermills require relatively large motors and frequent rebuilding or replacement of the hammers and grate. For this reason, operating and maintenance costs for hammermills are quite high. The operating capacity is 15 tons per hr or less for most units. In some plants, two hammermills are used, the first being a coarse grinder and the second producing a finer particle size.

The flail mill is a horizontal, single-rotor unit with a studded shell and hammers attached to the rotor by means of chains. Replacements of hammers and chains are reported (Ref. 2-35) to be relatively easy with new shell studs moved into position by simple rotation because of the unique design. For this reason, maintenance costs for flail mills are relatively low. Flail mills are designed specifically to handle the reduction of municipal solid wastes when shock loads and abrasion predominate.

The vertical-axis rasp was developed in the Netherlands. It consists of a vertical axle carrying eight horizontal arms hinged to rotate upward. The axle rotates at about 15 to 25 rpm, sweeping the arms over the upper bottom, or grinding floor, of the unit. The grinding floor is made up of alternate plate sections containing either perforations or welded extensions. The material to be reduced is dropped into the unit where the rotating arms move it into contact with the protruding rasping pins. The material that is sufficiently reduced then drops through the holes in the following plate.

Rasping units are approximately 16 ft to 17 ft in diameter by about 7 ft in height. Capacities are in the 5 to 15 tons per hr range, depending primarily on the size of the holes in the perforated plates. Compared to a hammermill, a rasping unit has a higher first cost and is larger. The advantages of a rasp include reduced power requirements and much less maintenance.

Another type of rasping unit is the drum type; it has been developed in the form of a unit known as a pulverator. This is a 6 ft diameter by 16 ft long inclined drum covered on the inside wall with triangular steel plates. This unit has been used to tear open bags, break up large agglomerations, and mix the incoming refuse before it is sent to the flail-type hammermills.

The Dano grinder has a screen within the outer drum. Material falling through the screen is passed to the grinders or directly to compost piles. It rotates at 12 rpm and is reported to require about 6 kilowatts per ton of refuse.

A crusher that apparently has had limited application to composting operations is the roller type. Limited mention of its use is found in the literature (Ref. 2-35).

Wet pulpers (such as the one periodically used at Altoona, Pennsylvania, where cans, bottles, and other noncompostable items are not normally received in the garbage) are also used to shred refuse. The wet pulper at Altoona consists of a 6 ft diameter steel bowl with a rotating steel plate mounted in the bottom. During operation, the bowl is first partially filled with water, which is followed by the solid waste material to be composted. The steel plate rotates at 650 rpm and in 5 minutes produces a slurry containing about 5 percent solids. The slurry is then discharged through a screen to a screw-type dewatering press where the moisture content is reduced to 75 to 80 percent. A second press follows which reduces the moisture content to 50 to 60 percent.

2-7.6 Finishing Equipment

To improve the uniformity, composition, and general salability of the final product, it is necessary to regrind, screen, and perhaps even pelletize and bag the compost, especially for the horticultural market. The mill used for this final grinding process is usually a small hammermill. After the material is reground, it is usually screened to assure that all foreign objects which might detract from the appearance of the compost are removed. There are two types of screens in general use: vibrating screens and rotary drum screens.

The rotary drum screen is housed to prevent dissemination of dust. The drum slopes slightly from inlet to outlet and, where dual screening is provided, the section nearest the inlet is provided with coarse perforations (1 to $1\frac{1}{2}$ in.) and that nearest the outlet, with fine perforations (about $\frac{3}{8}$ in.). Separate hoppers beneath each screen receive the screenings and discharge them to trucks or conveyors to be disposed of in another manner.

Vibrating screens may be either single or double deck types. Usually single deck screens are provided with uniform perforations throughout the length but they may have sections with two sizes of holes. A double deck unit has coarse perforations (1 to $1\frac{1}{2}$ in.) in the upper screen and finer ones ($\frac{1}{4}$ to $\frac{1}{2}$ in.) in the lower. Decks are inclined downward from the receiving end to the discharge end so that the tailings bounce along to the outlet end. In twin deck screening, the lower deck is longer than the upper.

2-8. REGULATORY CONSTRAINTS

Most state solid waste disposal laws do not consider regulatory constraints specifically for the operation of a composting plant. Therefore, the operation of a composting system is almost always governed by general guidelines applicable to what is usually termed "other methods." In general, the "other method" guidelines require that: 1) plans and specifications be submitted before a permit for construction and operation is granted, 2) engineering reports be submitted with said plans and specifications, stating the site for, and method of, secondary disposal, and 3) the grounds surrounding the solid waste disposal facility are maintained in an acceptable condition and are not allowed to become a storage area for salvage material or a public health nuisance. With these purposes in mind, it would be safe to assume that there would be no grounds for objection to the operation of a modern composting operation providing that it was constructed in an industrially zoned area and properly supervised.

Presently, there are no apparent federal constraints upon the disposal of solid waste by composting if the plant is properly operated. Any projected constraints would most likely come in the area of compost utilization in terms of its application as a building material and possibly as an agricultural additive. Such constraints would probably focus upon pathogenic bacteria survival in composted material.

2-9. CASE HISTORIES

Despite considerable investment and technical know-how, not one large-scale composting plant has operated economically long enough in the United States to indicate that the process is feasible. As of 1972, there had been 18 attempts at composting in the United States and of those 18 only one or two were operating, and they existed under special circumstances. Composting has failed in the United States for four main reasons: 1) no steady market for the end product has been found, 2) initial investment and operating costs are generally high compared to other disposal methods, 3) a high quality end product usually cannot be derived from refuse in the United States without excessive expense, and 4) the separation of noncompostables requires a secondary method of disposal.

Many composting plants have been operated by private contractors who in turn charge the city an amount that defrays operating costs. Profits for the contractor are hoped to be realized from the sale of the end product. Operators usually charge the city from $1.00 to $5.00 (1967) per ton, while the costs for composting refuse usually range from $5.00 to $10.00 per ton. The operators must make up the balance of the costs and any profits from the sale of the compost. To date this has not been possible.

In order to better understand the problems associated with composting in this country, some case histories of attempts at composting in the United States will be presented. The plants considered will be those at Altoona, Pennsylvania, Houston, Texas (United Compost Services and Lone Star Organics), Mobile, Alabama, St. Petersburg, Florida, Johnson City, Tennessee, and Gainesville, Florida.

The plant in Altoona, Pennsylvania is the oldest and the most successful composting plant in the United States. For a number of years the plant, owned by Altoona FAM, Inc., composted garbage and rubbish by the windrow method. Then, in December, 1963, the Fairfield Engineering Company began operating a Fairfield-Hardy composter at the Altoona plant. The Altoona FAM plant disposes of Altoona's garbage for $2700 (1964) per month (see Table 2-2 for cost breakdown). The yearly average for 1964 was 26.25 tons per day, ranging from 44.3 tons per day in August to 23.2 tons per day in February. The plant presently operates at an average between 20 to 50 tons per day. The success of the Altoona plant is surrounded by circumstances which make the operation unique, and should not be considered as indicative of chances for composting on a large scale across the nation.

TABLE 2-2   Estimated Cost for Fairfield-Hardy Composting Process

| Capacity | | | | Operating Costs | | | | | Amortized Capital Cost[a] (Excluding Land) and Interest | Total Processing Cost |
| | | Payroll | | Utilities and Supplies | | Administration Including Interest | | Total | | |
Tons/ Day	Tons/Yr	$/Yr	$/Ton	$/Yr	$/Ton	$/Yr	$/Ton	$/Ton	$/Ton	$/Ton
100	28,600	81,000	2.83	74,400	2.60	145,800	5.11	10.54	2.41	12.95
200	57,200	126,400	2.21	141,700	2.48	214,630	3.76	8.45	1.75	10.20
300	85,800	172,400	2.01	189,000	2.21	270,435	3.20	7.42	1.46	8.88
400	114,400	204,800	1.79	262,000	2.29	336,750	3.02	7.10	1.40	8.50

[a] Assuming 20-year life.

Source: Drobny, N. L., Hull, H. E. and Testin, R. F., "Recovery and Utilization of Municipal Solid Waste," USPHS Publ. No. 1908, Govt. Printing Office, 1971, pp. 55–65.

First of all, the Altoona FAM plant operates at a very low capacity. A plant processing only 25 tons per day could only serve a total population of 20,000. Secondly, Altoona has separate garbage and rubbish collection, which reduces the cost of separation and decreases the C/N ratio, thus increasing the speed of digestion. Finally, the plant has had a variety of contracts with private distributors for all of its compost and the finished product is sometimes shipped in bulk a distance of more than 500 miles, which would not be economical for most cities.

In 1966, three composting plants were planned for Houston, Texas. One of those plants used the Snell digester and was owned by United Compost Services, Inc., which built the plant on city-owned land in a residential neighborhood despite vigorous protests by the residents. The company invested from $1.5 to $2 million in the plant. The original agreement provided for the city to pay $3.74 per ton of refuse up to 300 tons per day. The contract stipulated that the city accept the plant if it satisfactorily completed a 6 day trial period. The plant was closed before the trial period was completed (and the city was released from the contract) because of complaints of a consistent discharge of bad odors.

TABLE 2-3 Operating Cost for the USPHS–TVA Composting Plant at Johnson City, Tennessee

Activities	Operations								Maintenance						Total
	Salaries and Benefits	Super-vision	Electric Power	Utilities (Excluding Electricity)	Truck Use	Supplies and Materials	Miscel-laneous	Total	Salaries and Benefits	Super-vision	Supplies and Materials	Repairs	Miscel-laneous	Total	
Receiving[a]	$6,905	$1,181	$59			$28		$8,173	$974	$250	$661	$480		$2,365	$10,538
Picking and sorting	8,116	1,388	17			319		9,840	305	78	30			413	10,253
Disposal of rejects[b]	7,351	1,258			$3,524	38		12,171	620	159	56			835	13,006
Grinding (rasper)	3,211	549	770					4,530	2,134	548	2,925			5,607	10,137
(hammermill)	39	7	17					63	231	59				290	353
Composting															
Hauling and handling	5,930	1,015	28		5,764	327	$50	13,114	3,073	789	677	140		4,679	17,793
Turning and wetting	4,615	790				291	21	5,717	2,342	600	1,548	783		5,273	10,990
Curing	982	168			224	9		1,383	45	12		15		72	1,455
Storage	2,477	424			577		48	3,526							3,526
Operation and maintenance															
Grounds, buildings, and utilities			1,165					1,165	4,506	1,156	509		$68	6,239	7,404
Cleanup of process and receiving buildings	9,556	1,635				123		11,314							11,314
Office and laboratory	6,123	1,048	378	$614		800	144	9,107							9,107
Other	1,712	293		134	1,044	4,519	624	8,326							8,326
Regrinding and screening	4,293	734	130		1,230			6,387	846	217			409	1,472	7,859
Sewage sludge processing	3,350	573	54			782		4,759	2,970	762	1,395	218		5,345	10,104
Total	64,660	11,063	2,618	748	12,363	7,236	887	99,575	18,046	4,630	7,801	1,636	477	32,590	132,165

[a]At plant site.
[b]Includes cost of haulage to landfill (no landfilling costs).

Source: Breidenbach, Andrew W., "Composting of Municipal Solid Wastes in the United States," U.S. Environmental Protection Agency Publ. No. SW-47r, Govt. Printing Office, 1971.

The second Houston plant used the MetroWaste Conversion process and was operated by Lone Star Organics, Inc. This plant was the largest in the country, with a capacity of 360 tons per day. The capital costs for the plant were 1.7 million dollars, exclusive of land rental. The contract with the city provided that Houston must deliver 300 tons per day at $3.50 per ton. However, the company only needed to accept 150 tons per day. The major markets for the compost were in bulk to agriculture and in small quantities to the Texas Valley fruit growers, the Brazos River bottom planters, and to homeowners. Despite its initial signs of success, the Lone Star Organics plant was recently closed because of a lack of a compost market. The third plant planned for Houston was not constructed because of a lack of financing.

The fourth plant to be discussed is the windrow plant in Mobile, Alabama which was constructed in 1966 and operated by the city of Mobile. The capital cost for the plant was $1.4 million. After only 8 months the plant was closed because very little of the compost was sold. Maintenance and breakdowns on the machinery was excessive and labor costs were double the original estimates. The plant suffered from failure of overworked motors, the need to rebuild the crushing mechanisms, and the necessity to redesign the assembly line of the facility. The building costs estimated at $980,000 actually exceeded $1.4 million and the profits from sale of the end product were only $1000. An annual pay-

TABLE 2-4 Approximate Capital Costs of Windrow Composting Plants

Item of Cost	Daily Plant Capacity in Tons per Day (T/D)				
	52 T/D (Johnson City Plant–1 Shift)[a]	50 T/D (1 Shift)[b]	100 T/D (50 T/D 2 Shifts)[c]	100 T/D (1 Shift)[c]	200 T/D (100 T/D 2 Shifts)[c]
Buildings	$368,338	$210,000	$231,000	$231,000	$251,000
Equipment	463,251	482,700	482,700	607,100	607,100
Site improvement[d]	126,786	126,800	126,800	152,000	152,000
Land cost[e]	7,600	8,400	12,400	12,400	21,200
Total cost	965,980	827,900	852,900	1,002,500	1,031,300
Total cost per ton daily capacity	18,580	16,560	8,530	10,020	5,156

[a]Actual cost of the research and development USPHS–TVA composting plant at Johnson City, Tennessee.
[b]Based on Johnson City cost data adjusted for building and equipment modifications.
[c]Estimates based on actual Johnson City cost data projected to the larger daily capacity plants.
[d]Includes preparation of composting field with crushed stone and needed utility lines.
[e]Land costs are estimated based on approximate land values near Johnson City, Tennessee, of $800 per acre.

Source: Breidenbach, Andrew W., "Composting of Municipal Solid Wastes in the United States," U.S. Environmental Protection Agency Publ. No. SW-47r, Govt. Printing Office, 1971.

roll of $90,000 was estimated for a crew of 20 men, but the total work force actually totaled 53, with a payroll of $182,000. Inadequate experience in predicting the requirements for plant operation and inability to sell the finished product were the major factors that forced the closing of the Mobile, Alabama installation.

A plant using the Naturizer system was constructed in St. Petersburg, Florida in 1966 by the Westinghouse Corporation, which later combined with the Salvage and Conversion Systems Corporation and the All State Insurance Company to form the International Disposal Corporation (IDC). The capital cost for this plant was $1.5 million. The contract with the city required that the IDC plant dispose of 31,200 tons of refuse per year (100 tons per day, 6 days per week) in a "nuisance-free" manner. The city was to pay $3.24 per delivered ton with allowances for some fluctuation in future years (Ref. 2-71). IDC attempted to sell the compost for $9.00 per ton to commercial agriculture (peat moss costs $4.95 per bale—6 cu ft) and for $1.98 per 10 lb bag to homeowners. The plant employed 26 people. Nearly $500,000 were spent by the firm to

TABLE 2-5 Approximate Investment Costs of Windrow Composting Plants

Item of Cost	Plant Capacity in Tons per Day (T/D)				
	52 T/D (Johnson City, 1 Shift, 7,164 Tons, 1968)[a]	50 T/D (1 Shift, 13,000 T/ Year)[b]	100 T/D (2 Shifts, 26,000 T/ Year)[b]	100 T/D (1 Shift, 26,000 T/ Year)[b]	200 T/D (2 Shifts, 52,000 T/ Year)[b]
Construction	$958,380	$819,500	$840,500	$989,100	$1,071,100
Land costs	7,600	8,400	12,400	12,400	21,200
Total	965,980	827,900	852,900	1,001,500	1,092,300
Depreciation/year[c]	47,920	41,000	42,000	49,500	53,550
Interest/year[d]	45,080	38,600	39,800	46,200	51,000
Cost per ton daily capacity	18,580	15,560	8,530	10,020	5,460
Cost per ton refuse processed	12.98	6.12	3.15	3.68	2.01
	(6.88)[e]	(5.38)[f]	(2.76)[f]	(3.28)[f]	(1.73)[f]

[a]Actual costs of plant as built at Johnson City. Plant operates on 1-shift day. Cost per ton based on 1968 level of 7,164 tons of refuse processed.
[b]Based on Johnson City plant cost data adjusted for less elaborate equipment, buildings, and modifications.
[c]Straight line depreciation over 20 years of buildings and equipment, excluding land.
[d]Bank financing at $7\frac{1}{2}$ percent over 20 years. Yearly figure is average of 20-year total interest charge. Land cost included.
[e]Cost of Johnson City plant adjusted to design capacity of 13,520 tons refuse processed per year.
[f]Estimated cost without sludge processing equipment.

Source: Breidenbach, Andrew W., "Composting of Municipal Solid Wastes in the United States," U.S. Environmental Protection Agency Publ. No. SW-47r, Govt. Printing Office, 1971.

TABLE 2-6   Estimated Investment Costs for Windrow and Enclosed Composting Plants

	150 Ton/Day Capacity	
Item of Cost	Windrowing	Enclosed
Construction and equipment	$185,500.00[a]	$300,800.00[b]
Depreciation[c]	9,280.00	15,040.00
Interest $(7\frac{1}{2}\%)$[d]	8,660.00	14,040.00
Capital cost per ton daily capacity	1,237.00	2,005.00
Total cost per ton refuse processed	0.46	0.75
Land	9,300.00	2,640.00
Interest $(7\frac{1}{2}\%)$	430.00	120.00
Cost per ton daily capacity	62.00	18.00
Cost per ton of refuse processed[e]	0.01	<0.01 (0.003)
Total cost		
Per ton of daily capacity	1,300.00 (1,550.00)[f]	2,023.00
Per ton of refuse processed	0.47 (0.52)[f]	0.75

[a]Based on costs from USPHS–TVA composting plant at Johnson City, Tennessee, and land at $800 per acre.
[b]Based on costs from composting plant at Gainesville, Florida, and land at $4,000 per acre.
[c]Straight line depreciation of equipment and buildings over 20 years.
[d]Average yearly interest, bank financing over 20 years.
[e]Computed from interest only; land is assumed not to depreciate.
[f]Computed with comparable land values estimated at $4,000 per acre.

Source: Breidenbach, Andrew W., "Composting of Municipal Solid Wastes in the United States," U.S. Environmental Protection Agency Publ. No. SW-47r, Govt. Printing Office, 1971.

operate on a nuisance-free basis; of this $110,000 was spent on a dryer. However, despite this expenditure, the plant was closed by the city council as a "public nuisance." The reason—uncontrolled odors. This plant was located in a resort and retirement community. A three-man committee for the Florida city declared that the operation was a public nuisance for four basic reasons: 1) a lack of preventative maintenance causing frequent breakdowns of essential equipment, requiring stockpiling of collected refuse, 2) the building was poorly designed and incoming trucks had to be unloaded in the open, 3) winds carried odors to a nearby residential area and a golf course, and 4) the dryer which was installed to alleviate odor repeatedly malfunctioned. An IDC spokesman explained the stench as arising when bacterial action was interrupted by a change in temperature, or an insufficient amount of oxygen or carbon dioxide, or when there was a mechanical breakdown. The lack of a compost market meant that the necessary funds were not available for needed maintenance, thus leading to the closing of the plant.

The final two plants to be discussed were constructed under grants from the

United States Public Health Service as research facilities to study the feasibility of composting in the United States. Therefore, very careful records have been kept of the operation of these two installations and much information as to operating requirements and problems is available in the literature. The first plant is the windrow plant at Johnson City, Tennessee.

The plant at Johnson City was constructed in 1967 as a joint United States Public Health Service–Tennessee Valley Authority venture. The plant capacity was 52 tons per day. The capital cost for the plant, including land, was $1 million. Extensive cost data are given in Tables 2-3 through 2-9 (pages 43–49). Problems were reported with odors and the control of fly populations. Another drawback to windrow composting was the excessive amount of land required; in this case, 30 acres to serve a population of 100,000. This demonstration project has since been completed.

The Gainesville plant was a joint project of the USPHS and the University of Florida and the municipality of Gainesville. It was a MetroWaste Conversion System plant using the Metro digester. The capital cost of this plant was $1.25 million for a capacity of 150 tons per day (see Table 2-10). Problems at this installation included the inability to handle bulky items or rubbish mixed with

TABLE 2-7 Approximate Yearly Operating Costs for Windrow Composting Plants (Various Capacities)

Plant Capacity (Tons of Refuse Processed/ Day) (T/D)	Number of Shifts	Plant Operating Costs ($)			Operating Costs per Ton Refuse Processed ($/Ton)
		Operations	Maintenance	Total	
52 T/D 1968 Johnson City (7,164)[a]	1	$99,575 (139,817)[b]	$32,590 (41,200)[b]	$132,165 (181,017)[b]	$18.45 (13.40)[b]
50 T/D (13,000)[c]	1	133,950	43,700	177,650	13.65
100 T/D (26,000)[c]	2	213,795	59,150	272,945	10.50
100 T/D (26,000)[c]	1	197,850	59,850	257,700	9.90
200 T/D (52,000)[c]	2	357,015	95,400	452,415	8.70

[a] Figure in parentheses is total tons of raw refuse processed in 260-day work year.
[b] Costs projected for operating USPHS–TVA composting plant at design capacity of 52 tons per day (13,520 T/year) in 1969.
[c] Estimated costs based on USPHS–TVA composting project operating cost data.

Source: Breidenbach, Andrew W., "Composting of Municipal Solid Wastes in the United States," U.S. Environmental Protection Agency Publ. No. SW-47r, Govt. Printing Office, 1971.

TABLE 2-8   Total Costs of Composting Plants[a]

Capacity (Tons/Day)	Number of Shifts	Type Plant[b]	Capital Cost (per Ton/Day)	Cost per Ton Refuse Processed		
				Capital	Operating[c]	Total
50	1	W	16,560	6.12	14.53	20.65[d]
100	1	W	10,000	3.68	10.62	14.30[d]
100	2	W	8,530	3.15	11.22	14.37[d]
100	1	HR	5,400	1.66	—	—
157	1	HR	8,830	2.97	7.56	10.53[e]
200	1	HR	4,800	1.48	—	—
200	2	W	5,460	2.01	9.22	11.23[d]
300	1	HR	8,600	2.76	—	—
300	?	W	5,000	1.53	5.00	6.53[f]
300	1	HR	5,000	1.45	2.40	3.85
300	1	HR	4,500	1.38	5.12	6.50
346	2	HR	4,420	1.64	6.94	8.58[g]

[a]Cost data provided for plants other than Johnson City and Gainesville were used without adjusting to current economic conditions.
[b]W, windrowing; HR, enclosed high-rate digestion.
[c]In the case of the 50, 100, and 200 tons per day windrowing plants, an estimated cost of $0.88, $0.72, and $0.52 per ton of refuse received has been included for landfilling rejects.
[d]Projected from Johnson City composting project data, at 26,000 tons per year per 100 tons per day capacity (260 days), straightline depreciation of equipment and buildings over 20 years. Bank financing at $7\frac{1}{2}$ percent for 20 years. Includes disposal of rejects into landfill.
[e]Actual data from Gainesville plant with interest at $7\frac{1}{2}$ percent over 20 years, at 45,000 tons per year (286 workdays). Includes sludge handling equipment and disposal of non-compostables remaining after paper salvage.
[f]Actual data from Mobile, Alabama, composting plant. Components of costs not known.
[g]Gainesville plant at 90,000 tons processed per year.

Source: Breidenbach, Andrew W., "Composting of Municipal Solid Wastes in the United States," U.S. Environmental Protection Agency Publ. No. SW-47r, Govt. Printing Office, 1971.

the garbage. There were also some problems associated with odor and disease vector control. This plant has also ceased operation. The Gainesville and Johnson City experiences cannot be directly applied to the general experience because of their experimental nature.

2-10.   ECONOMIC ANALYSIS OF COMPOSTING AS A SOLID WASTE DISPOSAL ALTERNATIVE

Composting is currently one of the most expensive of the acceptable systems of solid waste disposal. Costs per ton for composting exceed costs per ton for landfilling by approximately 400 to 500 percent. There are, however, inherent

TABLE 2-9   Estimated Net Costs of Composting

Plant Capacity (Tons/Day)	Type of Plant[a]	Number of Shifts	Total Cost per Ton Refuse Processed[b]	Estimated Potential Income per Ton Processed		Estimated Net Cost per Ton Refuse Processed[c]
				Compost Sales	Salvage Sales	
50[d]	W	1	$20.65	$2.00–$3.50	–	$17.15–$18.65
100[d]	W	1	14.30	2.00– 3.50	–	10.87– 12.30
100[d]	W	2	14.37	2.00– 3.50	–	10.94– 12.37
157[e]	HR	1	10.53	2.00– 3.50	1.63	5.40– 6.90
200[d]	W	2	11.23	2.00– 3.50	–	7.73– 9.23
300[f]	HR	?	7.88	3.24	3.50	1.14
300[g]	W	?	4.03	–	–	–
300[h]	HR	9	3.85	3.00– 6.00	–	–
346[i]	HR	2	8.58	2.00– 3.50	1.63	3.45– 4.95

[a]W, windrowing; HR, high-rate digestion, enclosed type.
[b]Estimated costs included sludge processing.
[c]No credit has been estimated for handling sewage sludge.
[d]Estimated from 1968 data from the Johnson City composting project. Costs include depreciation, interest, land, and operations.
[e]Projected costs for Gainesville plant at 39,000 tons per year.
[f]A hypothetical plant.
[g]Based on data from Mobile, Alabama, plant; all components of cost are not known. There was no sludge processing.
[h]Reported costs do not include administration.
[i]Projected costs for operating Gainesville plant at 90,000 tons per year.

Source: Breidenbach, Andrew W., "Composting of Municipal Solid Wastes in the United States," U.S. Environmental Protection Agency Publ. No. SW-47r, Govt. Printing Office, 1971.

advantages to composting not obtained in other systems. The absence of obnoxious by-products in a properly managed system is not the least of these. The production of a potentially useful by-product and the advantage of a smaller economical plant size contribute to the long-range desirability of composting as a permanent solid waste disposal system. On the other hand, high per-ton processing costs and the uncertainty of the market for compost as a fertilizer or soil conditioner undermine the desirability of composting, especially in larger population centers.

About 70 to 80 percent of the solid waste generated in the U.S. (10 lb per person per day) is compostable, or about 7.0 to 8.0 lb per person per day. Therefore, in a municipality of 350,000 persons, approximately 1200 to 1400 tons of compostable refuse is generated daily. The expected economic life of a composting plant is estimated to be 20 years and financing through tax-free municipal bonds is assumed to be available at $5\frac{1}{2}$ percent per annum. Fixed as well as operating costs of composting operations are difficult to ascertain because almost all of the actual experience in composting in this country has been

TABLE 2-10   Estimates of Capital Costs, Energy, and Labor Requirements for Metro Compost Systems

Capacity (Tons/Day)	Capital Cost (Dollars)	Power Requirements (hp)	Labor Requirements (Men)
100	900,000[a]	1,250[a]	12[a]
150 (Gainesville)	1,250,000[b]	1,100[b]	—
200	1,200,000[a]	1,700[a]	17[a]
300	1,500,000[a]	1,900[a]	25[a]
360 (Houston)	1,700,000[c]	—	—
400	1,600,000[a]	2,000[a]	30[a]

[a]Harding, C. I., "Recycling and Utilization," In: *Proc.*, the Surgeon General's Conference on Solid Waste Management for Metropolitan Washington, July 19–20, 1967, USPHS Publ. No. 1729, Govt. Printing Office, 1967, p. 115.
[b]Personal communication from H. W. Houston, Project Director, Gainesville Municipal Waste Conversion Authority, Inc., August 8, 1967.
[c]Personal communication from G. Vaughn, Plant Manager, Lone Star Organics, Inc., August 8, 1967.
Source: Drobny, N. L., Hull, H. E. and Testin, R. F., "Recovery and Utilization of Municipal Solid Waste," USPHS Publ. No. 1908, Govt. Printing Office, 1971, pp. 55–65.

on a proprietary basis. The following estimated cost schedules reflect an adjustment of available data based on inflation and local labor costs. Because of the lack of adequate cost breakdowns in the literature, it is unwarranted to attempt to estimate more specific figures than those presented. Total costs are, however, believed to be sufficiently accurate for use in comparative evaluations with other systems of solid waste disposal.

Three basic composting systems have been included in the following estimated cost summary (see Table 2-11). The windrow composting method has been omitted from consideration because of the land required and the length of the operating cycle.

It is assumed that the daily solid waste will be divided into three processing plants because of the practical limitations of plant size. The basic assumptions used to develop these estimates include:

Population	350,000
Domestic waste generated per day	1750 tons
Compostable waste generated per day	1225–1400 tons
Compostable refuse per plant (3 plants)	408–467 tons per day

2-11. SUMMARY—COMPOSTING

Although composting plants have operated successfully in Europe for many years, composting operations have a history of failure in this country. Me-

TABLE 2-11   Cost Analysis for Various Composting Systems

I. Metro Compost System

Plant Size	Total Cost	Amortization		Land[a]	Operating[b] per Ton	Total per Ton
		Annual	Per Ton			
300 tons/day (85,800 tons/yr)	$1,875,000	$154,800	$1.80	$0.02	$8.32	$10.14
400 tons/day (114,400 tons/yr)	2,000,000	165,095	1.43	0.02	7.96	9.41

II. IDC-Naturalizer System

Plant Size	Total Cost	Amortization		Land[a]	Operating[b] per Ton	Total per Ton
		Annual	Per Ton			
300 tons/day (85,800 tons/yr)	$3,375,000	$278,000	$3.73	$0.02	$8.19	$11.44
400 tons/day (114,400 tons/yr)	4,000,000	330,192	2.87	0.02	7.64	10.53

III. Fairfield-Hardy

Plant Size	Total Cost	Amortization		Land[a]	Operating[b] per Ton	Total per Ton
		Annual	Per Ton			
300 tons/day (85,800 tons/yr)	$3,137,500	$259,920	$3.01	$0.02	$8.32	$11.35
400 tons/day (114,400 tons/yr)	4,012,500	331,400	2.88	0.02	7.96	8.26

[a]Land requirements are not expected to exceed 20 acres. This figure represents an estimated per ton payment to service the debt incurred for land acquisition.
[b]Includes labor, maintenance, utilities, and miscellaneous operating expenses.

chanical malfunction and the lack of adequate compost markets have adversely affected almost every compost facility built in the U.S. to date. Europeans have accused U.S. composting attempts of being too complicated and too profit-oriented. It seems apparent that composting will not be accepted as a viable disposal process in this country until a large-scale plant operates economically over a long period of time. Consequently, the success or failure of the proposed 150 ton per day FAM plant in Altoona, Pennsylvania, may well determine whether composting will be utilized in the U.S. in the foreseeable future.

REFERENCES

2-1. Aerojet-General Corporation, "A Systems Study of Solid Waste Management in the Fresno Area; Final Report on Solid Waste Management

Demonstration," USPHS Publ. No. 1959, Govt. Printing Office, 1969, p. 411.

2-2. American Public Works Association, *Municipal Refuse Disposal*, 2nd ed., Public Administration Service, Chicago, 1966.

2-3. Anon. "Can Composting Ease the Solid Waste Load?" *Chemical Engineering*, Vol. 74, March 13, 1967, p. 98.

2-4. Anon. "Completely Controlled Compost Plant Opens," *Engineering*, April 1, 1971, p. 28.

2-5. Anon. "Composting Gets Try-Out," *American City*, Vol. 80(a), April, 1965, pp. 99–102.

2-6. Anon. "Composting Operation Handles Refuse and Sewage Sludge," *Public Works*, Vol. 99, March, 1968, p. 84.

2-7. Anon. "Compost Plant for Refuse and Sewage Sludge (Johnson City, Tennessee)," *Civil Engineering*, Vol. 38, October, 1968, p. 87.

2-8. Anon. "Engineering Equipment Used in Fully-Mechanized Composting," *Compost Science*, Vol. 7(3), Winter, 1967, pp. 22–25.

2-9. Anon. "Fertilizer From Domestic Waste," *Engineering*, Vol. 205, January 12, 1968, p. 67.

2-10. Anon. "Refuse-Reclamation Plant for St. Petersburg," *Civil Engineering*, Vol. 36, October, 1966, p. 93.

2-11. Association of Bay Area Governments, "Bay Area Regional Planning Program, Refuse Disposal Needs Study, Supplemental Report," Berkeley, California, July, 1965.

2-12. Banse, H. J. and Strauch, P., "Importance of Prefermentation in Composting," *Compost Science*, Vol. 6(3), Autumn–Winter, 1966, pp. 17–23.

2-13. Beckerman, G., "Mechanical Composting Method Turns Solid Waste to Advantage," *Product Engineering*, Vol. 40, September 8, 1969, pp. 13–14.

2-14. Behe, R. A., "Disposal of Municipal Garbage by Composting," *Journal of Environmental Health*, Vol. 27, March–April, 1965, pp. 824–829.

2-15. Breidenbach, A. W., "Composting of Municipal Solid Wastes in the United States," U.S. Environmental Protection Agency Publ. No. SW-47r, Govt. Printing Office, 1971.

2-16. Brown, J., "How Much Does Composting Cost Per Ton?" *Compost Science*, Vol. 8(1), Spring–Summer, 1967, pp. 16–17.

2-17. California Department of Public Health, Bureau of Vector Control, "California Integrated Solid Wastes Management Project, A Systems Study of Solid Wastes Management in the Fresno Area," A Report Under Public Health Service Grant No. D01-UI-00021-01, June, 1968.

2-18. California, University of, "Reclamation of Municipal Refuse by Composting," Technical Bulletin No. 9, Series 37, Sanitary Engineering Project, Berkeley, University of California, June, 1953, p. 89.

2-19. Cservenyak, F. J. and Kenahan, C. B., "Bureau of Mines Research and Accomplishments in Utilization of Solid Wastes," Bureau of Mines Information Circular 8460, Govt. Printing Office, 1970.

2-20. Davies, A. G., *Municipal Composting*, Faber and Faber, London, 1961, p. 203.

2-21. Drobny, N. L., Hull, H. E. and Testin, R. F., "Recovery and Utilization of Municipal Solid Waste," USPHS Publ. No. 1908, Govt. Printing Office, 1971, pp. 55–65.

2-22. Elam, R. J., "The Reduction and Utilization of Solid Waste, A Study of the Continued Development of a Solid Waste Disposal System for the County of Bergen, New Jersey," The Office of the County Engineer, County of Bergen, New Jersey, November, 1964.

2-23. Furlow, H. G., and Zollinger, H. A., "Westinghouse Enters Composting Field," *Compost Science*, Vol. 4(4), Winter, 1964, pp. 5–10.

2-24. Gainesville Municipal Waste Conversion Authority, Inc., "Gainesville Compost Plant; An Interim Report," U.S. Department of Health, Education and Welfare, Cincinnati, 1969, p. 345.

2-25. Galler, W. S., "Animal Waste Composting With Carbonaceous Material," Summary Progress Report—June 1, 1968 to May 31, 1970, Grant No. EC 00270 03, North Carolina State University, Raleigh, North Carolina, 1970.

2-26. Gilbertson, W. E., "Animal Wastes: Disposal or Management," *Proc.*, National Symposium of the American Society of Agricultural Engineers, May 5, 6, and 7, 1966, pp. 144–5.

2-27. Golueke, C. G., "Composting Refuse at Sacramento, California," *Compost Science*, Vol. 1(3), Autumn, 1962, pp. 5–8.

2-28. Golueke, C. G. and McGavhey, P. H., "Comprehensive Studies of Solid Waste Management—First and Second Annual Reports," USPHS Publ. No. 2039, Govt. Printing Office, 1970.

2-29. Golueke, C. G., "Comprehensive Studies of Solid Waste Management—Third Annual Report," U.S. Environmental Protection Agency Report SW-10rg, Govt. Printing Office, 1971.

2-30. Golueke, C. G., and Gotaas, H. B., "Public Health Aspects of Waste Disposal by Composting," *American Journal of Public Health*, Vol. 44(3), March, 1954, pp. 339–348.

2-31. Golueke, C. G., "San Fernando Plant Uses 'Total System Concept'," *Compost Science*, Vol. 3(3), Autumn, 1962, pp. 5–8.

2-32. Gotaas, H. B., "Composting; Sanitary Disposal and Reclamation of Organic Wastes," World Health Organization Monograph Series, 31, WHO, Geneva, 1956, p. 205.

2-33. Gotaas, H. B., "Compost Plant Design and Operation," *The American City*, Vol. 82, July, 1967, p. 237.

2-34. Gray, K. R., "Accelerated Composting," *Compost Science*, Vol. 7(3), Winter, 1967, pp. 29–32.

2-35. Hagerty, D. J., Pavoni, J. L., and Heer, J. E. Jr., *Solid Waste Management*, Van Nostrand Reinhold, New York, 1973.

2-36. Harding, C. I., "Recycling and Utilization," *Compost Science*, Vol. 9(1), Spring, 1968, pp. 4–9.

2-37. Harding, C. I., "Recycling and Utilization in Proceedings; the Surgeon General's Conference on Solid Waste Management for Metropolitan Washington, July 19–20, 1967," USPHS Publ. No. 1729, Govt. Printing Office, 1967, pp. 105–119.

2-38. Hart, S. A., "Solid Waste Management/Composting; European Activity and American Potential," USPHS Publ. No. 1826. Govt. Printing Office, 1968, p. 40.

2-39. Hart, S. A., "Solid Wastes Management in Germany; Report of the U.S. Solid Wastes Study Team Visit, June 25–July 8, 1967," USPHS Publ. No. 1812, Govt. Printing Office, 1968, p. 18.

2-40. Howard, A., "The Manufacture of Humus by the Indure Process," *Journal of the Royal Society of Arts*, Vol. 84, November 22, 1935, pp. 26–59.

2-41. Jamison, H. M., "Several Methods Available for Solid Waste Control," *Water and Sewage Works*, Vol. 116, July, 1969, pp. 14–15.

2-42. Jensen, M. E., "Observations of Continental European Solid Waste Management Practices," USPHS Publ. No. 1880, Govt. Printing Office, 1969, p. 46.

2-43. John Carollo Engineers, "Solid Wastes Disposal Report," A Report for the Maricopa County Health Department, Maricopa County, Arizona, 1968.

2-44. Kehr, W. Q., "Microbial Degradation of Urban and Agricultural Wastes," Environmental Quality: Now or Never, SIM Special Publ. No. 5, Michigan State University, East Lansing, 1972, p. 184–191.

2-45. Knuth, D. T., "Composting of Solid Organic Waste," *Battelle Technical Review*, Vol. 17, March, 1968, pp. 14–20.

2-46. Kupchik, G. J., "Economics of Composting Municipal Refuse," *Public Works*, Vol. 97, September, 1966, pp. 127–128.

2-47. Kupchick, G. J., "Economics of Composting Municipal Refuse in Europe and Israel, with Special Reference to Possibilities in the USA," *Bulletin of the World Health Organization*, Vol. 34, 1966, pp. 798–809.

2-48. Lossin, R. D., "Compost Studies; Part 1," *Compost Science*, Vol. 11(6), November–December, 1970.

2-49. McGauhey, P. H., "American Composting Concepts," USPHS Publ. No. 2023, Govt. Printing Office, 1969, p. 48.

2-50. McGauhey, P. H., "Refuse Composting Plant at Norman, Oklahoma," *Compost Science*, Vol. 1(3), Autumn, 1960, pp. 5–8.

2-51. National Academy of Engineering–National Academy of Sciences, "Policies for Solid Waste Management," USPHS Publ. No. 2018, Govt. Printing Office, 1970.

2-52. Obrist, W., "Additives and Windrow Composting of Ground Household Refuse," *Compost Science*, Vol. 6(3), Autumn–Winter, 1966, pp. 27–29.

2-53. Prescott, J. H., "Composting Plant Converts Refuse into Organic Soil Conditioner," *Chemical Engineering*, Vol. 74, November 6, 1967, pp. 232–234.

2-54. Progress Report (June, 1967–September, 1969); Joint USPHS–TVA Composting Project, Johnson City, Tennessee, unpublished report, p. 80.

2-55. Reimer, L. G., "Refuse Reclamation; A Solution to a Growing Urban Problem," *Westinghouse Engineering*, Vol. 26, November, 1966, pp. 175–177.

2-56. Schulze, K. L., "Fairfield-Hardy Composting Plant at Altoona, Pa.," *Compost Science*, Vol. 5(3), Autumn–Winter, 1965, pp. 5–10.

2-57. Shell, G. L. and Boyd, J. L., "Composting Dewatered Sewage Sludge," USPHS Publ. No. 1936, Govt. Printing Office, 1969, p. 28.

2-58. Snell, J. R., "On the Basis of a Dumping Fee Only," *Compost Science*, Vol. 8(1), Spring–Summer, 1967, p. 17.

2-59. TV-27246A-Cooperative Project Agreement, Joint U.S. Public Health Service–Tennessee Valley Authority Composting Project, Johnson City, Tennessee, February, 1966.

2-60. Varro, S., "Composting as a Feasible Waste Disposal System," *Proc.*, New Directions in Solid Waste Processing, Institute held at Farmingham, Massachusetts, 1970.

2-61. Westerhoff, G. P., "Current Review of Composting," *Public Works*, Vol. 100, November, 1969, pp. 87–90.

2-62. Wiley, J. S., "A Discussion of Composting Refuse with Sewage Sludge," Presented at the 1966 APWA Public Works Congress, Chicago, Illinois, September 13, 1966.

2-63. Wiley, J. S., *Composting of Organic Wastes; An Annotated Bibliography*, Technical Development Laboratories, Savannah, February, 1958, p. 126. Supplement 1, June, 1959, p. 65. Supplement 2, April, 1960, p. 64.

2-64. Wiley, J. S. and Kechtitzky, O. W., "Composting Developments in the United States," *Compost Science*, Vol. 6(2), Summer, 1965, pp. 5–9.

2-65. Wiley, J. S., Gartrell, F. E. and Smith, H., "Concept and Design of the Joint USPHS-TVA Composting Project, Johnson City, Tennessee," *Compost Science*, Vol. 7(2), Autumn, 1966, pp. 11–14.

2-66. Wiley, J. S., ed., *International Research Group on Refuse Disposal (IRGRD), Information Bulletin Numbers 1–12, November 1956 to September 1961*, U.S. Department of Health, Education and Welfare, 1969, p. 308.

2-67. Wiley, J. S., ed., *International Research Group on Refuse Disposal (IRGRD), Information Bulletin Numbers 13–20, December 1961 to May 1964*, U.S. Department of Health, Education and Welfare, 1969, p. 274.

2-68. Wiley, J. S., ed., *International Research Group on Refuse Disposal (IRGRD), Information Bulletin Numbers 21–31, August 1964–December 1967*, U.S. Department of Health, Education and Welfare, 1969, p. 387.

2-69. Wiley, J. S., "Pathogen Survival in Composting Municipal Wastes," *Journal of the Water Pollution Control Federation*, Vol. 34(1), January, 1962, pp. 80–90.

2-70. Wiley, J. S., "Some Specialized Equipment Used in European Compost Systems," *Compost Science*, Vol. 4(1), Spring, 1963, pp. 7–10.

2-71. "$1½ Million Composting Plant Closes," *Refuse Removal Journal, Solid Waste Management*, Vol. 11, May, 1968, p. 23.

3

Incineration

Incineration is one of the methods used most extensively in the United States for disposal of solid waste materials. The technique of waste incineration has developed over many years, largely through work by municipal and county governments. Additionally, advances in the incineration art have been made by the designers and developers of commercial and industrial incinerators.

3-1. DEFINITION OF INCINERATION

Incineration is a controlled combustion process for reducing solid, liquid, or gaseous combustible wastes primarily to carbon dioxide, other gases, and a relatively noncombustible residue. The residue is usually deposited in an accompanying landfill (generally located in a distant spot) after recovery of any valuable materials. The carbon dioxide and other gases generated through the combustion process are released to the atmosphere.

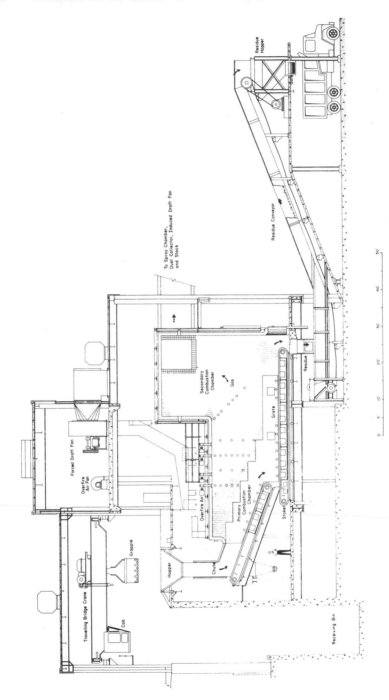

Fig. 3-1   Cross section through a typical municipal incinerator (*courtesy* Office of Solid Waste Management Program, USEPA).

3-2. DEVELOPMENT OF THE INCINERATION PROCESS

Incineration generally is considered to be the second oldest form of waste disposal practiced today (only landfill could be considered to be of greater longevity). Undoubtedly, with the discovery of fire by stone age man, the obvious advantages of obtaining heat while disposing of litter and debris from a cave led to the use of primitive incineration as a disposal practice. Even up to the present day, primitive cultures and especially nomadic groups rely heavily on waste fires, especially fires kindled with the ordure from pack animals and herd animals. These people, because of the open air environment they inhabit, generally ignore the consequences of burning wastes. However, established communities no longer can ignore the consequences of open burning because rather significant production of air pollutants is associated with uncontrolled burning (Ref. 3-6 and Ref. 3-9).

For many years, incineration of waste materials consisted essentially of open burning on level ground or in pits. One of the earliest innovations in incineration was the use of a fire wagon, a simple rectangular wagon with a frame protected by a clay coating. The wagon was pulled by horse through medieval towns so that the city dwellers could toss wastes from their homes through upper story windows onto the moving bonfire in the wagon. Interestingly enough, a similar idea resurfaced during the late 1950s, when a proposal was made to construct an enclosed mobile incinerator. The incinerator was to be used for the reduction of domestic and demolition wastes in a number of different localities (Ref. 3-6).

Two American engineers, Hering and Greeley, have described the historical development of incinerators (Ref. 3-12). According to their account, the first municipal incinerator, called a destructor, was built by Alfred Fryer in 1874 for the city of Nottingham, England. Early incinerators were loaded with a charge of wastes, fired, and then cleaned out; this kind of operation is called "batch" processing. Simple batch destructors became very popular in the early years of the twentieth century; by the early 1920s there were several hundred such plants in Great Britain and Western Europe. The potential value of this method of disposal was recognized then by engineers who noted that combustion temperatures in the incinerators were sufficiently high to generate steam and power. Incineration was the only large-scale method of waste disposal in use in England during the 1920s (Ref. 3-12). Hering and Greeley have stated also that the first incinerator in the United States, a government facility, was constructed on Governor's Island in New York Harbor in 1885. Many other incinerators followed soon after. A municipal incinerator was constructed in 1885 in Alleghany City, Pennsylvania. This plant and the Governor's Island plant were both designated "garbage crematories"; they burned domestic garbage with coal as an auxiliary fuel. Many of the early incinerators were designed to burn the nonorganic portion of municipal refuse and were not designed to handle the organic portion, the garbage. It was somewhat common for a municipal area to have separate plants for the incineration of garbage and the incineration of rubbish.

The first grate-furnace type incinerator built in the United States was constructed in Des Moines, Iowa in 1887. Similar plants were built in Ellwood, Indiana in 1893 and in Minneapolis, Minnesota in 1901. In almost all of these plants, wet refuse was dried, prior to its combustion, in basket-type grates. The stationary basket grate or flat grate was used for over 50 years in these early batch-fed incinerators. In a later modification, wet refuse was supported on air-cooled, cast iron fingers that rotated to release the dried material to a lower grate for final burning (Ref. 3-6).

Because of the seeming advantages of creating steam and power from the heat produced during incineration, early attempts were made to recover the heat associated with refuse burning. A small experimental plant was built in New York City in 1903, which incorporated steam production with refuse burning. Hering and Greeley have stated that more than 200 plants (many with steam recovery) were operating in the United States by 1921, but poor design and unskilled operation of many of the "garbage crematories" yielded poor results during this period.

Incinerators grew larger and more mechanized as time passed. A plant with a capacity of 300 tons per day was operating on mixed refuse in Milwaukee, Wisconsin by 1910. Between the years of 1910 and 1920, mechanical charging apparatus were developed and were installed at plants in Savannah, Georgia, Atlanta, Georgia, and Toronto, Canada. Almost all of the incinerators built in the United States during this period were equipped to produce steam (Ref. 3-12).

Automation increased in municipal incinerators during the second quarter of the twentieth century. Four specific types of automatic stoking grates and furnaces evolved during the period between 1920 and 1950: the "beehive" or hemispherical furnace; the multiple-hearth, mutual assistance furnace; the traveling grate furnace; and the rotary kiln incinerator (first constructed in Atlanta, Georgia about 1925). All four of these designs are in current use in the United States. Additionally, control mechanisms for these automated units are becoming more sophisticated. Observation of grate operation and refuse travel within incinerators by remote television systems is becoming more common (Ref. 3-5).

Several types of new incinerators are in developmental stages at the present time (1974). One of these types is the "elevated temperature" incinerator. High-temperature incinerators are designed to melt all metal and similar materials within the refuse and to reduce all of the refuse charged into the units (Ref. 3-15). Examples of such high-temperature incinerators are the "Melt-Zit" incinerator (Ref. 3-13), the Torrax System in Erie County, New York (Ref. 3-17), and the General Electric incinerator in Shelbyville, Indiana. High-temperature incinerators are described and discussed in more detail in Chapter 5 of this book. The Torrax System is classified as a pyrolysis system and is discussed in detail in Chapter 7.

The bulk of incinerators now in use (in the United States) are of the refractory-lined type with no heat recovery. Many of the difficulties associated with incinerator operation are connected with failure of refractory-lined furnaces caused by cyclic heating and cooling of the furnaces over short periods of time. During recent years, double furnace walls of silicon-carbide with an air cooling system for the air in the space between the walls have been developed; one of these systems has been installed in the Long Island, New York incinerator (Ref. 3-22). Also, "water-wall" furnaces equipped with heat recovery systems consisting of water-filled tubes along the walls of the combustion chambers have begun to be used in North America. One of the early installations of such systems was in Montreal, Canada (Ref. 3-19). These systems have long been used in Europe (Ref. 3-2), and formerly were used in the United States around the turn of the century. A more recent development in the incinerator art is less well-tested in use for combusting municipal refuse—the fluidized bed incinerator. This type of incinerator requires no grates and appears to be very promising for use in municipal incineration (Ref. 3-15).

The concept of using solid waste as a fuel for power production or as a fuel supplement for power production is also being developed. For example, the Combustion Power Company has built a pilot plant in Menlo Park, California for combustion of wastes and power generation. The proposed prototype plant based upon this pilot plant would be designed to produce 5 to 10 percent of the power consumed by the population area it served (Ref. 3-25). In St. Louis, Missouri, the Union Electric Company is at present using refuse as a supplementary fuel with coal for the production of power (Ref. 3-24). In Nashville, Tennessee (Ref. 3-23), the use of refuse as a fuel is being incorporated into a heating and cooling facility for the central city area. These systems and plants are discussed in more detail in Chapter 7 of this book.

Several other new approaches to municipal incineration are being attempted. For example, in Harrisburg, Pennsylvania, better incineration is being sought by placing responsibility for refuse combustion and effluent control on a contractor (Ref. 3-18). This type of "turn key" operation is attracting widespread attention at the present time.

Other developments in the incinerator art include provisions of special units for burning of wastes which are not amenable to combustion in a standard municipal incinerator. Bulky refuse is a problem in incineration because of incomplete combustion of slow-burning materials contained in automobile bodies, refrigerators, etc. Special incinerators for bulky wastes are being developed. A special driftwood incinerator has been constructed in New York Harbor (Ref. 3-26) and a log incinerator of a special type has been built in Baltimore, Maryland (Ref. 3-10). These are only two specific examples of units designed to process bulky refuse.

In another innovative experiment, underground incineration of refuse has been attempted (Ref. 3-27). A test facility for such underground incineration was built in California and preliminary test results were considered in a study of the feasibility of a commercial incinerator for Santa Clara, California. Test results indicated that the system may be technically feasible; however, the process has not been proven to be foolproof or economical as of this date (1974).

Coming years promise to hold new developments in refuse incineration with expanding American technology; however, increasingly strict air pollution control laws have militated against the use of incinerators for refuse disposal. There is some question at the present time whether developing technology can overcome the serious limitation of air pollutant generation which exists for this disposal method.

3-3. INCINERATION METHODOLOGY

The basic principle of refuse incineration is the conversion of a solid waste material into gaseous constituents which are released to the atmosphere and a relatively incombustible solid residue. As a result of the operation of most present day incinerators, a portion of the incombustible residue from the waste is entrained in the exit gas stream and, if not removed, constitutes an air pollutant. The operation of an incinerator must be designed to process the maximum amount of wastes which are anticipated to be generated in the area serviced by the incinerator. Furthermore, to obtain maximum efficiency and economy from the incinerator operation, the burning process must be regulated to produce as little residue as possible and must be controlled to prevent emissions of harmful effluents. Also, the operation of the incinerator must be regulated to insure long life of the facility itself; i.e., poor operation of an incinerator can lead to damage in the combustion chambers and in the residue removal system, for example, and necessitate costly repairs or curtail the effective life of the facility. The next section of this discussion is devoted to a description and evaluation of the operational parameters which affect the efficiency of an incinerator facility.

3-3.1 Operational Parameters

Some of the operational parameters which affect the efficiency of an incineration operation are not subject to control by the managers of the incinerator itself. For example, the quantities and characteristics of the wastes which are brought to the incinerator cannot be controlled by the incinerator operators. On the other hand, the combustion process may be controlled rather closely through the regulation of grate speeds, air supply, etc. All of these factors, however, have considerable influence on the success of the incinerator operation.

Refuse Characteristics. One of the most important quantities which must be determined for the proper design and operation of incinerators is the total amount of wastes which are to be handled by the facility. For efficient operation of such a facility, it must be designed and planned to serve a particular waste-generation area for a number of years; the large sums required for construction of the physical plant of an incinerator necessitate a long life for the facility. In order to ascertain the quantity of wastes to be anticipated for processing in the incinerator, it is necessary to forecast the population of the area to be served by the facility and to anticipate the per capita generation rate for solid waste. Table 3-1 contains some information on the quantities of wastes which are to be expected in metropolitan areas in the United States today. As is

TABLE 3-1   Solid Waste Collected for Incineration, lbs/cap/day

Item	Quantity
Residential (domestic)	1.5–5.0
Commercial (stores, restaurants, businesses, etc.)	1.0–3.0
Incinerable bulky solid wastes (furniture, fixtures, brush, demolition, and construction wastes)	0.3–2.5

evident from this table, the quantities of wastes generated per capita are quite variable. This variation in waste quantities has been observed in many different areas of the United States and among metropolitan zones within a given geographic area of this country. Additionally, an estimate must be made of the number of commercial and industrial establishments which will contribute solid waste for processing in the incinerator facility. Table 3-2 includes some information on the quantities of wastes which could be generated from various sources. All of the information shown in these tables must be considered to be approximate data and should be used only in formulating preliminary estimates of refuse quantities. In the planning for any facility such as an incinerator which requires the expenditure of large amounts of public funds, it behooves the design engineer or waste manager to conduct a thorough survey of the waste generation practices in the community to be served by the facility. The variability of wastes has been mentioned. Table 3-3 shows the range in composition of incinerator charge which has been observed at a number of existing facilities in the United States. As is obvious is this table, the individual constituents of the incinerator charge vary greatly. The variations in the relative amounts of the constituents such as food waste and paper products have great influence on the heating value of the wastes. In other words, a significant variation in the quantity of paper in the incinerator charge will significantly alter the gross heat content of the waste and will have an important influence on the operation of the incinerator. Because of the variability of the wastes, it is desirable to incorporate into the

TABLE 3-2   Refuse Quantities for Incineration

Classification	Building Types	Quantities of Waste Produced
Industrial buildings	Factories	Survey must be made
	Warehouses	2 lb per 100 ft^2 per day
Commercial buildings	Office buildings	1 lb per 100 ft^2 per day
	Department stores	4 lb per 100 ft^2 per day
	Shopping centers	Study of plans or survey required
	Supermarkets	9 lb per 100 ft^2 per day
	Restaurants	2 lb per meal per day
	Drug stores	5 lb per 100 ft^2 per day
	Banks	Study of plans or survey required
Residential	Private homes	5 lb basic and 1 lb per bedroom
	Apartment buildings	4 lb per sleeping room per day
Schools	Grade schools	10 lb per room and ¼ lb per pupil per day
	High schools	8 lb per room and ¼ lb per pupil per day
	Universities	Survey required
Institutions	Hospitals	8 lb per bed per day
	Nurses or interns homes	3 lb per person per day
	Homes for aged	3 lb per person per day
	Rest homes	3 lb per person per day
Hotels, etc	Hotels—1st class	3 lb per room and 2 lb per meal per day
	Hotels—Medium class	1½ lb per room and 1 lb per meal per day
	Motels	2 lb per room per day
	Trailer camps	6 to 10 lb per trailer per day
Miscellaneous	Veterinary hospitals	Study of plans or survey required
	Industrial plants	
	Municipalities	

Note: Do not estimate more than 7 hours operation per shift for industrial installations.
Do not estimate more than 6 hours operation per day for commercial buildings, institutions, and hotels.
Do not estimate more than 4 hours operation per day for schools.
Do not estimate more than 3 hours operation per day for apartment buildings.
Whenever possible an actual survey of the amount and nature of refuse to be burned should be carefully taken. The data are of value in estimating capacity of the incinerator where no survey is possible and also to double check against an actual survey.

Source: Incinerator Standards, Incinerator Institute of America, November 1968, p. 4A.

design of the incinerator provisions for utilizing temporary storage in a refuse pit to dry out excessively wet refuse, to utilize auxiliary fuels to aid in the combustion of low-heat value refuse, etc.

The composition of municipal solid waste tends to vary as a function of time throughout the year, locality, or geographic location in the United States, and general income level of the community to be served. Table 3-4 shows some of the variations in the physical and chemical characteristics of the refuse charged into municipal incinerators. In general, solid wastes in garbage cans or receptacles

TABLE 3-3   Range in Composition of Incinerator Charge

	Minimum	% (by Weight) Maximum	Average
Food waste	7.5	34.6	22.3
Garden waste	1.6	12.1	4.4
Paper products	29.8	61.8	47.6
Metals	6.6	10.9	8.3
Glass and ceramics	4.6	11.0	6.8
Plastics, rubber, leather	2.5	5.8	2.8
Textiles	1.4	4.8	2.2
Wood	0.4	2.2	1.4
Rocks, dirt, ashes, etc.	0.2	12.5	4.2

Source: Office Solid Waste Management Programs.

at the household or on the street curb weighs between 200 and 300 lb/cu yd. After compaction in a standardized collector truck, the density of the wastes increases to about 500 to 700 lb/cu yd. However, after the refuse is dumped into a storage pit at an incinerator facility, the density of the material is likely to decrease to approximately 400 to 500 lb/cu yd. Therefore, in addition to variations in the composition of the wastes, there are variations in the densities of the waste. This variation in density has some influence on the operation of materials transfer devices such as charging hoppers at the incinerator. Moreover, at times items are brought to a municipal incinerator which cannot be characterized as "standard" municipal refuse. These items generally are quite bulky and are not amenable to incineration in a municipal incinerator. At times, in some localities, bulky items have accounted for as much as 20 percent of the volume of the wastes brought to particular facilities. The management of bulky wastes by incineration is discussed in a later section of this chapter.

TABLE 3-4   Physical and Chemical Characteristics of Charge

Proximate Analysis Constituents	% by Weight	Ultimate Analysis Constituents	% by Weight
Moisture	15–35	Moisture	15–35
Volatile matter	50–65	Carbon	15–30
Fixed carbon	3–9	Oxygen	12–24
Noncombustibles	15–25	Hydrogen	2–5
		Nitrogen	0.2–1.0
		Sulfur	0.02–0.1
		Noncombustibles	15–25

"Higher" heating value
3,000–6,000 BTU/charged lb

Source: Office Solid Waste Management Programs.

To regulate the operation of an incinerator properly, it is customary to determine a materials balance on the refuse input to the incinerator. Of particular interest in calculating a materials balance of the incoming and outgoing matter in the combustion operation are the quantities shown in Table 3-4 under the heading "Ultimate Analysis." The ultimate analysis shows the total percentage by weight of the various combustible constituents in the refuse and also shows the percentages by weight of moisture and oxygen present. The moisture content is very important and the variations shown in moisture content has a great effect on the operation of incinerators. If the moisture content is high, a great deal of heat must be supplied to the refuse in order to evaporate the moisture, dry the refuse, and vaporize the combustible constituents. In many cases, it is necessary to supply additional auxiliary fuel with the incinerator charge of wastes in order to accomplish this drying-out of the moisture present. The information on higher heating value shown in Table 3-4 is also of considerable interest. Higher heating value is the total amount of heat which is released during the combustion process per unit weight of combusted material. The heating value of metropolitan refuse in the United States has increased in recent years because of the increasing percentages of paper and plastics in the refuse. Table 3-5 shows some comparative data on combustion of paper, wood, and garbage. It is obvious from this table that an increase in the amounts of paper or wood in municipal refuse would lead to a much higher overall heat content in that refuse. Formerly, the relative percentage of garbage in municipal refuse was much higher and the heat content of the refuse was correspondingly lower. In order to simplify the planning and design of incinerators, the Incinerator Institute of America has classified the various types of wastes which are likely to be brought to incinerator facilities. These waste classifications, Type 0 to Type 6, are shown in Table 3-6. In addition to classifying the types of waste which are likely to be brought to a municipal incinerator, the personnel of the Incinerator Institute have also categorized the various types of incinerators which are likely to be used in burning municipal refuse. These classifications are useful primarily in the application of data concerning combustion parameters which will be shown in the following paragraphs. Table 3-7 lists the classifications of incinerators developed by the Incinerator Institute of America. Municipal incinerators have been placed in Class V of this system. The primary information for design of a combustion process includes the value of heat which is released during combustion of a particular material and the requirements for air to supply the oxygen necessary for the combustion operation. The approximate combustion characteristics of various types of material which are commonly found in municipal refuse are shown in Table 3-8. These data should be considered only approximate in connection with the loose densities listed and the moisture contents listed in the table. However, they give some indication of the heat release values which would apply in the combustion of municipal wastes.

TABLE 3-5   Combustion Data Paper, Wood, and Garbage

Material	Sulfite Paper,[a] %		Average Wood, %		Douglas Fir, %		Garbage,[b] %	
Carbon, C	44.34		49.56		52.30		52.78	
Hydrogen, H	6.27		6.11		6.30		6.27	
Nitrogen, N			0.07		0.10			
Oxygen, O	48.39		43.83		40.50		39.95	
Ash	1.00		0.42		0.80		1.00	
Gross BTU/lb, dry basis	7,590		8,517		9.050		8,820	
Constituent (Based on 1 lb)	scf[c]	lb	scf	lb	scf	lb	scf	lb
Theoretical air	67.58	5.16	77.30	5.90	84.16	6.43	85.12	6.50
40% sat at 60°F	68.05	5.18	77.84	5.93	84.75	6.46	85.72	6.53
Flue gas with theoretical air								
CO_2	13.99	1.62	15.64	1.81	16.51	1.91	16.66	1.93
N_2	53.40	3.94	61.10	4.51	66.53	4.91	67.23	4.97
H_2O formed	11.78	0.56	11.48	0.54	11.84	0.56	11.88	0.56
H_2O (air)	0.47	0.02	0.53	0.02	0.58	0.02	0.59	0.02
Total	79.65	6.15	88.77	6.90	95.46	7.42	96.37	7.49
Flue gas with % excess air as indicated								
0	79.65	6.16	88.77	6.91	95.47	7.43	96.38	7.50
50.0	113.44	8.74	127.42	9.86	137.55	10.64	139.24	10.77
100.0	147.23	11.32	166.07	12.81	179.63	13.86	182.00	14.04
150.0	181.26	13.91	204.99	15.78	222.01	17.09	224.86	17.21
200.0	215.28	16.51	243.91	18.75	264.38	20.12	267.72	20.58
300.0	283.33	21.70	321.75	24.68	349.13	26.58	353.44	27.12

[a]Constituents of sulfite paper, %
Cellulose	$C_6H_{10}O_5$	84
Hemicellulose	$C_5H_{10}O_5$	8
Lignin	$C_6H_{10}O_5$	6
Resin	$C_6H_{10}O_5$	2
Ash	$C_{20}H_{30}O_2$	1

[b]Estimated.
[c]Measured at 60°F and 14.7 psia.

Source: Air Pollution Engineering Manual, U.S. Department of Health, Education, and Welfare, Public Health Service, 1967, p. 422.

Consideration of the data shown in the preceding tables leads to the conclusion that the quantities of wastes which will be brought to an incinerator will be quite variable, and the combustion characteristics of those wastes will also be quite variable. In planning an incinerator facility, these variations must be taken into account. Much of the variation in quantities and composition can be attributed to seasonal "weather" fluctuations. Maximum quantities of solid wastes will be generated during summer months; in many localities, the heat value of the wastes will also decrease drastically in the summer months. However, the

TABLE 3-6 Classification of Wastes to be Incinerated

Classification of Wastes	Principal Components	Approximate Composition, % by Weight	Moisture Content, %	Incombustible Solids, %	BTU value/lb of Refuse as Fired	BTU of Aux. Fuel per lb of Waste to be Included in Combustion Calculations	Recommended Min BTU/hr Burner Input per lb Waste
Trash,[a] Type 0	Highly combustible waste, paper, wood, cardboard, cartons, including up to 10% treated papers, plastic or rubber scraps; commercial and industrial sources	Trash, 100	10	5	8,500	0	0
Rubbish,[a] Type 1	Combustible waste, paper, cartons, rags, wood scraps, combustible floor sweepings; domestic, commercial, and industrial sources	Rubbish, 80 Garbage, 20	25	10	6,500	0	0
Refuse,[a] Type 2	Rubbish and garbage; residential sources	Rubbish, 50 Garbage, 50	50	7	4,300	0	1,500
Garbage,[a] Type 3	Animal and vegetable wastes, restaurants, hotels, markets, institutional, commercial, and club sources	Garbage, 65 Rubbish, 35	70	5	2,500	1,500	3,000

Type	Principal components						
Animal solids and organic wastes, Type 4	Carcasses, organs, solid organic wastes, hospital, laboratory, abattoirs, animal pounds, and similar sources	Animal and human tissue, 100	85	5	1,000	3,000	8,000 (5,000 primary) (3,000 secondary)
Gaseous, liquid or semi-liquid wastes, Type 5	Industrial process wastes	Variable	Dependent on predominant components	Variable according to wastes survey	Variable according to wastes survey	Variable according to wastes survey	Variable according to wastes survey
Semi-solid and solid wastes, Type 6	Combustibles requiring hearth, retort or grate burning equipment	Variable	Dependent on predominant compo-	Variable according to wastes survey	Variable according to wastes survey	Variable according to wastes survey	Variable according to wastes survey

aThe figures on moisture content, ash, and BTU as fired have been determined by analysis of many samples. They are recommended for use in computing heat release, burning rate, velocity, and other details of incinerator designs. Any design based on these calculations can accommodate minor variations.

Source: *Incinerator Standards*, Incinerator Institute of America, November 1968, p. 5A.

TABLE 3-7   Classification of Incinerators

Class 1—Portable, packaged, completely assembled, direct-fed incinerators, having not over 5 ft^3 storage capacity, or 25 lb per hour burning rate, suitable for Type 2 waste.

Class 1A—Portable, packaged or job assembled, direct-fed incinerators 5 ft^3 to 15 ft^3 primary chamber volume; or a burning rate of 25 lb per hour up to, but not including, 100 lb per hour of Type 0, Type 1, or Type 2 waste; or a burning rate of 25 lb per hour up to, but not including, 75 lb per hour of Type 3 waste.

Class II—Flue-fed, single chamber incinerators with more than 2 ft^2 burning area, for Type 2 waste. This type of incinerator is served by one vertical flue functioning both as a chute for charging waste and to carry the products of combustion to atmosphere. This type of incinerator has been installed in apartment houses or multiple dwellings.

Class IIA—Chute-fed multiple chamber incinerators, for apartment buildings with more than 2 ft^2 burning area, suitable for Type 1 or Type 2 waste. (Not recommended for industrial installations.) This type of incinerator is served by a vertical chute for charging wastes from two or more floors above the incinerator and a separate flue for carrying the products of combustion to atmosphere.

Class III—Direct-fed incinerators with a burning rate of 100 lb per hour and over; suitable for Type 0, Type 1 or Type 2 waste.

Class IV—Direct-fed incinerators with a burning rate of 75 lb per hour or over, suitable for Type 3 waste.

Class V—Municipal incinerators suitable for Type 0, Type 1, Type 2, or Type 3 wastes, or a combination of all four wastes, and are rated in tons per hour or tons per 24 hours.

Class VI—Crematory and pathological incinerators, suitable for Type 4 wastes.

Class VII—Incinerators designed for specific byproduct wastes, Type 5 or Type 6.

Source: Incinerator Standards, Incinerator Institute of America, November, 1968.

magnitude of the fluctuations in per capita quantities of wastes to the incinerator vary significantly from one community to another. As a rule of thumb, the 4-week average of waste generation quantities within any given area will range from −10 to +10 percent of the average weekly quantity. These significant fluctuations in quantities of solid waste have considerable bearing on the planning of the incinerator. The incinerator must be sized, for economy, on the basis of a certain weekly quantity which is not likely to be exceeded very often during the year. Sizing of the incinerator to hold maximum expected weekly quantities would result in an oversize facility and would cause the expenditure of large amounts of money to provide capacities needed very infrequently. Several techniques have developed in incineration practice for estimating incinerator sizes. According to one method, the highest estimated 4-week period quantity is determined by a survey of the local community for the design year; the average weekly quantity of waste to be delivered to the incinerator can then be estimated as a certain percentage of the average weekly quantity for the highest 4-week period. The incinerator then can be designed to accomodate that average weekly quantity. Other methods have been developed; these methods include the use of frequency diagrams, so that the incinerator is planned in such a manner that its capacity will be exceeded only during a certain percentage of time throughout the year. In any event, the incinerator should be planned so that it is possible to conduct preventive or routine maintenance and repairs frequently. The time necessary for repair of newer incinerators is considerably less than that required for older models. Newer incinerators which permit continuous operation generally require repairs only about once every 90 days. At the end of the 90 day period, the average down time for such facilities will be approximately 15 percent of the total operating time for the period. In good incinerator design, provisions will be made to include a number of furnaces and combustion units in the facility so that the requirement for 15 percent down time may be accomodated by the use of an extra furnace. If this choice of including an extra unit appears to be excessively expensive, the incinerator designer

TABLE 3-8(a)   **Approximate Combustion Characteristics of Various Kinds of Materials and Amounts of Air Needed for Combustion**

Material	High Heat Value,[a] BTU per lb of MAF[b] Waste	Air Needed for Complete Combustion lb per lb of MAF Waste[c]
Paper	7,900	5.9
Wood	8,400	6.3
Leaves and grass	8,600	6.5
Rags, wool	8,900	6.7
Rags, cotton	7,200	5.4
Garbage	7,300	5.5
Rubber	12,500	9.4
Suet	16,200	12.1

[a]Values are necessarily approximate, since the ultimate composition of the combustible part of the materials varies, depending upon sources. The heating value of the material as it is received is obtained by multiplying the moisture-free and ash-free BTU value of the materials by $1 - (\% \text{ moisture} + \% \text{ ash})/100$. For example, garbage with an MAF value of 7,300 and containing 35 percent moisture and 5 percent ash or other noncombustible material will have an "as-fired" heating value of 4,380 BTU per lb.

[b]MAF means moisture-free and ash-free if ash refers to total noncombustible materials.

[c]These values are also approximate and are based on 0.75 lb of air per 1,000 BTU for complete combustion. For various percentages of excess air, multiply these values by (100 plus percent of excess air)/100. For example, if paper is burned with 100 percent excess air (5.9) 200/100 = 11.8 lb of air per lb of moisture-free and ash-free paper will be required.

Source: Municipal Refuse Disposal, Institute for Solid Wastes, Am. Publ. Works Assn., 1970, p. 174.

TABLE 3-8(b) BTU Values[a]

Waste	BTU Value/lb as Fired	Weight in lb/ft³, Loose	Weight in lb/ft³	Content by Weight in Percentage Ash	Content by Weight in Percentage Moisture
Type 0 waste	8,500	10		5	10
Type 1 waste	6,500	10		10	25
Type 2 waste	4,300	20		7	50
Type 3 waste	2,500	35		5	70
Type 4 waste	1,000	55		5	85
Kerosene	18,900		50	0.5	0
Benzene	18,210		55	0.5	0
Toluene	18,440		52	0.5	0
Hydrogen	61,000		0.0053	0	0
Acetic acid	6,280		65.8	0.5	0
Methyl alcohol	10,250		49.6	0	0
Ethyl alcohol	13,325		49.3	0	0
Turpentine	17,000		53.6	0	0
Naphtha	15,000		41.6	0	0
Newspaper	7,975	7		1.5	6
Brown paper	7,250	7		1.0	6
Magazines	5,250	35		22.5	5
Corrugated paper	7,040	7		5.0	5
Plastic coated paper	7,340	7		2.6	5
Coated milk cartons	11,330	5		1.0	3.5
Citrus rinds	1,700	40		0.75	75
Shoe leather	7,240	20		21.0	7.5
Butyl sole composition	10,900	25		30.0	1
Polyethylene	20,000	40–60	60	0	0
Polyurethane, foamed	13,000	2	2	0	0
Latex	10,000	15	45	0	0
Rubber waste	9,000–11,000	62–125		20–30	
Carbon	14,093		138	0	0
Wax paraffin	18,621		54–57	0	0
1/3 wax–2/3 paper	11,500	7–10		3	1
Tar or asphalt	17,000	60		1	0
1/3 tar–2/3 paper	11,000	10–20		2	1
Wood sawdust, pine	9,600	10–12		3	10
Wood sawdust	7,800–8,500	10–12		3	10
Wood bark, fir	9,500	12–20		3	10
Wood bark	8,000–9,000	12–20		3	10
Corn cobs	8,000	10–15		3	5
Rags, silk or wool	8,400–8,900	10–15		2	5
Rags, linen or cotton	7,200	10–15		2	5
Animal fats	17,000	50–60			0
Cotton seed hulls	8,600	25–30		2	10
Coffee grounds	10,000	25–30		2	20
Linoleum scrap	11,000	70–100		20–30	1

[a]The chart shows the various BTU values of materials commonly encountered in incinerator designs. The values given are approximate and may vary based on their exact characteristics or moisture content.

Source: *Incinerator Standards*, Incinerator Institute of America, November 1968, p. 1C.

should provide some alternative method of disposal to handle the wastes which will be brought to the incinerator when some of the combustion units are shut down for repairs.

Combustion Controls. Combustion of refuse within an incinerator is controlled through regulation of the so-called "three T's" of combustion—time, turbulence, and temperature. Combustion time is governed in an incinerator by adjusting the travel rate of refuse through the combustion chambers. The chambers themselves must be made large enough to retain liberated gases a sufficient time to allow complete combustion. In practice, if an incinerator is overloaded, combustion time is shortened at the expense of complete combustion. Uninformed incinerator operators thus maintain that they consistently process the required quantities of solid waste which have been assigned to their particular facility. However, the too-rapid passage of wastes through the combustion chambers may lead to increases in air pollution and an increase in the amount of residue remaining after incineration. This practice, obviously, should be avoided.

Turbulence is achieved in an incinerator by physical tumbling of refuse, primarily on grate surfaces, and by blowing air through the burning refuse and the primary combustion gases created during the initial phases of the combustion process. Turbulence is also attained in the primary gases by passing these gases over a series of baffles or constrictions; changes in gas directions (caused by baffles) and increases or decreases in velocity (caused by constrictions or expansions in gas passages) mix the primary products of combustion with the additional air necessary for complete combustion.

Turbulence is desirable and necessary. The tumbling of the refuse on grates accelerates the combustion of the solid materials by breaking up agglomerations of the waste and allowing easier escape of moisture and easier access of air to the heated refuse. Furthermore, without turbulence in the primary and secondary combustion chambers, the primary gases created during the first phases of combustion would tend to separate or stratify on the basis of differences in composition and density and some of the gases would pass through the incinerator at least partially unburned. The passage of the primary combustion products over baffles and through constrictions, and the addition of air through ports in secondary combustion chambers creates turbulence and insures complete combustion of the primary gases.

Temperature is one of the most important operational parameters in the entire combustion process. Temperature gradient is the driving force which causes heat transfer to incoming solid refuse and primary gases to sustain combustion. Temperatures in the combustion chambers of an incinerator must be maintained sufficiently high at all times to reduce the refuse and to insure that all of the primary combustion products have been heated to temperatures above the ignition temperature of the various refuse constituents listed in Table 3-4. Additionally, the temperature at the exit port of the combustion chambers should be sufficiently high to insure complete oxidation of gases such as aldehydes so that noxious odors will not be generated. Additionally, higher temperatures are required to create complete burnout of carbon particles suspended in the gas stream; these suspended carbon particles produce smoke if they exit from the incinerator stack. Also, the temperature should be maintained sufficiently high to insure the complete combustion of carbon monoxide to carbon dioxide. All of these requirements can be satisfied if the temperature at the exit port of the combustion chambers is maintained at at least $1400°F$.

For any given configuration of incinerator furnaces and grates, an optimum furnace temperature exists. This temperature can be estimated, if not calculated, on the basis of the characteristics of the refuse being charged into the incinerator. The temperature should not be maintained above a level where the ash and residue from the combustion process fuse and clog the incinerator grates. Also, the moisture content and heating value of the refuse will have an important effect upon the operation of the incinerator and will govern, to a certain degree, the optimum operating temperature. The determination of an optimum temperature for incinerator operation is performed in the same way as the determination of optimum operating temperatures in power plants or steam boilers. The calculations for such a determination are contained in any standard handbook on power plant design. In many instances, such calculations will show that it is necessary to supply additional heat to the combustion process if the refuse has a low heat value or a high moisture content. The combustion process can be altered by the addition of auxiliary fuels to the solid waste, or by the use of an auxiliary burner to preheat the combustion air. In general, the air supplied to the combustion process is the one ingredient in the operation which is easiest to control, and which therefore can be used to control the entire combustion process.

Air Supply. Air, obviously, is needed as a source of oxygen to complete the combustion reaction. Air is also used to control the temperature in the incinerator during combustion and to create necessary turbulence. Air can be drawn into an incinerator through the creation of a natural draft in a high chimney or stack. Under normal conditions, the higher the stack the more air that will be drawn into the incinerator. In most cases, forced-draft and induced-draft fans are also used in refuse incinerators. Natural draft systems have been used for many years and continue to be popular; forced-draft units also are popular because of their overall reliability. Induced-draft systems have not found as much favor as the forced-draft systems because the induction fans must be located between the incinerator combustion chambers and the chimney or stack for the unit. In such a situation, the hot effluent gases from the incinerator must be cooled to protect the induction fans. For this reason, induced-draft fans generally can be used only on incinerators having air pollution control devices which cool the effluent gas stream to prevent damage to the induction fan; wet scrubbers are typical of this kind of air pollution control device.

Regardless of the means for creating the draft, the air is generally added to combustion chambers through openings in the bottom of the furnace area, through interstices in the grates, and also through openings in the upper parts of the combustion chambers. Air admitted in the lower parts of the combustion chambers is called underfire air; air admitted above the grates is called overfire air. Underfire air generally is used to supply the necessary oxygen for the combustion process, to cool the grates which transfer the wastes through the combustion chambers, and to create turbulence. In fluidized-bed incinerators, underfire air is the means for suspending refuse and fuel materials in the combustion bed. Overfire air is added primarily to provide necessary temperature control and additional turbulence. The required volumes of underfire air and of overfire air necessary for a given combustion operation depend upon the characteristics of the refuse (moisture content, heat value, etc.) and on the design of the incinerator. Generally, underfire air amounts to roughly one-half of the total amount of air entering the combustion chamber. It is not advisable to supply the major part of the required air for combustion as underfire air. Air entering under the refuse grates tends to entrain light particles of refuse, such as bits of paper, in the mass of exiting primary combustion products (gases). Particulate loadings in effluent gas streams have been found to vary in direct proportion approximately as the square of the velocity of the exit gas stream. Thus, air entering through the bottom portion of the combustion chambers may greatly increase the particulate loadings in the effluent gases. Overfire air, introduced above the refuse fuel bed, produces much less entrainment of small particles of refuse in the exit gas stream. In general, it is advisable to supply overfire air by means of forced-draft fans with sufficient capacity to create turbulence and mixing of the primary products of combustion.

Auxiliary Fuels. In incinerator operation, auxiliary fuels are used for several purposes: 1) warm-up of combustion chambers, 2) promotion of combustion when the charged solid waste is wet or does not contain an adequate heat value for good combustion, 3) completion of combustion in secondary combustion chambers to promote odor and smoke control, and 4) to furnish supplemental heat for energy-recovery units during periods when the refuse supply (or heat value of the solid waste) is not sufficient to satisfy energy production demands.

Auxiliary fuels, usually gas or oil, may be mixed with the waste, inserted into the combustion chambers in the grate areas, or added elsewhere; insertion locations depend upon the purpose of the auxiliary fuel. If the incinerator is charged with low moisture content, high heat value waste, the auxiliary fuel may be needed only in the secondary combustion chamber for odor and smoke control.

Extra fuel is added in the grate areas of the primary combustion chambers when extra heat is required for moisture evaporation and drying of low heat value wastes.

Waste Gases. Control of waste gases is another important operational consideration for an incinerator. Carbon dioxide, water vapor, and nitrogen are liberated during incineration. These gases ordinarily are not considered air pollutants. Additionally, because of waste composition and because, at times, incomplete combustion occurs, nitrogen dioxide, nitric oxide, sulfur dioxide, sulfur trioxide, and carbon monoxide are also produced. These gases are air pollutants and must be treated or eliminated from the waste gas stream. The incineration process must be designed so that complete combustion will occur and emission of the gases will be controlled by air quality control mechanisms. The control of air quality in incinerator effluents will be discussed in much greater detail in a later section of this chapter.

Summary on Incinerator Operational Parameters. The operational parameters mentioned above are essential to the efficient processing of solid wastes in a municipal incinerator. It is absolutely necessary to form a reliable estimate of the quantities and the range and characteristics of the refuse to be processed in a given facility. Because the quantities and characteristics of the refuse cannot be controlled by the incinerator operator, it is necessary to design the incinerator to accommodate variations in waste quantities and in waste characteristics. The variation in waste characteristics such as moisture content and heating value can be accommodated through the control of the combustion parameters mentioned above: combustion time, combustion temperature, turbulence in combustion chambers, and supply of air. Because of high moisture contents or low heat values in the waste charged into an incinerator, it may be necessary to use auxiliary fuel in addition to the charged solid waste. Finally, it is of the utmost importance to control the effluents from the incinerator which consist primarily of the gases produced during combustion.

3-3.2 Incinerator Components

The number, type, and arrangement of the parts of refuse incinerators vary from one facility to another, but a general description of the types of components used in incinerators may be given here. After the operational requirements of the incineration operation have been determined on the basis of waste characteristics and the other parameters described in the preceding section, the components of the incinerator may be designed to fit the conditions and needs of the particular facility.

Scales. An incinerator scale is needed in order to establish the weights of incoming solid waste. Scales also should be used to weigh outgoing residue, including flyash, grit, and ashes. A weight tally also should be kept for salvaged materials such as ferrous metals. Accurate weight records can be used to improve incinerator operation, to assist in management control of the facility, to assist in planning for new facilities, and to provide an equitable means for the assessment of processing fees. Accurate weights of incoming refuse and outgoing residues are needed in any attempt to rate the effective reduction of refuse in the incinerator and for determining the materials balance for the facility.

In small incinerators of 50 to 100 tons per day capacity, the scales may be manually-operated mechanical devices. On the other hand, large incinerators such as municipal refuse incinerators require automatic scale systems, incorporating load cells, electronic circuitry, and other sophisticated devices. The scales at an incinerator should have sufficient capacity to weigh the largest vehicle which is anticipated to use the incinerator. The dimensions of the scale should be suited to the largest vehicle to be weighed. The scale platform should be long enough to accommodate all of the axles on every incoming vehicle simultaneously; i.e., it should be possible to place all of the wheels of the largest tandem trailer vehicle using the incinerator on the scale platform at one time. Most collection compactor trucks could be accommodated on a scale platform 10 ft wide and 34 ft long; a 50 ft long platform would be sufficient to accommodate most trailers and semitrailers used in the United States for the transfer of solid wastes. Scale capacity should be adequate for the largest incoming vehicles (up to 30 or more tons). Incinerator scales should be tested periodically for accuracy. Wood, steel, or concrete may be used in the construction of the scale platform. The scale pit should have well-founded walls, constructed to resist settlement. The scale pit floor should be paved to facilitate cleaning and maintenance. Sufficient depth should be provided in the pit to allow maintenance and inspection of the scale. If periodic inspection and cleaning of the scale is performed, maintenance requirements can be minimized (Ref. 3-9).

Storage Pit and Tipping Area. The storage pit into which refuse trucks deposit their loads should be large enough to hold much more than the daily collection of refuse. In continuous-operation incinerators (three 8-hr shifts per day) the pit volume may be somewhat smaller than for batch-fed plants or plants which do not operate on weekends. A rule of thumb for pit capacity calls for a volume in the storage pit equal to 5 to 7 days input of refuse. The design and planning for the storage pit must include consideration of the approaches to the pit and the overall dimensions of the tipping area. If the tipping area is cramped or causes delays in the maneuvering of refuse trucks, the operation of the incinerator facility will be greatly hampered. The tipping area should be sufficiently large to permit rapid unloading of about two-thirds of the trucks expected at the peak unloading hour. The anticipated flow of trucks should be carefully estimated so that the area of the tipping floor and the entrances and exits from the tipping floor at the incinerator are adequate. Sufficient vertical clearance should be provided in the tipping areas to allow unloading of the highest trucks which will use the incinerator facility.

Storage pits and tipping areas should be enclosed against wind and weather. Dumping doors or gates should be used to confine odors, dust, and noise produced during the unloading of collection vehicles. An impervious, sloped surface should be constructed on the tipping floor (and in the storage pit) so that periodic washing and cleaning can be accomplished easily. Refuse dumping and the transfer of refuse from the storage pit into the charging mechanism for the incinerator combustion chambers are dusty operations, particularly when the transfer of refuse is done with an overhead crane and grapple. Therefore, due consideration should be given to providing dust control by means of water sprays in the storage pit area.

The storage pit should be arranged so that the overhead crane and grapple combination can be lowered into all points of the storage pit with ease. Operation of the overhead crane and grapple requires some training and expertise on the part of the incinerator personnel. At times, serious damage is done to the concrete walls of the storage pit through careless maneuvering of the heavy transfer grapple. In some incinerator facilities, the sidewalls of the storage pit have been given additional reinforcement to counteract such damage. Also at some facilities, the refuse grapples have been equipped with pads or bumpers to prevent damage of the storage pit walls. Storage pits, for the most part, are of concrete construction and generally are from 20 to 40 ft wide and from 40 to 80 ft deep. Storage pits may be several hundred feet long in large multifurnace incinerator facilities serving high-population metropolitan areas. Water supply fixtures should be installed during construction of the storage pit to facilitate flushing and cleaning of the bottom and side walls; these water pipes also may be used to extinguish fires in the storage pit caused by spontaneous combustion in the deposited refuse. Such fires may be prevented, in large part, through frequent turning and mixing of the deposited refuse by the crane operator. Furthermore, the crane operator may improve operating conditions in the combustion phase of the incinerator operation by mixing the refuse in the storage pit. For example, incoming wet refuse may be mixed with stockpiled dry refuse already in the storage pit; in this way, a more uniform feed is obtained for the combustion chambers.

Incinerator Cranes. In most incinerators, the storage pit is deep and generally is located at a much lower level than is the entrance to the charging facility for the combustion chambers. Therefore, some type of overhead crane is required for transfer of refuse. In large incinerators, it is advisable to have more than one crane available so that periodic maintenance on the cranes may be accomplished with no absolute shutdown of the facility. The provision of more than one crane also provides some security against breakdowns of the equipment. The cranes used in municipal incinerators generally are equipped with clamshell buckets or similar type grapples. In some small incinerator facilities, rail cranes may be used to transfer wastes from narrow storage pits into the entrance to the combustion chambers. However, for large incinerators the use of bridge cranes is preferred. Bridge cranes, moving on parallel rails located on both sides of the storage pit, are capable of reaching into all corners of the storage pit and rapidly transferring the refuse to any of the charging hoppers or charging chutes in multifurnace incinerators because the bridge configuration permits rapid horizontal and vertical movements of the refuse grapple. In most cases, the transfer cranes in an incinerator are powered by heavy duty electric motors. The operation of the overhead cranes may be controlled by operators sitting in small cabs on the crane bridge itself or in a central station. In any event, the operator compartment should be designed to permit easy visibility into all portions of the storage pit; the operator cabin or compartment also should be designed to furnish a comfortable working environment for the operators. The number of cranes and the capacity of the cranes used at an incinerator should be determined on the basis of the capacity of the incinerator and the capacity of the combustion units to be served by the cranes. Generally, crane buckets and grapples are constructed of steel with hardened teeth and edges, and they range in size from approximately 1 to 5 cu yd capacity.

Charging Mechanisms. In many older incinerators and in many small incinerators, the solid waste is charged into the combustion chambers in batches. The solid waste is fed by means of the overhead crane from the storage pit into a charging gate or hatch; the gate is maintained closed except when the waste is being charged. The material in the combustion chamber is allowed to be consumed to a great degree before the charging gate is opened and a fresh batch of refuse is deposited within the combustion chamber. In general, operation of batch-fed incinerators is not as easy to control and is not as efficient as operation of an incinerator combustion chamber in a continuous feed mode. In the continuous feeding systems, solid waste is fed directly into the combustion chambers, generally through a vertical or inclined rectangular chute that is maintained full at all times in order to create an air seal at the entrance to the combustion chamber. Continuous feeding of refuse is desirable in that it minimizes the irregularities in the combustion process; the thermal shock associated with the input of a batch of wet refuse into a combustion chamber wherein the refractory walls and grate systems have attained a high temperature is very significant. In batch-fed incinerators the fluctuation in temperatures caused by such feeding is a serious cause of deterioration of the grates and the refractory walls in the combustion chambers. Additionally, the feeding of the solid waste material on an intermittent basis causes an introduction of large quantities of cooler air whenever the charging gate or hopper is opened. This intermittent operation is much harder to control than is a continuous feeding system.

In most incinerators, some sort of hopper is used to accept the solid waste dropped by the transfer crane. Charging hoppers are particularly advantageous in continuous feed incinerators. In batch-fed furnaces, the charging hopper furnishes temporary storage of the waste on the charging gate until the gate is opened and the refuse falls into the combustion chamber. In continuous feed systems, the charging hopper furnishes temporary storage for the refuse and also creates an air seal for the combustion chamber, in that the entrance to the combustion chamber is completely filled with refuse.

In most incinerators, each furnace or combustion chamber is provided with at least one charging hopper. Most charging hoppers are built in the shape of an inverted truncated pyramid. The size of the hopper opening into the combustion chamber depends to a great extent upon the size of the combustion chamber; in any event, this opening should be large enough to prevent arching of the material which is being deposited in the hopper. The problem of arching of bulky wastes is a significant difficulty associated with the incineration of bulky refuse. In most cases, however, a hopper opening approximately 4 ft wide and from 4 to 10 ft long should be sufficient to pass most items of solid waste fed into municipal incinerators. The charging hopper should be deep enough so that a full grapple load of waste may be deposited in the hopper with no spillage. Charging hoppers may be constructed of reinforced concrete or steel. In many cases, because of the abrasion and wear from the solid waste which is dropped from the crane grapple, the charging hopper is constructed of steel because steel is, in general, more durable than the reinforced concrete. Moreover, the steel has proven to be more durable with respect to the heat from the combustion chamber which enters the charging hopper. Finally, construction of a charging hopper from steel will facilitate repair and replacement of damaged segments.

In some cases, a straight-sided chute is provided between the charging hopper and the combustion chamber. The chute, if provided, should be designed so that the refuse will pass easily through the chute without clogging. In most cases, transfer of refuse into the combustion chambers is accomplished under the influence of gravity; however, in some large incinerators, vibrating or reciprocating feed mechanisms have been installed in the charging chutes. It is of the utmost importance that the refuse passes easily through the chute without clogging. Because of this requirement, chutes should be constructed of materials which will resist corrosion caused by proximity to the combustion chamber and the presence of moisture in the refuse. Also, the inside surfaces of the charging chutes (and the charging hoppers) should be smooth. To facilitate passage of refuse, some charging chutes are designed with increasing cross section from top to bottom. Because of its location close to the combustion chamber, a charging chute should be made of fire resistant material or should be provided with some protection against heat. In many cases, charging chutes are designed with water jackets for cooling.

Furnaces. The furnaces used in refuse incinerators are built in various configurations and shapes, may be equipped with refractory or water-tube walls, and may be equipped with various systems of grates. The most commonly used furnaces in refuse incinerators are the rectangular furnace, the multicell rectangular furnace, the vertical circular furnace, and the rotary kiln furnace. (In this discussion, the term "furnace" is taken to include the primary combustion chamber equipped with grates and the secondary combustion chamber wherein more complete burning of primary combustion gases and entrained particulate matter takes place.) All of the common types of furnaces used in refuse incinerators consist of apparatus for transferring the wastes through the combustion chamber, the grates, and an enclosed refractory or water-wall structure wherein air is mixed with the solid waste materials and the primary gases produced during volatilization of the solid materials.

Furnace design and configuration is based upon the maintenance of a burning rate calculated to create temperatures in the range between approximately 1300 and 1800°F. If the burning rate is the critical factor in the operation of a furnace, then it follows that the design of the furnace must be based upon a consideration of waste characteristics, the travel time for the passage of the wastes through the furnace, and the volumes of air supplied to the wastes during their passage through the combustion chambers. The configuration of the furnace itself then can be modified to furnish sufficient volume for the combustion of the gases produced after drying of the wastes. Additionally, the grate system used for transfer of wastes through the furnace area can be designed on the basis of the heat release value of the refuse materials; heat release rates can be determined by actual tests or they can be estimated through reference to compiled data. Table 3-9(a) shows the design conditions which are pertinent to the incineration of three grades of refuse. Reference to the grate loadings and required volumes shown in this table will furnish an estimate of the required loadings and volumes for a refuse incinerator. If reference is made to Table 3-6 which shows the classifications of wastes established by the Incinerator Institute of America, the designer may obtain an estimate of maximum burning rate of any given type of waste through reference to Table 3-9(b). The information contained in Table 3-9 is useful in the formation of preliminary estimates and may furnish backup information for checking design calculations. However, the quantities shown in that table should be considered estimates only. Figure 3-2 also indicates estimated grate areas required for incineration of refuse, based upon average heat content values of the refuse.

TABLE 3-9(a) Parameters of Design for Refuse Furnaces

Type of Refuse[a]	Grate Loadings in lb of Refuse per Hour of Operation per Square Foot of Grate Area	Volume in Cubic Feet per Ton of Refuse per 24 Hours	
		Furnace Primary Chamber	Secondary Combustion Chamber
Range of values			
M	58–109	8.5–25.0	12.1–28.0
R	50–72	13.4–14.5	26.6–31.8
C	54–98	9.9–13.8	17.2–28.3
Average values			
M	77	12.7	18.5
R	58	13.6	29.9
C	77	11.5	21.3

[a]M—Mixed refuse made up of garbage, rubbish, and noncombustibles.
R—Refuse comprised of burnable rubbish only.
C—Refuse containing combustibles only, such as garbage and burnable rubbish.

Source: Municipal Refuse Disposal, Institute for Solid Wastes, Am. Publ. Works Assn., 1970, p. 167.

The required area of grates and the required combustion volume for incineration of refuse can be furnished through a variety of configurations. The common furnace configurations used in refuse incinerators are shown in Figs. 3-3, 3-4, 3-5, and 3-6.

The most common type of furnace installed in refuse incinerators in the United States at the present time is the rectangular furnace. This type of furnace can be adapted to function with any one of several different types of grate systems. The refuse may be charged into the furnace on an intermittent or a batch-fed basis, or, as is more commonly done at the present time, the refuse may be charged continuously. The refuse moves through the series of combustion areas from the primary or drying area, through the ignition zone and into the primary combustion chamber. The wastes are agitated through the mechanical motion of the transfer grates and through the passage of air coming through the grates and through entry ports in the sides of the combustion chambers. Primary combustion gases pass into the secondary combustion chamber where more complete oxidation takes place; unburned residue passes out of the furnace on a residue removal conveyor or similar apparatus. Figure 3-3 shows the principal components of a rectangular furnace. In large urban areas, it appears advantageous to modify the single rectangular furnace to create the multicell, or "mutual assistance" rectangular furnace. In the multicell furnace, two or more primary combustion chambers, complete with their own individual sets of grates,

TABLE 3-9(b)   Maximum Burning Rate of Various Type Wastes, lb/ft^2/hr

Burning rates are calculated as follows:

Maximum burning rate in lb/ft^2/hr for types #0, #1, #2, and #3 wastes, using factors as noted in the formula:

BR = Factor for type waste $\times$ log of capacity/hr

#0 waste factor = 13
#1 waste factor = 13
#2 waste factor = 10
#3 waste factor = 8
BR = Maximum burning rate in lb/ft^2/hr

Example: Assume incinerator capacity of 100 lb/hr for type #0 waste
BR = 13 (factor for #0 waste) $\times$ log 100 (capacity/hr) = 13 $\times$ 2 = 26 lb/ft^2/hr

Capacity lb/hr	Logarithm	#0 Waste[a] Factor, 13	#1 Waste[a] Factor, 13	#2 Waste Factor, 10	#3 Waste Factor, 8	#4 Waste[b] No Factor
100	2.00	26	26	20	16	10
200	2.30	30	30	23	18	12
300	2.48	32	32	25	20	14
400	2.60	34	34	26	21	15
500	2.70	35	35	27	22	16
600	2.78	36	36	28	22	17
700	2.85	37	37	28	23	18
800	2.90	38	38	29	23	18
900	2.95	38	38	30	24	18
1,000	3.00	39	39	30	24	18

[a]The density of the mixture and therefore the burning rate in lb/ft^2 of Type 0 waste, or Type 1 waste, are affected if the trash or rubbish mixture contains more than 10% by weight of catalogues, magazines, or packaged papers.

[b]The maximum burning rate in lb/ft^2/hr for Type 4 waste depends to a great extent on the size of the largest animal to be incinerated. Therefore, whenever the largest animal to be incinerated exceeds $\frac{1}{3}$ the hourly capacity of the incinerator, use a rating of 10 lb/ft^2/hr for the design of the incinerator.

Source: Incinerator Standards, Incinerator Institute of America, November 1968, p. 6A.

are placed side by side. Refuse is fed into each furnace cell individually through the top of the cell at one end; however, the primary combustion products (gases and entrained particulates) enter a common secondary combustion chamber. The individual rectangular furnace cells are serviced by a common residue removal apparatus. The advantage of the multicell configuration is the capability for maintenance of individual primary combustion chambers without interruption of operations in the remaining furnaces. Figure 3-4 shows the basic elements of the multicell rectangular furnace.

The vertical circular furnace, or cylindrical furnace, is of older vintage than the

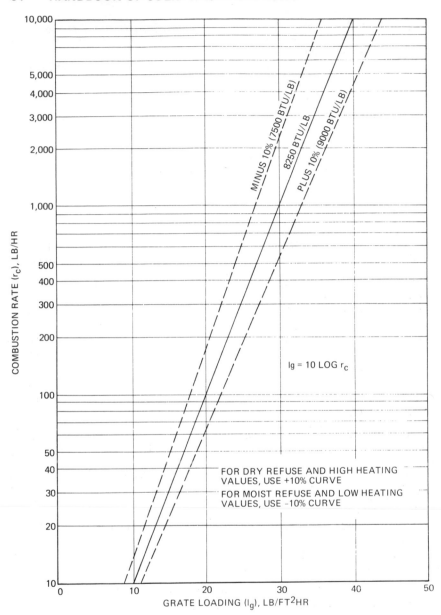

Fig. 3-2 **Relationship of grate loading to combustion rate for multiple chamber incinerators (redrawn from** *Air Pollution Engineering Manual*, **U.S. Dept. of Health, Education, and Welfare, USPHS, 1967, p. 419).**

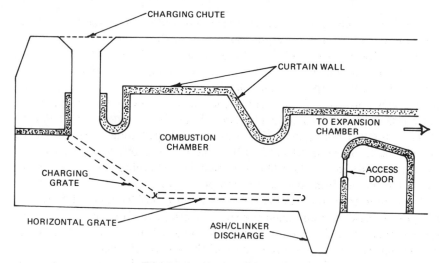

Fig. 3-3 Rectangular furnace.

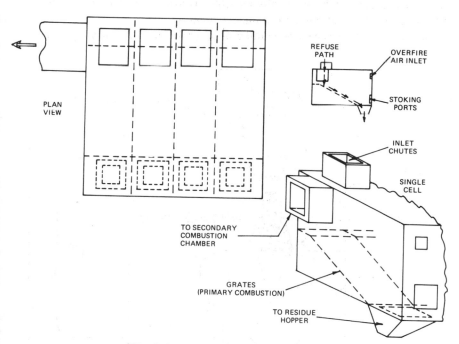

Fig. 3-4 Multicell rectangular furnace.

multicell rectangular furnace described in the preceding paragraph. The cylindrical furnace has been used to a great extent in batch-fed incinerator operations. In the cylindrical furnace, refuse is charged intermittently through a gate or door in the upper portion of the furnace; when the charging gate opens, the refuse drops onto a cone-shaped central grate. The configuration of the cylindrical furnace is shown in Fig. 3-5. The central cone-shaped grate in the vertical circular furnace is surrounded by a circular grate. Underfire air enters the combustion chamber through both grates. The central cone-shaped grate rotates during operation and rabble arms attached to the central cone move slowly over the surrounding circular grate. The rabble arms tumble the burning refuse and gradually sweep the residue from the combustion process to the outsides of the circular grate where it drops through the grate system into a residue removal device. The hot primary gases and entrained particulate matter exit from the primary combustion chamber and are more completely oxidized in an adjacent secondary combustion chamber. In general, most of the vertical circular furnaces now in operation in refuse incinerators are refractory lined. The principal means for combustion control is the supply of underfire air in the primary combustion chamber and the supply of overfire air which is inserted in the secondary combustion chamber to sustain more complete combustion.

The rotary kiln furnace is something of a hybrid construction in that it con-

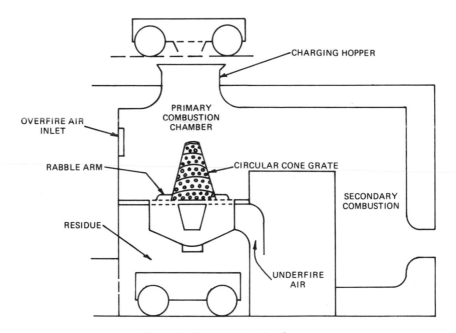

Fig. 3-5   Vertical circular furnace.

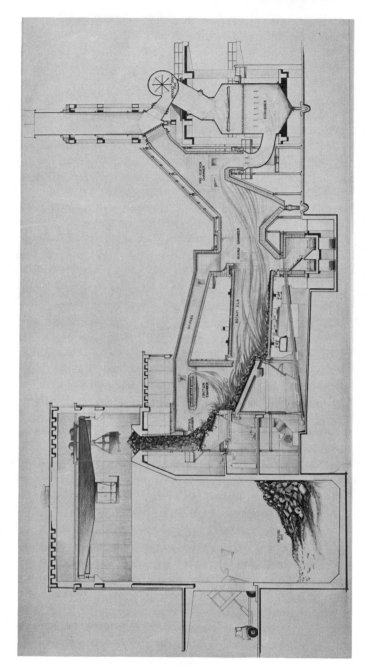

Fig. 3-6 A rotary kiln furnace incinerator (*courtesy* International Incinerators, Inc.).

sists of a rectangular drying and ignition zone coupled with a large kiln oriented at a slight slope to the horizontal; wastes enter the higher end of the kiln and the slow revolving movement of the kiln tumbles the waste and propels it toward the lower end of the kiln. Ideally, the refuse is dried and partially burned in the primary rectangular combustion chamber and then completely combusted in the rotating kiln where the tumbling action creates necessary turbulence. The hot gases and entrained particulates created within the kiln are completely combusted in a secondary combustion chamber which follows the rotating kiln in this type of incinerator. The unburned residue gradually moves along the bottom surface of the rotating kiln and falls from the end of the kiln into a residue removal apparatus. The general constituents of a rotary kiln furnace are shown in Fig. 3-6.

A large variety of different types of grates are in use in refuse incinerators today. A number of these grates are shown in Figs. 3-7, 3-8, and 3-9. Different designers and manufacturers make conflicting claims for the various types of grates and furnaces which have been shown in the accompanying figures and described in the preceding paragraphs. It is interesting to note in Table 3-9(a) the large variations in volumes required for combustion which are shown in this table. This table was prepared by personnel of the American Public Works Association and is based upon an analysis of the designs used in 24 large incinerators which were built in 18 cities in the United States. As a rule of thumb, the area of grates and the corresponding grate loading can be found on the assumption of a heat release value for the refuse equivalent to approximately 300,000 BTU per hr per sq ft of grate. The combustion volumes generally correspond to a calculated heat release value of approximately 12,500 BTU per hr per cu ft. Although the volumes of combustion chambers shown in Table 3-9(a) vary considerably, it was found upon analysis of the individual incinerators included in the APWA study that the total combustion volume varies very little from one incinerator to another; i.e., the combined volumes of the primary and secondary combustion chambers vary insignificantly from one incinerator facility to the other for the incinerators studied (Ref. 3-27). The variation in configurations

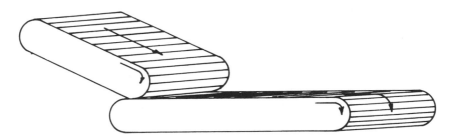

Fig. 3-7   Traveling grates.

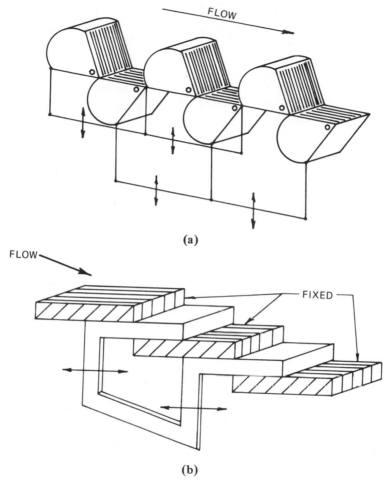

Fig. 3-8   (a) Rocking grates; (b) reciprocating grates.

and the allotment of different volumes in the primary and secondary combustion chambers in the different types of furnace-grate combinations indicates that caution should be exercised in the design of these combustion facilities. The data shown in Table 3-9 should be taken to indicate only approximate values. Any detailed final design of an incinerator facility should be based upon complete calculations of heat balance, materials balance, air supplied (volumes and velocities), grate transport rates, and combustion temperatures. This design calls for a comprehensive application of engineering thermodynamics.

As an indication of the areas and volumes required in refuse incinerator combustion chambers and as an aid in preliminary estimate and design of such

Fig. 3-9   Dusseldorf roller grates (*courtesy* Vereinigte Kesselwerke AG).

facilities, Table 3-10 has been included. This table has been developed by the Incinerator Institute of America. The waste classifications and incinerator classifications shown in Tables 3-6 and 3-7 furnish the starting point for use of Table 3-10. With a preliminary estimate of the characteristics and quantities of wastes to be processed in a given facility, and with a knowledge of the type of facility to be built, a waste manager may use the information in Table 3-10 to form an estimate of the required configuration, volume, etc. of the combustion system for the incinerator. It should be remembered, however, that detailed, comprehensive calculations are required for the actual design of such a facility.

In most of the incinerators in use in the United States today, no matter what the configuration of the combustion chambers, the walls of the furnaces are built of refractory bricks. The refractory brick used in refuse incinerators and in similar combustion chambers are produced from combinations of earth materials

TABLE 3-10   Multiple Chamber Incinerator Design Factors

Item and Symbol	Recommended Value	Allowable Deviation
Primary combustion zone		
Grate loading, L_G	$10 \log R_c$; lb/hr ft^2 where R_c equals the refuse combustion rate in lb/hr	±10%
Grate area, A_G	$R_c \div L_G$; ft^2	±10%
Average arch height, H_A	$4/3 (A_G) 4/11$; ft	—
Length to width ratio (approx)		
Retort	Up to 500 lb/hr, 2:1; over 500 lb/hr; 1.75:1	—
In-line	Diminishing from about 1.7:1 for 750 lb/hr to about 1:2 for 2,000 lb/hr capacity. Over-square acceptable in units of more than 11 ft ignition chamber length.	—
Secondary combustion zone		
Gas velocities		
Flame port at 1,000°F, V_{FP}	55 ft/sec	±20%
Mixing chamber at 1,000°F, V_{MC}	25 ft/sec	±20%
Curtain wall port at 950°F, V_{CWP}	About 0.7 of mixing chamber velocity	—
Combustion chamber at 900°F, V_{CC}	5 to 6 ft/sec, always less than 10 ft/sec	—
Mixing chamber downpass length. L_{MC}, from top of ignition chamber arch to top of curtain wall port	Average arch height, ft	±20%
Length to width ratios of flow cross sections		
Retort, mixing chamber, and combustion chamber	Range—1.3:1–1.5:1	—
In-line	Fixed by gas velocities due to constant incinerator width	—
Combustion air		
Air requirement batch-charging operation	Basis: 300% excess air. 50% air requirement admitted through adjustable ports; 50% air requirement met by open charge door and leakage	
Combustion air distribution		
Overfire air ports	70% of total air required	—
Underfire air ports	10% of total air required	—
Mixing chamber air ports	20% of total air required	—
Port sizing, nominal inlet velocity pressure	0.1 inch water gage	—

(Continued)

TABLE 3-10 (*Continued*)

Item and Symbol	Recommended Value	Allowable Deviation
Combustion air (*Continued*)		
Air inlet ports oversize factors		
Primary air inlet	1.2	
Underfire air inlet	1.5 for over 500 lb/hr to 2.5	
	for 50 lb/hr	
Secondary air inlet	2.0 for over 500 lb/hr to 5.0	
	for 50 lb/hr	
Furnace temperature		
Average temperature, combustion products	$1,000°F$	$\pm 20°F$
Auxiliary burners		
Normal duty requirements		
Primary burner	3,000–10,000 } BTU/lb of moisture	
Secondary burner	4,000–12,000 } in refuse	
Draft requirements		
Theoretical stack draft, D_T	0.15–0.35 inch water gauge	—
Available primary air induction draft, D_A. (Assume equivalent to inlet velocity pressure.)	0.1 inch water gauge	
Natural draft stack velocity, V_S	Less that 30 ft/sec at $900°F$	—

Source: Air Pollution Engineering Manual, U.S. Department of Health, Education, and Welfare, Public Health Service, 1967, p. 418.

such as clay minerals, silica, alumina, and other substances which are highly resistant to heat. Some of the properties of typical refractory materials are shown in Table 3-11. The designations "super-duty," "high-duty," "medium-duty," and "low-duty" refer to various grades of refractories produced from clay. The refractory linings in incinerator combustion chambers serve several important purposes: they protect the furnace structure and the surrounding portions of the incinerator from the high temperatures generated in the combustion area; they function as a storage reservoir for heat, so that heat is emitted from the refractories during combustion phases when the temperature drops and is absorbed by the refractories when the temperature rises again (the refractories thus sustain combustion); and the refractory walls physically contain the combustion process.

The choice of refractory type is based upon consideration of a number of factors including the abrasion resistance, heat resistance, physical strength, cost, and resistance to spalling and slagging shown by the different grades of refractory materials. Table 3-12 gives an indication of the types of refractories which should be placed at various locations within the furnace walls, ceiling, and floor on the basis of resistance to heat, abrasion, slagging, spalling, mechanical shock, and adherence of particulate matter. In former years, the refractory walls in

TABLE 3-11 Typical Physical Properties of Refractories[a]

Type	Pyrometer Cone Equivalent	Density, lb/ft³	Modulus of Rupture, lb/in.²	Cold Crushing Strength, lb/in.²	Thermal Conductivity at 1800 BTU/hr f	Thermal Expansion at 1,800°, % lineal	Porosity, % Vol	Absorption, % by Wt	Resistance to Spalling
Super-duty	33–34	140–145	700–850	1,200–2,400	10.6	0.64	10–18	4–7.5	Excellent
High-duty	31–33	140–144	800–1,000	2,100–2,800	10.6	0.64	12–19	5–8	Good
Intermediate-duty	29–31	136–140	1,100–1,400	4,000–5,500	10.3	0.64	13–16	6–7	Good
Low-duty	19–26	160–165	1,000–2,300	2,000–6,000	0.12	0.64	12–25		Fair
Insulating brick	17–32	40–46	60–320	115–380	3.0	0.50	22–26		Good
Alumina, 50%	34–35	130–135	700–1,000	1,400–2,000	10.1	0.60	27–34	10–12	Good
Alumina, 60%	35–37	130–140	850–1,150	2,100–2,500	10.1	0.60	29–36	12–17	to
Alumina, 70%	36–38	125–130	700–1,000	1,400–2,000	9.5	0.60	23–27	13–18	Excellent
Silica	31½–32½	103–107	600–1,200	1,500–3,000	12.4	1.3		13–16	Excellent above 1,200°F; poor below
Silicon carbide	38	157	2,000		108.0	0.44	13.2		Excellent; slagging resistance excellent

[a]Data, obtained from representative manufacturers' catalogs, determined through tests made according to recommendations of the American Society for Testing and Materials.

Source: *Municipal Refuse Disposal*, Institute for Solid Wastes, Am. Publ. Works Assn., 1970, p. 177.

TABLE 3-12 Suggested Refractory Placement in Incinerators

Incinerator Part	Temperature Range, °F	Abrasion	Slagging	Mechanical Shock	Spalling	Flyash Adherence	Recommended Refractory
Charging gate	70–2,600	Severe, very important	Slight	Severe	Severe	None	Super-duty
Furnace walls, grate to 48 in. above	70–2,600	Severe	Severe, very important	Severe	Severe	None	Silicon carbide or super-duty
Furnace walls, upper portion	70–2,600	Slight	Severe	Moderate	Severe	None	Super-duty
Stoking doors	70–2,600	Severe, very important	Severe	Severe	Severe	None	Super-duty
Furnace ceiling	700–2,600	Slight	Moderate	Slight	Severe	Moderate	Super-duty
Flue to combustion chamber	1,200–2,600	Slight	Severe, very important	None	Moderate	Moderate	Silicon carbide or super-duty
Combustion chamber walls	1,200–2,600	Slight	Moderate	None	Moderate	Moderate	Super-duty or 1st quality
Combustion chamber ceiling	1,200–2,600	Slight	Moderate	None	Moderate	Moderate	Super-duty or 1st quality
Breeching walls	1,200–2,400	Slight	Slight	None	Moderate	Moderate	Super-duty or 1st quality
Breeching ceiling	1,200–2,400	Slight	Slight	None	Moderate	Moderate	Super-duty or 1st quality
Subsidence chamber walls	1,200–1,600	Slight	Slight	None	Slight	Moderate	Medium-duty or 1st quality
Subsidence chamber ceiling	1,200–1,600	Slight	Slight	None	Slight	Moderate	Medium-duty or 1st quality
Stack	500–1,000	Slight	None	None	Slight	Slight	Medium-duty or 1st quality

Source: Municipal Refuse Disposal, Institute for Solid Wastes, Am. Publ. Works Assn., 1970, p. 179.

incinerator combustion chambers were built as gravity walls; i.e., the wall was constructed of thick courses of brick and designed to be self-supporting. As incinerators became larger to meet the demand of growing urban areas, the sizes of the combustion chambers necessarily increased. The thickness of the gravity walls and their corresponding weights increased to the point where gravity wall construction was quite expensive. The maintenance of large expanses of masonry gravity walls also is a difficult problem. To counter the rising cost and increased difficulties in maintenance associated with gravity walls, suspended or hung walls were developed. In the suspended wall, the basic structure consists of structural steel columns placed 3 to 5 ft apart on the outside of the combustion chamber. Along the inside surfaces of these steel columns are heat resistant fastenings to which are fastened horizontal shelves. The horizontal shelves are used to support low sections of refractory wall; in this way a series of low walls placed vertically on top of each other is created. The structural steel framework supports the entire weight of the refractory walls. Because the steel framework supports the weight of the refractories, the thickness of the wall need not be increased to support incumbent weight as is the case in the gravity type walls. Because the individual low sections of wall are entirely supported by the steel framework, a small gap may be created between the top of one section and the bottom of the next higher section to provide for vertical expansion of the refractories upon heating. Because the suspended wall has the capability for vertical expansion, damage due to rapid or unequal expansion is virtually eliminated. Such damage causes severe maintenance problems with gravity walls. An additional advantage of the suspended wall construction is the ease with which individual sections of the refractory walls may be repaired or replaced. Thus, the suspended wall may be less expensive in terms of initial costs because of smaller thickness (generally from one-half to one-third the thickness of a gravity wall, on the average), and easier and cheaper to maintain and repair.

In a small number of American incinerators and in a large number of European incinerators, the walls of the refuse combustion chambers consist of closely spaced metal tubes containing water. Heat is transferred through the metal tubes to vaporize the water and the resulting steam is used to create power or to provide space heating capability. In the event that water-tube walls are installed in a refuse incinerator, only a very thin refractory outer wall is required for the combustion chamber. However, thick refractory linings are needed at the exit areas of the combustion chambers.

In addition to the advantages of a lower requirement for refractory in the combustion chamber walls, and in addition to the capability for power production associated with steam production, water-tube walls in refuse incinerator furnaces also possess another advantage. Because of more effective transfer of heat from the combustion chamber, air requirements for combustion are considerably reduced as compared to the requirements for excess air in refractory wall combustion chambers. With reduction in required air in the combustion chambers, the gas velocity at the air entry ports and throughout the furnace itself may be reduced. Smaller volumes and lower velocities of air and combustion gases reduce the amount of particulate matter entrained in the exit stream. This reduction in particulate loading of the effluent gas stream is a significant advantage of water-tube walls because of the consequent savings in air pollution control equipment associated with the lower particulate loadings.

However, water-tube walls possess certain disadvantages or limitations. The temperature of the water in the walls must be maintained at 300°F or higher in order to prevent pitting of the metal tubes on the external surfaces of the tubes adjacent to the combustion area. This often requires preheating of the water in the tubes. Furthermore, the water in the tubes must be deaired and treated to remove hardness. If the air is not removed from the water in the tubes, as is done in standard boiler operations, internal oxygen corrosion on the insides of the tubes will result. If the water in the tubes is not softened, scale will form on the insides of the tube and lead to the formation of hot spots on the tubes and eventual tube failure. Most of the problems associated with the use of water tubes in the walls of American incinerators have resulted from the use of untreated water (water which has not been deaired and softened).

As mentioned previously, the operational parameters which control combustion in a refuse incinerator are combustion time, turbulence, and combustion temperature. Combustion time may be regulated by the speed of transfer of the refuse on the system of grates used in the incinerator. Additionally, these grates can be used to promote combustion by increasing the agitation and turbulence in the burning wastes. However, the advantages in complete combustion gained through tumbling the burning wastes on transfer grates must be balanced against the disadvantages of entraining particulate matter in the effluent from the combustion chambers. The agitation produced by the grate system should provide some turbulence but should not be so severe that excessive amounts of particulates are entrained in the exit gas stream. A continuous gentle tumbling or agitation of the wastes appears to be most satisfactory. This tumbling is achieved through a mechanical action of the grates and through the passage of underfire air through the grate systems. Various types of grates have been shown in Figs. 3-7, 3-8, and 3-9. The choice of grate system for use in a particular combustion chamber is governed to a great extent by the characteristics of the wastes to be incinerated and by the provisions for air supply in the combustion chamber. In any event, the grate system selected for use must be capable of withstanding high combustion temperatures, thermal shocks which occur when wet or low temperature wastes encounter grates previously heated, abrasion and wear caused by dense metallic or mineral wastes, wedging and clogging of the interstices between grates caused by small particles of inert material dropping out of the burning wastes, and generally detrimental conditions created by the necessity for transferring heavy loads of abrasive and sometimes corrosive hot materials. Many different types of grates have been designed and a number of different grates systems are currently available commercially. The selection of a grate system for use in a particular incinerator should be done carefully because proper functioning of the grates in transferring the wastes through the combustion chambers is essential for proper overall functioning of the incinerator.

Combustion also may be controlled, in addition to the control capabilities furnished by the grate systems, through the regulation of air supply in the combustion chambers. Air supply is regulated through the design and operation of the chimney system for the incinerator and the induced or forced-draft fans used in the facility. Natural draft is provided in incinerators through the construction of an upright chimney or flue. A chimney functions because of the fact that the temperature of the gases within the chimney is higher than the temperatures in the surrounding ambient air. Because of the difference in the temperature from inside to outside, the gases inside the chimney are less dense than those outside. Thus, the hot gases tend to rise and pass out of the chimney and create a draft through the chimney and also through the combustion chambers which precede the chimney structure. In general, the design of natural draft chimneys is such that the pressure differential from inside to outside ranges from approximately 2 to 4 in. of water; the gas velocities in natural draft chimneys range between approximately 25 to 50 ft per sec. In general for incinerator facilities, masonry chimneys have been favored over steel chimneys because of the unsightly appearance and high maintenance costs of the steel chimneys. Also, the metal chimneys require heat protection in the form of heat-resisting alloys and special protective coatings. Masonry chimneys require no unsightly supports such as guide wires and other exterior reinforcements, and, therefore, are more aesthetically pleasing than steel chimneys. They also have a longer life than do metal chimneys and maintenance costs generally are lower for the masonry chimneys. However, because of their great bulk and weight, masonry chimneys require very large and strong foundations. Moreover, masonry chimneys are subject to severe cracking and deterioration if the temperature of the chimney is not slowly increased. Thus, extreme care should be taken during furnace warmup to avoid rapid heating of masonry chimneys, especially newly installed masonry chimneys.

In recent years, because of the disadvantages associated with high chimneys (large foundation requirements, high construction costs, hazard to aircraft, etc.), there has developed a trend towards shorter stub stacks. Because of the lower height of the stub stacks, the natural draft created in such stacks is not sufficient, in general, to provide air supply pressures for incinerator operations. Therefore, induced-draft fans or forced-draft fans must be used where such stub stacks are installed on incinerators. As mentioned previously, if induced-draft fans are installed, the gases which exit from the secondary combustion chamber must be cooled sufficiently to prevent damage to the induced-draft fan, located between the secondary combustion chamber and the incinerator chimney. Tables 3-13 and 3-14 list estimated requirements for both forced-draft fans and induced-draft fans for refuse incinerators. Theoretical volumes of air required for complete combustion of refuse may be obtained if the composition of the refuse is known (see Table 3-8). However, the composition of municipal refuse is variable and quite heterogeneous. Therefore, it is impossible to calculate exact combustion air requirements. Additionally, air is used to control the combustion process, as mentioned previously. For these reasons, it is quite common for the operation of a refractory wall refuse incinerator to include the provision of as much as 200 percent excess air in the combustion chambers. On the basis of average compositions for American refuse, the requirement of 200 percent excess air amounts to about 8 lb of air per lb of refuse. The suggested provisions of draft fans shown in Tables 3-13 and 3-14 are based upon similar estimates of required amounts of excess air. With greater emphasis in recent years upon recovery of heat and power through the incineration of refuse, and with the consequent installation of water-tube walls in refuse incinerators, excess air requirements should correspondingly decrease. Approximately 50 percent excess air may be used in water-tube wall incinerators, as opposed to approximately 200 percent excess air required for complete combustion and temperature regulation in refractory wall units. The advantages of lower entrainment loadings of particulates and smaller volumes of effluent gases requiring treatment have been mentioned previously.

In addition to the gases and entrained particulate matter which are removed through the fan and chimney systems mentioned above, residue produced during the incineration process also requires removal. Incinerator residue, all the refuse material remaining after burning, can include such diverse items as pieces of rock, bits of glass, tin cans, ash, clinkers, and various amounts of unburned combustible materials. If the incinerator is not operated properly, a significant amount of combustible items may pass through the combustion chambers when combustion time is excessively reduced. Rapid transfer of wastes through the combustion chamber, with insufficient combustion time, is a common failing of ill-informed incinerator operators who are concerned principally with processing a stated volume of input solid wastes per day. The rapid processing of low heat value or wet refuse in order to meet a production quota for input refuse leads to poor reduction of the refuse and considerable amounts of unburned residue. However, even in adequately operated incinerators, a significant fraction of the input wastes will be practically incombustible at temperatures below 2000°F. This residue requires removal.

Residue removal may be accomplished on an intermittent basis or as a continuous operation. The residue consists of the material which falls off the end of the last transfer grate in the primary combustion chamber, the small particles of unburned material which drop through the transfer grates, and any particulate

TABLE 3-13 Recommended Natural Draft Stacks or Chimneys

Incinerator Capacity, lb/hr	Class III Incinerators — Type 0 Waste Air Supply[a]	Type 0 Stack Diameter, in.	Type 0 Stack Height, ft[b]	Type 1 Waste Air Supply[a]	Type 1 Stack Diameter, in.	Type 1 Stack Height, ft[b]	Type 2 Waste Air Supply[a]	Type 2 Stack Diameter, in.	Type 2 Stack Height, ft[b]	Class IV Incinerators Type 3 Waste Air Supply[a]	Type 3 Stack Diameter, in.	Type 3 Stack Height, ft[b]	Class VI Incinerators Type 4 Waste Air Supply[a]	Type 4 Stack Diameter, in.	Type 4 Stack Height, ft[b]
50													90	9	25
100	445	16	30	350	14	30	250	12	25	200	12	30	180	10	25
150	670	18	30	525	16	30	375	14	30	300	14	30	270	12	25
200	890	20	35	700	18	35	500	16	30	400	14	35	360	14	25
300	1,335	22	35	1,050	20	35	750	18	35	600	16	40	540	16	30
400	1,780	24	40	1,400	22	40	1,000	20	40	800	18	40	720	18	30
500	2,225	26	40	1,750	24	40	1,250	22	40	1,000	20	45	900	20	30
600	2,670	28	40	2,100	26	40	1,500	24	40	1,200	22	45			
700	3,115	30	45	2,450	28	45	1,750	26	45	1,400	24	50			
800	3,560	32	45	2,800	30	45	2,000	28	45	1,600	26	50			
900	4,000	34	45	3,150	32	45	2,250	30	45	1,800	28	50			
1,000	4,450	36	45	3,500	34	45	2,500	32	45	2,000	30	50			

[a]Air supply is given in ft³/min at 70°F and is the minimum which must be available at all times in the incinerator room at atmospheric or a slight positive pressure. The incinerator room or rooms should never be under a negative or minus pressure. If the incinerator is charged from a room other than the incinerator room the quantity of air shown must be available in both rooms.
The quantity of air shown must be increased to satisfy the following:
1. If stack or chimney is higher than minimum to satisfy the larger barometric damper involved.
2. If any other equipment requiring air supply is located in the incinerator room or charging room.

[b]The stack heights are based upon the following:
1. Installation made at or near sea level.
2. Stack heights measured from base of the incinerator.
3. Incinerator is side charged.
4. Breeching or flue connection not exceeding 10 ft in length in a straight run or 3 ft including not more than one 90° bend or two 45° bends.
5. Stack extends not less than 3 ft above any roof within 75 ft of the top of the stack.
The stack heights must be increased or may be decreased as follows:
1. Increase height 5% per 1,000 ft above sea level.
2. Decrease height 25% if stack is directly on top of incinerator eliminating any breeching or flue connection.
3. Increase height 15% if incinerator is top charged.
4. Increase height 15% for each additional 10 ft of straight breeching and 15% for each additional 90° bend.

Source: Incinerator Standards, Incinerator Institute of America, November 1968, p. 4C.

matter collected from the effluent gas stream. Although these materials may not be combustible at the operating temperatures of the refuse incinerator, they are very hot when they leave the combustion chambers and therefore require quenching and cooling. In batch-operated furnaces, residue falls through the systems into storage hoppers and is then periodically removed. Upon removal, the

TABLE 3-14   Recommended Induced Draft Fans

Incinerator Capacity, lb/hr	Type 0 Waste					Type 1 Waste				
		lb/hr		Fan			lb/hr		Fan	
	Air Supply[a]	Flue Gases	Cooling Air	ft³/min at 700°F	Cold Static Pressure	Air Supply[a]	Flue Gases	Cooling Air	ft³/min at 700°F	Cold Static Pressure
Class III Incinerators										
100	1,060	1,380	2,760	2,060	0.7	850	1,080	2,160	1,630	0.7
150	1,600	2,060	4,120	3,100	0.7	1,275	1,620	3,240	2,445	0.7
200	2,100	2,750	5,500	4,130	0.72	1,700	2,160	4,320	3,260	0.72
300	3,200	4,140	8,280	6,200	0.75	2,550	3,240	6,480	4,890	0.75
400	4,250	5,520	11,040	8,300	0.75	3,400	4,320	8,640	6,520	0.75
500	5,300	6,900	13,800	10,300	0.8	4,250	5,400	10,800	8,150	0.8
600	6,350	8,280	16,560	12,400	0.8	5,100	6,480	12,960	9,780	0.8
700	7,350	9,650	19,300	14,500	0.85	5,950	7,560	15,120	11,410	0.85
800	8,450	11,000	22,000	16,500	0.85	6,800	8,640	17,280	13,040	0.85
900	9,500	12,400	24,800	18,600	0.85	7,650	9,720	19,440	14,670	0.85
1,000	10,600	13,800	27,600	20,600	0.85	8,500	10,800	21,600	16,300	0.85
Class III Incinerators, Type 2 Waste										
100	600	768	1,540	1,130	0.7					
150	900	1,152	2,310	1,700	0.7					
200	1,200	1,536	3,080	2,250	0.72					
300	1,800	2,304	4,610	3,380	0.72					
400	2,400	3,072	6,150	4,500	0.75					
500	3,000	3,840	7,680	5,680	0.75					
600	3,600	4,608	9,220	6,750	0.75					
700	4,200	5,376	10,750	7,880	0.8					
800	4,800	6,144	12,290	9,000	0.8					
900	5,400	6,912	13,830	10,130	0.8					
1,000	6,000	7,680	15,360	11,250	0.8					
Class IV Incinerators, Type 3 Waste										
50										
100	485	625	1,250	920	0.68					
150	728	938	1,875	1,380	0.7					
200	970	1,250	2,500	1,840	0.7					
300	1,455	1,875	3,750	2,760	0.72					
400	1,940	2,500	5,000	3,680	0.75					
500	2,425	3,125	6,250	4,600	0.75					
600	2,910	3,750	7,500	5,520	0.75					
700	3,395	4,375	8,750	6,440	0.8					
800	3,880	5,000	9,000	7,360	0.8					
900	4,365	5,625	10,250	8,280	0.8					
1,000	4,850	6,250	12,500	9,200	0.8					

TABLE 3-14   *(Continued)*

| Incinerator Capacity, lb/hr | Type 0 Waste | | | | | Type 1 Waste | | | | |
| | lb/hr | | Fan | | | lb/hr | | Fan | | |
	Air Sup-ply[a]	Flue Gases	Cool-ing Air	ft³/min at 700°F	Cold Static Pressure	Air Sup-ply[a]	Flue Gases	Cool-ing Air	ft³/min at 700°F	Cold Static Pressure
			Class VI Incinerators, Type 4 Waste							
50	200	262	525	385	0.68					
100	400	523	1,050	770	0.68					
150	600	785	1,570	1,155	0.68					
200	800	1,046	2,100	1,540	0.68					
300	1,200	1,569	3,050	2,310	0.7					
400	1,600	2,092	4,200	3,080	0.7					
500	2,000	2,615	5,250	3,850	0.7					

[a]Air supply is given in ft³/min at 70°F and is the minimum that must be available at all times in the incinerator room for combustion and fan cooling air.
The total flue gases or total products of combustion are given in lb per hour.
The cooling air is given in lb per hour and is the air required to be bled into and mixed with the flue gases before entering the induced draft fan and unlined breeching section.
The fan capacity is given in ft³/min at 700°F, which is the anticipated temperature of the air-gas mixture entering the induced draft fan.
The static pressure of the fan is given as the cold (70°F) static pressure and with the installation made at or near sea level. The static pressure at 700°F is 45% of the "cold" static pressure. Increase the cold static pressure 3.5% for every 1,000 feet above sea level.
Water sprays or a combination of water and air may be used to cool the flue gases before they enter the fan. The ft³/min of the fan reduces but the static pressure of the fan increases to overcome the resistance created by the gas washer or scrubber used.

Source: Incinerator Standards, Incinerator Institute of America, November 1968, p. 6C.

residue is quenched and cooled and transferred to a residue hopper. In most batch-fed incinerators, the residue is transferred from the storage hopper beneath the furnace area to the residue hopper by means of a transfer tunnel beneath the furnace floor. This tunnel should be given adequate attention during design stages, since clogging of residue transfer facilities could lead to interruption in service of the incinerator. The tunnel should be wide enough so that an employee can walk safely beside a transfer vehicle. It should also be tightly paved, well-drained, and adequately lighted. Provisions should also be made either in the tunnel or in the residue hopper for removal of excess quenching water since such water represents unnecessary extra weight in trucks used to transfer the residue to a final disposal site (Ref. 3-9). Also, leakage of such water from transfer trucks can lead to community dissatisfaction with the operation of the incinerator. In most of the newer incinerators built for refuse disposal in the United States, refuse is charged into the combustion chambers on a continuous basis and, likewise, residue is removed from the combustion area on a continuous basis. In most incinerators, siftings and small particles which drop through the grate systems are carried by a conveyor or a similar device to the quenching tank. Unburned materials falling from the terminal grate in the combustion chamber also fall into the quenching tank. The residue is removed from quenching tanks in most installations through the use of a drag conveyor which is submerged in the water of the quenching tank. The residue is carried upward along the inclined drag conveyor and excess quenching water drains back down the slope of the conveyor into the quenching tank. In most incinerator facilities, flyash and particulate matter are collected from the effluent gas stream in a different location than the residue quench tank area. In electrostatic precipitators, for example, entrained particulate matter is collected in the dry form. However, in some incinerator installations, collected particulate matter is mixed with water to form a slurry for ease of transporting. The water-particulate slurry can be pumped into the same transfer trucks which are used to remove residue from the quench tank. However, more recent practice favors separate removal of flyash and residue. With increasing emphasis upon materials recycling, residue removal systems may require modification to allow separation of valuable constituents from the incinerator residue. The United States Bureau of Mines has developed a complex, sophisticated system for the reclamation of valuable constituents of incinerator residue. This system consists of a number of separation screens and mills or grinders to reduce the size of the individual particles of residue to aid in further separation. Magnetic separators are also included in this system. The USBM system is described in more detail in Chapter 6.

The provision for removal of the particulate matter from the effluent gas stream in a refuse incinerator requires comprehensive analysis and planning. Furthermore, the control of air pollutants in the effluent from an incinerator is a complex and difficult operation. Therefore, a comprehensive description of such control measures will be given in a later section of this chapter, section 3-4, in which the products of this disposal process are discussed. Certainly the air pollution control facilities which are required for environmental protection at refuse incinerators constitute vital components of the incinerator system. However, because of their complexity and importance, they merit special detailed description and evaluation, as is found in section 3-4.

In any discussion of the components of a refuse incinerator, the treatment would be incomplete without a description of the controls and instrumentation required for monitoring and management of the combustion operation. The apparatus used to control the combustion process in a refuse incinerator will consist of essentially the same elements and components as any other type of control system. These basic components would include a sensor or monitor designed to determine some operational parameter such as temperature, a standard for proper operation against which the monitored value may be compared, and a control device which may be manipulated to create a change in process conditions so that monitored parameter values approach and equal the desired standard values. In most refuse incinerators built to date, the control systems consist of various monitoring devices such as temperature thermocouple and associated readout or display devices so that an operator may compare the performance of the incinerator unit against the desired standards of performance. The operator then makes adjustments in the operation of the incinerator (increase in air supply, increase in refuse feed rate, etc.), to effect a desired change if such a change is necessary to alter the process toward optimum conditions. However, in newer incinerators, closed-loop control systems are being installed in which the entire operation is automatic. Signals sent from monitoring devices to a central control facility, usually computerized, are mechanically compared with the standards of performance for the unit and the automatic control system then effects the proper changes in operational parameters to create optimum combustion conditions. The types of instruments and monitoring devices used in control of incinerator combustion processes generally consist of devices to measure temperature or pressure. Temperatures are measured at various places in the combustion chamber and gas transfer zones. For example, gas temperatures are measured in the incoming air passage, the passage from the primary to the secondary combustion chamber, in the primary combustion chamber and the secondary combustion chamber, in various air pollution control devices, and in the incinerator chimney. Optical pyrometers are suitable for use in the combustion chamber if the temperatures are expected to exceed $2000°F$. For most conventional incinerators, shielded thermocouples manufactured from chrome-alumel materials are suitable for temperature monitoring in combustion chambers. In other areas of the incinerator, temperatures may be monitored with standard bulb thermometers or with other types of thermocouples. It is also convenient to monitor supplies of air and volumes of effluent gases indirectly through measurement of draft pressures or gas pressures. Manometers, bourdon gages, and diaphragm-actuated sensors are suitable for measurement of pressures in incinerator applications. A number of other devices should be used in monitoring incinerator performance: Venturi meters or Pitot tubes for measurement of gas flows; dynamic flow meters or weirs for measurement of liquid flows; tachometers for conveyor or fan speeds; volt meters, ammeters, and watt meters for measurement of electrical characteristics; photoelectric devices for measurement of effluent gas densities (smoke density); and closed-circuit television systems or simple visual ports for observation of furnace interiors and chimney interiors (Ref. 3-9).

The trend in modern incinerator design is toward the complete control of incinerator facilities through a centralized, computerized, closed-loop control system. Such control systems include monitors for the measurement of incinerator performance in all areas of the plant. Information from all of the monitored devices is fed to the central control computer where the incoming data is used in a control program to determine required changes in order to alter the combustion process for optimum burning conditions. The computerized control system may be designed so that consideration is given to optimizing the operation of individual components of the incinerator system, such as the furnace itself or the air pollution control devices, and the system may be designed through the use of weighting factors to arrive at an overall optimal operating situation. In other words, the designers of the incinerator facility may program the control computer to insure that certain operating conditions are always satisfied; i.e., minimum generation of air pollutants, for example, may be a primary goal in the operation of the facility. On the other hand, the operation may be designed to attain maximum burning rates with little consideration given to the generation of air pollutants. In any case, computerized, fully automatic monitoring and control systems using closed-loop logic should find extensive application in large refuse incinerators in the near future.

3-3.3   Process Requirements

In order to accomplish the desired combustion of refuse at an incinerator facility, some obvious requirements must be met. These requirements include the provision of a physical plant including combustion chambers, the provision of various utilities such as electrical power and water, and the employment of skilled and semiskilled personnel to operate the facility.

Physical Plant. The physical plant itself is an important consideration in the location and design of any incinerator facility. Site selection is one of the most important operations in planning and developing an incinerator facility. Perhaps the most important criterion to use in selecting a site for an incinerator is the acceptance by the general public of an incinerator at that particular spot. The importance of public acceptance of an incinerator facility will be discussed in a later section of this chapter dealing with limiting parameters for incinerator operations. If the site is judged to be suitable and acceptable to the public, on the whole, various other physical characteristics about the site must be examined. These physical factors or considerations include: the foundation conditions, topography, drainage provisions, the availability of utilities, the building code and zoning code restrictions for that area and site, and the general meteorological conditions of the proposed site. Topography, for example, can be very important in affecting the design of the structure itself and it can be important in the dispersal of gases and particulate effluents into the atmosphere. The meteorological conditions at a particular site must be considered together with the topography of the site in assessing the air pollution hazard associated with constructing a refuse incinerator at that particular site. Additionally, the structural design of the incinerator will be affected by foundation conditions and drainage conditions, since an incinerator with its furnaces and other apparatus creates heavy, concentrated loads on foundation materials. Availability of utilities is important and will be discussed in detail in the next paragraph. In addition to these particular physical properties of a given site, the overall location of the site with respect to the collection area is very important. The traffic flow pattern of collection vehicles coming to the incinerator location must be determined so that the travel distances for the collection vehicles are minimized. In other words, the collection pattern for the given area to be serviced must be determined in advance and the incinerator must be located as close as possible to the weighted center of the flow pattern. The term "weighted center" is used to indicate that not only must the geographic location of the collection routes be considered, but also the relative amounts of waste generated at various spots on the collection routes must be given full consideration. The incinerator should be located so that the number of ton-miles of collector truck travel is minimized. At the incinerator itself, the traffic flow pattern must be designed so that collection trucks have easy access to and from the facility. An incinerator may be used by hundreds of collection vehicles per day, both public and private; accessibility for these vehicles must be maximum. Furthermore, the physical plant of the incinerator should be designed so that it will harmonize with the overall character of the surrounding community and the facility should be oriented so that any unsightly operations carried out at the facility would be visible to a minimum number of persons. In addition to the external characteristics of the plant, certain internal facilities must be incorporated into the design of the physical plant. For example, dining facilities, locker rooms, showers, and toilet areas must be provided for the labor force of the incinerator. Office space must also be provided within the incinerator to house a centralized control room, a monitoring laboratory, maintenance and repair facilities, administrative offices, and a scale masters office. The provision of service facilities such as toilets and washrooms for incinerator personnel should be considered a high priority item in the planning of the incinerator facility since the operational efficiency of the work force at the incinerator will have an important bearing on the overall success of the disposal facility; the efficiency of the work force will be significantly affected by the quantity and quality of service facilities provided at their place of work. Outside the physical plant, other considerations must be made also. Adequate lighting must be provided for the entire area. The incinerator building should be surrounded by well-planned and landscaped areas with perimeter planting of shrubs, if possible. The perimeter of the facility should be fenced and large signs and lights should be provided to indicate the function of the facility and any restrictions on access to the plant. The necessity for easy access of collection vehicles to the site often requires the establishment of one-way traffic flow. This one-way traffic flow may also be desirable in that it permits careful scrutiny of incoming collection vehicles and eliminates promiscuous dumping or dumping without payment of required fees. In summary, the planning and design of the physical plant for a refuse incinerator should include every design consideration made for any other multimillion dollar industrial or commercial facility. Unfortunately, in the past adequate provision of the physical plant requirements listed above has not been guaranteed at all refuse incinerators.

Utilities. As mentioned previously, certain utilities must be readily available to an incinerator operation. These utilities, required for the operation of a refuse incinerator, include: supplies of electricity for operation of charging and control facilities and for lighting, supplies of potable water for incinerator personnel, wastewater removal systems for incinerator process water and wastewater from personnel sanitary facilities, storm sewers and drains, telephone service, and heating facilities for the incinerator plant and hot water supplies. The electrical power required in a refuse incinerator varies according to the types of equipment used in the plant. Electricity is used to power such items as fans, pumps, hoists, cranes, air pollution control devices, stoker mechanisms, etc. Electrical power costs may amount to as much as 75 cents per ton of wastes incinerated (Ref. 3-9). Since at the present time in the United States waste heat recovery is not generally practiced, all the electrical power required in an incinerator operation must be supplied from outside sources.

The water requirements for a refuse incinerator will vary with the type of combustion chambers and residue quenching facilities incorporated into the incinerator design. For units operating in the United States at the present time, water requirements vary from approximately 300 to 2000 gallons per ton of waste incinerated. The cost for this amount of water, on the basis of a water cost of 25 cents per 1000 gallons, would amount to between $7^{1}/_{2}$ and 50 cents per ton of waste incinerated. Because the water which is supplied to an incinerator plant will be changed in chemical characteristics, solids content, and temperature, it will require subsequent treatment, either on site at the incinerator facility or in a wastewater treatment system. Because of the requirements for wastewater treatment and because of rising costs in water supply, a trend is presently developing toward greater recycling and reuse of water in incinerator operations.

Additional facilities which may not be considered generally to be utilities but which are necessary for the good operation of an incinerator include: an internal communications system within the incinerator plant consisting of public address systems, intercoms, bells, or other suitable apparatus capable of operation in areas with high levels of background noise; storm drains and sanitary sewers to carry away natural drainage and wastewaters; and, finally, a fuel supply or energy source for space heating within the incinerator facility and for heating of water for washrooms and shower rooms. The last-named requirement for fuel may also be necessary whenever auxiliary fuels are required to sustain the combustion process. Because of the heterogeneous nature of municipal solid waste and because of large seasonal variations in water content, it is virtually mandatory to have auxiliary fuels available to aid in the combustion of wet or low heat value wastes. Therefore, the provision of oil or natural gas as an auxiliary fuel for an incinerator facility should be considered a necessary "utility"; in some circumstances, the provision of auxiliary fuels would be much more important than the provision of some of the other utilities listed in the paragraph above.

Personnel. The combustion of a variable material such as municipal refuse without creating environmental degradation is a complex operation which requires the employment of trained capable personnel. Among the personnel required for incinerator operation are: a general supervisor, a scalemaster, one or more crane operators, charging-floor operators, general process controllers, residue removal operators, and maintenance men. The supervisor generally is in charge of the entire facility and has the responsibility of the overall coordination of the individual efforts of the entire work force. It is desirable that the supervisor have adequate knowledge of all of the apparatus and techniques used to control the incinerator process. Additionally, the supervisor should have at least minimal training in the mechanical trades. Obviously, it is desirable that the supervisor have labor management skills, be capable of leading his own labor force, and be able to communicate effectively with the public. Finally, the supervisor should have a working knowledge of business management and records.

The scalemaster has an important function at an incinerator facility in that he regulates the flow of traffic through the incinerator. He comes into contact with the public frequently since he must regulate the dumping of materials at the incinerator by private haulers and citizens who occasionally visit the disposal facility. Finally, the scalemaster records the amounts of refuse delivered to the site and, therefore, is responsible for the maintenance of important records concerning refuse input to the facility. When a dumping fee is charged for refuse brought to the incinerator facility, the records kept by the scalemaster form the basis for the assessment of fees. Therefore, the scalemaster should have a basic knowledge of arithmetic and business records.

The operators within the incinerator facility have various important functions to perform. For example, crane operators have the responsibility to maintain a uniform flow of materials to the combustion chamber, while striving to thoroughly mix the solid wastes in the storage pit so that fluctuations in waste composition will be minimized. An experienced and competent crane operator can do much to improve the overall operation of an incinerator facility. Moreover, the operators who directly supervise the control of the charging and burning apparatus must be well-trained and must accept significant responsibility. The charging floor operator must maintain a steady flow of refuse through the charging system and must be able to perform basic maintenance and cleaning operations in that area. The combustion chamber operator regulates the stokers and fans which charge wastes into the furnaces and which supply required combustion air. Thus, this operator is in direct control of the basic combustion process. In order to maintain adequate control of that process, he must periodically inspect the furnace through viewports or similar devices and he must constantly be aware of the operating conditions within the furnace as shown by various instruments and gages used to measure temperature, pressure, and refuse transfer rate. The residue control system operator also is quite important to the overall success of the operation in that he is responsible for maintaining the flow of residue through the quenching system and away from the combustion chamber. However, this job requires a minimum amount of experience and very little training. Nevertheless, this operator, like every operator in the facility, must accept his responsibility as part of the entire work force charged with the operation of a very important municipal service facility.

Maintenance personnel are responsible for periodic inspection and repair or replacement of the apparatus in the incinerator facility. They must also keep adequate records showing the date and extent of required repairs and replacements. A competent maintenance man will instruct the personnel who are operating the equipment in the incinerator facility as to the proper operating procedures for their equipment and will inform them about lubrication and maintenance requirements for that equipment. It is essential that the maintenance personnel in municipal incinerators be experienced and qualified mechanics.

It is highly desirable that all of the operating personnel of the incinerator facility be very familiar with the overall operation of the plant and be aware of the importance of their work to the general community well-being. The importance of personnel morale in the proper operation of an incinerator facility should not be underestimated. Provision of adequate salaries, job tenure, and in-plant promotions should be made whenever possible in order to insure optimal performance by all operating personnel.

3-3.4 Process Output

It must always be remembered that a municipal incinerator is a facility for the processing of refuse and not for ultimate disposal. In other words, a certain amount of residue will always be produced from the incineration of solid wastes; this residue will require ultimate disposal. Therefore, the residue, consisting of unburned material from the grate area, siftings which have dropped through the grates, and collected particulate matter from the effluent gas stream should be considered products from the processing of the solid waste. The gases produced during incineration of solid waste may contain gaseous air pollutants also. Finally, the water used in the incinerator processes such as the quenching of residue, unless recycled in the incinerator plant, will require treatment and disposal. All of these materials and quantities should be considered products of the incineration operation. Because these products require additional processing, and because the success of the overall incineration operation depends upon adequate treatment and disposal of these products, the next section in this chapter is devoted to a comprehensive discussion of the by-products of incineration, particularly the generation and control of the air pollutants produced during refuse combustion.

3-4. PROCESS PRODUCTS

One of the most serious difficulties associated with the operation of municipal refuse incinerators is the prevention of environmental pollution through the emission of various types of effluents from the incinerator facility. A municipal incinerator is a processing facility wherein much of the solid waste brought to the facility is transformed to gaseous materials which are released to the atmosphere. Improper design or inefficient operation of the refuse incinerator may create pollution of air, water, or land.

3-4.1   Effluent By-Products

During the dumping of refuse from collector trucks and during transfer of the refuse from the storage pit to other parts of the incinerator, dust and litter may be generated. This type of pollution may appear to be minor but it is very troublesome to the operation of an incinerator facility and has a significant detrimental effect on the public image of the incinerator. Housekeeping procedures should be designed to eliminate dust and litter as much as possible. In this connection, special provisions may be made to completely enclose the dumping area so that dust does not leave the area of the dumping zone and the storage pit. Furthermore, the incinerator facility may be designed so that air intakes are placed in the area of the storage pit; air drawn through these entry ports into the combustion chambers of the incinerator will satisfy the oxygen requirements for the combustion process and will draw dust from the dumping area into the combustion chambers and consequently into collection facilities designed for the effluent gases from the combustion chambers. Litter control, on the other hand, must be accomplished through periodic sweeping and cleaning of the dumping area. If litter is allowed to accumulate in the dumping area, odors may be produced and disease vectors may find sustenance in the refuse. Also, odor problems may be created in the storage pit area if the materials are not turned frequently in the storage pit; putrefaction of organic materials can occur in deposited refuse if it is allowed to remain in a storage pit for too long a period of time. These problems of dust, litter, and odor can have a significant influence on the overall success of the incinerator operation since they create unpleasant working conditions and impair the efficiency of the operating personnel. For this reason, frequent cleaning and disinfection of the dumping areas are essential. Additionally, dust sprays and disinfectant sprays can be installed in the storage pit area to control dust and vectors.

The problems associated with dust and litter caused by dumping operations are rather minor, however, in comparison to the problems associated with the production of air pollutants. The air pollutants consist of solid, unburned particulate matter and gaseous air pollutants.

Particulate matter, generally termed "flyash," is entrained in the combustion chambers and carried out of those chambers in the exit stream of combustion gases. Flyash consists of very small particles of ashes, cinders, mineral dust, charred paper and other partially burned materials, and soot. Table 3-15 shows some of the properties of samples of flyash collected in municipal incinerators. As shown in this table, flyash particles range in size from as much as 120 to less than 5 microns (μ) in equivalent diameter. The distribution of particle sizes within this range is variable from time to time in a given incinerator, and also varies erratically from one incinerator to another. Much of the flyash created in municipal refuse incinerators is inorganic and consists of the oxides of aluminum, calcium, iron, and silicon. Flyash must be collected from the exit gas stream in

TABLE 3-15(a)   Particle Size of Flyash Emitted from Incinerators, per cent by weight

Size, μm	Incinerator Guidelines[a]	Gansevoort Incinerator, New York City[b]	South Shore Incinerator, New York City[c]
<2	13.5		
<4	16.0		
<5		12.0	
<6	19.0		
<8	21.0		
<10	23.0	17.8	
<15	25.0	39.0	
<20	27.5	42.4	
<30	30.0	44.3	
<40		56.8	
<44			28.1
<60		70.0	
<74			66.5
<90		87.7	
<120		94.2	
>120		5.8	
<149			86.7
<250			95.5
<841			99.6
>841			0.4

Sources:
[a]*Incinerator Guidelines*, U.S. Department of Health, Education, and Welfare, Public Health Service, 1969, p. 52.
[b]*Municipal Refuse Disposal*, Institute for Solid Wastes, Am. Publ. Works Assn., 1970, p. 201.
[c]*Municipal Refuse Disposal*, Institute for Solid Waste, Am. Publ. Works Assn., 1970, p. 202.

TABLE 3-15(b)   Chemical Analysis of Incinerator Flyash

Component	Gansevoort Incinerator, New York City,[a] % by Weight	South Shore Incinerator, New York City,[b] % by Weight	Arlington, Va.,[c] %	Jens-Rehm Study,[d] %	Kaiser Study,[e] %
Carbon					
Organic	14.5	10.4	11.62		
Inorganic	85.5	89.6			
Silicon as SiO_2	36.0	36.1			36.3
Si			18.64	5+	
Aluminum as Al_2O_3	27.7	22.4			25.7
Al			10.79	1–10	
Iron as Fe_2O_3	10.0	4.2			7.1
Fe			2.13	0.5–5.0	
Sulfur as SO_3	9.7	7.6			8.0
S			Small or trace		
Calcium as CaO	8.5	8.6			8.8
Ca			4.70	1.0+	
Magnesium as MgO	3.4	2.1			2.8
Mg			0.98	1–10	
Titanium as TiO_2					0.9
Ti			2.24	0.5–5.0	
Ni			Small or trace	1–10	
Na			Small or trace	1–10	
Zn			Small or trace	1–10	
Ba			Small or trace	0.1–1.0	
Cr			Small or trace	0.1–1.0	
Cu			Small or trace	0.1–1.0	
Mn			Small or trace	0.1–1.0	
Sn			Small or trace	0.05–0.5	
B				0.01–0.1	
Pb			Small or trace	0.01–0.1	
Be				0.001–0.01	
Ag			Small or trace	0.001–0.01	
V				0.001–0.01	

(Continued)

TABLE 3-15(b) (*Continued*)

Component	Gansevoort Incinerator, New York City,[a] % by Weight	South Shore Incinerator, New York City,[b] % by Weight	Arlington, Va.,[c] %	Jens-Rehm Study,[d] %	Kaiser Study,[e] %
Sodium and potassium oxides	4.7	19.0			
Na as Na$_2$O					10.4
K as K$_2$O					
Na			Small or trace		
K			Small or trace		
Ga			Small or trace		
Hg			Small or trace		
Mo			Small or trace		
Ta			Small or trace		
Apparent specific gravity	2.58				
Ignition loss			14.45		

Sources:
[a]*Municipal Refuse Disposal*, Institute for Solid Wastes, Am. Publ. Works Assn., 1970.
[b]*Municipal Refuse Disposal*, Institute for Solid Wastes, Am. Publ. Works Assn., 1970.
[c]*Municipal Refuse Disposal*, Institute for Solid Wastes, Am. Publ. Works Assn., 1970.
[d]W. Jens and F. R. Rehm, "Municipal Incineration and Air Pollution Control," *Proc. Natl. Incin. Conf.*, Am. Soc. Mech. Eng., 1966, p. 74.
[e]E. R. Kaiser, "Refuse Composition and Flue-Gas Analyses from Municipal Incinerators," *Proc. Natl. Incin. Conf.*, Am. Soc. Mech. Eng., 1964, p. 35.

order to prevent particulate air pollution. The collected particulate matter constitutes a by-product solid waste. Moreover, the collected ash is very difficult to handle in its dry state and frequently it is transported from the incinerator in the form of a water slurry. If the flyash is not removed in a water slurry, care must be exercised in the storage of the dry flyash. The dry ash is subject to scattering by the wind, and precipitation falling on stockpiles of flyash may leach soluble constituents from the ash. Therefore, the ash must be carefully removed to an ultimate disposal site. If the flyash is transported in water slurry form, the slurry may be treated together with the process water from the incinerator. The process water comes, for the most part, from quenching operations designed to cool incinerator residue. Also, however, water may be used in the incineration process to cool various parts of the incinerator plant. As mentioned previously, water requirements for an incinerator will vary tremendously on the basis of whether or not water is recirculated in the plant.

A number of investigations have demonstrated that incinerator process water contains significant amounts of inorganic and organic materials in solution or in suspension. Table 3-16 shows the characteristics of wastewaters obtained from a number of incinerators. In addition to the inorganic and organic constituents shown in this table, several studies have indicated that bacteria may be present in wastewaters from a municipal incinerator. The most frequently used disposal method for wastewaters from an incinerator is discharge of those waters into a sanitary sewer. However, this practice may be unwise. The added amount of wastewater should be estimated and the possible effects on the sewer system

TABLE 3-16   Incinerator Wastewater Data

Characteristic	Plant 1 Max.	Plant 1 Min.	Plant 1 Avg.	Plant 2 Max.	Plant 2 Min.	Plant 2 Avg.
pH	11.6	8.5	10.4	11.7	6.0	10.5
Diss. solids, mg/l	9,005	597	3,116	7,897	1,341	4,283
Susp. solids, mg/l	2,680	40	671	1,274	7	372
Total solids % volatile	53.6	18.5	36.3	51.6	10.5	31.2
Hardness (CaCO$_3$) mg/l	1,574	216	752	1,370	112	889
Sulfate (SO$_4$) mg/l	430	110	242	780	115	371
Phosphate (PO$_4$) mg/l	55.0	0.0	23.3	212.5	1.0	23.5
Chloride (Cl) mg/l	3,650	50	627	2,420	76	763
Alkalinity (CaCO$_3$) mg/l	1,250	2.5	516	1,180	292	641
5-day BOD @ 20°C	–	–	–	–	–	–

Characteristic	Plant 5 Max.	Plant 5 Min.	Plant 5 Avg.	Plant 6 Max.	Plant 6 Min.	Plant 6 Avg.
pH	6.5	4.8	5.8	4.7	4.5	4.6
Diss. solids, mg/l	1,364	7,818	8,838	6,089	5,660	5,822
Susp. solids, mg/l	398	208	325	2,010	848	1,353
Total solids, % volatile	–	–	–	24.69	23.26	23.75
Hardness (CaCO$_3$) mg/l	2,780	2,440	2,632	3,780	3,100	3,437
Sulfate (SO$_4$) mg/l	1,350	1,125	1,250	862	625	725
Phosphate (PO$_4$) mg/l	15.0	11.5	13.0	76.2	32.2	51.5
Chloride (Cl) mg/l	3,821	3,077	3,543	2,404	2,155	2,297
Alkalinity (CaCO$_3$) mg/l	28	16	23	4	0	1.33
5-day BOD @ 20°C	13.5	6.2	8.8	–	–	–

Sources: Plants 1 and 2 USPHS *Report on the Municipal Solid Wastes Incinerator System of the District of Columbia,* 1967.
Plants 5 and 6 USPHS unpublished data (SW-11ts) (SW-12ts).

Plant 1 110TPD Residue Quench (Batch).
Plant 2 425TPD Residue Quench (Batch).
Plant 5 300TPD Cont. Feed-Flyash Effluent.
Plant 6 300TPD Cont. Feed-Flyash Effluent.

should be anticipated and taken into account. The wastewater may be acidic and may have detrimental effects upon the sanitary sewer system. If a sanitary sewer system is not available for discharge of the process waters from the incinerator, a separate wastewater treatment plant must be provided for the facility.

Air Pollutants. Of much more importance than the generation of wastewater from the incinerator facility, however, is the creation of air pollutants during combustion. Such pollutants include the entrained particulate matter which was discussed previously, and various gaseous by-products. As mentioned before, the particulate matter must be collected from the exit gas stream. In connection with the collection of particulates, the important characteristics of the material include the sizes of the particles, their specific gravity, chemical composition, and electrical resistivity. Obviously, the quantity of particles entrained in the gas stream per unit volume of effluent gas is of primary importance. The particle size distribution and specific gravity of the flyash are of great importance in the design of collection systems because most particulate collectors are designed to operate on particles in a rather small size range. In general, larger, denser particles are much easier to collect, especially through the use of gravimetric methods. Coarse, dense particles can be collected easily in devices utilizing inertial characteristics of the materials; settling chambers and cyclones utilize inertia forces in the collection of particulates. Smaller, lighter particles remain entrained in exit gas streams more easily and, therefore, require more sophisticated procedures for collection. High-energy wet scrubbers, fabric filters, and electrostatic precipitators are used to collect very small-sized particles. As a rule of thumb, an equivalent diameter of 10 μ is a boundary line for particles with respect to the type of collection equipment which should be used to remove the particles from the effluent gas stream. In other words, if most of the particulate matter in the exit gas stream is smaller than 10 μ in equivalent diameter, sophisticated collection devices must be utilized.

Because the size and specific gravity of entrained particulates will vary with the characteristics of the input solid wastes, the characteristics shown in Table 3-15 should be considered only generally indicative of the characteristics of the particulate matter. Because of the variability in the particulate size and loading anticipated in refuse incinerators, collection systems should be designed conservatively to operate on a wide range of particulate sizes. Additionally, other measures should be used to obtain particulate control. Figure 3-10(a) shows the relationship between the rates of particulate entrainment and the supply of underfire air in incinerator installations. As mentioned previously, the provision of underfire air will have a great influence on the entrainment of unburned particles of waste in the primary combustion products. If it is at all possible to reduce the amount of underfire air supplied to the combustion process, such reduction will significantly reduce the amounts of entrained particulate matter in the effluent gas stream. Figure 3-10(b) shows general combustion products which could be anticipated from refuse incineration. The products shown in Fig. 3-10(b) were calculated for various percentages of excess air supplied to the combustion process on the basis of an assumed composition of refuse. The refuse included over 50 percent paper, had an assumed moisture content of 20 percent, and had an assumed heating value of approximately 5450 BTU per lb (Ref. 3-13). Thus, the curves shown in Fig. 3-10(b) should be considered approximate indicators only. In most cases, over 99 percent of the effluent gases from municipal incinerators consist of carbon dioxide, nitrogen, oxygen, and water vapor; all of these gases are generally not considered air pollutants. Other trace gases are present in incinerator stack effluents. Table 3-17 shows some of these trace elements.

In comparison to the other combustion processes that occur daily in the United States, the contribution of pollutants from the incineration of solid wastes generally is quite insignificant. For example, the total amount of nitrogen oxides and sulfur oxides generated in the combustion of municipal refuse is only about 1 to 10 percent of the amount generated per ton during the combustion of fossil fuels in power-generating plants.

In addition to the particulate matter and gaseous products mentioned above, water vapor is entrained in effluent gas streams from municipal refuse incinerators. The generation of water vapor plumes from incinerator facilities is highly undesirable. If water is not used in cleaning or cooling the effluent gases from an

TABLE 3-17 Trace Elements in Incinerator Effluents

Gaseous Emissions	Analysis A[a] (ppm)	Analysis B[b] (lb/Ton)	Analysis C[c] (lb/Ton)
Aldehydes	1–22	23.6×10^{-4}	1.1
Nitrogen oxides	58–92	2.7	2.1
Organic acids			
(expressed as acetic acid)	–	–	0.6
Ammonia	–	–	0.3
Hydrocarbons			
(expressed as CH_4)	none	0.8	1.4
Sulfur oxides	–	–	1.9
Carbon monoxide	1000–3000	–	–

Sources:
[a]Hein, G. M. and Engdahl, R. B., *A Study of Effluents from Domestic Gas-Fired Incinerators*, AGA, New York, 1959. Glendale, Cal. municipal incinerator.
[b]Stenburg, R. L., *et al.*, "Field Evaluation of Combustion Air Effects on Atmospheric Emissions from Municipal Incinerators," *J. Air Pollution Control Assoc.*, Vol. 12, No. 2, pp. 83–89. Theoretical analysis of "typical" refuse.
[c]Stanford Research Institute, *The Smog Problem in Los Angeles County*, Los Angeles, Western Oil and Gas Assoc., 1954. Mostly twigs and branches, no garbage.

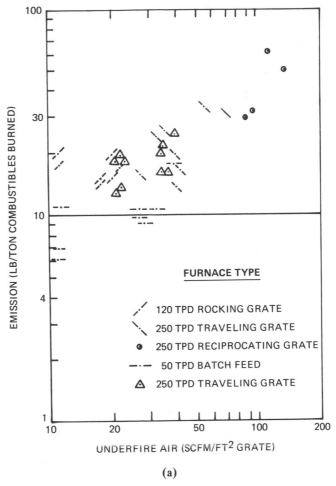

Fig. 3-10   (a) Particulate emissions; (b) combustion products per pound of solid waste.

incinerator, the occurrence of water vapor plumes from that facility will be limited and generally will occur only during periods of very low outside ambient temperatures. On the other hand, if water is used to cool effluent gases, the occurrence of such plumes is much more likely. Even in this case, however, the temperature of the effluent gases in the stack of an incinerator will generally be sufficiently high to prevent the generation of vapor plumes except during conditions of very low temperature in the outside air and high relative humidity in the outside atmosphere. If water is used in gas cooling and in cleaning of the effluent gas stream, however, the occurrence of water vapor plumes may be a

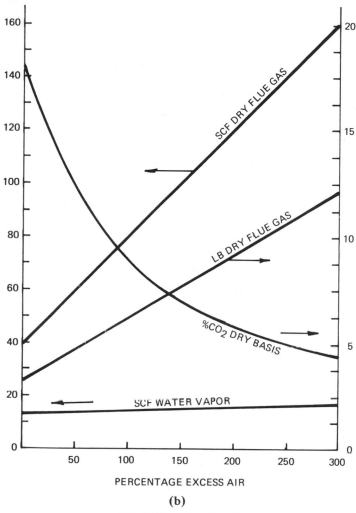

(b)

Fig. 3-10 (*Continued*)

regular feature of the incinerator facility. Under these circumstances, the exit gases will be saturated with moisture and the absolute humidity of the effluent gas stream will be quite high. Also, the temperature of the effluent gases will be low, dispersion of the gases will be reduced, and condensate plumes will occur under almost all atmospheric conditions. Water vapor plumes are usually not inherently harmful; however, they constitute a major cause of poor public relations for incinerator facilities. In some cases there have been public complaints of material corrosion as a result of condensate fallout, hazardous decrease in ground level visibility on surrounding terrain, and some other rather particular complaints. Whether or not these complaints have always been based on fact and have been justified is of minor importance. The undesirable end results in poor public relations for the incinerator facilities have been quite significant. Thus, every effort should be made to prevent the generation of water vapor plumes.

In order to evaluate the performance of any incinerator facility, it is necessary to employ monitoring devices and control systems such as were described in section 3-3.2. Of particular interest in recent years, however, is the necessity to monitor effluents from incinerators in order to comply with federal air pollution control legislation. Under provisions of the Clean Air Act Amendments of 1970, the Administrator of the United States Environmental Protection Agency has been given the authority to demand records of emission from stationary sources of air pollutants such as refuse incinerators. In order to comply with this requirement, the operators or managers of a refuse incinerator must sample the effluent gas stream from their facilities and obtain actual test data showing concentrations of gaseous and particulate constituents in the exit gas stream. The emissions from any incinerator can be estimated on the basis of material balances using an assessment of the composition of the wastes and an assumption concerning the relative degree of combustion in the incinerator. However, such data obtained from "desk calculations" is not sufficient to comply with the federal legislation, nor is it adequate to provide comprehensive information on the performance of the incinerator facility and the air pollution controls at that facility. Therefore, it is necessary to obtain samples in the effluent gas stream through standard testing methods in order to determine the gaseous and particulate constituent concentrations therein. The sampling of incinerator effluents has been standardized by the American Society of Mechanical Engineers and by the United States Environmental Protection Agency. Sampling systems, called "trains," have been developed by these organizations. In general, these trains include some sort of nozzle or inlet structure, a filter, a collector for gaseous pollutants, a condenser or similar cooling device, a gas flow meter, and a pump or aspirator to activate the entire system. The sampling methods used in monitoring effluents from incinerator stacks are referred to as isokinetic sampling techniques. In such techniques, the withdrawal of a sample of stack gases is done in such a manner that the flow conditions for the sample entering the collector train are identical to the flow conditions within the flue or stack. In this way, representative results are obtained. Maintaining appropriate velocities in the sampling train and in the stack or flue is a very complex procedure which varies in proportion to the gas velocity in the stack, the temperature of the gases in the stack, and the configuration of the sampling train. Various techniques have been developed in order to insure the procurement of a truly representative sample. A comprehensive discussion of sampling techniques cannot be given here and is not appropriate within the scope of this book. R. E. George and J. E. Williamson have prepared an excellent discussion of sampling procedures for testing incinerator performance in Chapter 10 of Reference 3-6.

3-4.2 Effluent Controls

The major concern of effluent control efforts carried out at a municipal incinerator is the prevention of air pollution. Wastewaters are produced in the operation of an incinerator and an incombustible residue is also produced, but disposal or treatment of these by-products is generally much easier to accomplish than is control of gaseous and particulate air pollutants. Water pollution can be prevented effectively by the transmission of all wastewaters to municipal wastewater treatment systems, or by the provision of a wastewater treatment plant for the incinerator. Solid residues may contain constituents which could be hazardous if they were liberated in the atmosphere or in water. The disposal of incinerator residue will be discussed in the next section.

The gaseous emissions from refuse incinerators consist primarily of carbon dioxide, nitrogen, oxygen, and water vapor. All of these gases are normally present in the atmosphere and generally are not considered to be air pollutants. However, other gases may be produced during refuse combustion which are deleterious substances. Examples of these gases are sulfur oxides, nitrogen oxides, carbon monoxide, hydrogen chloride, and other hydrocarbon gases. These air pollutants generally constitute only a very minor portion of the effluent gas stream from refuse incinerators. A comparison of measured amounts of these materials in incinerator effluents and currently proposed criteria for air emissions indicate that the emission levels for these gases are well below the emission limits presently under consideration. Competent operation of a refuse incinerator with maintenance of sufficient combustion time and sufficiently high temperatures will eliminate many of the pollutant gases mentioned above. The composition of municipal solid waste generally is such that generation of sulfur oxides is minimal. Finally, nitrogen oxides are not produced in any significant quantities in refuse incinerators unless the operating temperatures remain above 2000°F for a considerable period of time.

With respect to particulate matter entrained in the effluent gas streams from incinerators, the need for removal and control is much greater than the case with gaseous air pollutants. This control of particulate matter has been attempted through regulation as well as technology. In many cities and communities throughout the United States, control of suspended particulate matter from incinerators has been attempted through the enactment of local ordinances pertaining to the amount of suspended particulate matter which would be considered permissible in the atmosphere of the subject communities. Permissible levels of particulate matter generally are expressed in terms of pounds of suspended solids per thousand pounds of dry flue gas with reference to a standard condition of 50 percent excess air. Particulate matter concentrations have also been expressed in grains of particulate matter per standard cubic foot of dry flue gas corrected to a datum condition of 12 percent carbon dioxide content. Standard conditions for volume measurements consist of an atmospheric pressure of 29.92 in. of mercury and an ambient temperature of 70°F. In most cases, for the incineration of municipal solid wastes, these two standard reference conditions are approximately equivalent. If the effluents from an incinerator are measured and it is desired to compare particulate concentrations with emission criteria, it will be necessary to correct the obtained sample value to the basic conditions of 50 percent excess air content or 12 percent carbon dioxide content. This conversion or correction can be accomplished through reference to Fig. 3-11. In this illustration, an investigator would enter the figure at the appropriate value of excess air in the flue gas sample obtained. For example, if the flue gas sample had contained approximately 200 percent excess air and the quantity of particulate matter was to be expressed on the basis of 12 percent carbon dioxide

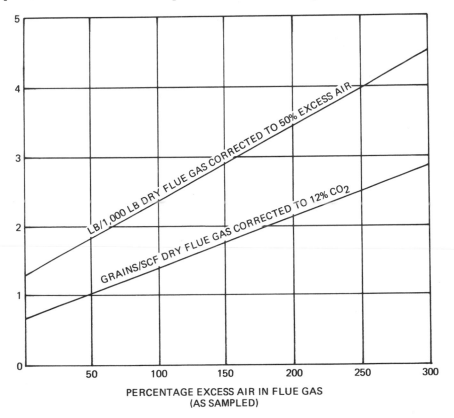

Fig. 3-11   Dust concentration equivalents.

content as a reference condition, the investigator would begin at the abscissa of Fig. 3-11 at the value of 200 percent excess air. Proceeding upward in the figure to the line designated for corrections to 12 percent carbon dioxide content, the investigator would obtain a value of approximately 2.05 on the ordinant scale at the left of the figure. Therefore, for every grain of particulate matter found in a standard cubic foot of the dry flue gas as sampled, there would be approximately 2.05 grains of particulate matter in a standard cubic foot of flue gas on the basis of 12 percent carbon dioxide content. If 5 grains of particulate matter had been obtained in the sample at 200 percent excess air, 10.2 grains of particulate matter would be found in each standard cubic foot of flue gas corrected to 12 percent carbon dioxide. When the corrected value of particulate concentration in a flue gas stream is obtained, it can be compared to the allowable emission standards for the community in which the incinerator facility is located.

In addition to quantitative particulate emission codes, many communities have attempted to control particulate emissions through the use of a visual measurement of smoke or opacity in the effluents from an incinerator stack. Many air pollution control codes are based upon so-called visual emission levels. The basis for the determination of visibility in many of the existing air pollution control codes is the Ringelmann number. This number is obtained through a visual observation of the stack plume at an incinerator or similar facility. An observer compares the opacity of the stack plume with a series of reference grids of black lines on white. The observer stands at some distance from the reference grids. When the reference grids are properly positioned, they appear as shades of gray to the observer, and the observer selects the proper shade of gray in the chart to match the color of the incinerator stack plume. Obviously, this determination is very subjective. The interpretation of color tone in a moving dispersing stack plume is very difficult. Additionally, the correlation between the quantitative values of particulate concentrations in the flue gases and the existing visibility criteria is rather complex. It is definitely preferable to base air pollution control regulations upon quantitative values of particulate emissions. As an approximate correlation, however, between visibility levels and quantitative particulate concentrations, the following is offered: a particulate concentration between 0.01 and 0.02 grains per standard cubic foot of effluent gas would correspond to a Ringelmann visibility number of less than 1, indicating an effluent which is optically clear.

Because of the measured quantities of particulate matter found in all samples of uncontrolled incinerator effluents, and because of the presence of trace amounts of pollutant gases in effluent samples, it should always be anticipated that air pollution control devices will be required on incinerators. In choosing a particular method or device for control of air pollutants, consideration must be given to the type of pollutant which must be controlled; i.e., gaseous emissions or suspended particulate emissions. As mentioned previously, the gaseous emissions from combustion of municipal solid wastes generally consist of materials which are not considered to be air pollutants. However, it is possible under certain circumstances to generate nitrogen oxides and sulfur oxides during refuse incineration and these gases definitely must be considered air pollutants. In most cases, the amounts of sulfur oxides produced per ton of solid waste burned are quite low compared to the amounts produced in fossil fuel combustion. Solid wastes generally have sulfur contents on the order of about 0.1 to 0.2 percent by weight; fossil fuels such as coal and oil in use in the United States today have sulfur contents which range from about 1 to 3 percent. Thus, the amount of sulfur in a fossil fuel, on a weight basis, is from 5 to 30 times the amount in solid waste. It has also been found that much of the sulfur content of solid waste charged into refuse incinerators is retained in the residue from the incinerator rather than being discharged as an oxide in the effluent gas stream. Similarly, little difficulty should be anticipated in the control of nitrogen oxide emissions because the emissions of these compounds at normal operating temperatures in refuse incinerators are insignificant.

Some constituents of the solid waste stream may produce detrimental effluent gases. The plastics which are appearing in ever greater amounts in the solid waste stream decompose to a variety of hydrocarbon gases during primary decomposition. For the most part, incineration of the plastics used today leads to the ultimate formation of innocuous gases. However, some plastics, such as polyvinyl chloride, produce deleterious end products such as hydrogen chloride gas. However, PVC packaging materials and other forms of this plastic have come under severe restriction in recent months. Also, it appears that hydrogen chloride emissions will be relatively easy to control since HCl is highly soluble in water and can be removed from the effluent gas stream through the use of wet scrubbers. (Removal of PVC materials from the waste stream is to be preferred over the control of emissions through the use of wet scrubbers, since the use of wet scrubbers is expensive and may lead to the generation of water vapor plumes.)

The major emphasis in control of effluents from refuse incinerators is in the control of particulate emissions. A number of different techniques and devices have been used in this control effort and new designs and methods are continually being developed. The earliest type of particulate collector consisted of a settling chamber which had either a wet or a dry bottom. For the most part during the first 50 years of the twentieth century, simple settling chambers were the only types of particulate collection devices used in incinerators in the United States. The efficiency of simple gravity chambers in collecting particulate matter is only about 33 percent, and, therefore, the use of this type of device has been virtually discontinued under the impact of increasingly stringent air pollution control regulations. Within the last 15 years, a small percentage of the incinerators built in the United States have been equipped with single and multiple cyclones. The collection efficiency of these devices has been determined at be-

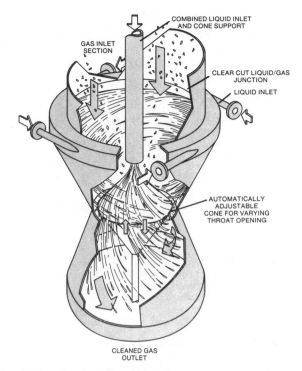

Fig. 3-12   Flyash wet scrubber (*courtesy* American Air Filter).

tween 60 and 65 percent in practice. Therefore, this type of device is not wholly satisfactory. The installation costs for cyclone collectors are in the range of 15 to 25 cents per cfm of treated gas. The gases are slowed appreciably during the passage through a cyclone collector; the gage pressure drop in this type of collector is in the range of $2\frac{1}{2}$ to 4 in. of water.

In a large number of incinerators built in the United States, wetted baffle spray collection devices have been used. In this type of collector, vertical baffle screens are placed in the path of the effluent gases; the screens are wetted by flushing sprays or by overflow weirs. This type of device has been installed on over half of the incinerators built in the last 15 years. The efficiency for multiple baffle units has been measured at as much as 50 percent; the removal of particulate matter necessary to attain Ringelmann number 1 visibility level has been calculated to be approximately 99 percent and, therefore, this type of device would not be adequate to remove particulate matter from the effluent gas stream in a municipal incinerator. However, these devices are rather inexpensive; installation costs for this type of device range from about 2 to 4 cents per cfm of treatment capacity.

More efficient collection devices have been developed through modifications of the wetted baffle spray collector mentioned in the previous paragraph. One type of collector developed is the wet scrubber, which has been installed in about 20 percent of the incinerators built since 1960. In this type of collector, the cleaning of the effluent gas stream is accomplished through direct contact of the entrained particles and water droplets rather than contact of particles with a wetted screen surface. Particles of entrained solid material collide with sprayed water droplets and the larger droplet–particulate mass can be collected in various types of inertial collection devices. The efficiencies of this type of apparatus have been measured at from 94 to 96 percent, with accompanying pressure drops in the range of 5 to 7 in. of water. As mentioned previously, a disadvantage of the use of wet scrubbers is the fact that effluent gases are saturated with water vapor and the creation of water vapor plumes from the stack of an incinerator equipped with wet scrubbers is much more likely than for an incinerator equipped with dry collection devices. Water requirements for wet scrubbers range from about 5 to 15 gal per 1000 cu ft of gas treated per minute. The water used in the scrubbing process is highly corrosive and, therefore, the scrubber system must be corrosion-proof. If the process water from a scrubber is dis-

Fig. 3-13 Typical installation of electrostatic precipitators (*courtesy* American Air Filter).

charged to sanitary sewer facilities, deterioration of sewer lines may result. Some form of pretreatment for the corrosive process water from wet scrubbers may be necessary before release of scrubber water to sewer facilities. Because of the more complex construction of this type of apparatus, the installation cost for these collectors is somewhat higher, from $0.25 to $1.25 per cfm, as compared with the cost of simpler collector devices.

Because of the problems associated with wet collection of particulate matter (corrosive process water and water vapor plumes), in recent years attention has centered on dry collection of particulates. One of the more effective techniques for dry collection of particulates is the use of electrostatic precipitators. Electrostatic precipitators operate by electrically charging the suspended particulate matter and then causing the charged particles to be deposited on the surface of an oppositely charged or grounded electrode. The deposited particulate matter

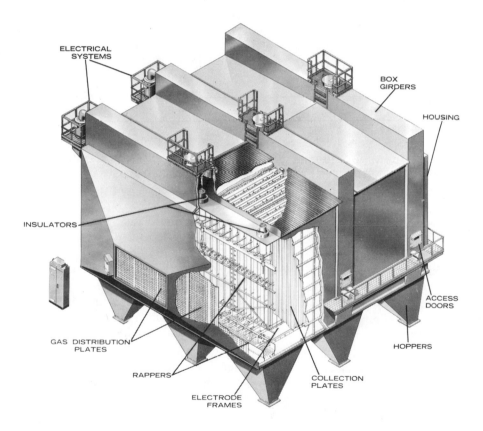

Fig. 3-14   Electrostatic precipitator (*courtesy* Wheelabrator-Fray, Inc.).

is removed from the collection surface by either vibrating the collector surface or by washing the surface with water. The vibration collection method is much more advantageous than the introduction of water into the collection process. Several variable conditions in the particulate matter determine the efficiency of electrostatic precipitators: the electrical resistivity of the particulate matter, the moisture content of the effluent gas stream, and the temperature of the latter. Gas temperature is very important; precipitators operate most efficiently in a temperature range from about 450 to 520°F. The operation of an electrostatic precipitator is more effective than most of the other collection devices described above; collector efficiencies for precipitators have been measured at from 96 to 99.5 percent. Also, the gage pressure drop across the precipitators is generally less than $1/2$ in. of water and the velocity of the effluent gas stream, consequently, is not significantly reduced by passage through an electrostatic precipitator. Electrical power is required for the operation of a precipitator, however; from 200 to 400 watts of electrical power are required per 1000 cu ft of gas treated per minute. The advantages of electrostatic precipitators are several: the precipitator has a low operating cost and maintains a high collection efficiency; as mentioned, it operates at a low pressure drop; the electrical power requirements for the precipitator are not high; and, no use of water is required in collection by the precipitator. Installation costs for electrostatic precipitators range from about $0.85 to $1.50 per cu ft of gas treated per minute (basic cost with no auxiliaries).

Another type of collection device has been used to a limited extent in incinerators in the United States; this is the fabric filter. Fabric filters operate in much the same manner as does a vacuum cleaner. A dust cake is formed by the passage of polluted gases through the fabric filter, and further filtering is done predominantly by the cake itself. High collection efficiencies are obtainable with fabric filters; as much as 99.9 percent removal of particulate matter. The utilization of fabric filters in refuse incinerators has been retarded by the fact that the filter material must be capable of withstanding high temperatures and corrosive constituents in the effluent gas stream. Additionally, the creation of a dust cake on the filter creates a large pressure drop in the gas stream after passage through the filter. Pressure drops in fabric filters have been measured at from 4 to 7 in. of water. The installation and capital cost for fabric filters range from $0.75 to $1.50 per cu ft of gases treated per minute. As mentioned above, fabric filters have not been installed to any great extent in American incinerators.

In summary, simple collection devices such as settling chambers and wetted baffles generally are inadequate for collection of particulate matter from incinerator effluent gas streams. Cyclone collectors are somewhat more efficient, but with the application of more stringent air pollution emission controls, it is quite likely that in the near future cyclone collectors will not be sufficient for use in incinerators. In all probability, future collection requirements will have to be met through the use of wet direct impaction scrubbers, fabric filters, or electrostatic precipitators.

Two additional topics which must be considered under the heading of emission control are the control of odors and the prevention of water vapor plumes. Odors generally may be effectively controlled through the oxidation of all hydrocarbons produced in the primary combustion chamber of an incinerator. These hydrocarbons can be converted to carbon dioxide and water; a 0.5 second retention time at a temperature of 1500°F or greater generally will be sufficient to eliminate by-product hydrocarbons and thus eliminate most odors.

As mentioned previously, water vapor plumes are caused when the relative humidity of the effluent gas stream is significantly higher than the relative humidity of the ambient air around an incinerator chimney or stack. This is caused quite frequently by the use of wet scrubbers as collection devices for particulate matter. If high efficiency wet scrubbers are used to control particulate emissions, vapor plumes may be eliminated in the following way. Heat may be removed from the hot furnace gases before they are introduced into the wet scrubber, and then this heat may be reintroduced into gases after they exit from the scrubber. The relative and absolute humidities of the stack effluent are reduced by the reintroduction of heat and the creation of vapor plumes is minimized. Effluent gases saturated with water vapor from wet scrubbers may also be processed in dehumidifiers. In any case, the use of heat exchangers or dehumidifiers will add to the cost of effluent treatment. Such additional cost should be considered in the planning process when air pollution control systems are chosen.

3-4.3 Residue Recovery

In addition to the by-products listed above, residue from an incineration operation amounts to approximately 15 to 25 percent of the original input volume of solid wastes. Residue includes such items as tin cans, rocks, glass, ashes, clinkers, and unburned organic materials. Table 3-18 shows some typical residue compositions. Ultimate disposal of residue from an incinerator can be accommodated in a landfill operation. The incinerator residue must be treated with the same amount of consideration as any of the other solid waste materials which enter a disposal landfill. Incinerator residue may contain soluble organic and inorganic constituents which could be removed in a leaching action. Such constituents may pose a threat to quality of groundwater resources in the area of the disposal landfill. Care must be exercised to prevent the occurrence of such leaching. This admonition is necessary because many persons assume that the residue remaining after incineration is completely inert and innocuous. In a competently operated incinerator, very little combustible material remains in the residue coming from the combustion chambers. However, it is quite possible that portions of the incinerator residue are soluble in water and could be removed from a disposal site to

TABLE 3-18 Incinerator Residues

Material	Washington, D.C.,[a] Metro-Average Grate-Type Municipal Incinerators, % Dry Weight	Rotary-Kiln Incinerators,[b] % Dry Weight
Tin cans	17.2	19.3 + 6.5 (nonmetallics)
Mill scale and small iron	6.8	10.7
Iron wire	0.7	0.5
Massive iron	3.5	1.9
Nonferrous metals	1.4	0.1
Stones and bricks	1.3	
Ceramics	0.9	0.2
Unburned paper and charcoal	8.3	3.4 (charcoal)
Partially burned organics	0.7 ⎫	
Ash	15.4 ⎬	57.0
Glass	44.1 ⎭	

Sources:
[a]Kenahan, C. B. and Sullivan, P. M., "Let's Not Overlook Salvage," *APWA Rep.*, Vol. 34, No. 3, 1967, p. 5.
[b]Rampacek, G., "Reclaiming and Recycling Metals and Minerals Found in Municipal Incinerator Residues," *Proc.*, Mineral Waste Utilization Symp., March 27–28, 1968, IIT Research Institute, p. 129.

cause degradation of the environment. In any case, a more suitable method of disposal for residues from incineration may be the processing of such residues for material recovery. Such recovery is discussed in detail in Chapter 6.

3-4.4 Waste Heat

The combustion of the solid wastes in a refuse incinerator creates a significant amount of heat. To the present time in the United States, this heat has been wasted through dispersal to the atmosphere. With the increasing shortage of energy supplies such as fossil fuels in recent years, the economic value of the heat created during incineration has increased tremendously. In the United States today, less than a score of municipal incinerators are operated so that heat recovery is practiced. On the other hand, heat recovery from incineration of refuse is widely practiced in Western Europe. The recovery of the energy content of solid wastes through waste heat recovery is discussed in detail in Chapter 7 and waste heat recovery operations in European incinerators are described in Chapter 8.

3-4.5   Summary—Incineration Process Products

In summary, the processing of solid waste in an incinerator creates a number of products, including effluent gases, particulate matter entrained in the effluent gas stream, solid residues from quenching operations and from air pollution control operations, wastewater resulting from cooling and air pollutant control operations, and a great amount of waste heat. Control of air pollutants is a primary operation in the overall processing of solid waste in a refuse incinerator. Techniques and devices exist for such control and these techniques must be used in order that emissions from incinerators will comply with emission standards promulgated by federal, state, and local governments. Processing of solid residue and the heat generated during incineration can lead to significant recovery of much-needed resources. These latter two operations are discussed in later chapters of this book.

3-5.   CRITICAL AND LIMITING PROCESS PARAMETERS

There are a number of limiting process factors associated with an incineration process. Some of these parameters are subject to technical control or analysis, while others are essentially non-technical in nature.

3-5.1   Technical Parameters

The technical parameters that limit the applicability of incineration as a waste disposal operation include the variable quantities and characteristics of refuse, the detrimental production of air pollutants during air incineration, and the characteristics of the wastewaters and residues produced during incineration which require treatment and careful disposal. As mentioned previously, the quantities and characteristics of the refuse vary with climate, season, community characteristics, the relative prevalence of commercial and industrial facilities in the service area, and a number of other factors such as the amount of usage of food grinders, home compactors, etc. The variations in solid waste quantities and characteristics were given in Tables 3-1 and 3-2. These variations indicate that a local study of waste quantities and characteristics will be necessary for the adequate design of an incinerator. In that variations in these quantities and characteristics require provision of extra capacity and auxiliary fuels, and because of the fact that the characteristics of the refuse may limit the recovery of materials and energy from the incineration process, these variations must be considered limiting factors in the application of refuse incineration as a disposal method. However, as indicated, these variations may be accommodated through competent and conservative design of a facility.

The characteristics of the residue from an incineration process also slightly limit the applicability of incineration as a disposal method, because the residue material requires disposal. In addition, the stability of the residue varies with the variation in solid waste characteristics, and the disposal method selected for this residue must accommodate such variations in the qualities of the residue. Moreover, any salvage or materials recovery operation designed to operate on the residue from a refuse incinerator must be capable of accepting the variations in composition and quantity of the residue which have been indicated as characteristic of refuse incineration. These variations in the character and quantity of the residue from an incinerator may be accommodated through competent design and planning of a residue disposal and/or recycling operation.

As demonstrated in Table 3-16, the wastewaters from an incinerator operation may resemble strong industrial wastewaters and obviously require treatment and disposal. However, such disposal can be easily accomplished using standard techniques of wastewater treatment. Therefore, the production of wastewaters in an incinerator operation can be considered to be of only minor importance in limiting the applicability of incineration as a waste disposal method. Of much greater importance in the success of any incineration operation will be the production of air pollutants and the required control and elimination of those pollutants. In refuse incineration, unfortunately, complete combustion does not always take place and unburned hydrocarbons may be produced. In addition to these gaseous constituents such as aldehydes and carbon monoxide, slight amounts of other gaseous pollutants may be produced. Significant quantities of particulate matter are entrained in the effluent gases from most incinerators. All of these air pollutants must be collected and removed from the effluent gas stream in an incinerator. The necessity to remove such pollutants and to satisfy ever-stricter air pollution control laws tends to severely limit the applicability of incineration as a waste disposal method. The cost of air pollution control operations significantly increases the overall cost of refuse incineration. Because of this increase in cost, the use of an incinerator in a given community may not be as advantageous as the disposal of refuse through composting or through sanitary landfilling. Furthermore, the elimination of smoke and water vapor plumes from the stack of an incinerator requires extensive control effort and the necessity for these efforts must be considered a severe limitation on the practical applicability of incineration for refuse processing. The need for air pollution control in incinerator facilities is the most severely limiting or critical parameter associated with this method of waste disposal.

3-5.2 Nontechnical Parameters

The principal parameters of a nontechnical nature that tend to limit the applicability of incineration for refuse disposal are public acceptance of the method and the availability of land for the facility.

Public acceptance of an incinerator can be achieved through proper selection of the site for the facility and an effective public relations effort intended to educate the public concerning the advantages of waste disposal by incineration. An incinerator should be located so that it generally is compatible with the existing character of the community at that locality. Since an incinerator resembles an industrial plant, areas of industrial and commercial activity are much more suitable for the sites of refuse incinerators than are residential areas. To insure continued public acceptance of an incinerator facility, it must be located in an area which is consistent with its industrial character, and if vacant land is found around the incinerator site, that vacant land should be developed for commercial or industrial facilities only. The later construction of residences or institutions such as hospitals or schools near an existing incinerator should be discouraged. The large amount of truck traffic associated with an incinerator and the industrial operation at the incinerator may conflict with such facilities. In order to eliminate such conflicts, any undeveloped land surrounding an incinerator should be allocated to industrial uses through appropriate zoning. The incinerator facility itself should be designed so that it is aesthetically pleasing; every effort should be made to screen undesirable aspects of the facility from public view and landscape design professionals should assist in the overall layout of the facility. If these recommendations are carried out, much of the public hostility toward incinerator facilities, developed through long years of contact with unsatisfactory incinerators, may be eliminated.

Secondly, before the incinerator facility is built, and even during preliminary design stages, it is advisable to begin an effective public relations effort aimed at educating the public about the advantages of refuse incineration. Input from the citizens living in the vicinity of the proposed incinerator should be actively encouraged. To secure such citizen participation, public meetings should be held in which the proposals and design for the incinerator are described and presented. The full cooperation of newspapers, radio, and television communication facilities should be sought in this effort. The public relations program should include some brief review of the alternatives available to the proposed incinerator and should indicate the reasons for the choice of incineration as the preferred disposal method and the choice of the particular site as the location for the proposed facility. This type of activity can do much to assuage citizen doubts concerning the impact of the proposed facility and can transform public hostility into public acceptance.

The comments in the foregoing paragraphs are based on the assumption that a site is available for a proposed facility and that the major difficulty associated with developing that site is public acceptance. In some communities throughout the country, sites may not be available at first glance for incineration facilities. However, this limitation can be overcome through the expenditure of sufficient funds; i.e., every piece of land is available for purchase at a sufficiently high price. The limitation of available land enters the decision-making process concerning the choice of disposal method in the economic value of the land required for the incinerator facility. In communities where land is not readily available near the center of the waste generation area, the high prices required for purchase of suitably located land may make the disposal of waste by incineration inordinately expensive and, therefore, may preclude the use of incineration as a disposal technique.

3-5.3 Summary–Incineration Critical and Limiting Process Parameters

Certain factors limit the applicability of incineration as a waste disposal method; some of these factors are amenable to technical control and some are not. Of those factors amenable to technical control, the most significant is the necessity for control of air pollutants produced during the combustion of refuse. The most important factor which limits the applicability of incineration, but which is not amenable to technical control, is the necessity for public acceptance of such a facility. The gaining of public acceptance for a disposal facility of any kind will require a comprehensive public relations program and a full consideration of aesthetic and social factors in the design of the incinerator facility.

3-6. PROCESS ECONOMICS

The costs associated with refuse disposal through incineration include capital cost for the construction of the facility and operating costs which, in turn, include both direct and indirect operational and maintenance expenditures.

3-6.1 Capital Costs

Capital costs for incinerator facilities include expenditures for such items as furnaces, cranes, scales, air pollution control devices, fans, process residue treatment and recycling equipment, instrumentation, waste heat recovery equipment, etc. Table 3-19 shows some project captial costs established in a study of incineration in the District of Columbia. The relative expenditures associated with each of the items listed above can be seen through reference to Table 3-20. Air pollution control equipment can account for a significant portion of the capital expenditures for any incinerator facility. Additionally, a major item of expense is the set of combustion chambers or furnaces to be used in the incinerator. Commonly, furnaces and their appurtenances account for 60 to 65 percent of all capital costs. Tables 3-21, 3-22, and 3-23 show cost ranges for the various items necessary for the construction of an incinerator and also indicate the relative importance of operational cost. In the 1968 National Survey of Community

TABLE 3-19   Comparative Capital Cost Estimates

Four Unit Incinerator Plant	Type of Air Pollution Control Equipment	
	Electrostatic and Mechanical	Wet Scrubber
General building contract		
Incinerator building and foundations	$ 606,700	$ 447,000
River pump house		7,200
Clarifier basins		45,000
	606,700	499,200
Mechanical contract		
Refractory furnaces and flues	204,000	248,400
Spray cooling chamber	221,900	
Steelwork	324,000	228,000
Insulation	133,700	26,500
Instrumentation	60,000	42,000
Installation of purchased equipment	132,000	60,000
Piping		58,500
	1,075,600	663,400
Purchased equipment		
Fans and drives	135,900	125,900
Pumps and drives		19,000
Air pollution control equipment	396,000	337,300
Clarifier equipment		64,800
	531,900	547,000
Electrical contract: power and lighting	195,000	129,000
Subtotal physical cost	2,409,200	1,838,600
Engineering and field supervision	169,000	130,000
Contingency	241,000	185,900
Escalation to December 1968	120,500	92,900
Total incremental physical cost	2,939,700	2,247,400

Source: Special Studies for Incinerators for the Government of the District of Columbia, U.S. Department of Health, Education, and Welfare, Public Health Service, 1968, p. 18.

Solid Waste Practices conducted by the United States Public Health Service, the average capital costs for incinerators operating at that time in the United States were approximately $6200 per ton of incinerator capacity. In addition to the furnaces which account for approximately two-thirds of the capital cost in an incinerator, the physical plant accounts for 20 to 30 percent and air pollution control equipment commonly accounts for 8 to 10 percent of total capital expenditures.

TABLE 3-20 Distribution of Incinerator Construction Costs for Equipment and Building, dollars per ton per day[a]

	Architectural and Structural	Mechanical	Electrical	Heating and Ventilating, Plumbing and Miscellaneous	Total Cost	% of Total
Delivery of refuse tipping floor, scales and accessories	337	4	14	18	373	8
Handling of refuse (storage pit, crane hoppers and accessories)	658	141	14	27	840	17
Burning of refuse (furnaces, flues, chimneys, combustion chamber and accessories)	955	1,321	121	55	2,452	51
Residue removal (ash cellar, conveyors, controls and accessories)	309	108	66	41	524	11
Flyash removal (subsidence chambers, screens and accessories)	113	15	9	9	146	3
Miscellaneous (utilities, grading, landscaping demolition, furniture, tools)	313	127	30	16	486	10
Total costs	2,685	1,716	254	166	4,821	100
Percent distribution	55%	36%	5%	4%	100%	

[a]1958 cost index.

Source: Rogus, C. A., "Municipal Incineration of Refuse," *J. Sanit. Eng. Div.*, Am. Soc. Civ. Eng., *90*, June, 1964.

TABLE 3-21   Approximate Costs of Recommended Multiple Chamber
Incinerators and Scrubbers in 1968[a]

Size of Incinerator, lb/hr	General Refuse Incinerators	Scrubbers[b]	Pathological Incinerators
50	$ 1,200	$ 2,200	$2,000
100	1,700	3,000	2,700
150	2,000	3,600	4,000
250	2,700	4,400	5,500[c]
500	5,000	6,200	
750	9,500	7,600	
1,000	12,500	8,800	
1,500	20,000	11,200	
2,000	25,000	13,200	

[a]Incinerator costs are exclusive of foundations.
[b]Scrubber costs are exclusive of foundations but include reasonable utility connections.
[c]For a 200 lb/hr incinerator.

Source: Interim Guide of Good Practice for Incineration at Federal Facilities, U.S. Department of Health, Education, and Welfare, Public Health Service, 1969.

3-6.2 Operational and Other Costs

Figures 3-15 and 3-16 contain comparative cost data for incinerators as opposed to other disposal facilities, and show costs for operation of incinerators versus the age of the incinerators. The average operating cost for municipal incinerators was found to be $5.00 per ton of incoming refuse in 1968. On the basis of the

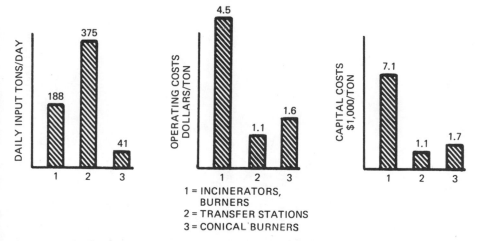

1 = INCINERATORS, BURNERS
2 = TRANSFER STATIONS
3 = CONICAL BURNERS

Fig. 3-15   Characteristics of disposal facilities (Study of Community Solid Waste Management, HEW, USPHS, 1968).

TABLE 3-22   Estimated Capital Investments and Operating Costs,[a] Incinerator and Incinerator Boiler Plants, Refuse Design Capacity—4000 Tons per Week

	Case I Refractory Furnace	Case II Refractory Furnace with Boiler	Case III Waterwall Furnace with Boiler	Case IV Waterwall Furnace
Capital costs:				
General building construction	$ 930,000	$1,215,000	$1,030,000	$ 938,000
Equipment delivered to site	1,340,000	2,620,000	2,230,000	2,150,000
Mechanical contract	1,075,000	1,580,000	800,000	983,000
Electrical contract	238,000	322,000	237,000	232,000
Total incremental cost	$3,583,000	$5,737,000	$4,297,000	$4,303,000
Annual operating expenses:				
Operating days per week	5	7	7	5
Maintenance labor and supplies	140,000	161,000	147,000	123,000
Operating labor	326,000	495,000	495,000	365,000
City water	28,000	25,000	43,000	30,000
Auxiliary fuel	—	140,000	187,000	10,000
Electric power	161,000	184,000	174,000	161,000
Operating supplies and chemicals	1,000	2,000	3,000	1,000
Subtotal	$ 656,000	$1,007,000	$1,049,000	$ 690,000
Fixed charges on investment	276,000	442,000	331,000	331,000
Estimated total annual expense	$ 932,000	$1,449,000	$1,380,000	$1,021,000
Estimated value of steam per 1000 lb	—	1.06	0.49	—

[a]Capital investments and operating expenses include only those variables affected by plant design. They are not intended to include all costs of operation or construction at the incinerator plant (1968 cost figures).

Source: Day and Zimmerman, Inc., *Special Studies for Incinerators for the District of Columbia*, U.S. Department of Health, Education, and Welfare, Cincinnati, Ohio, 1968.

TABLE 3-23   Typical Comparative Cost Figures for Furnace Systems, 3 units at 300 ton/day

	Water Furnace with Waste Heat Boiler	Refractory Lined Furnace	Refractory Lined Furnace with Waste Heat Boiler	Refractory Furnace with Rotary Kiln
Incinerator structure	$2,240,000	$2,290,000	$2,240,000	$2,290,000
Utility construction	259,000	265,000	256,000	265,000
Furnace components	2,933,000	903,000	2,375,000	3,253,000
Air pollution control equipment	690,000	1,500,000	900,000	1,500,000
Estimated project cost	$6,122,000	$4,958,000	$5,771,000	$7,308,000
Amortization of capital cost based on 4½% annual interest, 20 yr bonds	$ 472,000	$ 380,000	$ 447,000	$ 563,000
Plant labor	520,000	495,000	520,000	496,300
Utilities	15,000	35,000	15,000	36,500
Building supplies, operation and maintenance	193,000	312,000	280,000	350,400
Estimated annual cost	$1,200,000	$1,220,000	$1,262,000	$1,446,200

Source: Proposals for a Refuse Disposal System in Oakland County, Michigan, U.S. Department of Health, Education, and Welfare, Public Health Service, 1970, p. 30.

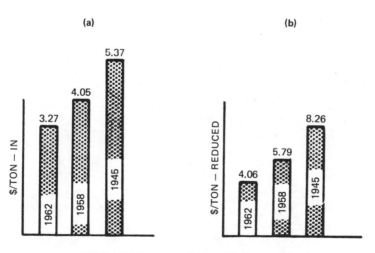

NOTE SCALE CHANGE, PART (A) TO PART (B).

Fig. 3-16   Incinerator operating costs vs. "age" in the U.S. (Study of Community Solid Waste Management, HEW, USPHS, 1968).

reduced wastes, however, this figure is much higher, as shown in Fig. 3-16. Although most of the incinerators surveyed in the 1968 survey had operating costs below $5.00 per ton, some of the incinerators investigated had operating costs above $10.00 per ton. This variation in operating costs can be attributed to differences in the amount and type of pollution control equipment installed in the various incinerators and the differences in labor and utility rates from one locality to another.

In addition to the capital and operating expenses associated with equipment, many indirect costs are incurred during the operation of an incinerator. Table 3-24 shows estimated direct and indirect costs for incineration. It should be

TABLE 3-24   Cost of Refuse Disposal by Incineration[a]

	Continuous Type (5 Plants)
Direct costs	
Operating labor (cranemen, firemen, stationary engineers, laborers)	$2.20
Maintenance labor (masons, mechanics, laborers)	.13
Residue and flyash disposal:	
Hauling to sanitary landfills (average 5 miles)	.21
Landfilling disposal operations	.53
Overhead and administration (departmental only)	.25
Total direct costs	$3.32
Indirect costs	
Fringe benefits (pensions, vacations, sick leave)	.72
Vehicle maintenance (residue loading and hauling vehicles)	.08
Utilities (telephone, electricity, gas—no water or sewage charge)	.07
Materials and supplies (for operation and maintenance)	.09
Capital amortization at 3% (45 yr life for plant and 15 yr life for	
incinerator equipment)	.71
Total indirect costs	$1.67
Total operating cost per ton processed	$5.00

[a]Unit costs per ton of refuse processed—1960 cost index.

Source: Rogus, C. A., "Municipal Incineration of Refuse," *J. Sanit. Eng. Div.*, Am. Soc. Civ. Eng., *90*, June, 1964.

noted that the costs shown in Table 3-24 are based on the 1960 Cost Index and should be multiplied by an appropriate factor to transform them to current costs.

The discussion in this chapter has indicated that significant economies can be realized through the recovery of energy from an incinerator. Some indication of the income which can be obtained through heat recovery in a municipal incinerator can be obtained from Table 3-25. The economics of waste heat recovery are more fully discussed in Chapter 7 of this book.

TABLE 3-25   Summary of Operating Costs for Chicago Southwest Municipal Heat-Recovery Incinerator

	Operating Cost, $/Ton
Cost item	
Labor (62 men)	$1.92
Repairs	0.73
Electrical	0.19
Supplies	0.60
Residue hauling	1.05
Steam production (4000 lb at $0.50/1000 lb)	2.00
	6.49
Income item	
Sales of steam (4000 lb at $0.625/1000 lb)	2.70
Profit from salvage	0.14
	2.84
Net operating cost	3.65
Amortization of capital cost	1.26[a]
Total operation and maintenance cost	4.91[b]

[a]These calculations assume 100 percent steam utilization. ASME states that actual operating cost is $3.98 per ton implying that only 86 percent $\left(\dfrac{2.70 - 0.38}{2.70}\right)$ steam utilization is practiced.

[b]ASME gives $5.24 per ton, again reflecting less than 100 percent steam utilization.

Source. Am. Soc. Mech. Eng., Incinerator Committee, June 1966.

3-6.3   Summary—Incinerator Process Economics

In comparison to other methods of refuse disposal, municipal incineration appears to be rather expensive in terms of both capital costs and operating costs. However, in some localities land costs associated with sanitary landfilling or composting operations may be sufficiently high that centralized incineration is by far the most economical method of refuse disposal in that community. More extensive data on actual incinerator costs and costs for projected incinerators are contained in section 3-8 wherein case histories of municipal incinerators are given.

3-7.   LEGAL RESTRICTIONS OF REFUSE INCINERATION

The application of refuse incineration as a disposal method will, of necessity, be subject to certain federal, state, and local laws relating to refuse disposal and environmental protection.

3-7.1   Federal Laws Pertaining to Incineration

At the present time in the United States, the use of incinerators in disposal of municipal refuse is not specifically regulated by federal legislation. However, existing federal laws requiring various measures for environmental protection will have important influences on incinerator operation. As mentioned previously, the Clean Air Act Amendments of 1970 require monitoring and compilation of data concerning emissions from incinerators into the atmosphere. Emission levels established by the Administrator of the Environmental Protection Agency pertain to the operation of a refuse incinerator since a refuse incinerator must be considered a stationary source of possible air pollution. Furthermore, federal water pollution control laws call for the elimination of all pollutant discharges into the nation's waters by the early 1980s. This requirement will certainly have an indirect effect upon the disposal of wastewaters from incinerator facilities. Although the present state-of-the-art of noise pollution abatement is not highly advanced, it is highly likely that within the near future emission levels for noise will be established by the Environmental Protection Agency for facilities such as refuse incinerators. These regulations together with pertinent air pollution control and water pollution control laws will have important effects upon the operation of any incinerator facility. Certain state and local laws will also further restrict the operation of incinerators.

3-7.2   Local Laws Pertaining to Incinerators

In selecting any method of waste disposal, local laws or ordinances governing the operation of any of the processes should be determined. The incinerator, for example, is especially confined by such items as pollution laws and zoning regulations.

Air Pollution Laws. Almost every city and state now has under review some type of air quality standard for its particular area. The federal government has outlined air quality standards which were to be met by 1973. The elements of air pollution to be considered are particulate matter, sulfur dioxide, carbon monoxide, hydrocarbons, and nitrogen dioxide. Table 3-26 shows a comparison of the air quality standards as presently set forth by the federal government, the state of Kentucky, and Jefferson County, Kentucky.

Noise Laws. Noise laws in most cities are not enforced and the interpretation of these laws varies widely from occurrence to occurrence. The decision as to when a noise law is broken is made first by the citing officer and then by the courts. Section 86-135 of the General Ordinances for the City of Louisville, 1954, a Revised Compilation is a sample of one city's ordinance pertaining to noise. It reads:

NOISE (1) It shall be unlawful for any person within the corporate limits of the City of Louisville to make, continue or cause to be made or continued any loud, unnecessary or unusual noise which either annoys, injures, or endangers the comfort, repose, health or safety of others unless the making and continuing of the same be necessary for the protection or preservation of property or the health, safety, life or limb of said person. Any person, firm or corporation violating any provision of this subsection of this section shall be fined not less than one dollar ($1.00) nor more than twenty-five ($25.00) for each offense.

(2) It shall be unlawful to:

(a) sound any horn or signal device on any vehicle not in motion, except as a danger warning if another vehicle is approaching apparently out of control;

(b) sound any horn or signal device on any vehicle in motion except as a danger warning after or as an attempt is made to decelerate the vehicle by the application of brakes;

TABLE 3-26   Air Quality Standards

Pollutant	Federal (1973) primary	(1973) secondary	State of Kentucky	Jefferson County Kentucky
Particulate	$\mu g/m^3$			
Annual geom. mean	75	60	65	65
24 hr[a]	—	—	180	150
24 hr[b]	260	150	220	180
SO_2	ppm			
Annual geom. mean	0.03	0.023	0.02	0.015
max one month	—	—	0.05	0.030
24 hr[b]	0.14	0.10	0.08	0.10
2 hr avg.	—	—	0.15	0.30
1 hr avg.	—	—	0.20	0.40
CO	ppb			
Any 8 hr period	8.7	8.7	8.0	8.0
max 1 hr	13.1	13.1	30	20
Total oxidant	ppm			
max 24 hr	—	—	0.02	0.02
max 1 hr	0.064	0.064	0.05	0.03
Total HC	$\mu g/m^3$			
any 3 hr period	125	125	100	100
NO_2	$\mu g/m^3$			
Annual arith. mean	100	100	—	—
24 hr avg.	250	250	—	—

[a]Not to be exceeded 1% of the time.
[b]Not to be exceeded more than once per year.

(c) sound any horn or signal device on a vehicle for an unnecessary and unreasonable period of time or in such manner as to create an unreasonably loud or harsh sound;

(d) use or operate any vehicle which produces, or use or operate any vehicle so out of repair or so loaded with any materials as to cause any loud and unnecessary grating, grinding, rattling or other loud and excessive noise;

(e) discharge into the open air the exhaust of any vehicle except through a muffler or other device which will effectively prevent loud or explosive noises therefrom;

(f) create any loud and excessive noises in connection with loading or unloading any vehicle;

(g) use any mechanical loud speaker or amplifier on any moving or standing vehicle for advertising or other purposes.

As used in this section, the word vehicle shall include any device in, upon or by which any person or property is or may be transported or drawn upon any street in the City of Louisville. Any person, firm or corporation violating any provision of this subsection shall be fined not less than five dollars ($5.00) nor more than twenty-five dollars ($25.00) for each offense.

Zoning Laws. Incinerators preferably should be built on land zoned for industrial use; however, in some instances commercially-zoned land may be used. Most local governments have a zoning classification that pertains to land used for an incinerator. As a typical example, an incinerator built in Jefferson County, Kentucky would require land that is zoned M-3 with a special condition permit. These special conditions pertain to the operation of the plant. The regulation governing the M-3 zone is as follows:

The following uses having accompanying hazards such as fire, explosion, noise, vibration, dust, or the emission of smoke, odor, or toxic gases may, if not in conflict with other laws or ordinances, be located in the M-3 Industrial District by conditional use permit after the location and nature of such use shall have been approved by the Louisville and Jefferson County Board of Zoning Adjustment. The Board of Zoning Adjustment shall review the comprehensive plan, the plans and statements of the applicant, and shall not permit such buildings, structures, or uses until it has been shown that the public health, safety, morals, and general welfare will be properly protected, and that necessary safeguards will be provided for the protection of surrounding property and persons.

The special conditions provided by this regulation are:

(2) Incinerators, public and private
    (a) No incinerator building or structure shall be located closer than 200 feet from any site boundary line, and no other building or

structure used in connection with the operation shall be located closer than 30 feet from any site boundary line. Access to the incinerator shall be located so as to require a minimum of travel on a public way.

(b) The entire site shall be enclosed with fencing and gates as required in Section 3c above.

(c) All materials delivered to the site which are organic or of organic origin or other combustible materials such as paper, cardboard, rubber, plastic, wood fiber, sawdust, floor sweepings, plaster board, framing, lumber, laths, tree stumps, trunks, branches, foliage, furniture, rags, garbage and industrial wastes and including metal and glass containers shall be burned in the incinerator.

(d) All residue resulting from the burning operations and other fill materials which are inorganic or substances which are not subject to decomposition combustion or the production of odors shall be spread and thoroughly compacted as they are deposited.

(e) All materials which are to be burned shall be placed on or in a concrete slab or hopper enclosed by a building, masonry walls, or chain link type fencing at least 6 feet high provided with doors or gates which shall be securely locked when the incinerator is not in operation. The materials shall be transferred from the slab or hopper into the incinerator as soon as they are received but in any case all combustible materials shall be burned during the same day that they were delivered. The slab or hopper shall be kept clear of all materials when not in active use.

(f) There shall be no separation or picking of materials or storage for salvage thereof on the site.

(g) All deliveries of materials to the site, loading of the incinerator filling, spreading, compact and grading shall be done between the hours of 7:00 A.M. and 5:00 P.M. on weekdays only.

(h) A watchman shall be stationed at the site at all times for whom a suitable shelter or living quarters shall be provided.

(i) Sanitary toilet facilities shall be provided on the site in accordance with the requirements of the Department of Health.

3-7.3 Summary—Legal Restrictions of Refuse Incineration

The planning, design, construction, and operation of a refuse incinerator should be carried out with a full knowledge of existing federal, state, and local legislation regulating the incineration of refuse. For the most part, of any federal legislation, air pollution control legislation will have the most important effect on incineration. On the other hand, zoning regulations will be most influential in the location and siting of incinerator facilities as far as local or state laws are concerned.

3-8. PROCESS CASE HISTORIES—TECHNICAL AND ECONOMIC

Several comprehensive studies have been made in the investigation of existing incinerators or proposed incinerator facilities in cities or communities in the United States. A review of certain of these case histories will present valuable information concerning operational characteristics and process economics. For this reason, several case histories are presented here.

3-8.1 Seven Incinerators

The case histories of seven incinerators (incinerators A through G) are chronicled in the publication, "Seven Incinerators," by W. C. Achinger and C. E. Daniels (Ref. 3-1). This publication gives data on the quality and quantity of waste incinerated, residue, and gasborne particulate emissions, the quality of flyash collected and the wastewater produced, as well as the economics involved in incineration.

Incinerator A is a 300 ton per day plant built in 1966. The waste is charged by dumping it onto an enclosed tipping floor and transporting it to the charging hoppers by a front end loader. Waste is fed continuously from the hoppers by conveyors. The plant uses two refractory-lined, multiple-chambered furnaces with inclined, modified reciprocating grate sections followed by stationary grate sections. An 11,000 cu ft per minute forced-draft underfire air fan and a 57,000 cu ft per minute induced draft fan per furnace provide the components of the air-draft system. No overfire air is provided for the furnace. The residue is handled by a common chain flight conveyor for both furnaces. Partial spray quenching of the residue is practiced. Air pollution is controlled by a wet scrubber of the impingement type. Impingement occurs on forty-one 12-in. diameter wetted columns. Residue quenching water flows to complete retention lagoons. Flyash scrubbing water flows to settling basins and then to the retention lagoons. The plant is located in the western United States.

Incinerator B is a 300 ton per day plant located in the eastern United States and was constructed in 1966. The waste is dumped onto an enclosed tipping floor and then stored in a 3000 cu yd storage pit. The waste is conveyed from the storage pit by a bridge crane with a grapple bucket into two charging hoppers. Two refractory-lined, multiple-chambered furnaces with three sections of inclined rocking grates burn the waste. The air draft system consists of two 19,000 cu ft per minute forced-draft fans for each furnace and one 200 ft tall stack for natural draft. The residue handling system has a quench tank with a chain flight conveyor with a duplicate system also available. A wet scrubber of the flooded baffle wall type is used as the air pollution control equipment. The flyash scrubbing water receives pH adjustment, is ponded in a settling basin, and is then discharged weekly to the sewerage system. Residue-quench water is detained in the basin and is discharged weekly to the sewerage system.

Incinerator C is located in the southern United States and is a 5 to 6 ton per day pilot plant constructed in 1967. There is no permanent storage and charging is achieved by a screw conveyor from the hopper. The furnace is a conical burner with double metal walls and fixed grates. The air is supplied to the system by an 1800 cu ft per minute forced-draft underfire air fan and a 3600 (when a water scrubber is used) or 5000 cu ft per minute (when an electrostatic precipitator is used) induced draft fan. The residue is cleaned out manually after a cooling period. The air pollution controls used are a water scrubber of the centrifugal type, an afterburner and water scrubber, and an electrostatic precipitator. The flyash scrubbing water flows to a settling basin before final discharge to an open watercourse.

Incinerator D is a 500 ton per day plant located in the midwestern United States; it was constructed in 1965. The plant has an open pit floor, two storage pits, two bridge cranes, and two charging hoppers. Incineration is achieved in two refractory-lined, multiple-chambered furnaces with one inclined section and one horizontal section of traveling grates. A forced-draft fan and natural draft from a 200 ft tall stack for each furnace supply the air to the system. A quench tank with a chain flight conveyor handles the residue. A duplicate backup system is also available. Air pollution control is achieved by using a wet scrubber of the flooded baffle wall type. All process water flows through a settling basin and then to the sewerage system.

Incinerator E was built in 1963 in the southern United States. It has a capacity of 500 tons per day. The plant has an open 5150 cu yd storage pit, two bridge cranes, and two charging hoppers. Incineration is achieved in two furnaces with three reciprocating grate sections followed by a rotary kiln. Air is supplied by one 25,000 cu ft per minute forced-draft underfire air fan per furnace, and one 200 ft tall stack for natural draft. Residue is handled by a residue-quench tank with a chain flight conveyor. A duplicate system is available. Wet scrubbers of the water spray and baffle wall type are the pollution control devices used for upgrading the air quality. Flyash scrubbing water is used for quenching residue. The water then flows through a grit chamber before being discharged to an open watercourse.

Incinerator F is a 600 ton per day plant located in the southern United States and was constructed in 1967. The tipping floor is open and the waste is stored in a 2430 cu yd storage pit. Two bridge cranes feed the two charging hoppers. Two furnaces with three reciprocating grate sections followed by a rotary kiln are used to burn the waste. Air is supplied by a 25,000 cu ft per minute forced-draft underfire air fan (per furnace) and one 200 ft tall stack for natural draft. A residue-quench tank with a chain flight conveyor is used to handle residue. A duplicate system is available. Air pollution is controlled by a wet scrubber of the water spray and baffle wall type. Flyash scrubbing water is used for residue quenching. The water flows to a lagoon and then to an open watercourse.

Incinerator G was built in 1967. It is a 400 ton per day plant located in the southern United States. The plant has an open tipping floor, a 1750 cu yd storage pit, one bridge crane, and two charging hoppers. Incineration is achieved in two furnaces with four sections of inclined reciprocating grates. In each, furnace air is supplied by a 20,000 cu ft per minute forced-draft underfire air fan, a 24,000 cu ft per minute forced-draft overfire air fan, and a 120,000 cu ft per minute induced draft fan. Residue is handled in a quench tank with a chain flight conveyor. A duplicate system is available. Air pollution control is achieved by the use of a multitube dry cyclone following a wet-baffle wall. All process waters enter the residue quench tank and then go to a lagoon with subsequent discharge to a canal.

Waste Characteristics. The solid waste received by the incinerators during the study was composed of 79 percent combustibles and 21 percent noncombustibles in general. Incinerators C and G showed lower contents of combustible materials; therefore, their volatile and heat contents were lower than in the other five incinerators. Tables 3-27, 3-28, and 3-29 give a breakdown of the characteristics and quality of the waste coming into each incinerator.

Residue Characteristics. Tables 3-30 and 3-31 show the composition and analyses of the residue from each incinerator. The higher percentages of fines found in the residue from incinerators E and F were probably caused by the reduction of the size of glass and rocks due to the tumbling action in the rotary kilns. The lowest percentage of fines was produced by incinerators C and D, a conical burner and a traveling grate incinerator.

Flyash. Flyash samples were obtained from only four of the seven plants. Table 3-32 shows the type of pollution control used and the analyses of the flyash collected from each control.

Wastewater. Wastewaters from the plants that were analyzed were scrubber water, quench water, and effluent (after treatment). Table 3-33 shows the characteristics of the wastewaters from each incinerator. The scrubber water generally was found to be acidic with increases in chloride, hardness, sulfate, and phosphate concentrations. Where quench waters were not added in with the scrubber water, it was found to be alkaline. The total solids content of the quench water in all the incinerators was found to be high. The data show that

TABLE 3-27 Heat-Release and Burning Rates—Seven Incinerator Case Histories

| Incinerator | Capacity (Tons/Day) | | Burning Rate per Unit Area of Grate (lb/ft²/hr) | | Rate of Heat Release per Unit Volume (BTU/ft³/hr) | | | |
| | | | | | Primary Chamber | | Total Furnace | |
	Design	Actual	Design	Actual	Design	Actual	Design	Actual
A	300	281	45	42	23,000	19,000	14,300	11,800
B	300	308	52	53	28,600	25,300	13,800	12,300
C	1000[a]	1444[a]	3	5	—[b]	—[b]	2,400	2,600
D	500	[b]	[b]	[b]	[b]	[b]	[b]	[b]
E	500	660	45	59	23,300	31,000	13,900	18,600
F	600	645	47	50	21,900	26,000	14,400	17,000
G	400	482	51	62	23,600	22,000	14,400	13,400

[a]lb/hr.
[b]Information not available.

Source: Achinger, W. C. and Daniels, C. E., "Seven Incinerators," paper presented to 1970 National Incinerator Conference, Cincinnati, Ohio, May 17–20, 1970 (EPA Publ. No. SW-51ts.lj).

TABLE 3-28 Solid-Waste Composition
(Percent by Weight)

Component	Incinerator						
	A	B	C	D	E	F	G
Combustibles:							
Food waste	7.4	6.1	20.3	8.5	12.2	18.3	11.0
Garden waste	3.4	8.4	11.1	0.5	1.6	0.6	9.8
Paper products	62.5	58.0	30.2	60.4	58.7	60.6	44.9
Plastic, rubber,							
leather	2.8	3.3	3.1	5.4	3.0	2.1	3.5
Textiles	2.4	3.1	5.2	2.4	1.8	1.8	3.2
Wood	2.4	1.4	1.7	5.4	0.4	2.3	3.1
Total	80.9	80.3	71.6	82.6	77.7	85.7	75.5
Noncombustibles:							
Metal	9.0	8.2	6.8	9.0	8.6	8.5	8.1
Glass, ceramics	4.2	8.1	10.5	3.5	10.3	5.4	9.5
Ash, rock, dirt	5.9	3.4	11.1	4.9	3.4	0.4	6.9
Total	19.1	19.7	28.4	17.4	22.3	14.3	24.5

Source: Achinger, W. C. and Daniels, C. E., "Seven Incinerators," paper presented to 1970 National Incinerator Conference, Cincinnati, Ohio, May 17–20, 1970 (EPA Publ. No. SW-51ts.lj).

TABLE 3-29 Solid Waste Analyses

Incin-erator	Moisture, as Sampled (%)	Heat, as Sampled (BTU/lb)	Ash, Dry Basis (%)	Volatiles, Dry Basis (%)	Density, as Sampled (lb/yd^3)
A	20.0[a]	4410	34.2	65.8	[b]
B	20.0[a]	4320	31.8	68.2	[b]
C	26.5	3770	47.1	52.9	[b]
D	20.7	4520	35.6	64.4	[b]
E	20.2	5030	29.9	70.1	200
F	21.0	5530	22.7	77.3	140
G	28.2	3870	42.3	57.7	230

[a] Assumed.
[b] No measurement made.

Source: Achinger, W. C. and Daniels, C. E., "Seven Incinerators," paper presented to 1970 National Incinerator Conference, Cincinnati, Ohio, May 17–20, 1970 (EPA Publ. No. SW-51ts.lj).

TABLE 3-30   Residue Composition
(Percent by Weight)

Component	Aa	B	Ca	D	E	F	G
Fines	44.9	52.5	38.9	36.4	74.5	79.4	52.6
Unburned combustibles	b	b	1.3	35.8	0.1	0.7	1.1
Metal	23.9	14.6	13.0	14.5	21.4	16.8	20.0
Glass, rock	31.2	32.9	46.8	13.3	4.0	3.1	26.3

[a]Dry samples.
[b]Unburned combustibles included with fines.
Source: Achinger, W. C. and Daniels, C. E., "Seven Incinerators," paper presented to 1970 National Incinerator Conference, Cincinnati, Ohio, May 17–20, 1970 (EPA Publ. No. SW-51ts.lj).

there is a need for treatment of incinerator wastewaters before they are discharged into open waterways.

Particulate Emissions. Particulate emissions of each incinerator are given in Table 3-34. The data were collected and calculations were based upon standard conditions of 29.92 in. of mercury and 70°F. The incinerators failed to meet all but the weakest standards with the existing air pollution control equipment then employed by the seven units.

TABLE 3-31   Residue Analyses

Incinerator	Moisture as Sampled (%)	Heat, Dry Basis (BTU/lb)	Ash, Dry Basis (%)	Volatiles, Dry Basis (%)	Density as Sampled (lb/yd³)
A	15.0a	170	97.4	2.6	b
B	24.5	200	98.4	2.0	b
C	0.3	180	98.0	2.0	b
Dc	–	–	–	–	–
E	21.8	520	97.0	3.0	1490
F	24.8	940	92.7	7.3	1620
G	10.5	70	99.4	0.6	1600

[a]Assumed.
[b]No measurement made.
[c]No laboratory analysis performed.
Source: Achinger, W. C. and Daniels, C. E., "Seven Incinerators," paper presented to 1970 National Incinerator Conference, Cincinnati, Ohio, May 17–20, 1970 (EPA Publ. No. SW-51ts.lj).

TABLE 3-32   Flyash Analyses

Incinerator, Type of Air Pollution Control Equipment	Moisture, as Sampled (%)	Heat, Dry Basis (BTU/lb)	Volatiles, Dry Basis (%)	Ash, Dry Basis (%)
A, wetted-column water scrubber	64.9	180	14.0	86.0
B, flooded baffle-wall water scrubber	a	1290	13.9	86.1
C-1, centrifugal water scrubber	a	a	16.4	83.6
C-3, electrostatic precipitator	52.4	3400	27.5	72.5
G, multitube cyclones	0.3	440	4.2	95.8

[a]No measurement made.

Source: Achinger, W. C. and Daniels, C. E., "Seven Incinerators," paper presented to 1970 National Incinerator Conference, Cincinnati, Ohio, May 17–20, 1970 (EPA Publ. No. SW-51ts.lj).

Costs. The annual costs of the incinerators ranged from $171,838 to $675,864. Unit costs ranged from $4.02 to $6.69 per ton of waste processed. Since all of the incinerators were operated under design capacity, the adjusted annual cost was computed. The annual costs of the incinerators consisted of repairs and maintenance cost, operating cost, and capital cost. Tables 3-35 to 3-39 give a breakdown of these costs for the incinerators.

3-8.2   Chicago Calumet Incinerator

Another case study of interest is the Calumet Incinerator in Chicago, Illinois. The plant consists of six furnaces capable of incinerating 200 tons per day. The furnaces are divided into three cells with individual charging and ash removal for each cell. Three 5-ton P&H cranes transfer refuse from the 1200-ton capacity storage pit to the charging hoppers. Solid waste enters the furnace through pneumatic-powered Beaumont Birch gates 5½ ft by 4 ft in size. The refuse is agitated by hydraulically-operated inclined Flynn and Emrich stokers 22½ ft wide and 12 ft deep. Each pair of furnaces is served by a 12 ft diameter stack 250 ft in height. Dual submerged Link-Belt dragchain conveyors running the length of the building remove and discharge the ash into hoppers for removal by truck (Ref. 3-20).

Air pollution control requirements are met by the removal of particulates.

TABLE 3-33 Wastewater Analyses

Incinerator, Sample Source	pH	Temperature (°F)	Suspended Solids (mg/l)	Dissolved Solids (mg/l)	Total Solids (mg/l)	Alkalinity (mg/l CaCO$_3$)	Chlorides (mg/l)	Hardness (mg/l CaCO$_3$)	Sulfates (mg/l)	Phosphates (mg/l)	Conductivity (μmhos/cm)
A, quench water	8.4–11.2	b	1860	1280	3140	120	420	460	230	0.5	3000
A, scrubber water	3.8–4.2	b	1350	5820	7170	1.0	2300	3430	720	51	7100
B, quench water	11.2–11.5	110	1300	2660	3960	720	680	980	120	38	–
B, scrubber water	4.8–6.5[a]	165	320	8840	9160	23	3540	2630	1250	13	–
C-1, scrubber water	2.6	b	110	540	650	0	270	110	110	4.4	1800
C-1, settling-tank water	2.6	b	120	500	620	0	280	110	80	4.1	970
C-2, scrubber water	2.6–3.4	b	90	450	540	0	200	150	100	4.1	1000
C-2, settling-tank water	2.4–3.6	b	180	480	660	0	230	120	70	6.0	850
C-3, precipitator drain water	3.6–4.0	b	1720	7360	9080	0	3200	1890	460	54	6000
C-3, settling-tank water	3.4–4.2	b	600	1300	1900	0	470	400	100	24	1600
D, quench water	5.9–7.1	b	460	2040	2500	600	360	550	280	21	2020
D, scrubber water	1.8–7.6	b	280	1740	2020	80	700	900	220	19	3640
E, tap water	8.4	b	0	55	56	100	7	33	1.0	0.1	46
E, quench water	3.9–7.0	120	900	590	1490	240	200	290	25	21	810
E, scrubber water	2.5–3.0	150	90	750	840	0	300	260	28	13	1360

TABLE 3-33 (Continued)

Incinerator, Sample Source	pH	Tempera- ture (°F)	Suspended Solids (mg/l)	Dissolved Solids (mg/l)	Total Solids (mg/l)	Alkalinity (mg/l CaCO₃)	Chlorides (mg/l)	Hardness (mg/l CaCO₃)	Sulfates (mg/l)	Phosphates (mg/l)	Conductivity (μmhos/cm)
E, final effluent water	4.5–6.9	110	85	570	655	110	200	270	33	4.9	750
F, tap water	5.9	b	0	75	75	74	4.0	46	5.0	0.2	46
F, quench water	5.4–7.1	68	760	360	1120	140	98	180	45	14	530
F, scrubber water	3.0–5.0	82	90	520	610	29	180	190	24	8.8	630
F, lagoon effluent water	5.8–7.9	65	580	320	900	140	94	180	–	9.3	430
G, well water	7.0–8.4	75	0	950	950	350	420	30	20	0.9	1550
G, spray water	6.6–10.3	104	740	2350	3090	260	1050	400	210	43	3780
G, flyash wash water	10.9–12.5	57	3180	890	4070	720	240	340	89	160	1690
G, quench water	9.4–10.9	88	450	1200	1650	470	450	95	53	16	1940
G, lagoon effluent water	9.4–10.3	70	40	1210	1250	310	450	100	70	3.1	1960

[a]Sample was obtained after soda-ash neutralization.
[b]No measurement made.

Source: Achinger, W. C. and Daniels, C. E., "Seven Incinerators," paper presented to 1970 National Incinerator Conference, Cincinnati, Ohio, May 17–20, 1970 (EPA Publ. No. SW-51ts.li).

TABLE 3-34 Particulate Emission Data

Incinerator	Particulate Emissions				CO_2 %	Stack Temp. (°F)	Excess Air (%)	Moisture (%)	Gas-Flow Rate (ft³/min)
	gr/st ft³ at 12% CO_2	lb/1000 lb at 50% Excess Air	lb/hr	lb/Ton of Waste Charged					
A	0.55	1.06	122	10.4	4.6	455	270	16.3	69,800
B	1.12	–	186	14.5	3.5	585	–	16.1	131,000
C-1	0.56	0.75	3.2	4.1	2.8	138	370	18.0	3,890
C-2	0.41	0.46	2.4	3.4	3.3	158	220	25.6	3,990
C-3	0.30	0.52	1.7	2.9	3.3	325	410	15.5	4,460
D	0.46	0.85	173	8.8	5.0	485	260	18.1	120,000
E	0.73	1.19	238	8.6	5.0	305	220	26.6	186,000
F	0.72	1.18	–	12.5	3.9	365	320	16.0	165,000
G	1.35	2.70	386	20.4	3.2	500	500	14.3	130,000

Source: Achinger, W. C. and Daniels, C. E., "Seven Incinerators," paper presented to 1970 National Incinerator Conference, Cincinnati, Ohio, May 17–20, 1970 (EPA Publ. No. SW-51ts.1)).

TABLE 3-35 Annual Cost Data

Item	Incinerator A Normal Capacity Actual	Normal Capacity Adjusted	Design Capacity Projected	Design Capacity Adjusted	Incinerator B Normal Capacity Actual	Normal Capacity Adjusted	Design Capacity Projected	Design Capacity Adjusted	Incinerator D Normal Capacity Actual	Normal Capacity Adjusted	Design Capacity Projected	Design Capacity Adjusted
Operating costs:												
Direct labor	$ 75,184	$ 73,703	$112,776	$110,555	$197,500	$185,730	$197,500	$185,730	$193,138	$186,301	$193,138	$186,301
Utilities	17,352	17,352	35,135	34,135	20,000	20,000	25,850	25,850	18,000	18,000	19,301	19,301
Parts and supplies	12,509	12,509	24,608	24,608	32,950	32,950	42,580	42,580	0	0	0	0
Vehicle operations	1,739	1,739	3,421	3,421	7,200	7,200	9,300	9,300	7,670	7,670	8,225	8,225
External repairs	7,346	7,346	14,451	14,451	6,250	6,250	8,080	8,080	22,339	22,339	23,954	23,954
Disposal charges	700	700	1,377	1,377	2,000	2,000	2,580	2,580	32,232	32,232	34,562	34,562
Overhead	11,326	11,326	11,326	11,326	52,800	52,800	52,800	52,800	32,959	32,959	32,959	32,959
Total operating cost	126,156	124,675	202,094	199,873	318,700	306,930	338,690	326,920	306,338	299,501	312,139	305,302
Operating cost/ton	2.95	2.92	2.41	2.38	4.90	4.72	4.03	3.89	2.35	2.29	2.23	2.18
Financing and ownership costs:												
Plant depreciation	23,581	21,651	23,581	21,651	80,149	84,842	80,149	84,842	200,000	142,572	200,000	142,572
Vehicle depreciation	6,042	6,042	6,042	6,042	9,675	9,675	9,675	9,675	0	0	0	0
Interest	16,059	32,477	16,059	32,477	64,558	127,262	64,558	127,262	70,448	213,858	70,448	213,858
Total financing and ownership cost	45,682	60,170	45,682	60,170	154,382	221,779	154,382	221,779	270,448	356,430	270,448	356,430
Financing and ownership cost/ton	1.07	1.41	0.54	0.72	2.38	3.41	1.84	2.64	2.07	2.73	1.93	2.55
Total cost	171,838	184,845	247,776	260,043	472,082	528,709	492,072	548,699	576,786	655,931	582,587	661,732
Total cost/ton	4.02	4.33	2.95	3.10	7.28	8.13	5.87	6.53	4.42	5.02	4.16	4.73

Item	Incinerator E Normal Capacity Actual	Adjusted	Design Capacity Projected	Adjusted	Incinerator F Normal Capacity Actual	Adjusted	Design Capacity Projected	Adjusted	Incinerator G Normal Capacity Actual	Adjusted	Design Capacity Projected	Adjusted
Operating costs:												
Direct labor	$202,407	$205,139	$202,407	$205,139	$165,684	$181,391	$165,684	$181,391	$150,949	$145,434	$160,949	$145,434
Utilities	65,260	65,260	90,418	90,418	67,632	67,632	70,500	70,500	31,952	31,952	75,777	75,777
Parts and supplies	57,332	57,332	79,433	79,433	51,540	51,540	53,725	53,725	2,700	2,700	6,403	6,403
Vehicle operations	4,188	4,188	5,802	5,802	9,600	9,600	10,007	10,007	13,968	13,968	33,127	33,127
External repairs	1,999	1,999	2,770	2,770	12,758	12,758	13,299	13,299	808	808	1,916	1,916
Disposal charges	0	0	0	0	10,364	10,364	10,803	10,803	27,720	27,720	65,741	65,741
Overhead	123,577	123,577	123,577	123,577	84,674	84,674	84,674	84,674	21,331	21,331	21,331	21,331
Total operating cost	454,763	457,495	504,407	507,139	402,252	417,959	408,692	424,399	259,428	243,913	365,244	349,729
Operating cost/ton	4.50	4.53	3.60	3.62	2.49	2.59	2.43	2.53	5.49	5.17	3.26	3.12
Financing and ownership costs:												
Plant depreciation	110,726	168,574	110,726	168,574	80,000	121,795	80,000	121,795	101,234	111,722	101,234	111,722
Vehicle depreciation	3,516	3,516	3,516	3,516	0	0	0	0	0	0	0	0
Interest	106,859	252,860	106,859	252,860	75,840	182,693	75,840	182,693	81,494	167,583	81,494	167,583
Total financing and ownership cost	221,101	424,950	221,101	424,950	155,840	304,488	155,840	304,488	182,728	279,305	182,728	279,305
Financing and ownership cost/ton	2.19	4.20	1.58	3.04	0.97	1.89	0.93	1.81	3.87	5.91	1.63	2.50
Total cost	675,864	882,445	725,508	932,089	558,092	722,447	564,532	728,887	442,156	523,218	547,972	629,034
Total cost/ton	6.69	8.73	5.18	6.66	3.46	4.48	3.36	4.34	9.36	11.08	4.89	5.62

Source: Achinger, W. C. and Daniels, C. E., "Seven Incinerators," paper presented to 1970 National Incinerator Conference, Cincinnati, Ohio, May 17–20, 1970 (EPA Publ. No. SW-51ts.1j).

TABLE 3-36 Analysis of Capital Investment

Incinerator	Actual Cost	Adjusted Cost	Adjusted Cost/Ton
A	$ 471,659	$ 541,276	$1804
B	1,848,240	2,121,040	7070
D	3,000,000	3,564,300	7129
E	3,321,779	4,214,341	8429
F	2,400,000	3,044,880	5075
G	2,530,855	2,793,052	6983

Source: Achinger, W. C. and Daniels, C. E., "Seven Incinerators," paper presented to 1970 National Incinerator Conference, Cincinnati, Ohio, May 17–20, 1970 (EPA Publ. No. SW-51ts.1j).

Large subsidence chambers and a spray chamber with a wet bottom are used for flyash removal.

A waste-heat boiler and a Spencer Turbine Company vacuum cleaning system for cleaning furnaces of flyash are also included in the plant.

The sizes of the various plant units are as follows:

Primary Furnace
12 ft deep, 22½ ft long, 10½ ft high
Volume 2835 cu ft
Cross-sectional area—270 sq ft

TABLE 3-37 Breakdown of Capital Investment

Plant	Adjusted Cost	Percent of Total
Incinerator A:		
Buildings	$ 191,979	35.5
Equipment	333,977	61.7
Miscellaneous	15,320	2.8
Total	541,276	100.0
Incinerator B:		
Buildings	1,428,119	67.3
Equipment	530,593	25.0
Miscellaneous	162,328	7.7
Total	2,121,040	100.0
Incinerator E:		
Buildings	1,312,506	31.2
Equipment	2,711,198	64.3
Miscellaneous	190,637	4.5
Total	4,214,341	100.0

Source: Achinger, W. C. and Daniels, C. E., "Seven Incinerators," paper presented to 1970 National Incinerator Conference, Cincinnati, Ohio, May 17–20, 1970 (EPA Publ. No. SW-51ts.1j).

TABLE 3-38 Repairs and Maintenance Cost Data

Item	Incinerator A Actual	Incinerator A Adjusted	Incinerator B Actual	Incinerator B Adjusted	Incinerator D Actual	Incinerator D Adjusted	Incinerator E Actual	Incinerator E Adjusted	Incinerator F Actual	Incinerator F Adjusted	Incinerator G Actual	Incinerator G Adjusted
					Expenditures							
Expenditure type:												
Labor	$10,442	$10,237	$29,625	$27,861	$53,590	$51,695	$61,335	$62,164	$31,679	$34,685	$44,003	$39,762
Parts	12,509	12,509	32,951	32,951	0	0	57,332	57,332	51,540	51,540	2,700	2,700
External charges	7,346	7,346	6,250	6,250	22,339	22,339	1,999	1,999	12,758	12,758	808	808
Overhead	1,574	1,574	7,919	7,919	9,145	9,145	37,447	37,447	16,189	16,189	6,799	6,799
Total	31,871	31,666	76,745	74,981	85,074	83,179	158,113	158,942	112,166	115,172	54,310	54,069
					Allocation							
Cost center:												
Receiving and handling	$ 6,824	$ 6,780	$ 9,112	$ 8,903	$13,766	$13,460	$ 39,345	$ 39,552	$ 17,960	$ 18,442	$18,642	$17,187
Volume reduction	21,641	21,502	56,825	55,518	60,109	58,770	85,597	86,045	77,804	79,888	17,834	16,441
Effluent handling and treatment	3,406	3,384	10,808	10,560	11,199	10,949	33,171	33,345	16,402	16,842	17,834	16,441
Total	31,871	31,666	76,745	74,981	85,074	83,179	158,113	158,942	112,166	115,172	54,310	50,059

Source: Achinger, W. C. and Daniels, C. E., "Seven Incinerators," paper presented to 1970 National Incinerator Conference, Cincinnati, Ohio, May 17-20, 1970 (EPA Publ. No. SW-51ts.1j).

TABLE 3-39   Operating Cost Breakdown by Cost Centers

Cost Center	Incinerator A Actual	Incinerator A Adjusted	Incinerator B Actual	Incinerator B Adjusted	Incinerator D Actual	Incinerator D Adjusted	Incinerator E Actual	Incinerator E Adjusted	Incinerator F Actual	Incinerator F Adjusted	Incinerator G Actual	Incinerator G Adjusted
Receiving and handling:												
Direct labor	$ 25,062	$ 24,568	$ 79,000	$ 74,290	$ 51,942	$ 50,103	$ 67,470	$ 68,381	$ 59,294	$ 64,915	$ 73,360	$ 66,288
Utilities	0	0	3,020	3,020	12,600	12,600	6,964	6,964	12,715	12,715	18,720	18,720
Vehicle operating expense	1,564	1,564	0	0	0	0	0	0	0	0	0	0
Repairs and maintenance	6,824	6,824	9,112	9,112	13,766	13,766	39,345	39,345	17,960	17,960	18,642	18,642
Overhead	3,775	3,775	21,119	21,119	8,864	8,864	41,192	41,192	30,302	30,302	9,081	9,081
Total	37,225	36,731	112,251	107,541	87,172	85,333	154,971	155,882	120,271	125,892	119,803	112,731
Volume reduction:												
Direct labor	35,504	34,805	59,250	55,720	48,451	46,736	30,667	31,081	34,226	37,471	14,260	12,885
Utilities	8,597	8,597	7,480	7,480	2,700	2,700	7,724	7,724	8,251	8,251	6,193	6,193
Repairs and maintenance	21,641	21,641	56,826	56,826	60,108	60,108	85,597	85,597	77,804	77,804	17,834	17,834
Overhead	5,348	5,348	15,839	15,839	8,268	8,268	18,725	18,725	17,492	17,492	1,819	1,819
Total	71,090	70,391	139,395	135,865	119,527	117,812	142,713	143,127	137,773	141,018	40,106	38,731
Effluent handling and treatment:												
Direct labor	4,176	4,094	29,625	27,860	39,156	37,770	42,935	43,515	40,485	44,323	29,325	26,498
Utilities	8,755	8,755	9,500	9,500	2,700	2,700	50,572	50,572	46,666	46,666	77,039	7,039
Vehicle operating expense	175	175	7,200	7,200	7,670	7,670	4,188	4,188	9,600	9,600	13,968	13,968
Disposal charges	700	700	2,000	2,000	32,232	32,232	0	0	10,364	10,364	27,720	27,720
Repairs and maintenance	3,406	3,406	10,809	10,809	11,199	11,199	33,171	33,171	16,402	16,402	17,834	17,834
Overhead	629	629	7,920	7,920	6,682	6,682	26,213	26,213	20,691	20,691	3,632	3,632
Total	17,841	17,759	67,054	65,289	99,639	98,253	157,079	157,659	144,208	148,046	99,519	96,691
Total	126,156	124,881	318,700	308,695	306,338	301,398	454,763	456,668	402,252	414,956	259,428	248,153

Source: Achinger, W. C. and Daniels, C. E., "Seven Incinerators," paper presented to 1970 National Incinerator Conference, Cincinnati, Ohio, May 17–20, 1970 (EPA Publ. No. SW51ts.1i).

Secondary Combustion Chamber
  Volume—6400 cu ft
  Cross-sectional area—124 sq ft (min.)

Subsidence Chamber
  Volume—8000 cu ft
  Cross-sectional area—200 sq ft (min.)

Flues (Cross-sectional area)
  Primary to combustion chamber—70 sq ft
  Combustion chamber to subsidence chamber—86 sq ft
  Spray chamber to stack—60 sq ft

The capacity of these units is based on a 46 percent combustible, 15 percent moisture, and 39 percent noncombustible solid waste. Chicago's refuse averaged 400 to 500 lb/cu yd and 4000 BTU per lb.

Tables 3-40 and 3-41 list various data obtained during the performance test.

TABLE 3-40   Performance Data

	Specification Requirements	Acceptance Test Results
Test duration	96 hours	96 hours
Refuse incinerated	4800 tons	5642.96 tons
Capacity percentage	100%	117.6%
Refuse analysis		
Combustible	46%	34.05%
Noncombustible	39%	37.39%
Moisture	15%	28.56%
Heat value	4000 BTU/lb	2724 BTU/lb

Source: Stellwagon, R. H., "Calumet Incinerator—Chicago's Second, Nation's Largest," *American City*, Vol. 75, No. 2, February, 1960, pp. 96–98.

3-8.3   District of Columbia Study

Another item of interest is a special study done for a proposed 800 ton per day incinerator for the District of Columbia. The report consists of six special studies (Ref. 3-8). These studies are of municipal incinerator effluent gases, control laboratory, size reduction of oversize burnable waste, size reduction of metal objects by compression presses, heat recovery, and can-metal recovery.

The purpose of the effluent gases study was to estimate the type of air pollutants associated with incinerators, investigate the effect of furnaces on the emission of pollutants, evaluate the performance of different types of air pollution control equipment, develop capital and operating cost estimates for

TABLE 3-41   Air Pollution Test Results (Expressed in grams per cubic foot)

Test Conditions	As Sampled	Corrected to SCTP[a] (60°F–30.0 in. Hg)	Corrected to SCTP[a] and 12% CO_2 Basis
Forced draft (all sprays on wet bottom used)	0.178	0.262	0.073
Natural draft (all sprays on wet bottom used)	0.034	0.450	0.151
Forced draft (all sprays off wet bottom used)	0.2178	0.5190	1.510
Natural draft (all sprays off wet bottom used)	0.2125	0.5150	1.140

[a]Standard conditions of temperature and pressure.

Source: Stellwagon, R. H., "Calumet Incinerator—Chicago's Second, Nation's Largest," *American City*, Vol. 75, No. 2, February, 1960, pp. 96–98.

acceptable air pollution control equipment, and to make recommendations for equipment to be installed at the proposed incinerator.

Two approaches were used to determine the amount and type of air pollution constituents. One was predictive, the other empirical, by review of actual test data. The pollutants were grouped into two categories: inorganic gases and particulates (such as the oxides of sulfur, nitrogen, ammonia, and metal oxides), and organic gases and particulates, which consist mainly of fatty acids, esters, aldehydes, hydrocarbons, and the oxides of carbon.

The effects of the furnace variables such as temperature, excess air, fuel bed agitation, and incomplete combustion on the amount and types of pollutants were reviewed. It was found that higher temperatures along with excess air tend to reduce the amount of organic pollutants. Particulate emission increased with more agitation, while incomplete combustion (based on furnace efficiency) was found to be inversely related to particulate size.

A particulate loading of 3 lb per thousand pounds of dry fluegas was used to evaluate the air pollution control equipment. The study showed that only two types of particulate control equipment would pass the current (1968) Federal Air Pollution Code; high energy scrubbers and electrostatic precipitators preceded by mechanical collectors.

The advantages of the high energy scrubber were found to be the relatively high efficiency of particulate removal obtained and the absorption of some of the gaseous pollutants. The disadvantages were the high water consumption rate, high power costs, and the presence of a water vapor plume at the stack discharge. Another disadvantage was the necessity for treatment of the effluent water for pH control and flyash removal.

The electrostatic precipitator was found to have a high particle removal efficiency for small particles. The disadvantages of this type of precipitator are the large space requirements, large capital cost, and a critical operating temperature range. In addition, some operating problems should be anticipated because of the lack of operating experience.

The cost for this particular plant with the electrostatic precipitator and a mechanical collector was estimated at $2,409,200 (1968) with an annual operating cost of $512,500 (1968). The plant with wet scrubber had an estimated capital cost of $1,838,600 (1968) and an annual operating cost of $401,000. The electrostatic precipitator unit was recommended because the plume created by the high energy scrubber was found to be objectionable at the location of the plant. Another reason for the rejection of the scrubber was that the temperature conditions in the nearby water source were not favorable for its use. Table 3-19 summarizes these costs.

The control laboratory study was made to itemize instrumentation and equipment and the estimated cost for their use in the monitoring of plant operation. The total cost of instrumentation and laboratory equipment was estimated to be $447,675 (Ref. 3-8).

The physical laboratory equipment is required for sample collection, preparation, and physical analyses of the samples of refuse, ash residue, flyash, and furnace slag. The cost of this equipment was estimated to be $25,615.

The monitoring equipment for test and development studies was to be used for the continuous indication and recording of the operating conditions of a single incinerator furnace unit. The estimated equipment cost for this function was $106,680 (Ref. 3-8).

Approximately 2400 sq ft of floor space would be needed for the control laboratory. Maintenance and operation of the laboratory would require specially trained personnel.

The study dealing with the size reduction of oversize burnable waste used two approaches for solving the problem. The first approach was to use size reduction equipment such as shredders and mills. The second approach was to use specially designed incinerators for burning dense or bulky objects.

Shredders were investigated and it was found that they could not handle all of the waste that would need processing. Impact mills, hammermills with grates, and knife hogs were also investigated. The hammermill type with grate bars selected to discharge a product approximately 1 in. by 8 in. by 6 in. as maximum dimensions was found to be the equipment best suited for the proposed incinerator. Its selection was based on the fact that it could handle bulky metal objects. The unit is estimated to cost $667,000 to install and $125,000 to operate annually (1968 costs).

The specially designed furnace would be constructed with refractory hearth, refractory walls, and roof arches from the charging doors to the wet scrubber. Charging would be done by a front-end loader with pusher blade attachment. The unit would be complete with both forced and induced draft fans and a wet scrubber for air pollution control. It would burn 16 lb of wastewood products per hour per square foot of hearth area. The capital cost of the furnace was estimated at $532,400 (1968) and the annual operating cost was estimated at $104,000. These costs do not include the independent air pollution controls needed for the incinerator. It was therefore determined that the use of a special bulky refuse incinerator does not seem feasible where shredder applications can be used.

The size reduction of bulky metal objects by compression presses was not necessary because of the hammermill installation that handles both metal objects and oversize burnable waste. The capital cost of such a press was estimated at $255,000 and the operating cost (1968) was estimated at $45,400 per year with the press being run 2 hr per day, $58,350 per year with the press being run 4 hr per day, $84,300 per year with the press being run 8 hr per day.

Heat recovery was studied for the proposed incinerator and the installation of water-cooled furnaces or steam boilers in the incinerator was not recommended because: there are higher operating expenses in a water-cooled furnace; the trend toward plastics and freons in the refuse will increase maintenance of the metallic boiler tubes because of the corrosive products of combustion; there is no market for the steam produced; and, up to 1968, successful operation of the water-cooled furnaces had not been experienced in the United States. Table 3-22, previously given, shows the cost estimates that were made for the four cases studied to determine the feasibility of steam production.

The can-metal recovery study determined that no satisfactory method exists for economically extracting scrap can metal from the refuse prior to incineration. Recovery after incineration can be achieved provided complete burnout of the residue is achieved. The study recommended that can-metal recovery not be practiced at the proposed incinerator site becasue of the lack of profits. The estimated investment cost (1968) of a can-metal recovery system for the proposed incinerator was $400,000. The operating costs (1968), including amortization, were estimated at $13.60 per ton of metal reclaimed. Freight charges (1968) for delivery of the metal to the processor was estimated at $83 per ton. Metal was presently selling for $75 per ton (1968) and there was no indication that the price would rise.

3-9. OTHER INCINERATORS

In addition to centralized municipal incineration, other types of incineration operations are carried out on certain solid wastes in the United States and in Western Europe. These types of incineration activities include on-site incineration, incineration of bulky wastes, incineration of hazardous wastes, and incineration of sewage sludge.

3-9.1 On-Site Incineration

On-site incineration of wastes has been practiced for many years in apartment houses and commercial and industrial facilities wherein large volumes of wastes are generated in a concentrated location. There are obvious economies to the incineration of such wastes at or near their source, but serious limitations exist for on-site incineration. For example, many on-site incinerators used in the past have utilized refuse charging through the incinerator flue and other features which have led to significant production of air pollutants. Because of the problems of air pollutant production, many apartment house and commercial facility incinerators have been abandoned. An additional factor in the evaluation of on-site incinerators is the necessity for automatic operation or for the provision of a skilled operator to supervise the incineration process. It is highly unlikely that skilled incinerator technicians can be employed in individual apartment houses or commercial facilities. Likewise, fully automated on-site incinerators designed to ignite periodically and to consume refuse deposited in haphazard fashion by apartment house dwellers or others have contributed significant amounts of air pollution and in general have performed very poorly in reducing deposited wastes. Because of these factors, the future of on-site incineration appears to be somewhat limited. However, in large commercial or industrial facilities, the use of on-site incinerators from which energy can be recovered should be encouraged. It will be necessary, in order to insure environmental protection, to closely supervise the operation of such incinerators. Nevertheless, because of savings in the collection and transport of wastes, and because of the possible recovery of energy in facilities where energy is in great demand, on-site incinerators in the future may prove to be very advantageous in the disposal of wastes from high density housing units, institutions, or commercial or industrial establishments.

3-9.2 Bulky Wastes

Bulky items such as mattresses, vehicle tires, tree stumps, logs, or discarded furniture may occasionally appear at a municipal incinerator. These items are very difficult to process in a conventional incinerator. They are collectively designated "bulky solid wastes." This type of waste does not burn sufficiently in the normal detention time used in standard municipal incinerators. Additionally, such bulky items may damage the charging and removal mechanisms of the incinerator or may be simply too large to enter the combustion chambers. In the past, bulky wastes have been processed by open burning or have been deposited in landfills. Open burning, with significant production of air pollutants, is not an acceptable solution to this problem at the present time. More-over, transport of bulky wastes to remote landfill sites may be very costly. In order to overcome these problems, two expedients have been developed for processing of bulky solid wastes: special incinerators and/or size reduction. Several cities in the United States today are using special incinerators to burn bulky refuse. These incinerators are generally batch-fed rectangular refractory-lined furnaces. In most of these installations, auxiliary fuel is used in order to sustain the early stages of the combustion process and to insure complete combustion of the wastes. On the other hand, size reduction of bulky wastes has

Fig. 3-17   Shears for bulky refuse, Edmonton incinerator, London.

not generally been successful in the United States. A number of size reduction devices such as hammermills and chippers have been used in these attempts. Generally these devices are successful with certain types of materials but are not capable of reducing large pieces of metal or other wastes. In European in-cinerators, large shears have been used to reduce bulky wastes. A typical example of such shears is shown in Fig. 3-17. Because of the hazards associated with operation of size reduction equipment and because of the generally poor performance of such equipment in the past, it appears likely that special in-cinerators will be favored in the future for the processing of bulky solid wastes.

3-9.3 Hazardous Wastes

In addition to the bulky wastes which were mentioned above, other materials may be charged into municipal incinerators and incineration of these materials may constitute a grave hazard to the operating personnel of the incinerator and to the community at large. Included in this category of materials are those which are highly flammable or explosive, those which are toxic chemical substances, and those which are radioactive. These materials must be disposed of in other facilities and should not be charged into municipal incinerators. However, a certain amount of such hazardous materials may be generated from domestic sources and it is virtually inevitable that these hazardous substances will arrive at municipal incinerators. If such wastes are detected at the incinerator facility in the dumping area or at the storage pit, they should be removed by the operating personnel. It may be possible to insert incendiary or explosive materials into the incoming waste stream in small quantities but extreme care should be exercised in such an operation. Under no circumstances should radioactive materials ever be processed in any incinerator. In addition to these highly dangerous materials, other wastes which are objectionable or unpleasant to process may be brought to a municipal incinerator. These wastes include hospital wastes, carcasses from slaughterhouses, carcasses from city streets, etc. Hospital wastes always should be incinerated in pathological incinerators located near the source of the wastes. It is very difficult to reduce hospital wastes such as excised flesh in conventional municipal incinerators because of the long detention time required to combust materials with very high moisture contents. In the case of slaughterhouses, butcher shops, or other facilities which generate highly putrescible organic materials, the best solution is special drying facilities and special incinerators. In many cases, however, it is desirable to make some provision at a municipal incinerator for animals which are collected from city streets. Some incinerators are equipped with a special hearth in a secondary combustion zone where such dead animals can be placed.

3-9.4 Sludge Incineration

Another special material which should be considered for incineration is sewage sludge. In many communities in the United States, such sludge is mixed with domestic and municipal refuse and incinerated. Considerable advantage is derived from such an operation. The excess heat generated when the solid wastes are burned is available for use in drying the partially dewatered sludge which then, in turn, burns more readily. Also, the provision of a single incinerator to handle sewage sludge and municipal solid wastes may produce significant economies in physical plant expenditures. At the present time, combined incinerator operations are receiving considerable attention in the United States.

3-10. SUMMARY—INCINERATION

The incineration of municipal solid wastes is a complex operation which is complicated by variations in the quantity and characteristics of the materials to be processed, the municipal refuse. It is virtually impossible to standardize refuse entering an incinerator. If recycling operations are carried out to any great extent in the future, the composition of municipal refuse may change considerably from its present already variable composition. An incinerator facility must be designed to accommodate the variations in the charged solid waste. Moreover, if long term changes in waste quantities and characteristics occur, such changes must be accommodated by changes in the design and operation of incinerator facilities.

The practice of centralized refuse incineration as a disposal method possesses certain distinct advantages:

1. If land is not available for sanitary landfills or composting facilities within economic haul distances from the center of waste generation, a centrally located incinerator may represent the most economical total system for collection and disposal of refuse.
2. It is possible to locate an incinerator plant in a central urban area if the facility is well-designed and landscaped.
3. The residue produced from incineration constitutes only a small fraction of the charged solid waste and contains only a negligible amount of decomposable materials.
4. A properly designed incinerator is capable of accommodating fluctuations of waste quantities and characteristics and also is free from interference by climate and weather.
5. Recovery of materials from incinerator residue and recovery of heat from the incineration process may produce significant incomes.

Although refuse incineration possesses these distinct advantages, it also has certain significant disadvantages:

1. A large expenditure of capital funds is required for the design and construction of an incinerator facility.
2. Operating costs for refuse incinerators commonly are much higher than operational costs for sanitary landfills because of the requirements for complex and detailed equipment and skilled personnel to operate the incinerator facility.
3. Location of an incinerator in the most desirable position with respect to the generation area for solid wastes may create detrimental conditions in an established neighborhood such as large volumes of truck traffic.
4. Most importantly, refuse incineration is not a complete and ultimate disposal method—residue from the combustion process will require further disposal.

Full consideration should be given to the above-mentioned advantages and disadvantages in any evaluation of central incineration as a disposal method for solid wastes.

REFERENCES

3-1. Achinger, W. C. and Daniels, C. E., *Seven Incinerators*," paper presented to 1970 National Incinerator Conference, Cincinnati, Ohio, May 17–20, 1972 (EPA Publ. No. SW-51ts.1j).

3-2. Bender, J., "Incineration Plant-Plus," *Power*, Vol. 111, No. 1, January, 1967, pp. 62–64.

3-3. Boucher, R. M. G., "Ultrasonics in Processing," *Chemical Engineering*, Vol. 68, No. 20, October 2, 1961, pp. 83–100.

3-4. Cassreino, J., Berry, E. E. and Donohue, P., "A High-Performance Incinerator," *American City*, Vol. 86, No. 1, January, 1971, p. 54–57.

3-5. Cerniglia, V. J. and Friedland, A., "Smile—Your Incinerator is on T.V.," *American City*, Vol. 84, No. 12, December, 1969, p. 96.

3-6. Corey, C., *Principles and Practices of Incineration*, Wiley-Interscience, New York, 1969.

3-7. Cross, F. C., Jr., *Handbook on Incineration*, Technomic Publishing Company, Westport, Connecticut, 1972, p. 64.

3-8. Day and Zimmerman, Inc., *Special Studies for Incinerators for the District of Columbia*, U.S. Department of Health, Education, and Welfare, Cincinnati, Ohio, 1968.

3-9. DeMarco, J., Keller, D. J., Leckman, J. and Newton, J. L., *Incinerator Guidelines—1969*, U.S. Department of Health, Education, and Welfare, Washington, D.C., 1969.

3-10. Dvirka, M. and Zanft, A. B., "Prototype Log Incineration Operates in a Baltimore Park," *Public Works*, Vol. 97, No. 11, November, 1966, pp. 72–73.

3-11. Hagerty, D. J., Pavoni, J. L. and Heer, J. E., Jr., *Solid Waste Management*, Van Nostrand Reinhold, New York, 1973.

3-12. Hering, R. and Greeley, S. A., *Collection and Disposal of Municipal Refuse*, McGraw-Hill, New York, 1921.

3-13. Kaiser, E. R., "Chemical Analyses of Refuse Components," ASME Paper No. 65WA/PID-9, November, 1965.

3-14. Kaiser, E. R., Halitsky, J., Jacobs, M. D. and McCabe, L. C., "Modifications to Reduce Emissions from Flue Fed Incinerators," *Journal of the Air Pollution Control Association*, Vol. 10, 1960, pp. 183–192, 207, 251.

3-15. Kramer, W. P., "The Capabilities and Limitations of Current Incinerator Designs," *Public Works*, Vol. 101, No. 7, July, 1970, pp. 57–59.

3-16. McDonough, P. A., "Design and Operation," *Solid Waste Management*, Vol. 5, National Association of Counties Research Foundation.

3-17. Modlin, R. A., "High Temperature Solid Waste Disposal," *Professional Engineer*, Vol. 41, No. 10, October, 1971, pp. 38–39.

3-18. Rogus, C. A., "Incineration with Guaranteed Top Level Performance," *Public Works*, Vol. 101, No. 3, September, 1970, pp. 92–97.

3-19. Spitzer, E. F., "Montreal's Combined Incinerator–Power Plant," *American City*, Vol. 85, No. 5, May, 1970, pp. 86–89.

3-20. Stellwagon, R. H., "Calumet Incinerator–Chicago's Second, Nation's Largest," *American City*, Vol. 75, No. 2, February, 1960, pp. 96–98.

3-21. Stenburg, R. L., Hangebrauck, R. P., von Lehmden, D. U. and Rose, A. H., Jr., "Field Evaluation of Combustion Air Effects on Atmospheric Emissions from Municipal Incinerators," *Journal of the Air Pollution Control Association*, Vol. 12, No. 2, February, 1962, pp. 83–89.

3-22. Wegman, L. S., "The Cleanest Incinerator Stack Gases . . . ," *American City*, Vol. 82, No. 5, May, 1967, pp. 89–91, 142–4.

3-23. Wilson, M. J. and Manmill, M., "Putting Solid Waste to Work in Nashville," *Professional Engineer*, Vol. 41, No. 10, October, 1971, pp. 41–44.

3-24. Wisely, F. E., Sutterfield, G. W. and Klumb, D. L., "St. Louis Power Plant to Burn City Refuse," *Civil Engineering*, Vol. 41, No. 1, January, 1971, pp. 56–59.

3-25. "Converting Solid Waste to Electricity," *Environmental Science and Technology*, Vol. 4, No. 8, August, 1970, pp. 631–633.

3-26. "New Driftwood Incinerator for New York Harbor," *Public Works*, Vol. 99, No. 9, September, 1968, pp. 99–100.

3.27. "Promising Future for Underground Incineration," *American City*, Vol. 85, No. 5, May, 1970, p. 48.

3-28. Institute for Solid Wastes, American Public Works Association, *Municipal Refuse Disposal*, Public Administration Service, APWA, Chicago, 1970, p. 167.

4

Sanitary landfill

4-1. INTRODUCTION

Landfilling of solid wastes is a method of waste disposal which has been practiced since very early times. As a general rule, it is still one of the most economical waste disposal techniques in current use. It is often referred to as the only *final* solid waste disposal method since, unlike incineration or composting, it is not a processing operation which yields a residue or end product which requires disposal. The wastes deposited in a sanitary landfill are considered to be ultimately "eliminated"; the landfill is their ultimate destination. Because of this, landfilling of solid waste material is in some ways a very undesirable procedure; many potentially useful materials which could be recycled are buried in the earth and lost.

In early fills, refuse was deposited in an open dump on a selected piece of land and allowed to decompose in the open air. However, the nuisances associated with such open dumping—odors, airborne litter and waste paper, the presence of disease vectors such as rats and mice, and other problems—soon caused an alteration in landfilling operations. In Champaign, Illinois (1904), Columbus, Ohio (1906), and Davenport, Iowa (1916), waste disposal managers began burying or covering refuse with earth (Ref. 4-1). The concept of "sanitary landfill" was first proposed in describing a cut-and-cover operation used for waste disposal in Fresno, California, in the 1930s. The name "sanitary landfill" has been applied mistakenly to many operations in the United States. Inherent in the designation "sanitary landfill" are three conditions of operation: daily cover of the refuse with earth; no open burning of deposited wastes; and no pollution of the surface or groundwaters around the site.

Today, sanitary landfills are widely used in the United States. In a survey conducted in 1968 by the United States Public Health Service, over 8800 landfills in 6259 U.S. communities were examined (Ref. 4-20). Of these more than 8800 landfills, 4811 were publicly owned and 4032 were privately owned. In these operations a total area of more than 195,000 acres were involved, approximately 116,618 acres of which were actually being used in the disposal operation. The landfilling sites were serving as the waste disposal facilities for a total population of 92.5 million people. Of this population, 75 percent would be classified as "urban" or city-dwellers, with the remainder inhabiting essentially

Fig. 4-1 Typical crawler for use at a sanitary landfill (*courtesy* John Deere Company).

rural communities. Obviously, sanitary landfilling is a technique of major importance in solid waste management in the United States at the present time.

There are several advantageous characteristics of the practice of waste disposal by sanitary landfilling which contribute to its popularity. Little or no capital investment is required for a physical plant, such as would be required for a solid waste incinerator. Land for the sanitary landfill site may be purchased, or it may be leased for the duration of the landfilling operation. The physical requirements for the landfilling operation also include a variety of trucks and earth movers which are virtually standard equipment in the construction industry. During operation of the sanitary landfill, no extremely high labor costs are incurred as would be the case where highly trained technical personnel are employed. As mentioned previously, there is no residue or by-product from a sanitary landfill operation such as the ash produced by incinerators, or the soil conditioner produced from a composting operation.

On the other hand, there are a number of difficulties associated with waste disposal in landfills. The most severe problem at the present time is that in urban areas in industrialized countries land suitable for sanitary landfills is becoming increasingly scarce. Because of this increasing scarcity of available

Fig. 4-2 Tracked crawler with special refuse blade (*courtesy* International Harvester Company).

sites, the practice of sanitary landfilling must be considered a temporary or stopgap measure in a long-term plan for waste disposal.

In addition to the problem of decreasing availability of landfill sites, difficulties may arise in the operation of a sanitary landfill because of the materials and conditions associated with waste decomposition within the landfill itself. If generated end products—gases and/or fluid leachate—are not contained within the landfill, they constitute the same sort of problem as do incinerator residue and compost. During the early stages of waste decomposition in a landfill, the degradation process is essentially aerobic and carbon dioxide is the principal gas produced. However, as oxygen is depleted within the deposited refuse, the decomposition process becomes anaerobic and other gases, principally methane, are created in significant quantities. Methane in proper proportion forms an explosive combination with air. Other generated gases, such as hydrogen sulfide, are toxic and lethal. Leachate produced by water moving through deposited refuse represents another environmental hazard. The pollution of groundwater and surface waters in the landfill area can result from the migration of leachate out of the refuse cells and into the environs of the landfill. Additionally, disease vectors such as rats, flies, and other vermin present difficulties to the operators of sanitary landfills. Improper operation of a landfill can create numerous environmental problems such as odors, blowing litter, and the other nuisances previously mentioned. Finally, a distinct disadvantage to the practice of sanitary landfilling is the present attitude of the general public toward this disposal method. The operation of a sanitary landfill is associated in the mind of the average citizen with an odorous, burning, open dump inhabited by rats, mice, and other vermin. The lack of public support for landfills has caused serious difficulties in upgrading landfilling operations. Continued open dumping is due in large part to public apathy (and in some cases hostility) towards the technique of sanitary landfill.

The future of the sanitary landfill as a waste disposal method is somewhat doubtful because of the ever increasing difficulty in finding land suitable for such an operation. However, the general economy associated with this rather simple operation makes the landfill an attractive method to many solid waste managers. Also, in recent years there have been some innovations in the operation of sanitary landfills such as the development of special types of compactors and earth movers which produce greater densities of compacted refuse. In addition, volume reduction at the source of the wastes by compaction in the home or in an industrial compactor has produced greater densities of refuse and more economy in landfilling. Compaction in transit, in standard compactor trucks and in transfer stations, has likewise yielded more economical operations. In the last few years, high-intensity compaction and baling operations at the landfill site or at an intermediate location have also been used to produce high density in the final deposit of refuse. Several other experimental techniques have also shown promise. For example, shredding and grinding of refuse at a landfill site before placement has been attempted with some success (Ref. 4-11). The action of shredding the refuse has minimized odor, has improved the handling characteristics of the refuse, and has limited severely the number of disease vectors which later are found in the deposited refuse (Ref. 4-13). Salvaging operations wherein certain components of the solid waste stream are reclaimed have also been added as accessory operations at many landfilling sites. The monies realized by reclamation of such materials as ferrous metals (through magnetic separation) obviously will make the operation of landfilling a more attractive disposal technique. Finally, the basic philosophy of the sanitary landfill has been that of a sequestration site or a refuse cache. However, in recent years several innovators have promulgated the concept that a landfill may be considered a waste treatment plant contained beneath the surface of the earth. In this operation of the landfill as a treatment plant, water is deliberately introduced into the refuse to aid in accelerating the decomposition process. As the water migrates through the refuse, it is collected and returned through the refuse again and again until the decomposition process is essentially complete. When indices such as the biochemical oxygen demand (BOD) indicate that the decomposition within the refuse is virtually complete, the leachate which is finally collected is removed from the site and treated in a standard wastewater treatment plant. By means of this process, the site is returned to a useful status at an earlier date and the pollutional hazard of the deposited refuse is minimized. The impact of this innovative technique has yet to be evaluated. An in-depth discussion of this new landfilling procedures is presented in Chapter 5, "Innovative Disposal Methods."

4-2. SANITARY LANDFILL METHODOLOGY

A sanitary landfill is a solid waste disposal facility wherein refuse is placed at the greatest possible density (in the smallest possible space) for final deposition; no open burning, no water pollution, and daily cover of deposited refuse with earth are requisites of the method. While the technique of sanitary landfilling is everywhere generally the same, there are three variations in refuse placement currently used in the United States: the trench method, the area method, and the ramp method (Ref. 4-12).

4-2.1 Filling Techniques

The Trench Method. In the trench method of sanitary landfilling, a long narrow excavation is made in the earth and the soil removed from this excavation is stockpiled. Wastes are then deposited at one end of the excavation on a sloped end of the trench. The refuse is spread on a rather shallow inclination

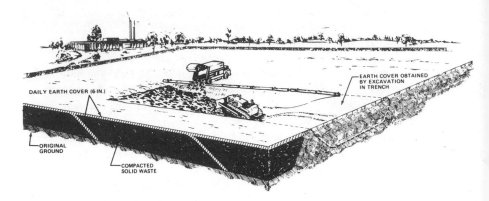

Fig. 4-3 Trench method of sanitary landfill (*courtesy* Office of Solid Waste Management Programs, EPA).

(usually about 3 horizontal to 1 vertical), and is then compacted by the placement/compaction equipment used at the site. At the end of the day's operation, the compacted layers of refuse are covered with a layer of soil taken from the stockpile of material removed in the original excavation. When the entire trench has been filled with refuse, a thicker final cover layer is placed over the completed deposit of refuse.

The trench method is most suitable for sites where the groundwater table is at significant depth and where there is a deep layer of suitable cover soil. Also, generally, the trench method of landfilling is most suitable where the site topography is rather regular.

The length and depth of the trench itself are variable from site to site; the width of the trench, however, is generally limited to a width approximately equal to $1^1/_2$ times the width of the blade of the excavation equipment bulldozer or the bucket of the loader used in creating the original trench and in spreading and compacting the refuse in the trench.

The Area Method. In the area method of sanitary landfilling, in contrast to the trench method, refuse is dumped on an undisturbed existing ground surface; the only prior operation in the area-method landfill may be surface removal of top soil and highly organic material (humus) suitable for final cover. After the refuse is dumped from collection and transportation vehicles, it is spread over the ground surface in a uniform layer and then compacted to a higher density. The compacted layer of refuse is covered with soil at the end of an operational day or when the deposition area is filled. When the final layer of refuse has been placed and the entire site is filled, a final cover layer of greater thickness is placed over the completed fill. Generally, in area-method

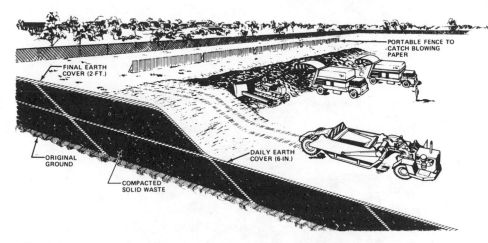

Fig. 4-4 Area method of sanitary landfill (*courtesy* Office of Solid Waste Management Programs, EPA).

landfills, the cover soil is transported onto the site from another location. The area method is used where the groundwater table is at or near the surface and on sites where the terrain is rough and irregular. The size of the working face in the area method is an important consideration for the overall success of the operation. The operating face should be large enough that no unnecessary delays for collection or delivery vehicles are caused; long lines of trucks waiting to dump their loads of refuse at a too-small operating face create an uneconomical operation. However, the operating or working face should not be so large that difficulties arise with blowing paper and litter, or the occurrence of disease vectors such as rats and mice, or other similar problems associated with poor control and management of the landfill site. Thus, the size of the working face will depend upon such operational variables as the site topography, the direction and velocity of prevailing winds, the quantities and characteristics of the refuse being brought to the site, and other factors.

The Ramp Method. The ramp method is a hybrid technique combining features of both trench and area methods of landfilling. Before refuse deposition is begun, a small excavation is made in front of the proposed face on an existing slope. The soil removed in this excavation is stockpiled nearby. Refuse is then deposited on the face of the slope, spread and compacted by standard landfilling equipment, and then covered with the soil which had been stockpiled from the preceding excavation. This process is repeated again and again at the face of the newly created slopes so that a succession of slopes are produced in a line across the landfill site. Because of this successive technique, the ramp method has also been called the "progressive slope" method.

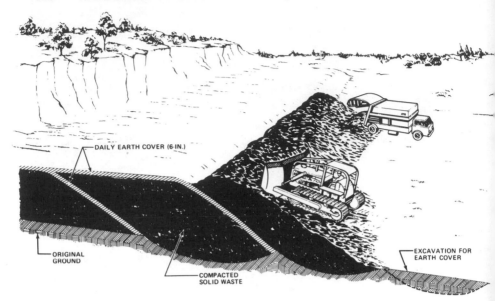

DAILY EARTH COVER (6-IN.)

ORIGINAL
GROUND

EXCAVATION FOR
EARTH COVER

COMPACTED
SOLID WASTE

Fig. 4-5 Ramp method of sanitary landfill (*courtesy* Office of Solid Waste Management Programs, EPA).

The same general considerations for size of working face are applicable in the ramp method as are applicable in the area method.

At any one given landfill site it is possible that more than one of the filling techniques will be used at the same time; use of more than one method of filling at the same site generally is associated with irregular or broken terrain.

4-2.2 Operational Parameters

Several factors will control or govern the operational methods used in sanitary landfilling at any given site. Included among these factors are the topography and geology of the site, the availability of cover material, the characteristics of the wastes to be deposited at the site, the surface and groundwater hydrology of the site, and the climatic characteristics of the site location area (Ref. 4-12).

Topography and Geology of the Site. If necessary, a sanitary landfill may be constructed on almost any topographic landform; however, while landfilling is an easily adaptable disposal operation, certain topographies are much more suitable than others for the deposition of wastes. Furthermore, the topography of the site, as mentioned previously, will govern the type of disposal technique employed; i.e., the trench method, the area method, or the ramp method. Flat land with a sufficient depth of usable cover soil is most suitable for the trench method of landfilling unless water occurs at or near the surface of the ground. Where the water table is high in flat land, the area method is more suitable; either the area or the ramp method is more suitable on irregular and rolling topography.

The overall geologic setting of the landfill site will be important to the operation. The presence of a large layer of suitable cover soil over deep bedrock is an ideal situation for sanitary landfill operation. The characteristics of the bedrock are also important in that certain types of rock are more pervious and therefore contain large quantities of rapidly-moving groundwater, offering significant possibliities of water pollution. Other types of rock (for example, shale) are typically very slowly pervious and groundwater moves very slowly and in only small quantities through such materials. The characteristics of the underlying bedrock are also important in that certain types of rock are more susceptible to chemical action, and may deteriorate rapidly under the attack of leachate produced in a sanitary landfill. Finally, geologic features such as active faults, landslides, or subsidence areas obviously will have detrimental effects upon a sanitary landfill.

Cover Material. The presence of suitable cover material at a particular site does much to create an efficient landfilling operation; if it is necessary to import cover soil from a source area outside the close environs of the landfill site, high transportation costs will be incurred. However, the quality of the available cover soil is just as important as is its quantity. The characteristics of a desirable cover material are easy workability, moderate cohesion, and significant strength. To fulfill these requirements, the most suitable cover soil is a mixture of sand, silt, and clay (Ref. 4-12). Generally, a sandy loam is a very desirable cover material (Ref. 4-1). Clean sands are somewhat unsuitable for cover material since they are readily permeable and allow large quantities of water to invade the deposited refuse. Fine-grained soils such as clays and silts are not ideal cover materials because of difficulties in working on and with such materials and because of the shrink–swell properties associated with cohesive soils. Pure clays in particular have a strong tendency to shrink and form deep cracks upon drying; such open cracks provide access to the deposited refuse for disease vectors such as rodents and insects, allow infiltration of significant amounts of surface water until they swell and close, and permit the escape of gases and odors from the decomposing refuse (Ref. 4-12).

General rules-of-thumb for operation of a refuse landfill state that the cover soil should be a well-graded mixture of fine and coarse soil components and that a sufficient amount of cover soil should be available to insure 1 part of cover soil for 4 parts of refuse in the completed landfill.

Refuse Characteristics. Physically, the overall success of the landfilling operation depends upon the ability of the operator to place the greatest amount of material in the smallest possible volume. Consequently, the efficiency of the landfill may be judged directly by the densities which are achieved in the compaction of the refuse. In common practice, a sanitary landfill operator must be able to achieve densities of 800 lb/cu yd or greater in the completed landfill site to remain in economic competition (Ref. 4-12). Obviously, then, those wastes which are easily compacted and densified are quite favorable to the operation of a sanitary landfill. Other waste types, particularly domestic and commercial wastes which contain large portions of paper and plastics, are difficult to compact and may be modified to only moderately high densities in a landfilling operation. The necessity to dispose of such low-density materials in a landfill will cause an increase in the operating costs per ton for the site.

Secondly, the characteristics of the wastes themselves may be of paramount importance in the landfilling operation if the wastes are hazardous materials. In other words, if the wastes present significant potential for important detrimental environmental effects, special treatment such as neutralization or containment may be necessary. The high costs of such special measures must be added into the cost of the landfilling operation and the overall economy of the site will be affected.

Hydrologic Considerations. The system of water flow on the land surface and through the materials of the sanitary landfill site is of paramount importance to the acceptability of the operation since water migrating through decomposing wastes will lead to the spread of pollutants into the general area and cause a severe environmental problem. In assessing the overall technical feasibility of sanitary landfilling as a disposal method it is necessary to consider the potential for the movement of water through the wastes deposited at the proposed landfill site. A surface drainage system must be incorporated into the landfill design to prevent contact of surface waters with deposited refuse. Likewise, the design and operation of the landfill must ensure that groundwater and decomposing wastes do not come into contact. The production of leachate and its migration into useable water supplies is one of the significant environmental problems associated with solid waste disposal by sanitary landfilling methods.

Climatic Conditions at the Site. The climate of the landfill location area will have an important influence on the landfilling operation. Since it is necessary to prevent, as much as possible, the contact of surface water with the deposited refuse, it is necessary to consider the amount, frequency, intensity, and duration of precipitation to be expected at the landfilling site. Likewise, since blowing paper and litter can be a significant nuisance at a landfill site, some consideration must be given to the velocity and direction of the local prevailing winds. Temperature ranges in the area for the proposed landfill are also important because of difficulties associated with the excavation and handling of frozen ground. At certain times during the year it may be necessary to curtail land-filling operations, and provisions must be made for temporary storage of accumulated solid wastes during such times.

4-2.3 Components of a Sanitary Landfill Operation

The Site Itself. As mentioned previously, a sanitary landfilling operation may be carried out at almost any site. However, this siting of a landfill is a very significant operation in itself. Certain areas may be used in other ways much more profitably than for sanitary landfilling and their use as landfill sites should be avoided. For example, tidal flats and marsh lands have been filled, in some cases, around urban areas in sanitary landfilling operations. The loss of the life-support capacity in these biologically active areas far outweighs the benefit gained by the availability of land as a disposal site. Such dumping of wastes in these ecologically prolific areas should be curtailed. The other characteristics of the sites such as topography, climate, geology, and hydrology have been mentioned previously. Therefore, no further mention or discussion of those variables need be made here.

Equipment. The equipment used in a sanitary landfilling operation will determine the efficiency and success of the waste disposal. The choice of type of equipment for use on a landfill site will depend upon a number of factors, including the size of the overall operation, the particular deposition technique employed (trench, area, or ramp method), the types of wastes deposited, and the water and soil conditions at the site. The size of the operation itself may govern the choice of equipment to a predominant degree (Ref. 4-27). Many operations, especially in small communities and in rural areas, are so small that only one piece of equipment is necessary or possible. The most common choice for this one piece of equipment is the tracked type front-end loader. These vehicles are usable in all kinds of weather and have excellent excavation characteristics. They are not especially maneuverable and are at some disadvantage in spreading the deposited refuse. However, the size of the working face in small operations is obviously not very large, so this disadvantage is of negligible importance in small fills. These tracked type excavators are particularly useful in trench-method operations since the working face is relatively narrow and confined. For larger landfilling operations and particularly for landfilling by the area method, the tracked dozer or bulldozer is more suitable than the front-end loader because of the capability of the dozer to spread the refuse quickly and efficiently. Moreover, the large tractors can be used to tow scrapers and other earthmoving equipment for large operations. As in the case of the front loaders, these tracked dozers operate in all kinds of weather and have good trafficability on all types of soils because of the low bearing pressures exerted by the tracks. Dozers are especially useful in site preparation, site finishing, and construction and maintenance of access roads (Ref. 4-8).

In large operations, specialized equipment may be used. Special steel-wheeled compactor-loader-dozer vehicles have been developed especially for sanitary landfill operations. They provide better compaction of heterogeneous refuse but lack the tractive ability of the dozers in wet-weather operations or on steep slopes. Likewise, they are not used optimally in excavation of undisturbed soil. For very large operations where cover material must be transported over distances of several thousand feet or more, larger earthmoving equipment such as self-propelled scrapers or rubber-tired vehicles may be used. Generally, rubber-tired loaders and other such vehicles are used where mobility is a prime consideration; for example, in irregular ground where small depressions or narrow sloughs are being filled. The mobility of such equipment offsets its inferior spreading and excavation capabilities (as compared to tracked vehicles) (Ref. 4-12).

All equipment used in sanitary landfilling operations must be fitted with special engine and radiator guards and screens, reversible fans to blow paper off radiators, underchassis guards to protect engines and transmissions, and special protective covers for hydraulic lines. Also, rubbertired vehicles must be equipped with specially strengthened tires to prevent blowouts. At sites where rainfall is

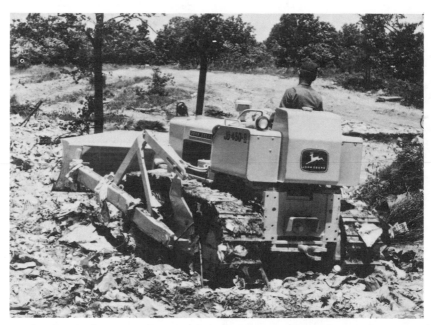

Fig. 4-6 Tracked crawler for spreading and compacting refuse and cover soil (*courtesy* John Deere Company).

Fig. 4-7   Tractor type spreader with special cleated wheels for fill compaction (*courtesy* John Deere Company).

low and dust is a problem, special air intake filters must be used and spraying by water trucks may be required to alleviate the dust problem. In very specialized situations, other types of equipment such as draglines have been used—a dragline may be used to excavate in very soft ground or below the water table. However, the use of such specialized equipment is very limited.

Physical Plant. Since the workmen who are operating the landfill require shelter and sanitary facilities, some sort of shelter or building must be constructed on the site as a headquarters for the operation. Housetrailers have been used successfully for this purpose; they are mobile and can be placed near the actual working operation. In very large operations, relatively permanent structures may be constructed at a central location within the large site. For the physical plant, whether it is a trailer or a permanent structure, certain utilities must be supplied. Electric power, potable water, and sanitary facilities must be furnished for the convenience and use of the employees. Furthermore, adequate access roads from the entrance of the site to the headquarters building and to the working face are absolutely necessary. Since much of the waste material deposited in a sanitary landfill is combustible, it is necessary to provide fire control capabilities for a landfilling operation, whether the fire control capabilities are achieved through subscription to local fire control authorities or provision of fire control equipment on the site itself. Finally, the exterior route or road to the landfill site must be adequately marked and notices should be posted for operational conditions such as the types of wastes accepted at the site, the hours of operation, etc. (Ref. 4-1).

Other items necessary at a sanitary landfill site include adequate fencing, sufficient lighting, and provisions for scales for weighing incoming refuse. Fencing is needed in order to prevent illegal dumping of wastes, to keep out unauthorized people and wandering animals, and to provide aesthetic control of the site to the operator. In other words, a well-run landfill will be operated in such a way that the working face and the open area of refuse are well-hidden from the eyes of the passerby so that the aesthetic insult of the operation is minimized. Additionally, light fences such as snow fences may be used to control blowing paper and litter on sites where wind velocities are significant. Sufficient lighting must be provided throughout the site to ensure safe operation of filling equipment, collection trucks, and other vehicles. Furthermore, the security of the site from unauthorized entrance necessitates the use of adequate lighting for the perimeter of the landfill area. Finally, it is necessary to install scales at a landfill site for proper assessment of dumping charges in operations where fees are levied on a per-ton basis. For example, in a municipal landfill, private haulers may be allowed to dump their loads of refuse in the landfill site providing they pay an appropriate fee. The fee is assessed on the basis of load weights determined at the site on truck scales installed at a central gate to the site. For best usage of the scales, it is recommended that adequate and comprehensive records be kept.

4-2.4 Auxiliary Operations

In the normal operation of sanitary landfills certain auxiliary operations such as waste processing by shredding or grinding, or volume reduction are commonly practiced. Some discussion should be devoted to these auxiliary operations at this point.

Shredding and Grinding. Shredding and/or grinding of refuse before landfilling is currently being investigated at several sites in this country. A comprehensive program of testing shredding of municipal refuse has been conducted at Madison, Wisconsin where a Heil-Gondard hammermill was used to grind the refuse. By reversing the rotation of the hammers within this mill, a ballistic ejection system was achieved that expels and rejects items which are normally difficult to landfill. This ejection of unsuitable material is being cited as one of the advantages of the shredding operation (Ref. 4-11). Further, it has been shown that the action of shredding the refuse intimately mixes the decomposable garbage fraction with nondecomposable refuse. Because of this intimate mixing, disease vectors such as rodents are unable to live in or on the exposed milled refuse. Direct tests at Madison and at the Purdue University Experimental Station have proven that rats cannot survive using the milled refuse as a food source (Ref. 4-13). Furthermore, the action of passing the refuse through the hammermill apparently destroys many of the larvae and the eggs which are laid in the refuse. The problem with mature insects emerging from the refuse at the landfill site has been minimized. Finally, the shredded refuse apparently is easier to handle, it is easier to compact, and it has little noticeable odor in the actual filling operation. Shredding has been attempted and evaluated at Bucyrus, Ohio, Tacoma, Washington, and Duchess County, New York. Shredding is widely used in Europe.

Volume Reduction. Volume reduction, or densification prior to landfilling, has been practiced in several different ways in this country up to the present time. First, refuse has been compacted at the source in stationary compactors in the home or in compactor trucks as soon as the refuse is brought from the home or source to the collection vehicle. Such compaction significantly increases the efficiency of the collection procedure, yielding higher truck loads and lower transportation rates for the refuse. Secondly, compaction of collected refuse has been practiced at central locations, i.e., in transfer stations where collected refuse is placed in high-energy compactors and compacted or pushed into special truck bodies which are then moved to the final deposition site. The high-pressure stationary compactors and balers being employed at the present time in transfer station operations achieve very high densities of the compacted refuse. Several American companies have built patented systems to produce bales of refuse in which the final bale average densities are on the order of 1700 to 2000 lb/cu yd. Baling of refuse and placement of the completed bales in a sanitary landfill has been investigated in field situations at San Diego, California and at Minneapolis, Minnesota. An earlier experiment in Los Angeles, California was unsuccessful because the baling equipment used was not satisfactory and the dozers at the landfill site were unsuitable for moving and the depositing the completed bales (Ref. 4-15). Finally, special densification and placement of refuse at landfill sites have been investigated in King County, Washington and in Niagara County, New York; at these sites, experimental vehicles that receive the refuse, compact the refuse, bale the refuse, dig a trench, place the bale in the trench, and cover the refuse with soil all in one operation have been tested and evaluated (personal communication, D. Keller, OSWMP, EPA, 1972). No comprehensive test data are available.

Other additional auxiliary operations currently being attempted with sanitary landfilling include the collection and treatment of leachate produced within the landfill refuse cells. A detailed discussion of such collection and treatment procedures will be given in Chapter 5 of this text, which will deal with landfilling with leachate recirculation.

4-2.5   Limiting Parameters

Parameters which limit the practice of sanitary landfilling include the decreasing availability of land, the production of leachate in landfills, the production of gas in decomposing refuse, the availability of cover material, the creation of nuisances by improper operations, and the public disregard for sanitary landfilling as a disposal method.

Land Availability. The most significant problem associated with the continuing disposal of solid wastes in sanitary landfills is the decreasing availability of sites suitable for such disposal operations. The ever-expanding urbanization of the land area of the United States, particularly of those areas of prime value and most useable topography, militates against the use of sanitary landfills. As land near urban areas becomes scarce, sanitary landfills will be located farther and farther from the source points of the solid wastes in the urban centers. The cost of transporting collected refuse long distances to remote landfill sites, added to standard landfill operating costs, will quickly transform the economics of refuse disposal in favor of methods other than sanitary landfill. One phase of this difficulty which may be altered is the fact that many sites are not now available for landfill operations because of public opposition to sanitary landfills near urban areas; the average citizen bitterly opposes the location of a sanitary landfill anywhere near his home. Reversal of public opinion could make many sites available which are not now in use.

Landfill "By-Products." The decomposition of refuse in a sanitary landfill produces carbon dioxide, methane, hydrogen sulfide, and other gases which, at best, are public nuisances, and at worst, are severe hazards. Such hazardous gases must not be allowed to escape from the landfill. Conventional control practices include venting and burning of collected gases. This type of control is costly but necessary.

In addition to gases, other decomposition products are generated as liquids and solids in landfill refuse cells. Passage of migrating water through such materials leaches out suspendable and soluble constituents and may lead to the introduction of pollutants into surface water and groundwater systems. Prevention of such leaching, or collection and treatment of generated leachate, must be accomplished to prevent water contamination.

The technical effort and attendant costs associated with preventing escape of decomposition products from sanitary landfills must be viewed as limiting factors to the practice of sanitary landfilling.

Other Limiting Factors. Other condition which may limit the successful use of sanitary landfilling as a disposal method include: the lack of adequate cover material at a given site; the creation of nuisances such as odors, blowing paper, and the prevalence of disease vectors by improper operation of the site; and the refusal of the public to accept the sanitary landfill as a suitable and sometimes necessary method of solid waste disposal. The need to import cover material to a site can drastically alter the profit–loss situation for a private operation or the total costs of waste disposal in a public operation. Creation of nuisances caused by improper landfill operation may make the landfill a poor disposal method. Good operation could eliminate such nuisances, obviously, but according to a 1968 national survey, only about 5 percent of all sanitary landfills are properly operated (Ref. 4-20). Finally, lack of public support for a landfill can cripple its operation through loss of long-term financial support (taxes), interruption of operation (injunctions against operation), and loss of available land (zoning changes in response to public demand).

4-2.6 Future Use of Sanitary Landfill Sites

An important consideration in the disposal of solid wastes in a sanitary landfill operation is the ultimate fate of the disposal site. Future use of a landfill site may yield both economic and social benefits; improper future use may lead to serious environmental problems.

Planning Future Use. Anticipation and planning of the future use of a landfill site should accompany the preliminary design and operational planning which takes place before the filling operation begins. However, in the aforementioned 1968 National Survey it was discovered that 82 percent of the more than 8000 landfills in operation at that time were being filled with no planned or anticipated future use (Ref. 4-20). This condition severely limits the overall utility of the method because if the future use of a site is anticipated, the filling operation may be tailored to future needs. For example, inert construction wastes may be segregated and placed at high density in selected portions of a site to form foundation layers for future structures.

Recreational Uses. Most completed sanitary landfill sites eventually are used as recreational areas. This occurs not as the result of long-range planning but simply because most sites are filled in a haphazard fashion and are unsuitable for any other use. Poor bearing capacity of surface layers, settlement over decomposing refuse, and collection of gases in refuse cells present little difficulty in a recreational facility such as a park or athletic field. In many cases, recreational use of completed landfill sites provides much-needed green areas in urban regions. Nevertheless, recreational use of a filled disposal site may be wasteful and inappropriate under certain circumstances.

Agricultural Use. In previous attempts to use completed landfill sites for agricultural purposes, deep layers of final cover soil have been placed over the completed refuse cells. Extra cover soil requirements have been balanced in an economic sense by the availability of new croplands in areas where urbanization is intense and arable land is scarce. However, major difficulties have been encountered because of the percolation of irrigation waters into underlying refuse with consequent leachate generation, and because of the detrimental effects on growing plants of the gases generated in decomposing waste (Ref. 4-2). The suspicion of pathogen migration from waste to crop has worked against public acceptance of agriculture over landfills. The continued use of completed landfill sites for agricultural purposes is doubtful.

Commercial and Industrial Uses. The founding of important, heavy, extensive structures over decomposing refuse cells is not feasible. Moreover, penetration of foundation elements such as piles through unsorted refuse in a conventional landfill is a chancy proposition (Ref. 4-28). Consequently, little important construction has been placed on completed sites. As a result, sites with central location, suitable zoning, and excellent accessibility have not been properly utilized. Adequate planning for future use before filling begins can alleviate these difficulties. Decomposable refuse can be placed in separate areas of a site; those areas may be finally used as parks or green borders. Future sites for structures may be demarcated as fill areas for inert wastes such as construction and demolition wastes, used foundry sand, etc. Bulky items which could ultimately collapse and cause surface settlement may be relegated to sections of the site destined for recreation or other light use.

4-3. CRITICAL METHOD PARAMETERS

In assessing the technical feasibility of any method for waste disposal, a significant amount of attention should be given to those conditions or parameters which severely limit the applicability or success of the process itself. In other words, in any given process, the condition or value of a certain parameter will give either a "yes" or "no" answer to the question "Can this process be used in this place at this time to dispose of solid waste?" In the following discussion, several parameters which furnish such as answer to this question will be described. The parameters may be grouped into two categories: 1) there are certain parameters or conditions which must be controlled technically in order to have a satisfactory sanitary landfill operation; 2) there are other factors or parameters which may not be controlled by the landfill designer or operator but which will have a very important influence on the success or applicability of the landfilling method. An example of a technical limiting parameter would be the necessity for control of gas generation and migration in a sanitary landfill. On the other hand, land availability is a limiting parameter which has an important influence (a predominant influence) on the applicability of sanitary landfilling but it is not amenable to control by the landfill operator or the landfill designer.

4-3.1   Limiting Parameters Subject to Technical Control

Those parameters or conditions which limit the applicability of sanitary landfill as a disposal method which are subject to technical control include the generation of gases in decaying refuse within the landfill, the generation of leachate in decaying refuse, and the creation of public nuisances such as disease vectors, litter and blowing paper, and odors by poor operation of a sanitary landfill.

Gases Generated. As mentioned previously in the discussion of the methodology of sanitary landfilling, the decomposition of solid wastes in a landfill produces gaseous end products. The exact character of the gases produced during the decay of the wastes depends upon the relative abundance of oxygen within the landfill refuse cells during the decomposition period. If oxygen is abundant within the refuse and aerobic microbes accomplish the biological phase of decomposition, the major gas which is produced is carbon dioxide. If the oxygen present in the landfill becomes depleted, as is the usual case in the common landfill operation, the aerobes die and a different type of microbe begins to digest the organic material. The anaerobes, microbes which flourish in the absence of oxygen, digest the organic material present in the solid wastes and create a number of by-products, including gases such as methane and hydrogen sulfide. These gases are the ones responsible for the odor problems associated with sanitary landfilling, and they may prove to be toxic if they are allowed to concentrate in an area wherein people live or work.

The generation of these gases can be prevented or controlled in several different ways. For example, the wastes which are deposited in the sanitary landfill may be selected so that during the decomposition phase certain gaseous products are selectively produced. Also, the decay process itself can be regulated by the addition of oxygen into the landfill refuse cells so that the decomposition process will be carried out by aerobic microbes and only carbon dioxide will be created. The carbon dioxide would constitute no significant air pollution problem. Finally, the problem of the gases which are generated in refuse decomposition could be controlled by regulating the migration and exit of these gases from the landfill and by disposing of the gases themselves. Such a control and disposal operation could be effected by providing impervious liners for the landfill site, collecting the gases in a central location, and burning the collected gases to create carbon dioxide during the oxidation process. Of the three means of control just mentioned, the easiest method of control is to limit the exit of the generated gases and dispose of the gases by venting and burning. The provision

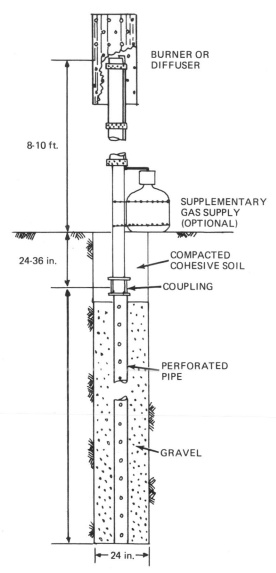

Fig. 4-8 Vertical vent gas control device (redrawn after Office of Solid Waste Management Programs, EPA).

of such control features does not present significant technical difficulty. Adequate design of a sanitary landfill would include a provision for the lining of the refuse cells to render them impervious to the travel of gases and would also include provisions for collection and burning of the gases which are produced. Thus, the limitations imposed upon the applicability of sanitary landfilling which arise from gas generation and decomposing refuse may be eliminated through 1) adequate design provisions incorporating liners and collection systems, and 2) operation of the site to include disposal by burning of the collected gases using devices such as that shown in Fig. 4-8.

Generation of Leachate. When the deposited wastes in a sanitary landfill degrade, the decomposition products may be removed from the landfill site by migrating water. The suspended and dissolved solids in the migrating water transform it from a harmless fluid to a leachate which has all of the characteristics of a strong industrial wastewater; the characteristics of this material will be discussed in detail in the following section describing the by-products of sanitary landfilling. The production and migration of leachate from a landfill site may be controlled and eliminated in a number of ways. As mentioned above in the case of the gases which are generated in a landfill, the production of leachate may be limited by selecting what goes into the landfill site. If only innocuous materials which would not be subject to decomposition were allowed to be deposited in a landfill site, no problem with leachate would result. However, only a very limited number of waste materials could then be disposed of in the sanitary landfill. As far as leachate is concerned, the limitation of materials deposited in a landfill site has only partial applicability as a means of control. Obviously, materials which are extremely hazardous in themselves (for example, a very toxic chemical waste) could and should be eliminated from the land site. However, a large portion of the materials which are found in ordinary domestic refuse decompose to form highly undesirable waste substances which may be removed from a landfill site by migrating waters. Thus, the limitation of what goes into the landfill site, as far as wastes are concerned, is only a partial remedy to the production of leachate. Likewise, an effort can be made to control the amount of water which either infiltrates from the surface into the landfill site or moves laterally as groundwater into the deposited refuse. Such a control procedure is also only a partial solution because the decomposition of solid materials in a landfill produces a certain amount of moisture and a certain amount of leachate without the movement of water from an outside source into the refuse cells. Furthermore, the likelihood of water moving into a landfill site must always be taken into account. For example, for many landfill sites, the probability of a flood condition may be very, very slight, but *some* probabilities or possibilities do exist that at some time in the future high water may cause either ground or surface water to flow into the deposited refuse which has decomposed to produce highly detrimental materials. Therefore, the control of deposited wastes and the limitation of infiltration of water into the refuse in a landfill site are only partial control procedures to prevent leachate contamination of water supplies.

Another means of control would be to control the decay process itself; in other words, oxygen could be pumped into the decaying refuse to create aerobic decomposition conditions. The end products produced during decomposition would be less harmful and toxic than the products of anaerobic decomposition but they would still be detrimental to such a degree that their exit from the landfill should be prevented in all cases. Therefore, controlling the decomposition process would have little influence as a solution to the problem of leachate migration.

A final step to control leachate problems in addition to the elimination of hazardous materials and the limitations of infiltration into deposited wastes is to provide impervious liners in the bottom and sides of the landfill site and to collect any migrating fluids and prevent their exit from the landfill site into the surrounding environment. Such prevention of leachate migration by collection would provide the necessary final step in the control of leachate. The collected fluid may be removed from the landfill site and transferred to a wastewater treatment plant wherein it can be rendered innocuous. Thus, the problem of leachate generation and migration with consequent contamination of subsurface water supplies can be controlled and eliminated through the proper technical design and operation of a landfill site. The additional costs attendant upon the installation of liners and a collection system must be borne as a necessary expense to proper sanitary landfilling. Furthermore, the limitation or elimination of hazardous materials (or special treatment such as containment in concrete vessels for hazardous materials), while expensive, may also furnish means of control for leachate. Thus, the problems associated with migration of water through decomposing refuse and the generation of leachate can be technically controlled.

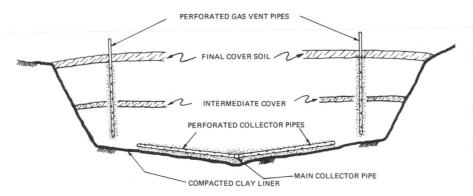

Fig. 4-9   Clay liner and collection pipe scheme.

Other Limiting Factors Amenable to Technical Control. As mentioned in a previous section, poor operation of a sanitary landfill site may lead to the presence of disease vectors such as rats, flies, and other vermin. Likewise, the improper operation of a landfill site wherein generated gases are allowed to escape may cause nuisance odors in the vicinity of the landfill. Finally, airborne litter and paper may constitute a significant nuisance problem at a landfill site which is not being properly operated. All of these factors which tend to limit the applicability of sanitary landfill as a disposal method may be overcome and eliminated by proper operation of the site. Disease vectors can be successfully eliminated by a program of pest control and the provision of daily cover for deposited refuse. In this regard, grinding and shredding of solid waste have been found to be effective means of rendering the material practically unusable as a food supply by rodents or birds. Such grinding and daily cover of the refuse will prevent the existence of conditions which support these vectors. Furthermore, shredding and grinding also appear to eliminate a large portion of the insects which are normally found in decaying refuse. Litter and blowing paper may be eliminated through the use of daily cover and light wind fences. Finally, odors are eliminated in a site which is operated properly where gases are not allowed to escape from decaying refuse but are collected and burned.

Summary—Limiting Parameters Subject to Technical Control. All of the conditions previously cited such as the generation of deleterious gases, the generation of leachate, and the prevalence of public nuisances may all be eliminated through the proper design and operation of a sanitary landfill. Such control and elimination will involve an additional expense to the operator and owner of the disposal facility; however, such expense should be considered a necessary expenditure for proper operation. Thus, these factors, while overtly limiting the operation of a landfill site, may be eliminated by technical solutions.

4-3.2 Limiting Factors Outside Technical Control

In addition to the limiting parameters discussed above, there are certain conditions or parameters which limit the applicability of a sanitary landfill as a waste disposal facility which are beyond technical control. The principal control parameters that are beyond technical control are the availability of land for landfill sites and the prevalent public opinion which is unfavorable to sanitary landfill as a disposal method.

Land Availability. As has been mentioned previously, the amount of land near centers of population which is suitable for sanitary landfilling operations is steadily decreasing. As the population explosion and move to the cities from rural areas continues, this decrease in the availability of land will accelerate. It must be remembered that a site for a proposed sanitary landfill must have the proper zoning and must be centrally located in order for it to be economically suitable for a landfilling operation. Approximately 80 percent of all monies expended for solid waste management in the United States each year are spent on collection of solid waste, not on disposal. Therefore, it is eminently important to locate a disposal facility in a central position which minimizes the travel distance which collection vehicles must traverse to the disposal site. Such centrally located sites are not plentiful. Moreover, to be suitable for a sanitary landfill, a site must be located in an industrial or commercial area and the land must be zoned for light-industrial or industrial use. If the site is zoned for commercial or residential use, a variance in zoning regulations must be obtained prior to the development of the site as a landfill. It has been the national experience that such variances are practically impossible to obtain. This combination of factors, suitable zoning and central location, is very rare within most of the metropolitan areas in the United States today. Such sites will become increasingly rare.

One of the factors which militates against the availability of existing sites which could be used for landfills is the opinion of the public which does not favor such use. Frequently, when a variance for use of a site as a landfill is requested from a zoning commission, citizen protest is a large factor in the commission's decision to turn down the request for such a variance. This factor of negative public opinion is sufficiently important that it must be discussed as a limiting parameter in itself.

Public Opinion. At the present time in the United States the designation "sanitary landfill" connotes for the average citizen a rat-infested, odorous, burning open dump. Naturally, the citizen is opposed to the operation of such a nuisance-ridden facility anywhere in the environs of his own home. Most often the citizen, when told that sanitary landfilling is a most economical method of waste disposal, reacts by saying that the metropolitan area should possess a landfill but that the landfill should be located in the section of town most removed from his own home. This factor of negative public opinion is prevalent nationwide and is a vital force in many efforts to limit and stop the practice of waste disposal by sanitary landfilling. A comprehensive program of education of the public is a requisite operation in any solid waste management program which envisions the use of a sanitary landfill as a disposal operation. A significant amount of professional assistance in overcoming the negative public outlook towards sanitary landfills will be required in any comprehensive solid waste management effort. The money, time, and effort associated with such an educational program is considerable, and every manager must plan for the expenditure of such time and money. In the past, such education programs have had only limited success. Furthermore, there is no hope for the success of such a program without a model landfill operation available for demonstration purposes. In many instances, the lack of such a model operation has prevented any educational program from being effective. As a consequence, in many instances a vicious cycle of no education program, no variance, therefore no model landfill, therefore no demonstration, therefore no education program, etc. results.

4-4. CHARACTERISTICS OF METHOD BY-PRODUCTS

In actuality, sanitary landfilling as a disposal method is intended to produce no end products. It has been called, for this reason, an ultimate disposal method. The very intent and idea of a sanitary landfill is that solid wastes will find a final deposition point from which they will not travel. However, imperfect operation of a sanitary landfill does produce undesirable by-products.

Actual By-Products. During the daily operation of a sanitary landfill, a decomposition process is initiated and proceeds until all degradable materials have been stabilized. This decomposition process produces by-products in the form of liquids and gases. These gases have been discussed in the previous section, as has the liquid which is produced and which may be removed from the landfill as leachate. Also, technical means of control of such by-products have been outlined in the previous section. Similarly, nuisance by-products such as odors from escaping gases, blowing paper and litter, and disease vectors such as rats and flies may be eliminated by adequate technical operation of the sanitary landfill site. Therefore, it is pertinent to consider that there are no actual by-products which leave a properly operated sanitary landfill. On the other hand, certain conditions will be created in the vicinity of an operating sanitary landfill which may be considered operational "by-products." Such operational conditions include the generation of large volumes of traffic; of necessity, collection vehicles will congregate at a landfill site. The concentration of traffic in the vicinity of a landfill will undoubtedly have an effect upon the surrounding neighborhood. If the surrounding area is a commercial or industrial zone, the detrimental quality of this effect will be minimal. However, if in any instance a sanitary landfill is located in a light commercial or residential area, the effects of this collection vehicle traffic can be quite detrimental. In addition to the obvious traffic of collection vehicles, secondary effects such as the noise associated with traffic and with the operation of earthmoving equipment in the site itself can be unfavorable to the surrounding environment. While these items and conditions are not, strictly speaking, actual by-products, they should not be forgotten in any assessment of the effects of a sanitary landfilling operation.

4-4.1 Potential By-Products

Because the very intent of a sanitary landfill is to furnish a site for the deposition and sequestration of solid waste materials, a potential will exist at any landfill

site for the removal of valuable constituents of the solid waste stream is a recycling operation. Therefore, portions of the solid waste deposited at any landfill site may be considered, in a sense, by-products in a potential recycling
operation.

Potential for Recycling Before Deposition. In many recycling operations in
the United States at the present time, constituents of the solid waste stream
which are high in resale value, such as aluminum and ferrous metals, are removed
before the remainder of the wastes are deposited in a sanitary landfill or before
the wastes are processed in an incineration operation. Thus, many landfill sites
appear to be centralized locations, ideal for a recycling operation. However, it
must be realized that a certain amount of processing is necessary for the complete
separation of valuable constituents of the wastes from the remainder of the
waste stream. For example, in order to separate ferrous metals, a magnetic force
field is created above a moving stream of wastes and the ferrous metal particles
are pulled by the magnetic forces from the remainder of the wastes. In order for
this magnetic separation to be effective, the ferrous metal pieces must be quite
small and must be separated from inert nonmagnetic material. Therefore, associated with any magnetic separation operation must be a previous grinding or
shredding operation which reduces the size of the metal particles and effectively
separates the metal from large pieces of inert material.

Such grinding and shredding operations are also necessary in other separation
activities such as flotation, air separation, and optical separation (as practiced for
glass particles). Thus, it appears that two facts must be recognized in viewing the
potential for recycling of solid waste in association with a sanitary landfill
operation:

1. By the time the wastes arrive at the landfill site they are intimately mixed,
   with desirable materials being closely intermingled with undesirable nonre-
   useable materials;
2. The possibility of separating and recycling desirable materials at any landfill
   site must include an evaluation of the costs in both labor and operational
   and capital expenditures for processing operations such as shredding and
   grinding which make feasible later separation operations. Therefore, the
   idea of a sanitary landfill as an ideal site for recycling must be tempered
   with a full realization of the cost in power, labor, and capital outlay associated with both processing and separation operations.

Potential for Mining Past Disposal Sites. Because solid wastes disposal in
sanitary landfills has been practiced in the United States for many years, there is
a significant deposition of valuable waste materials at numerous old landfill sites
throughout the United States today. Many of the former deposition sites were
not, in truth, sanitary landfills, and much of the material previously deposited

has been subjected to open burning. However, many of the most valuable portions of the solid waste stream, such as aluminum and ferrous metals, are not seriously harmed by the burning operation and hopefully may be removed from previous landfill sites in a profitable reclamation effort. The consideration of the potential for mining past disposal sites is somewhat extraneous to this chapter which considers methods to be used at the present time and in the future for waste disposal. However, in a comprehensive plan for solid waste management in any given region, some thought should be given to the possibility of obtaining the valuable constituents of the deposited wastes in a reclamation operation. In many cases old landfill sites are being used only for recreational purposes and no great inconvenience would be associated with the selective mining of portions of the old site to reclaim metals and other valuable constituents in the deposited wastes. In some instances, however, subsequent development of an old landfill site would preclude any such reclamation attempt. Nevertheless, a comprehensive waste management plan would include consideration of the potential for utilizing the former disposal sites as source areas for commodities which are extremely valuable in themselves or which are of limited supply in nature. At the present time, the feasibility of mining old landfill sites is being investigated by the U.S. Bureau of Mines.

4-5. REQUIREMENTS FOR A SANITARY LANDFILL

For a successful sanitary landfilling operation, certain practical requirements must be fulfilled. Chief among these requirements is available land. The next important requirement is adequate suitable cover soil. The amount and accessibility of the cover material in the area of the landfill operation will greatly influence the design of the refuse cells within the landfill, the method of operation of the landfill, and the amount and type of equipment needed at the landfill site. Less important method requirements for sanitary landfilling include fire protection, utilities such as potable water and sanitation for site personnel and electricity for both power and light, communications within the site boundaries and also with the area outside the site proper, and fuel for the equipment used to process the waste.

4-5.1 Amount of Land Required

National experience has shown that an optimal sanitary landfilling operation will result if the space or life expectancy of a particular site is on the order of 20 to 30 years. In order to estimate the amount of land required in a given situation for such a lifetime, several factors may be considered. For example, in standard practice the average compacted density of wastes in a completed landfill is on the order of 800 lb/cu yd. Secondly, the amount of wastes to be deposited in the given landfill site each day may be estimated on the basis that in urban areas approximately 8 lb per capita per day of domestic, commercial, and municipal wastes will be generated. Additional amounts of industrial, construction, and demolition wastes will also be generated and must be taken into account. Finally, the site topography and geology must be considered in order to arrive at a planned total depth of compacted refuse and cover material for the completed landfill site. An example of a determination of required land may be given below.

Example. A community of 100,000 people is to be serviced by waste disposal in a sanitary landfill operation. It is estimated on the basis of previously obtained information that approximately 400 tons of domestic, commercial, and municipal refuse is produced in this community each day. In addition, industrial operations and construction and demolition operations in this city produce an additional 150 tons of waste each day. It is assumed that an adequate amount of cover soil is available at the proposed landfill site. Because of topographical and geological limitations, it is planned that the final depth of deposited wastes plus cover soil will be 22 ft. It is desired that the landfill site have a minimum life expectancy of 30 years.

For these conditions, within a 30-year period, approximately 6.1 million tons of waste will be generated and must be deposited. This quantity of 6.1 million tons would occupy a space of approximately 15 million cu yd when compacted in the finished landfilling operation. As a rule of thumb, in cover construction it may be estimated that approximately 4 parts of refuse will be used with 1 part of cover soil in the filling operation. A final layer of cover material 2 ft thick is proposed for the given site. The total depth of compacted wastes at this site will be 16 ft; two 8 ft deep cells would require 2 ft of soil over each cell in daily cover and 2 ft of final cover to total 22 ft. If the waste volume of 15 million cu yd (405 million cu ft) is divided by the 16 ft depth of deposited refuse, an area of 29 million sq ft of land required for the proposed landfill is obtained. This corresponds to an area of approximately 666 acres to furnish a 30 year minimum life expectancy in this landfill.

Calculations similar to the example given above must be made in every operation wherein a sanitary landfill is planned to be a disposal operation for a given community.

4-5.2 Cover Material Needs

As mentioned in the previous section on methodology of sanitary landfilling, the material used to cover refuse in a landfilling operation ideally should be composed of soil types with a wide range in grain sizes; i.e., a mixed-grain soil such as a silty loam is an ideal material for covering solid wastes. Granular soils tend to be too pervious and admit large quantities of water into a landfill site. Furthermore, large granular material such as gravel may provide access to deposited wastes for small vermin and insects. Very fine-grained soils such as clays, on the other hand, may experience shrinkage and cracking during dry seasons and thus are not suitable as cover material because they likewise may allow large amounts of infiltration and may provide access to the deposited wastes for various types of vermin.

The amounts of cover material required in any given operation may be estimated quite easily using the previously mentioned rule of thumb of 4 parts of refuse to 1 part of cover soil. On this basis, in the preceding example approximately 6.4 million cu yd of cover material would have been required in the proposed landfilling operation. This quantity of cover soil would have to be available within a short distance of the proposed landfill for the landfilling operation to be economically attractive and technically feasible.

4-6. EQUIPMENT REQUIREMENTS

The equipment required for operation of a sanitary landfill generally may be categorized as either refuse-handling equipment or earthmoving and placement equipment. A few items not in these categories, such as watering trucks, may be desirable and/or necessary in some situations; such specialized items may be classed as equipment ancillary to the major refuse-handling and earthmoving equipment. In addition to the mobile materials-handling equipment at a landfill, a set of platform scales will be required at sites wherein private haulers are allowed to deposit refuse for a per-ton fee. Scales are required also at private landfills servicing public and private collection agencies. Physically, platform scales should be at least 10 ft by 35 ft with a capacity of 30 tons for standard operations; operations with transfer trailers will require 10 ft by 50 ft platforms.

4-6.1 Mobile Equipment

Table 4-1 given on the next page shows general recommendations for refuse-handling and earthmoving equipment for sanitary landfills. As mentioned in the preceding section on landfilling methodology, tracked vehicles are most popular. Tracked vehicles are effective in compacting refuse, suitable for moving cover soil, and operational in all extremes of weather. Tracked equipment is at a disadvantage in situations where vehicle speed and/or maneuverability are highly desirable, as in operations involving long haul distances for cover soil. Generally, a front-end loader is the most desirable tracked-vehicle configuration, but a dozer blade may be fitted to a tracked chassis in situations where spreading capability is critical (area fill method).

On sites where speed and machine "agility" are important, rubber-tired

TABLE 4-1 Average Sanitary Landfill Equipment Requirements

Population	Daily Tonnage	Quantity	Type	Equipment Requirements Size (lb)	Accessories[a]
0 to 15,000	0 to 40	1	Tractor-crawler or rubber-tired	10,000 to 30,000	Dozer blade Front end loader (1 to 2 cu yd) Trash blade
15,000 to 50,000	40 to 130	1 [a] [a] [a]	Tractor-crawler or rubber-tired Scraper Dragline Water truck	30,000 to 60,000	Dozer blade Front end loader (2 to 4 cu yd) Bullclam Trash blade
50,000 to 100,000	130 to 260	1 to 2 [a] [a] [a]	Tractor-crawler or rubber-tired Scraper Dragline Water truck	30,000 or more	Dozer blade Front end loader (2 to 5 cu yd) Bullclam Trash blade
100,000 or more	260 or more	2 or more [a] [a] [a] [a] [a]	Tractor-crawler or rubber-tired Scraper Dragline Water truck Road grader Steel-wheel compactor	34,000 or more	Dozer blade Front end loader Bullclam Trash blade

[a]Optional, dependent on individual need.
Source: "Sanitary Landfill Facts," HEW, USPHS, Publ. No. 1792.

vehicles may be used. Mobility is gained at the sacrifice of some compactive ability and trafficability.

Some specialized pieces of equipment such as "gear-tooth" steel-wheeled compactors have been developed. Higher compaction densities may be attained with this type of compactor, but the tractive capacity of such vehicles is somewhat lower than that possessed by tracked crawlers; consequently, they operate poorly on wet or soft ground.

Scrapers for long-hauling of soil and draglines for wet excavation work are similar specialized-use items of equipment.

4-6.2 Supporting Equipment

Certain types of equipment in addition to those items mentioned above are required for support operations in connection with sanitary landfilling. For example, road graders and rollers are required during construction of access

Fig. 4-10 Steel-wheeled front-end loader (*courtesy* International Harvester Company).

Fig. 4-11 Self-propelled scraper for longhaul of cover soil (*courtesy* WABCO, Inc.).

roads, and fire fighting equipment may be required at any time during the life of the filling operation. However, this type of support operation either is a "one-time" occurrence or involves the contracting of a standby service; i.e., incorporation of the landfill operation into a local fire prevention and control district. No special equipment should be purchased for this type of operation, but such a service should be secured in all cases.

4-6.3 Equipment Accessories

The equipment used in sanitary landfilling is subject to special hazards because of the nature of the wastes. Sharp or pointed objects in the fill may puncture tires, hydraulic lines, or other vulnerable parts of the vehicles. Consequently, "armored" tires and hoses must be installed. Likewise, engine screens and radiator guards, reversible fans to blow paper off radiators, and underchassis guards to protect engines should be utilized.

4-7. SANITARY LANDFILL PERSONNEL REQUIREMENTS

Approximately one-half the total operating costs of a sanitary landfill is attributable to labor costs. A rule of thumb for labor requirements is that 4 to 6 laborers are involved in the deposition of 1000 cu yd of refuse per day. Included in this group are the foreman or supervisor, equipment operators, gatemen, etc. Not included are the administrative personnel associated with a landfill operation (Ref. 4-1).

4-7.1 Administrative Personnel

The overall project supervisor is the key man in the administration of the land-fill. This individual should be acquainted with heavy equipment operation, general construction techniques, and grading and draining operations. He must also be an efficient and effective manager capable of maintaining good labor relations while operating a well-documented, tightly-controlled landfill. Finally, the supervisor must be adroit in maintaining good public relations with the general citizenry and, more especially, with the people living and working in the immediate neighborhood of the landfill.

In addition to the project supervisor, secretarial assistants, auditors, and bookkeepers are necessary members of the landfill administrative staff (Ref. 4-21).

4-7.2 Other Personnel

In addition to the personnel mentioned above, certain other types of individuals will be necessary members of the landfill labor force. For example, equipment mechanics and special equipment (shredders, grinders, etc.) operators may be desirable additions to the staff of a large landfill operation. The type and number of such artisans and operators will vary with the type of accessory operations used in conjunction with the basic landfill (Ref. 4-21).

4-8. FEDERAL LAWS PERTAINING TO SANITARY LANDFILLS

At the present time there are no specific laws or regulations governing the location and operation of sanitary landfills. The Rivers and Harbors Act of 1899 has been cited by the United States Army Corps of Engineers as authority for the regulation of landfills within the floodplains of navigable streams and rivers, but no special federal laws written for landfills themselves have been passed. However, a proposed set of regulations has been under study by a steering committee composed of members from the Environmental Protection Agency, the Council for Environmental Quality, and other environmental planning agencies. These regulations, not approved for publication at the time of this writing, are expected to appear in the *Federal Register*. The steering committee has gone to considerable trouble to elicit suggestions and comments from interested persons who wish to critically evaluate the proposed rules which have been aired informally at a number of conferences and meetings of the scientific and engineering community. The proposed regulations will govern the operation of sanitary landfills only on federal lands. However, it is expected that state and local regulatory agencies will use the federal regulations as guidelines in the formation of their own sets of rules and restrictions (personal communication, D. Brunner, OSWMP, EPA, 1972).

4-9. EXAMPLE LOCAL LAWS PERTAINING TO SANITARY LANDFILLS

As an example of state and county regulations on landfills, the regulations of the State of Kentucky will be presented here.

In June, 1972, the Kentucky General Assembly passed Senate Bill No. 1, which among other things created a "Department of Environmental Protection" for the State of Kentucky. The bill became a new section of Kentucky Revised Statutes Sect. 224 and stated, in part, that the new Department should

> ... issue, continue in effect, revoke, modify or deny under such conditions as the Department may prescribe, permits for ... the establishment or construction and the operation or maintenance of solid waste disposal sites and facilities; require that applications for such permits be accompanied by plans, specifications, and such other information as the Department deems necessary; ...

PERMIT APPLICATION CHECK LIST

Submit All Applications to the County Health Department

1. Name of operating permit applicant permit is issued to
2. Address of site
3. Mailing address and telephone number of applicant
4. Name of owner of property if different from applicant
5. Address and telephone number of owner
6. Site file number if known
7. Signed environmental impact statement from local health officer (Addendum #1)
8. Signed zoning conformance from county judge (Addendum #2)
9. United States Geographical Survey map with boundaries of area to be permitted plainly mapped
10. A 1" = 200' contour map showing EXISTING and PROPOSED construction and grading
11. Fencing, existing and proposed
12. Gate (one that can be locked)
13. Posting of hours of operation
14. Drinking water facilities and sanitary hand washing facilities
15. Toilet facilities
16. Equipment shed
17. Heated personnel shelter
18. Access road
19. Availability of firefighting equipment, if in fire district, statement from this district stating protection available
20. Communication equipment (radio or telephone)
21. Size and type of equipment to be used as well as quantity of equipment available
22. Backup equipment, size and type and how readily available
23. Control of water pollution, show all proposed drainage facilities on contour map
24. Control of air pollution, no burning whatsoever on a permitted site
25. Soils analysis
26. Covering program, daily for a sanitary landfill, availability of cover material
27. Method of operation, if trench method, show proposed location of trenches on contour map and if area fill method, be sure proposed contours are shown on contour map
28. Proposed use of completed landfill
29. Is site in flood plain? If so, has the Department of Natural Resources issued a permit for this operator
30. A list of types of material to be disposed of

COUNTY HEALTH DEPARTMENT WILL FORWARD PERMIT APPLICATION TO DIVISION OF SOLID WASTE, IF COMPLETE

Fig. 4-12 Jefferson County, Kentucky permit application for sanitary landfill.

Up to that time (June, 1972), the Division of Solid Waste Disposal of the Kentucky State Department of Health was the agency actually processing permit applications for, and the inspections of, sanitary landfills. Copies of the permit application used until 1972 and a sample site inspection report used by the Health Department in Jefferson County, Kentucky and the rest of the state are shown in Figs. 4-12 through 4-16. These applications are typical of state agency

<u>MEMORANDUM</u>

DATE:

TO: Health Division Director
          Division of Solid Waste

FROM:

SUBJECT: Addendum to Application, Number 2

I, _____, County Judge for

_____ County, certify that I have examined the

application for the establishment of a solid waste disposal

site or facility as proposed by the applicant and state that

this property is not located in an area where its operation

or construction is prohibited by planning or zoning.

                                    _____
                                             County Judge

                                    SW 12 (7-71)

Fig. 4-13 Memorandum (Part 1) to accompany permit application in Fig. 4-12.

MEMORANDUM

DATE:

TO: Health Division Director
        Division of Solid Waste .

FROM:

SUBJECT: Addendum to Application, Number 1

      I, _____, Local Health Officer

for _____ County, certify that in my opinion

and judgement after having made a thorough investigation am

of the opinion that the establishment of a solid waste dis-

posal site or facility as proposed by the applicant will not

create a nuisance nor detract from the beauty or quality of

the environment nor will the establishment of such site or

facility cause the spread of disease if operated pursuant to

the conditions set forth in the permit.

                              _____
                              Local Health Officer

                              SW 11 (7-71)

Fig. 4-14 Memorandum (Part 2) to accompany permit application in Fig. 4-12.

procedures. Creation of state "EPA" groups may change jurisdiction for approval and inspection, but procedures should remain constant. Local regulations in towns and cities also apply, in many cases, to location and operation of sanitary landfills.

Within the urban limits of the City of Louisville, Kentucky, solid waste disposal practices are regulated by provisions of the state statutes, Chapter 94 (shown in Fig. 4-17). Annual permits for sanitary landfilling are issued in Louisville by the Louisville and Jefferson County Board of Health, which also conducts all required site inspections.

The general experience in the United States is that landfills are inspected by local officials only semiannually or less frequently because of the small number of local government inspectors and the large number of facilities to be inspected,

MEMORANDUM

DATE:

TO: Health Division Director
           Division of Solid Waste
FROM:

SUBJECT: Addendum to Application, Number 3

        I, _____, Fire Chief

for _____ County and/or fire district,

certify that the solid waste disposal site or facility

as proposed by the applicant will be serviced by the

county fire department.

                                    _____
                                    Fire Chief

                                    SW 13 (1-72)

Fig. 4-15 Memorandum (Part 3) to accompany permit application in Fig. 4-12.

SOLID WASTE DISPOSAL SITE INSPECTION REPORT

Date _____ Sanitarian _____

Owner _____ Address _____

Operator _____ Operating Permit No. _____

Premise Address _____

Is the site adequately fenced? YES ☐ NO ☐ If no, explain: _____

_____

ANSWER YES OR NO TO THE FOLLOWING QUESTIONS:	YES	NO
1. Has an all weather road been provided?	☑	☐
2. Has a shelter been provided with drinking water?	☑	☐
3. Have toilet facilities been provided?	☑	☐
4. Have hand washing facilities been provided?	☑	☐
5. Is there a telephone at site? Number _____	☑	☐
6. Has fire protection been made available?	☑	☐
7. Was material being burned?	☐	☑
8. Was there any evidence of burning?	☐	☑
9. Is fill & surrounding area policed to collect all scattered materials?	☑	☐
10. Is waste being properly compacted?	☑	☐
11. Is the waste being placed in layers? At what depth? _____	☑	☐
12. Is waste being covered daily with six inches of soil?	☑	☐
13. Is final cover 24 inches in depth?	☑	☐
14. Is there evidence of sewage solids or liquid?	☐	☑
15. Is there evidence of septic tank pumpings?	☐	☑
16. Is there evidence of other hazardous liquids?	☐	☑
17. Are insects and rodents under control?	☑	☐
18. Is there a salvage operation?	☐	☑
19. Is there any feeding of farm or domestic animals?	☐	☑

What are the operating hours? _____

REMARKS _____

_____

Fig. 4-16 Jefferson County, Kentucky Health Department landfill inspection report.

CHAPTER 109
GARBAGE AND REFUSE DISPOSAL

109.010 Definitions. As used in this Chapter unless the context otherwise requires:

(1) "Garbage" means all putrescible animal and vegetable waste.

(2) "Refuse" means all putrescible and nonputrescible solid waste (except body waste) including, but not limited to, combustible trash, paper, cartons, boxes, barrels, woods, and noncombustibles including, but not limited to, material, dirt glass crockery and other mineral waste.

(3) "Disposal method" means any incinerator, sanitary landfill, or other method approved by the State Department of Health for disposing of garbage and refuse.

(4) "Collection system" means a system for collecting garbage and refuse by whatever method, and transporting said garbage and refuse from its source to the disposal site.

(5) "Person" means any individual, public or private corporation, political subdivision, copartnership, association, firm, estate, or other entity whatsoever.

(6) "Department" means the Kentucky State Health Department.

(7) "Board of directors" or "Board" means the governing body of the garbage and refuse district.

(8) "Chief engineer" means the engineer of the Kentucky State Health Department

designated to that position pursuant to the provisions of KRS Ch. 211 and the administrative orders of the department. (1966, c. 66, § 1)

109.020 Contracts between subdivisions for collection and disposal of garbage and refuse, contents. When the governing bodies of two or more cities or counties, or both, have declared by ordinance, order or resolution that it is in the best interest of such cities and counties to join with each other in the collection and disposal or solely in the collection or solely in the disposal of garbage and refuse they shall cause a contract to be prepared and executed which shall set forth:

(1) Whether and to what extent such cities and counties, or both, shall render services to the district and receive services from the district or from each other.

(2) Whether such cities, counties or any of them will contract with a private party or parties for the collection of garbage and refuse.

(3) The financial responsibilities and contributions of the respective cities and counties in the joint undertaking.

(4) The extent to which the cities and counties will furnish personnel and administrative services to the district.

Fig. 4-17 Louisville, Kentucky regulations for sanitary landfill, based on KRS 109 "Garbage Disposal Districts."

(5) The terms of the contract or agreement: Provided that such contract may be modified from time to time as conditions may warrant and provided one or more additional cities and counties desiring to participate in the joint undertaking may be permitted to become parties to the contract or that any participant may withdraw therefrom on such conditions as shall be agreed to by all participants.

(6) Such other provisions as may be necessary to execute a workable system of collection and disposal or solely of collection or solely of disposal of garbage and refuse. (1966, c. 66, § 3)

109.030 Garbage disposal districts, establishment by fiscal court and cities.

(1) The fiscal court of any county may be proper order or resolution lay out, establish and maintain one or more garbage and refuse disposal districts within the county, and may cause to be made surveys necessary to establish with reasonable accuracy a proper boundary for such districts. The district may acquire, construct, improve, enlarge, replace, maintain and operate such garbage and refuse collection systems within such district and such disposal methods within or without any such district as is necessary for the protection of the public health.

(2) The fiscal courts of two or more counties may so lay out, establish and maintain one or more districts to include and serve all or a designated portion of the territory within each such county upon such terms and conditions as may be agreed upon by the fiscal courts and set forth in the orders or resolutions adopted by each.

(3) Any city individually or jointly and cooperatively with any other one or more than one city or county may lay out, establish and maintain a garbage and refuse disposal district pursuant to the provisions of this chapter and exercise the power agreed herein.

(4) No city shall be made a part of a garbage and refuse disposal district without prior approval of its legislative body. (1966, c. 66, § 2)

109.040 Board of directors of garbage disposal district. The affairs of a district shall be controlled and managed by a board of directors consisting of three or more members as hereinafter provided. If the district lies wholly within one county, the board shall consist of three members and shall be appointed by the county judge of that county within thirty days after the establishment of the district. If the district lies within two counties, the county judge of the county in which the greater portion of the population of the district lies shall appoint two directors and the county judge of the other county shall appoint the third. If the district lies within more than two counties, in addition to the directors just enumerated, the county judge of each county exceeding two shall have the right to appoint one additional director. The legislative body of each city of the first three classes or if none each city of the highest class located within each county in the district, which city elects to operate as a part of the district, shall have the right to appoint one additional director. Each board member shall reside within the district and within the county or city by which he is appointed. (1966, c. 66, § 4)

109.050 Oath and bond of directors.

(1) Each director, before entering upon his official duties, shall take and subscribe to an oath that he will honestly, faithfully and impartially perform the duties of his office and that he will not be interested in any contract let for the purpose of carrying out any of the provisions of this chapter. The oath shall be filed with the records of the district.

(2) Each director shall give a good and sufficient bond, to be approved by his appointing authority, for the faithful and honest performance of his duties and as security for all monies coming into his hands or under his control. The cost of the

Fig. 4-17 (*Continued*)

bond shall be paid by the district. (1966, c. 66, § 5)

109.060 Directors, terms of office. The terms of office of the directors first appointed shall expire on the first Monday in January of the even numbered year next following their appointment. Thereafter, directors shall be appointed for terms of two years each, expiring on the first Monday in January of the next even numbered year or until their successors are appointed and qualified. Any vacancies shall be filled by the appointing authority for the unexpired term. (1966, c. 66, § 6)

109.070 Plans of district; operations.

(1) The board shall cause to be prepared a plan or plans for the improvements, methods and systems for which the district was created. The plans shall include such maps, profiles, plans, specifications and other data as are necessary to set forth properly the location and character of the work to be done.

(2) The board may devise, prepare for, execute, maintain and operate any or all works or improvements necessary or desirable to complete, maintain, operate and protect the works provided for in the plans. They may secure and use men and equipment under the supervision of the district engineer or other designated agents, or they may let contracts for such works, either in whole in or part. (1966, c. 66, § 9)

109.080 Plans to be approved by State Health Department; methods of disposal.

(1) Upon the completion of a plan the board shall submit it to the department for approval. If the department refers the plan back for amendment, the board shall prepare and submit an amended plan. If the department rejects the plan, the board shall prepare another plan. When the department approves the plan a copy of the action of the department shall be filed with the secretary of the board and incorporated in the records of the district.

(2) In carrying out the functions for which it was created a district shall provide such method or methods as shall be approved by the department for the collection and disposal of garbage and refuse or solely for the collection of garbage and refuse or solely for the disposal of garbage and refuse. The methods of disposal may include, but not be limited to, incinerators, sanitary landfills or reduction to fertilizer. All salvage or things of value resulting from the district operations may be sold and the proceeds used for the operation of the system. (1966, c. 66, § 10)

109.090 Collection and disposal before plans approved unlawful; abatement of operations. After the establishment of a district and the organization of the board, no person shall operate within the district any collection system or disposal method until the plans therefore have been submitted and approved by the board and by the department. Any operation or installation contrary to these provisions shall constitute a public nuisance and shall be abated by an injunction upon proper application by any one aggrieved including the district, the department or the local health department. (1966, c. 66, § 12)

109.100 Engineer and attorney for board; personnel; offices. The district may employ an attorney and a chief engineer who shall hold office at the pleasure of the board and who shall give such bond as required by the board. The board may employe such other administrative personnel as may be needed and may describe the duties and fix the compensation of all employes of the district. The board may maintain, fix and equip an office or offices as necessary to carry out the functions of the district. (1966, c. 66, § 8)

109.110 Chief engineer duties. The chief engineer of the department shall, in addition to his other duties, advise and assist all districts in the performance of their duties under the provisions of this chapter and in accordance with the requirements of the department. He shall be charged with other duties and services in relation thereto as the

Fig. 4-17 (*Continued*)

department may prescribe. (1966, c. 66, § 11)

109.120 Rules and regulations of board. The board may adopt such rules and regulations as are necessary to carry out the purposes for which the district was created and necessary to the adequate collection and disposal of garbage and refuse in a manner adequate to protect the public health. (1966, c. 66, § 7)

109.130 Contracts for service outside district. The board may make contracts or other income resulting from the operation outside the district and provided that no service may be given within the corporate boundaries of any city except with permission of the legislative body of the city. (1966, c. 66, § 13)

109.140 Funds, use of. All funds derived from taxes, service charges, sales or other income resulting from the operation of a district shall if collected or held by another person be paid over to and expended by the district to carry out its duties, functions and responsibilities. (1966, c. 66 § 14)

109.150 Property, district may acquire. The district may acquire by bequest, gift, grant or purchase such real and personal property or any interest therein as may be necessary to accomplish its purposes. Title to all property acquired by the district shall vest solely in the name of the district. With the approval of the department the district may apply for and receive funds made available under federal legislation in the form of grants and loans and for the purpose of improving collection systems and disposal methods of garbage and refuse. (1966, c. 66, § 15)

109.160 Board may condemn real property. The board may by resolution reciting the need, order the condemnation on behalf of the district of any real property or interest therein that in the opinion of the board is necessary for the proposed construction or use of a disposal method. Proceedings for condemnation shall be conducted in the manner prescribed in KRS 416.230 to 416.310. (1966, c. 66, § 16)

109.170 Revenue bonds, issuance, power of board. For the purpose of acquiring, creating and maintaining a collection system and disposal method any district may pursuant to the provisions of KRS Ch. 58 borrow money and issue negotiable revenue bonds. (1966, c. 66, § 17)

109.180 Financing; taxes; service charges.

(1) The district may levy an annual tax of not to exceed ten cents on each one hundred dollars of assessed valuation of real property within the district subject to taxation for county purposes. The proceeds of such tax shall be used for expenses of the district and for redemption of any bonds issued by the district pursuant to KRS 109.170.

(2) The district may in lieu of the tax provided in subsection (1) or in addition thereto finance the maintenance and operation of the district by service charges to be collected from all persons receiving services from the district. Such charges shall be fixed in such amounts as can be reasonably expected to yield revenues not in excess of the cost of operation and maintenance of the system and for an adequate depreciation fund.

(3) The service charges authorized by subsection (2) may be collected by the district directly or the district may enter into an agreement with other utilities either public or private where possible on such terms as may be acceptable to both for the collection of such charges. (1966, c. 66, § 18)

Fig. 4-17 (*Continued*)

including other public facilities such as restaurants, produce stores, etc. However, in Louisville, for example, if a specific complaint is registered against any given fill operation, a special inspection is made. Regulatory procedures usually consist of 1) enforcement of remedial clean-up measures as required, 2) frequent inspection to ensure improved operation, and 3) serious fines and possible closure for flagrant or major violations.

A special provision of note in Louisville restrictions and regulations is that combustible materials may not be deposited in a landfill within the city limits. Many other cities and towns have similar restrictions.

4-10. EXPERIENCE WITH SANITARY LANDFILLS

Sanitary landfilling is the most widely used method of solid waste disposal in the United States. Because of this existing nationwide experience with the method, it is inappropriate to discuss any single case history or example of the use of sanitary landfill. However, a compendium of data on the general economies of scale for sanitary landfills is given in Fig. 4-18 below, taken from *Sanitary Land-*

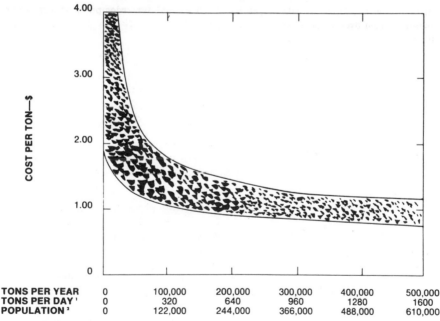

TONS PER YEAR	0	100,000	200,000	300,000	400,000	500,000
TONS PER DAY [1]	0	320	640	960	1280	1600
POPULATION [2]	0	122,000	244,000	366,000	488,000	610,000

[1] Based on 6-day work week.
[2] Based on national average of 4.5 lbs per person per calendar day.

Fig. 4-18 Operating costs for sanitary landfills. *Source: Sanitary Landfill Facts*, USPHS, 1968.

fill Facts, a publication of the Office of Solid Waste Management Programs, United States Environmental Protection Agency.

4-11.   ECONOMICS OF SANITARY LANDFILLS

Landfill is generally the least expensive method of solid waste disposal used in the United States today. The low operating costs shown in Fig. 4-18 are postulated on existing land costs. Population growth and urban sprawl are removing land from consideration as landfill sites. Thus, disposal costs in landfills are likely to increase drastically in the future. Long-range planning should consider this circumstance.

4-11.1   Cost Analysis

Landfilling involves a number of direct costs, indirect costs, and intangible expenditures. Intangible expenditures cannot be evaluated in quantitative terms but they represent significant costs. These intangible losses include the economic impacts on land values, rental rates, etc. of the detrimental by-products of improper landfilling such as odor, pollution of groundwater, and aesthetic degradation. Such costs can be avoided by utilizing proper management and control methods. Other intangible costs cannot be avoided; an example is the restriction of future use of landfill sites and any economic losses consequent therefrom.

4-11.2   Indirect Costs

Since materials are lost and energy is not recovered in sanitary landfilling, the value of all materials placed in sanitary landfills should be considered indirect costs of this method. These costs will increase in relative proportion to virgin material costs as natural resources are exhausted. Also, with better materials recovery methods landfilling will become less attractive.

4-11.3   Direct Costs

Direct costs incurred in sanitary landfilling include land costs; equipment purchase costs, operating costs, and refitting costs; labor costs; and infrequent miscellaneous expenditures.

To illustrate the economics of sanitary landfilling, an example will be given. This analysis was made by Mr. Edward Ranck of the University of Louisville as part of a study of waste disposal in Jefferson County, Kentucky.

The cost estimates below are based upon the following assumptions which reflect broad estimates of circumstances as they occur in Jefferson County. Land costs include acquisition and required preparation such as grading an access road, constructing a fence for permanent control, and other initial construction tasks which might be required. Two alternative cost structures have been included, one based upon the ability to dig down to 22 ft below the surface to begin landfilling, the other based upon the ability to dig down only 5 ft. It is believed that these two estimates cover the range of alternatives available in Jefferson County and immediately adjacent areas. These estimates are based upon the requirements for serving a population of 350,000 persons and utilizing a given landfill site for 25 years.

BASIC ASSUMPTIONS

Population	350,000
Waste generated per day (including industrial)	1,890 tons
Volume	4,725 cu yd
Waste generated per year	689,850 tons
Volume	1,724,625 cu yd
Required acreage (assume 25 year life of site, 800 lb/yd, 22 ft depth of landfill)	2,050 acres
Required acreage assuming 5 ft depth (see Note 1)	8,200 acres
Land cost and preparation (this figure represents the highest estimated cost)[a]	$3,000/acre
Interest rate (municipal bonds)	$5\frac{1}{2}\%$

[a]It can realistically be assumed that lower land prices can be obtained but only by trading off advantageous location and potential depth of operation so this cost would probably occur in some form anyway.

LAND

Land acquisition and preparation is estimated at $6,150,000. Amortized (including interest) over the 25 year life of the landfill site, annual payments will be approximately $417,840 or about $.60 per ton. Although the land will have recovery value at the end of the 25 years, this cannot be ascertained at this time. This recovery, however, does represent an asset to the municipality.

LABOR

On the basis of information on labor requirements furnished by the Environmental Protection Agency, a model labor force was assembled. Current (1974) national average wage rates were applied to this work force. The labor costs determined by this method were approximately $0.40 per ton.

EQUIPMENT

On the basis of Table 4-1 and current manufacturers' list prices (1974) for equipment, the total equipment costs for the various alternative plans were estimated. The equipment to be used in each plan is shown below. These pieces of equipment have an estimated life of 10,000 hr or about 5 years of full-time operation.

Alternative I (1 site)

1 steel-wheeled compactor	$ 78,500
1 tractor-loader	82,000
1 track-type vehicle	45,500
1 rubber-tired scraper (including 6% for specialized equipment)	101,760
Total initial equipment outlay expense	$331,020
Annual amortization expense	$ 76,000
Per ton	$ .11

Alternative II (1 site)

2 steel-wheeled compactors (smaller)	$ 90,000
1 tractor-loader	82,000
1 track-type vehicle	45,500
1 rubber-tired scraper	101,760
Total initial equipment outlay	$342,520
Annual amortization expense	$ 78,840
Per ton	$ .11

Alternative III (5 smaller sites, each to accept approximately 140,000 tons over a 25-year period).

5 tractor loaders	$188,750
5 track-type vehicles	190,000
1 motor grader	32,000
Total	$410,750
Annual amortization expense	$ 93,984
Per ton	$ .14

MAINTENANCE EXPENSE

Maintenance and repair averages 16 to 18 percent of the cost of equipment annually.

	Estimated Annual Expense	Per ton
Alternative I (1 site)	$55,000	$0.08
Alternative II (1 site)	56,500	0.08
Alternative III (5 sites)	73,000	0.11

FUEL AND LUBRICATION

Fuel and lubrication usage based upon manufacturers estimates are expected to range from $20,000 to $25,000 annually or about $0.03 per ton.

SUMMARY OF DIRECT COSTS

(Assuming that land and labor requirements are approximately the same under all alternatives—see Note 1)

Alternative I (1 site)

	Annual	Per ton (689,850 tons annually)
Amortization expense	$417,840	$0.60
Labor	273,000	0.40
Equipment	76,000	0.11
Maintenance	55,000	0.08
Fuel and lubricants	20,000	0.03
Utilities and miscellaneous	30,000	0.04
Total	$871,840	$1.26

Alternative II (1 site)

Amortization expense	$417,840	$0.60
Labor	273,000	0.40
Equipment	78,840	0.11
Maintenance	56,500	0.08
Fuel and lubricants	22,500	0.03
Utilities and miscellaneous	39,000	0.04
Total	$878,680	$1.26

Alternative III (5 sites)

Amortization expense	$417,840	$0.60
Labor	273,000	0.40
Equipment	93,984	0.14
Maintenance	73,000	0.11
Fuel and lubricants	25,000	0.03
Utilities and miscellaneous	30,000	0.04
Total	$912,824	$1.32

Note 1: The assumption here has been that it would always be possible to dig down to 22 ft to begin landfilling and that land acquisition would average $3000/acre. If these assumptions were modified to show the range of possibilities, another set of assumptions should be included. The assumptions of 5 ft depth to bedrock and land acquisition at $1000/acre have been chosen to illustrate how amortization expense and therefore total costs would be affected.

Land required	8,200 acres		
Price/acre	1,000		
Total price	$8,200,000		

		Total	Per ton
Annual amortization change (increase)		$604,320	$0.88
in amortization expense		185,480	0.28

4-12. SUMMARY—SANITARY LANDFILL

The techniques of sanitary landfilling have been presented. This method is relatively simple, requires little or no capital expenditure for physical plant, and may require only short lead time if zoning restrictions are met. It is the least expensive method of waste disposal in current use in the United States; this statement obviously does not apply in areas where land is not available at low or moderate prices.

Materials (and energy potentially available from those materials) are wasted when refuse is deposited in a sanitary landfill. For this reason, and because of growing shortages of raw materials, sanitary landfill will be less attractive in the future.

Limiting method parameters have been described; detrimental by-products such as noxious gases and leachate are prime factors limiting landfill useage. These factors are susceptible to technical control. Other factors, such as adverse public opinion, are not subject to technical remedy and may be much more critical to the overall success of refuse disposal by sanitary landfilling.

REFERENCES

4-1. American Public Works Association, *Municipal Refuse Disposal*, 2nd ed., Interstate Printers and Publishers, Inc., Danville, Illinois, 1966.

4-2. Anderson, D. R., "Gas Generation and Movement in Landfills," *Proc.*, National Industrial Solid Waste Management Conference, University of Houston, Texas, 1969.

4-3. Anderson, J. R. and Dornbush, J. N., "Influence of Sanitary Landfill on Ground Water Quality," *Journal of the American Water Works Association*, Vol. 59, No. 4, April, 1967, pp. 457–470.

4-4. Barnes, A. M. and Black, R. J., "Effect of Earth Cover on Housefly Emergence," *Public Works*, Vol. 87, March, 1956, pp. 109–111.

4-5. Bradley, C. L., "Operating Sanitary Landfills During Extremely Cold Weather," *Public Works*, Vol. 9, November, 1960, p. 153.

4-6. California State Water Pollution Control Board, "Report on the Investigation of Leaching of a Sanitary Landfill," Publication No. 10, Sacramento, California, 1954.

4-7. California State Water Pollution Control Board, "Effects of Refuse Dumps on Ground Water Quality," Publication No. 24, Sacramento, California, 1961.

4-8. Caterpillar Tractor Company, "Recommended Standards for Sanitary Landfill Operations," Caterpillar Tractor Company Bulletin No. TEO80014.

4-9. Committee of Sanitary Landfill Practice of the Sanitary Engineering Division, "Sanitary Landfill," *American Society of Civil Engineers Manual of Engineering Practice*, No. 39, New York, 1959.

4-10. Dobson, A. L., Wilson, H. A. and Burchinal, J. C. "Factors Influencing Decomposition in Sanitary Landfills," *Bacteriological Proceedings*, A77, 1965.

4-11. Duszynski, E. J., "Progress Report on Heil Gondard Refuse Reduction Mill in Madison, Wisconsin," paper presented to Institute for Solid Wastes, Boston, Mass., October 5, 1967.

4-12. Hagerty, D. J., Pavoni, J. L. and Heer, J. E., Jr., *Solid Waste Management*, Van Nostrand Reinhold, New York, 1973.

4-13. Ham, R. K., "A Study of the Relative Attractiveness of Milled vs. Unmilled Refuse to Rats and Flies," Report on the City of Madison—Heil Gondard Demonstration Project, University of Wisconsin, Madison, Wisconsin, 1969.

4-14. Hughes, G. M., Landon, R. A. and Farvolden, R. N., "Hydrogeologic Data From Four Landfills in Northeastern Illinois," Environmental Geology Notes, Illinois State Geological Survey, March, 1969.

4-15. "Landfill Operations by Los Angeles County," *Public Works*, Vol. 91, September, 1960, pp. 122–123.

4-16. Louisville, University of, Institute of Industrial Research, *Falls of the Ohio Metropolitan Region Solid Wastes Disposal Study*, Vol. I, April, 1969.

4-17. "Measuring Gas Escape From a Landfill," *Public Works*, Vol. 98, December, 1967, pp. 86–87.

4-18. Merz, R. C. and Stone, R., "Gas Production in a Sanitary Landfill," *Public Works*, Vol. 95, February, 1964, pp. 84–87.

4-19. Merz, R. C. and Stone, R., "Landfill Settlement Rates," *Public Works*, Vol. 93, September, 1962, pp. 103–106.

4-20. Muhich, A. J., Klee, A. J. and Gritton, P. W., *1968 National Survey of Community Solid Waste Practices: Preliminary Data Analysis*, USPHS Publ. No. 1867, Govt. Printing Office, 1968.

4-21. National Association of Counties Research Foundation, "Personnel," *Solid Waste Management*, Vol. 9 of 10, Washington, D.C., 1968.

4-22. "Pollution of Subsurface Water by Sanitary Landfills; a Summary on a Solid Waste Demonstration Grant Project," USPHS, HEW, Cincinnati, 1968.

4-23. Remson, I., Fungaroli, A. S. and Alonzo, A. W., "Water Movement in an Unsaturated Sanitary Landfill," *Proc., American Society of Civil Engineers, Journal of the Sanitary Engineering Division*, Vol. 94, No. SA2, April, 1968, pp. 307–317.

4-24. Rogus, C. A., "Use of Completed Sanitary Landfill Sites," *American Public Works Association Yearbook*, 1959.

4-25. Salvato, J. A., "Sanitary Landfill, Planning, Design, and Operation," *Public Works*, Vol. 101, February, 1970, pp. 93–97.

4-26. "Sanitary Landfill," *American Society of Civil Engineers, Manual of Engineering Practice*, No. 39, 1959.

4-27. Sorg, T. J. and Hickman, H. L., "Sanitary Landfill Facts," HEW, USPHS Publ. No. 1792, Govt. Printing Office, 1968.

4-28. Sowers, G. F., "Foundation Problems in Sanitary Landfills," *American Society of Civil Engineers, Journal of the Sanitary Engineering Division*, Vol. 94, No. SA2, February, 1968.

4-29.　Wolfe, H. B. and Zinn, R. E., "Systems Analysis of Solid Waste Disposal Problems," *Public Works*, Vol. 98, September, 1967.

4-30.　Zausner, E. R., *An Accounting System for Sanitary Landfill Operations*, Office of Solid Waste Management Programs, Cincinnati, 1969.

APPENDIX 4-1　BIBLIOGRAPHY

Because of the vast amount of literature available concerning the most widely used solid waste disposal technique, the sanitary landfill, a selection of landfilling references not previously cited in Chapter 4 is presented in this appendix.

Alexander, M., *Introduction to Soil Microbiology*, John Wiley and Sons, Inc., New York, 1961.

American Chemical Society, "Cleaning our Environment: The Chemical Basis Basis for Action," 1970.

Anderson, D. R., Bishop, W. D. and Ludwig, H. F., "Percolation of Citrus Wastes Through Soil," *Proc.*, 21st Industrial Waste Conference, Purdue University, Lafayette, Indiana, 1966.

Barnes, A. M. and Black, R. J., "Effect of Earth Cover on Fly Emergence from Sanitary Landfills," *Public Works*, Vol. 89, February, 1958, pp. 91–94.

Bishop, W. P., et al., "Gas Movement in Landfill Rubbish," *Public Works*, Vol. 96, November, 1965, pp. 64–68.

Bjornson, B. F. and Bogue, M. D., "Keeping a Sanitary Landfill Sanitary," *Public Works*, Vol. 92, September, 1961, pp. 112–114.

Bjornson, B. F., Pratt, H. D. and Littig, K. S., "Control of Domestic Rats and Mice," USPHS Publ. No. 563, Govt. Printing Office, 1968.

Bogue, D. and Boston, R. J., "Solid Waste Disposal; a New Area of Pollution," *Georgia Municipal Journal*, Vol. 18, February, 1968, pp. 14–15.

Bowers, N. A., "Sanitary Fill Proves Out for Plant Sites: Transforming San Francisco Tideland," *Engineering News-Record*, Vol. 161, December, 1958, pp. 44–46.

Brinkley, G. C., "Perimeter Sanitary Landfill for Dallas Refuse Disposal," *Public Works*, Vol. 87, September, 1956, pp. 102–103.

Bronow, John A., "U-Blade Dozer Speeds Refuse Handling," *Public Works*, Vol. 5, January, 1962, p. 82.

Burges, A., *Microorganisms in the Soil*, Hutchinson & Co., Ltd., London, 1958.

Burleson, R. A., "Landfill for a Wheel Loader," *Public Works*, Vol. 93, November, 1962, p. 113.

California State Water Quality Control Board, "Waste Water Reclamation in Relation to Ground Water Pollution," Publication No. 24, Sacramento, 1953.

California, University of, "Analysis of Refuse Collection and Sanitary Landfill Disposal," Technical Bulletin No. 8, Sanitary Engineering Research Project, Richmond, University of California, Section 37, December, 1952.

Cannella, A. A., "The Refuse Disposal Problem" *Public Works*, Vol. 99, February, 1968.

Chanin, G., "Decomposition Efficiency of Sanitary Landfills," *Public Works*, Vol. 87, February, 1956, pp. 102–103.

Clark, R. M. and Toftner, R. O., "Land Use Planning and Solid Waste Management," *Public Works*, Vol. 103, March, 1972, pp. 79–80, 98.

Committee on Sanitary Engineering Research, "Survey of Sanitary Landfill Practices: Thirtieth Progress Report," *Journal of the Sanitary Engineering Division, American Society of Civil Engineers*, Vol. 87, No. SA4, July, 1961, pp. 65–84.

Cook, H. A., Cromwell, D. L. and Wilson, J. A., "Microorganisms in Household Refuse and Seepage Water from Sanitary Landfills," *Proc.*, West Virginia Academy of Sciences, Vol. 39, No. 107, 1964.

"County Landfill Reclaims Marshlands," *Public Works*, Vol. 94, March, 1963, p. 117.

"County Runs Economical Landfills," *Public Works*, Vol. 92, August, 1961, p. 109.

Darnay, A. and Franklin, W. E., "The Role of Packaging in Solid Waste Management 1966–1976," Midwest Research Institute, HEW, USPHS, 1969.

"Dealing with Domestic Refuse," *Engineering*, Vol. 203, No. 6267, March, 1967, pp. 499–508.

Department of County Engineer, County of Los Angeles, "Development of Construction and Use Criteria for Sanitary Landfills," HEW, Cincinnati, 1969.

Diven, M. E. "A Study of Refuse Collection and Disposal Service," *Public Works*, Vol. 91, October, 1960, p. 129.

Division of Environmental Health, Solid Waste Program, Kentucky Department of Health, "Kentucky Solid Waste Disposal Laws and Regulations," Publication PAM SW-1, Frankfort, Kentucky, July, 1968.

Division of Sanitary Engineering, Indiana State Board of Health, *Manual for Storage, Collection, and Sanitary Landfill Disposal of Refuse*, Indianapolis, July, 1969.

Division of Sanitary Engineering, Indiana State Board of Health, *Standards for the Selection, Operation, and Maintenance of a Sanitary Landfill*, Indianapolis, October, 1968.

Dunn, W. L., "Landfill Burned for Odor Control," *Civil Engineering*, Vol. 27, November, 1957, pp. 790–791.

Dunn, W. L., "Storm Drainage and Gas Burning at Refuse Disposal Site," *Civil Engineering*, Vol. 30, August, 1960, pp. 68–69.

"Effects of Refuse Dumps on Ground Water Quality; Abstract," *Public Works*, Vol. 94, January, 1963, p. 142–143.

Ehlers, V. M., and Steel, E. W., *Municipal and Rural Sanitation*, McGraw-Hill Book Co., New York, 1965.

Engineering-Science, Inc., "In-Situ Investigation of Movements of Gases Produced from Decomposing Refuse," Fifth and Final Annual Report Prepared for State Water Quality Control Board of California, November, 1961.

"Fundamentals of Sanitary Landfill Operation," *Public Works*, Vol. 95, December, 1964, pp. 88–89.

Gilbertson, W. E., "Scope of the Solid Waste Problem," *Journal of the Sanitary Engineering Division, Proceedings of the American Society of Civil Engineers*, October, 1966.

Goodrow, T. E., "Sanitary Landfill Becomes Major League Training Field," *Public Works*, Vol. 96, August, 1965, pp. 124–126.

Hanna, M., "Municipal Golf Course Built on Waste Land," *Public Works*, Vol. 87, May, 1956, p. 155.

Harza Engineering Co., "Land Reclamation Project; An Interim Report," HEW, Cincinnati, 1968.

Hellbusch, R. A., "From Landfill to Landscape," *Public Works*, Vol. 100, June, 1969, pp. 102–103.

Holhuezer, O., "Steel Wheel Dozer Improve Landfill Compaction," *Public Works*, Vol. 98, April, 1967.

Horbitz, W. E., "Looking to the Future with a Regional Refuse Disposal Plan," *Public Works*, Vol. 98, June, 1967, pp. 120–121.

"How to Plan a Successful Landfill Program," *Public Works*, Vol. 87, November, 1956, p. 78.

Hughes, George M., "Selection of Refuse Disposal Sites in Northeastern Illinois," Environmental Geology Notes, Illinois State Geologic Survey, September, 1967.

Jamison, H. M., "Several Methods Available for Solid Waste Disposal," *Water and Sewage Works*, Vol. 116, No. IV, July, 1969, pp. 14–15.

Knudsen, E. J., "Village of 16,000 Saves $24,000 per Year With Sanitary Fill," *Public Works*, Vol. 86, February, 1955, p. 81.

Koch, A. S., "Sanitary Landfill Lives up to County's Expectations," *Public Works*, Vol. 96, July, 1965, pp. 70–71.

LeGrand, H. E., "Management Aspects of Ground Water Contamination," *Water Pollution Control Federation Journal*, Vol. 36, September, 1964, pp. 1133–1145.

LeGrand, H. E., "System for Evaluation of Contamination Potential of Some Waste Disposal Sites," *American Water Works Association Journal*, Vol. 56, August, 1964, pp. 959–974.

McKinney, R. E., "The Challenge of Solid Waste Research," *Journal of the Sanitary Engineering Division, Proceedings of the American Society of Civil Engineers*, October, 1966.

"Measuring Gas Escape From a Landfill," *Public Works*, Vol. 95, September, 1964, p. 163.

Merz, R. C. and Stone, R., "Progress Report on Study of Percolation Through a Landfill," *Public Works*, Vol. 98, December, 1967, pp. 86–87.

Merz, R. C. and Stone, R., "Sanitary Landfill Behavior in an Aerobic Environment," *Public Works*, Vol. 97, January, 1966, pp. 67–70.

Morris, W., "Burning Helps Conserve Landfill Area," *Public Works*, Vol. 92, September, 1961, p. 215.

National Association of County Engineers, "Location of Utilities-Refuse Disposal," *County Development*, Vol. 3 of 3, Washington, D.C., 1966.

National Association of Counties Research Foundation, "Design and Operation," *Solid Waste Management*, Vol. 5 of 10, Washington, D.C., 1968.

National Association of Counties Research Foundation, "Financing," *Solid Waste Management*, Vol. 6 of 10, Washington, D.C., 1968.

National Solid Waste Management Association and Federal Solid Waste Management Program, "Recommended Standards for Sanitary Landfill Design, Construction, and Evaluation & Model Sanitary Landfill Operation Agreement," Govt. Printing Office, 1971.

"Operation of Sanitary Landfills by Counties," *Public Works*, Vol. 90, February, 1959, pp. 125–127.

"Operation of Sanitary Landfills," *Public Works*, Vol. 89, September, 1958, pp. 115–117.

Paino, B. and Olsen, C. E., "Odor and Insect Control at an Open Landfill Operation," *Public Works*, Vol. 90, April, 1959, pp. 115.

Qasim, S. R., "Chemical Characteristics of Seepage Water from Simulated Landfills," Ph.D. Dissertation, West Virginia University, Morgantown, 1965.

Qasim, S. R. and Burchinal, J. C., "Leaching of Pollutants from Refuse Beds," *American Society of Civil Engineers, Journal of the Sanitary Engineering Division*, Vol. 96, No. 541, February, 1970, pp. 49–58.

Rawn, A. M., "Planned Refuse Disposal for Los Angeles County," *Civil Engineering*, Vol. 26, No. 4, April, 1956, pp. 41–45.

"Refuse Collection and Disposal in Albuquerque," *Public Works*, Vol. 89, June, 1958, p. 110.

"Refuse Volume Reduction in a Sanitary Landfill, 26th Progress Report of Commission on Sanitary Engineering," *Research of the Sanitary Engineering Division, American Society of Civil Engineers, Proceedings*, Vol. 85, November, 1959, pp. 37–50.

Rogers, P. A., "Status of Solid Waste Program in California," *Public Works*, Vol. 100, May, 1969, pp. 80–83.

Rogus, C., "Refuse Collection and Disposal in Western Europe; Port III—Salvaging, Landfilling, and Composting," *Public Works*, Vol. 93, June, 1962, pp. 139–143.

Rogus, C. A., "Sanitary Refuse Fills in Wet Areas," *Public Works*, Vol. 86, December, 1955, p. 67.

"Sanitary Fill Reclaims Valuable Land; Bordentown, New Jersey," *Public Works*, Vol. 91, April, 1960, p. 156.

"Sanitary Landfill Costs 68 Cents per Ton for Disposal," *Public Works*, Vol. 88, April, 1957, p. 177.

"Sanitary Landfill Operations Costs and Revenues," *Public Works*, Vol. 93, May, 1962, p. 191.

"Sanitary Landfill Projects Serve Both City and Contractor," *Public Works*, Vol. 89, August, 1958, p. 111.

Scates, C. H., "Sanitary Fill Ends 50 Years of Public Nuisance; Union City, Tennessee," *Public Works*, Vol. 88, October, 1957, p. 135.

Spornicha, J. E., "Is Sanitary Landfill Right for Your Community?" *Public Works*, Vol. 86, January, 1955, pp. 90–92.

Smith, C. D., "Municipal Refuse, A Low-Cost Resource: Sanitary Landfill to Construct Addition to Park," *Public Works*, Vol. 88, July, 1957, p. 94.

"Status of Refuse Collection and Disposal; Sanitary Engineering Research Commission Report," *American Society of Civil Engineers Proceedings, Journal of the Sanitary Engineering Division*, Vol. 83, February, 1957, pp. 1–7.

Ralph Stone and Company, Inc. Engineers, "Solid Wastes Landfill Stabilization," An Interim Report for the City of Santa Clara, California, 1968.

Stone, R. and Conrad, E. T., "Landfill Compaction Efficiency," *Public Works,* Vol. 100, May, 1969, pp. 111–113.

Stone, R. and Friedland, H., "National Survey of Sanitary Landfill Practices," *Public Works,* Vol. 100, August, 1969, pp. 88–89.

"A Study of a Metropolitan Solid Waste Program," *Public Works,* Vol. 100, March, 1969, pp. 78–79.

"Survey of Sanitary Landfill Practices: Progress Report, Commission on Sanitary Engineering Research," *American Society of Civil Engineers, Journal of the Sanitary Engineering Division,* Vol. 87, July, 1961, pp. 65–81.

Thompson, P., "Another County Operates a Sanitary Landfill; Douglas County, Washington," *Public Works,* Vol. 91, March, 1960, pp. 125–126.

"Transfer Station Assists Refuse Disposal," *Public Works,* Vol. 100, January, 1969, pp. 74–76.

Van Kleech, L. W., "Safety Practices at Sanitary Landfills," *Public Works,* Vol. 90, August, 1959, p. 113.

Vaughn, R. D., "Federal Solid Waste Program," *Civil Engineering,* Vol. 39, February, 1969, pp. 69–71.

Vaughn, R. D. and Black, R. J., "Public Health Service Solid Waste Program," *Journal of the Sanitary Engineering Division, American Society of Civil Engineers,* Vol. 95, April, 1969, pp. 233–238.

Williams, E. R., et al., "Effect of Adding Water to Municipal Refuse for Increased Compaction in Sanitary Landfills," *Public Works,* Vol. 90, October, 1959, p. 129.

Williams, E. R., et al., "Light Equipment for Small Town Sanitary Landfill Operations," *Public Works,* Vol. 89, February, 1958, pp. 89–91.

5

Innovations
in disposal

Although the traditional methods of solid waste disposal
(composting, incineration, and sanitary landfill) have been
practiced for many years, they are not without serious operational problems. Inadequate markets for compost have virtually
eliminated composting as a viable disposal technique in the
United States. With the passage of stringent air quality standards, incineration has become an increasingly expensive disposal technique. Furthermore, the volume reduction provided
by conventional incineration (approximately 80 percent)
necessitates the need for sizeable landfills for residue disposal.
The sanitary landfill process is also plagued by an increasing
scarcity of adequate land areas, groundwater pollution through
leachate generation, and excessive settlement precluding the
adoption of many valuable land uses following landfill
completion.

Consequently, a need exists for either upgrading the three
conventional disposal methods or developing innovations in
disposal techniques. Two recently developed innovations in

solid waste disposal which appear to hold some promise are the medium-temperature and high-temperature incineration process and the landfill with leachate recirculation process. This chapter briefly describes each process operation and outlines the advantages, disadvantages, and possible applications of the two systems.

5-1. MEDIUM-TEMPERATURE AND HIGH-TEMPERATURE INCINERATION

An innovation in disposal of solid wastes which has not yet developed to anticipated potential is incineration at elevated temperatures. The use of auxiliary fuel to create temperatures in the range of 2800 to 3200°F in a refuse combustion chamber originated in Western Europe. The purpose of this high-temperature environment is the achievement of so-called "total combustion" wherein bulky items, metals, glass, and other hard-to-reduce materials are either combusted or melted. Proponents of this technique claim that residue from high-temperature incineration amounts to only about 2–3 percent of the input refuse, as compared to residues of 20 percent (by volume) or more for conventional incineration practice. Moreover, the residue is said to be totally inert, as indicated by comprehensive laboratory tests by Fife (Ref. 5-10). Potential advantages outlined by Hagerty, et al. (Ref. 5-15) for high-temperature combustors (also known as "slagging" incinerators) include:

1. Acceptance of virtually all types of municipal refuse.
2. Reduction in generation of air pollutants.
3. Possible utilization of the residue produced during the combustion/fusion process.
4. Reduced process water requirements.

Disadvantages noted by Hershaft (Ref. 5-18) for developed processes include the need for auxiliary fuels, increased rate of refractory deterioration in the slagging units, and increased emission of NO_x gases as compared to conventional incinerators.

Several high-temperature incinerators are presently in various stages of development. The Ferro-Tech unit, the Sira unit, and the University of Hartford (Connecticut) method are commercially available (Ref. 5-18). Among the systems generally described as high-temperature incineration methods are the Torrax system, the Dravo/FLK system, and the American Thermogen system. The Torrax system includes a vertical "gasifier" wherein virtual pyrolysis occurs. Consequently, the Torrax system is discussed in the section of this work devoted to energy recovery (Chapter 7). The Dravo/FLK system includes a shredding operation wherein refuse is pre-sized before it is fed into a combustion chamber. A molten residue is produced in the Dravo/FLK unit, similar to the residue from the Torrax bottom zone (Ref. 5-18). A 100 ton per day Dravo plant has been in operation since 1972 at the Volkswagen works in Wolfsburg, West Germany. The American Thermogen, Inc. (ATI) unit was named the "Melt-Zit" incinerator. A pilot plant unit built in Whitman, Massachusetts, in 1966 by ATI was the first high-temperature incineration pilot plant built in the United States. It is discussed in detail in later paragraphs in this chapter.

Several other units have been termed "high-temperature" incinerators even though they operate at temperatures only slightly above the operating temperatures of conventional incinerators and well below the 3000°F range of the slagging incinerators. Typical of such units is the General Electric vortex pilot plant incinerator built at Shellbyville, Indiana. This unit also will be described in detail in this chapter.

5-1.1   Critical and Limiting Process Parameters

The application of high-temperature incineration to solid waste processing is limited by many of the same factors which limit conventional incineration. These factors include problems in the combustion process itself such as variable waste composition (variable heat content) and refractory wear as well as difficulties in controlling process emissions. Refractory wear is accelerated at the high temperatures employed in slagging incinerators. The operation of a high-temperature unit relies upon heating of refuse by extremely hot refractory walls while it is dropped into the furnace. Thermal shock is great, therefore, if wet solid wastes or similar low heat content material drops into the high-temperature furnace. Because of this problem of occasional low heat wastes, auxiliary fuels and the equipment and space for such fuels is a definite process limitation. Fuels such as oil or gas are easy to convey and control, but are expensive.

The need for high temperatures in the combustion chamber limit the utility of slagging incinerators in another way: waste heat cannot be recovered by means of efficient water-tube walls. Steam can be produced in "add-on" boilers attached to the basic system downstream from the secondary combustion chamber, but these boilers are less efficient than water-tube wall furnaces. Steam production in slagging incinerators has been estimated at 3.3 lb of steam per pound of refuse incinerated (Ref. 5-10).

The production of nitrogen oxides and increased particle emission as a result of high-temperature combustion are definite limiting factors in slagging incinerator application. These problems are described in more detail in the next section.

Finally, a limiting factor in the application of high-temperature incineration is the need for a number of skilled laborers to control the combustion process. Control of high-temperature incineration must be more precise than control of conventional incineration in order to obtain the maximum degree of combustion and the minimum production of residue. This close control requires skilled operators.

The above-mentioned limiting factors must be balanced against the advantages of high percentage volume reduction, elimination of incompletely oxidized hydrocarbons, production of inert residue, and the potential reclamation value of residue.

5-1.2 Components of Slagging Incinerators

In addition to the dumping floor, storage pit, etc. which are found in conventional incinerators, a slagging incinerator must also be furnished with facilities to convey and handle auxiliary fuel. Likewise, because the incineration process is zoned in most commercially available units, furnace construction is somewhat different than that of a conventional furnace. Commonly, the refuse is dropped into an upper ignition zone and falls into a "pyrolyzing" zone where oxygen is deficient to some extent and where combustion takes place for the most part; residue from the middle zone then falls into a lower high-intensity zone where it fuses. While these zones cannot be considered "components" of a high-temperature unit, the concept of zonation explains the configuration of such a unit.

The aforementioned auxiliary fuel systems may be conveyors or pneumatic feed lines if the fuel is coke or coal, or piping systems if the fuel is oil or gas. These auxiliary systems represent significant expenditures in addition to the furnace unit proper.

Residue recovery units for high-temperature incinerators differ considerably from the residue recovery devices employed in many conventional incinerators because:

1. The volume of slag produced in a high-temperature unit is only about ¼ to ⅙ of the volume of residue per ton of waste incinerated in a conventional unit.
2. The slag is in a molten liquid form rather than in solid form. (In medium-temperature incinerators such as the General Electric experimental vortex unit described in a later section, slag accumulates in the furnace and is removed periodically after the furnace is shut down and the material solidifies.)

Continuous casting of conventional shapes of molten slag has been proposed for some units. Residue processing systems will require engineering design and skilled operation if maximum returns are to be realized from recovery operations.

5-1.3 Process By-Products

As in conventional incineration, the by-products of high-temperature combustion are heat, particulate matter, effluent gases, and residue.

Heat recovery from slagging units has been discussed as has the effect of higher temperatures on refractory deterioration.

Because of the vertical configuration of some units and the drop-feeding of refuse into the furnaces, particulate emissions from a high-temperature furnace may exceed rates of particulate emission from conventional incinerators. Particulate collection may be doubly difficult in slagging units because of the heavy particulate loadings and the high temperatures of the effluent stream which render collection more difficult.

The nature of the gaseous combustion products change with increase in temperature. Combustion at temperatures in the range of 2800 to 3200°F will produce nitrogen oxides. NO_x emissions react with other atmospheric constituents in the presence of sunlight to create a number of serious secondary air pollutants. Therefore, the production of nitrogen oxides is a serious drawback to the use of high-temperature incinerators. Prevention of NO_x emissions from slagging units will necessitate the installation and use of complicated and/or expensive air pollution control devices.

The residue produced during slagging incineration has been described in previous paragraphs. In summary, this residue is formed as a granulate frit or as drops and globules of metal and glassy solid. The sale of this residue is expected to furnish income to the incinerator operation; however, as in the sale of other residues, markets for this material must be developed. Postquenching treatment such as grinding or pulverization may be required to render the residue suitable for secondary usage.

5-1.4 Economics

It is impossible to reliably forecast costs for high-temperature incineration operations at the present time because practically no commercial experience has been gained with such units. Preliminary tests indicate that capital costs for slagging units will be slightly higher than for conventional units because of the necessary auxiliary fuel system. Net operating costs will depend upon the resale value of residue and steam generated from recovered heat.

5-1.5 Process Case Histories

Two process case histories are described below to illustrate the characteristics of high-temperature and medium-temperature incinerators.

The Melt-Zit Incinerator. The "high-temperature incinerator" is typified by the Melt-Zit model of American Thermogen (American Design and Development Corporation). In this incinerator, the noncombustible fractions of refuse are melted in a bed of high-temperature coke and are then drained from the furnace area as molten slag and iron. The residue remaining after incineration is completely sterile, is inert, and possesses a high density. Figure 5-1 shows the features of the Melt-Zit pilot plant furnace built at Whitman, Massachusetts as a demonstration model. The components of the incinerator are the refractory-lined furnace, the refuse charging conveyor, the coke elevator, and the limestone and air charging mechanisms. The limestone, coke, and refuse are dropped into the furnace from a conveyor; each constituent is charged separately into the furnace as required by furnace performance. Air is supplied at two levels by means of air ports spaced around the furnace. Air at the lower level is fed into the coke bed and helps to produce the high temperatures required for the melting of slag from the refuse. The construction of the furnace is such that it induces a slight negative pressure at the input chute, and as a result additional air is drawn through the charging opening. The upward rush of hot gases causes partial entrainment of the solids which are dumped from the conveyor. The refuse is burned partly in suspension within the shaft of the furnace, and partly on the surface of the lower coke bed. Slag resulting from the melting of the noncombustible portions of the refuse is drained from the bottom of the incinerator through a taphole located at the level of a sand hearth which underlies the coke bed. The drained slag flows out of the furnace into a water reservoir where it immediately congeals into a black granulate frit. The furnace is put into operation in stages as follows:

1. A sand bed is laid at the bottom of the furnace and is sloped toward the slag spout;
2. Coke is laid over the sand hearth to a depth of 3–4 ft;
3. The coke is ignited and is gradually brought to operating temperature by means of forced air supplied through the air ports;
4. Congealed slag from a previous run is charged into the furnace to establish a flow of slag from the spout as an indication that the coke bed is at a sufficiently high operating temperature;
5. After the proper temperature has been developed in the coke bed, air is forced into the upper air ports over the coke bed and refuse combustion is begun.

The refuse charging is begun in small intermittent batches until the process has been well established, after which the rate of feed becomes continuous.

Coke is charged periodically into the furnace to maintain the desired thickness of 3–4 ft of the coke bed. The refuse feeding ordinarily is interrupted for approximately 5 minutes every half hour, at which time the coke bed is charged and

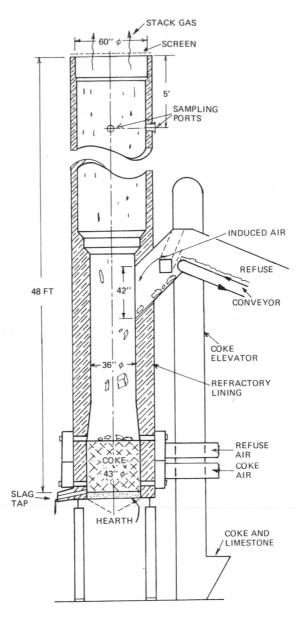

Fig. 5-1 Melt-Zit incinerator. *Source:* "Evaluation of the Melt-Zit High-Temperature Incinerator," USPHS, Cincinnati, 1969.

heated. When coke is charged into the underlying bed, the top of the coke bed ceases to radiate a large amount of heat, but radiation from the refractory lining reestablishes ignition promptly. The temperature of the coke bed is soon restored. The refuse layer is maintained on the coke at a thickness of only a few inches so that the flow of gases from the bed is not unduly interrupted. At the end of the incineration run, the bottom of the furnace is opened and the remaining coke, slag, and molten metal are dropped into a refuse container.

The USPHS supervised a series of tests of the Whitman, Massachusetts pilot plant. In a critical evaluation (Ref. 5-9, 5-27) of the Melt-Zit incinerator, several process problem areas were outlined, including:

1. A low refuse firing rate, approximately 1.6 tons/hr.
2. A relatively high consumption of coke, at least 1 ton of coke for every 6–8 tons of refuse incinerated.
3. An inability to drain molten slag and iron in order to maintain continuity of operation.
4. Consumption of limestone—1 ton for each 38 tons of refuse to promote fluidity of the slag.
5. Flyash which amounted to approximately 10 percent of the refuse charged.
6. A high labor requirement, at least 3 men per shift.
7. A high consumption of oxygen and iron pipe lances required to maintain the slag flow. In addition, the refractory lining in the furnace was attacked by the high temperatures in the slag ingredients.

Several distinct process advantages were also noted in this USPHS critique including:

1. A molten residue was received from the incinerator free of putrescible and combustible materials; thus, the residue could be deposited in a landfill without cover, would not attract vermin or rodents, and would be odorless and sterile.
2. A high yield of iron was obtained from the refuse as flattened or spherical drops if the molten slag was allowed to run into a water tank. During test runs, about 20 percent of the residual material consisted of iron.

The Melt-Zit by itself is only the primary combustion chamber for a total incineration plant. A secondary chamber would be required for complete combustion of the carry-over solids which move rapidly up in the hot gas stream. A dust collector would also be required after the secondary chamber in order to reduce the particulate load of the exhaust gases. The economics of this system are comparable to those of conventional municipal incinerators. A high-temperature plant requires virtually the same land area, buildings, roadways, handling equipment, secondary combustion chamber, and other appurtenances as a conventional incinerator. In addition, other units for receiving, storing, and conveying coke and limestone would be required. The cost of the stoker and hydraulic feed drive would be saved in the high-temperature apparatus. Only small costs would be incurred in a granulator and conveyor for the frit recovered from the quenching tank; this additional expense could be met through sale of the end products. An additional savings for the high-temperature incinerator would be the much smaller water requirement as opposed to conventional incinerators.

Capital costs for municipal incinerators are currently in the range of $7500 to $10,000 per ton/day of capacity. Present estimates for costs of high-temperature incinerators indicate that the cost for a comparable plant for the same site would be 90–100 percent of that of a conventional incinerator.

In respect to operating expenses, little savings are anticipated in operating a high-temperature incinerator. In addition, receiving, storing, and conveying coke and limestone would require additional manpower over and above the requirements for the conventional apparatus. Finally, the high-temperature incinerator incurs the cost of coke and limestone, which are not common to other incinerators. Disposal of the final residue of the Melt-Zit process would be less expensive than disposal of conventional incinerator residue. Since it is neither combustible nor putrescible, this residue could be deposited anywhere without a daily cover of soil. Sale of the metal pellets and the slag and frit at the present time is a highly questionable venture. The frit produced during high-temperature incineration is friable and crushes readily underfoot. Its strength is much lower than that of sand or stone and it could not be used to perform in structural functions.

In comparison with conventional incinerators, however, the Melt-Zit process is very economical as to water usage. Water used in the cleaning of flue gases would be slightly more than that for a conventional plant of the same tonnage because of the additional heat from the coke. However, runoff water from flue gas scrubbers could also be used for the quenching operation. Little or no process water would be discharged to sewers or streams.

As a result of preliminary tests and evaluations, the pilot plant unit was modified in 1969 to use gas or oil as auxiliary fuel to replace coke. Additionally, the primary combustion zone was enlarged and a steam injection system was installed to replace the original air pollution control equipment.

The Shelbyville, Indiana GE Pilot Plant. General Electric Company has built a pilot plant vortex incinerator at Shelbyville, Indiana. EPA officials have expressed considerable interest in this facility (Ref. 5-8). Many persons have referred to this device as a "high-temperature" incinerator. Actually, it is operated at temperatures only moderately higher than conventional incinerator temperatures. Also, the GE unit is designed to operate on the separated organic fraction only; inert materials such as glass and metals are air-separated from the combustible fraction of the refuse prior to burning. It is a tangentially-fired horizontal cylindrical vortex furnace similar to many industrial furnaces. No grates are used in the furnace; air flow is used to move unburned ash to the outlet end of the furnace where a transverse air stream blows the ash out the side of the furnace.

The refuse is delivered onto a dumping floor at the western end of the incinerator building. A front-end loader is used to place the refuse on a conveyor. A laborer must manually remove large, bulky, or hard-to-shred items from the first conveyor unit. The conveyor dumps the sorted refuse into an initial grinder which breaks up bundles, boxes, and bags of refuse. The refuse is then transported via an inclined conveyor into an inclined, rotating drum. Heavy materials travel down-slope and out of the drum onto a reject conveyor which carries them into a bin. Light materials "walk" up the inclined drum under the influence of an air draft and are dumped onto the main refuse conveyor. This conveyor carries the refuse to the principal shredder which reduces particle size to 3 in. or less. The reduced refuse leaves the shredder on a vibrating conveyor which carries it into a materials-handling fan. The fan blows the refuse up a curved, inclined conduit into the furnace unit. In the furnace, a vortex is created by air blown into the bottom of the furnace tangentially from a plenum supplied by an external FD fan. The refuse particles swirl around the chamber and burn quickly in the 1600-1800°F heat. Ash is blown out of the furnace at the bottom of the unit (tangentially) near the outlet to the stack. Air is added, by natural draft, to the ash stream to cool it and the cooler ash is carried into a cyclone separator. Air (and carry-over ash) leaves the cyclone and is returned to the furnace. Exhaust gases are vented from the furnace through a natural-draft stack.

The operation at Shelbyville has a capacity of 5 to 6 tons per hr. For each ton of shredded refuse which enters the furnace, approximately 20-25 lb of clinker are formed in the furnace. Residue ash amounts to 90-100 lb per input ton of shredded refuse. No air pollution control devices or techniques are being used at the present time (1974). A significant percentage of the rejected material is heavy food wastes such as potatoes, oranges, etc. All rejected material requires landfill or similar disposal. Approximately 30-40 percent of the delivered wastes are rejected in this system. The furnace requires as much as 1½ hr of warm-up in which oil and/or natural gas are used. Gas costs have been estimated at $2.85/ton of refuse burned ($0.85/1000 cu ft of gas).

At the present time the Shelbyville incinerator is considered an experimental device not as yet suitable for full-scale application. The General Electric Company is working to develop better materials handling systems and more reliable operating procedures. The existing facility seems to offer little advantage over conventional incineration installations.

5-2. LANDFILL WITH LEACHATE RECIRCULATION

Although solid waste disposal in sanitary landfills is one of the simplest and most economical methods of waste management, several distinct disadvantages of this disposal technique must be considered. First, the possibility of groundwater contamination exists through the leaching of deposited contaminants by percolating waters. Second, the slow rate of waste degradation and stabilization in a sanitary landfill may effectively preclude future use of the disposal site for long periods of time. In addition, the dry, loose condition of deposited refuse makes on-site compaction necessary and difficult, with resultant low densities (and high cost per ton in space used) in the completed fill. To overcome these disadvantages, several variant techniques of landfilling have been proposed. One of the most promising new techniques for improving the methods of waste disposal in sanitary landfills is the recirculation of leachate through deposited refuse with ultimate treatment and disposal of the collected leachate after waste stabilization.

5-2.1 Characteristics of Leachate from Conventional Sanitary Landfills

Numerous investigations have been conducted in laboratories and field situations concerning the generation of leachate from sanitary landfills. Deterioration of groundwater quality below and around landfills has been reported by Calvert (Ref. 5-2), Carpenter and Setter (Ref. 5-3), Lang (Ref. 5-23), Anderson and Dornbush (Ref. 5-1), Hughes, et al. (Refs. 5-20, 5-21, 5-22), Merz and Stone (Ref. 5-25), Davison (Ref. 5-6), Coe (Ref. 5-5), Hopkins and Papalisky (Ref. 5-19), and Fungaroli (Ref. 5-12).

In general, these studies have shown increases in mineral constituent concentrations (hardness, alkalinity, etc.) and organic content (BOD, COD, etc.) in groundwater which is in contact with deposited refuse. For example, Hughes (Ref. 5-20) reported that groundwater mounds formed below four investigated fills; he further found that groundwater quality increased with age of fill material and distance from the fill. Tremendous increases in the chloride, sodium, and specific conductance content of groundwater in close proximity to landfill sites were found by Anderson and Dornbush (Ref. 5-1). Coe (Ref. 5-5) reported the following percentage increases in pollutant parameters in the groundwater under a Southern California landfill:

BOD	2600%
Chloride	1000%
Sodium	900%
Sulfate	800%

Carpenter and Setter (Ref. 5-3) studied water quality immediately below a waste fill and found the following water quality parameter concentrations:

BOD	1987 mg/l
Alkalinity	3867 mg/l
Chlorides	3506 mg/l

The pollution of groundwater by leachate requires the existence of three conditions: the location of a landfill site near an aquifer, the movement of

TABLE 5-1 Leachate Composition as Reported by Various Sources

Analysis[a]	1	2	3	4	5	6	7	8	9	10	11	12	13	14	15	16	17
pH	5.6	5.9	8.3	–	–	–	–	–	7.63	5.60	7.4	6.4	4.9	8.4	6.3	6.48	5.88
Total hardness (as $CaCO_3$)	8,120	3,260	537	–	8,700	500	900	290	8,120	–	–	2,500	30	–	7,600	13,100	10,950
Total alkalinity (as $CaCO_3$)	–	–	–	–	–	–	–	–	650	–	–	–	–	–	–	–	–
Total iron	305	1,710	1,290	1,000	–	–	40	2	305	6	–	–	–	9,450	10,630	16,200	20,850
Sodium	1,805	336	219	–	–	–	–	–	1,805	85	206	152	300	100	175	546	860
Potassium	1,860	350	600	–	–	–	–	–	1,860	28	1,200	1,100	110	–	584	1,428	1,439
Sulfate	630	655	–	–	940	24	225	100	730	248	–	920	65	–	1,050	2,535	3,770
Chloride	2,240	1,220	99	2,000	1,000	220	–	–	2,350	90	1,845	970	485	12,300	615	1,002	768
NO_3 nitrogen	–	5	–	–	–	–	–	–	–	0.2	668	–	10	–	951	2,000	2,310
NH_3 nitrogen	845	141	–	–	–	–	160	100	845	2	940	196	–	–	473	756	1,106
TON	550	152	18	–	–	–	–	–	550	–	101	–	–	–	288	664	1,416
COD	–	7,130	–	750,000	–	–	3,850	246	–	–	35,700	21,120	282	–	–	26,940	–
BOD	32,400	7,050	–	720,000	–	–	1,800	18	33,100	81	5,491	–	–	7,330	14,760	–	–
TDS	–	9,190	2,000	–	11,254	–	–	–	–	–	11,254	15,830	1,740	–	–	–	33,360
Specific conductance	–	–	–	–	–	2,075	3,000	2,500	–	–	–	–	–	–	–	–	–

[a] All analyses in mg/l except pH and specific conductance.

Source: Pohland, F. G., "Landfill Stabilization with Leachate Recycle," Annual Progress Report, EPA No. EP-00658, July, 1972.

water into the fill producing leachate, and subsequent entrance of the leached fluids into the aquifier. If these requirements are satisifed, water pollution may be severe.

The amount of leachate introduced into an aquifier may be estimated by techniques of moisture routing described by Remson, Fungaroli, and Lawrence (Ref. 5-29) and Fungaroli (Ref. 5-12). Time lags between refuse placement and production of leachate are indicated for conventional landfills according to these techniques.

In a field investigation by Merz and Stone (Ref. 5-25), only a low level of leachate penetration into the subsoils below landfill cells was recorded. In this study, water was introduced into the study cells to simulate rainfall in Seattle, Washington; however, it is probable that this low precipitation model does not represent typical conditions in most of the United States.

The characteristics of leachate collected from actual operating fills and from simulated fills (laboratory studies) are summarized in Table 5-1. Leachate composition is variable, the character of generated leachate depending upon the deposited refuse, the quality of the groundwater and infiltrating precipitation, and the physiochemical character of the soils around and beneath the refuse cells (Ref. 5-14, 5-17, 5-30). Variations in the color and appearance of leachate were reported by Coe (Ref. 5-5) and Qasim and Burchinal (Ref. 5-4). Meade and Wilkie (Ref. 5-24) estimated on the basis of a California field study that continuous water flow for 1 year through 1 acre-ft of refuse would remove 1.5 tons of sodium and potassium, 1.0 tons of calcium and magnesium, 0.91 tons of chlorides, 0.23 tons of sulfates, and 3.9 tons of bicarbonate. Fungaroli and Steiner (Ref. 5-13) reported the results of a laboratory lysimeter study in which generated leachate possessed acidic characteristics, pH = 5.0 to 6.5, contained iron concentrations of more than 1600 mg/l at times, and possessed a COD initially of more than 50,000 mg/l. These organic and inorganic constituent concentrations were found to decrease with time. Merz (Ref. 5-30) conducted similar studies wherein high concentrations of iron, ammonia, phosphate, and organic nitrogen were found.

The operational parameters which appear to govern the characteristics of leachate are refuse moisture content, landfill cell temperature, and refuse composition. Refuse composition is a variable which cannot be controlled easily on-site; moisture content and temperature, however, are amenable to control. Control of these parameters may permit optimization of waste stabilization in landfills.

5-2.2 Leachate Recirculation

Increasing moisture content in a fill may be achieved by water addition at the time of solid waste placement; however, moisture in excess of the refuse field capacity will drain from the fill soon after addition. To increase moisture content permanently, water percolating through and out of the fill cells must be replaced. One way that this can be accomplished is by recirculating the leachate. The effect of a higher moisture content in landfill sites has been reported by Eliassen (Ref. 5-7), Salvato et al. (Ref. 5-31), and others (Ref. 5-30). Eliassen reported that for recently landfilled material, the optimum moisture content for biological decomposition was in the range of 50 to 70 percent. Such an optimum moisture content serves to accelerate microbial activity, thereby increasing the rate of waste degradation. Surface settlement for a fill with optimum water content was found to be 400 percent of the settlement for a fill without added water during studies by Merz (Ref. 5-30).

Temperatures in deposited refuse have been monitored by Carpenter and Setter (Ref. 5-3), Eliassen (Ref. 5-7), Merz and Stone (Ref. 5-25), and Fungaroli and Steiner (Ref. 5-13). The results of these studies indicate that the temperature in a conventional fill quickly rises to a range of 50–70°C following refuse placement, a range where thermophilic organisms thrive. In undisturbed landfill cells, the temperature apparently decreases slowly to about 60°C and then very rapidly to about 29°C at which point it stabilizes (Ref. 5-25).

Moisture content and temperature both can be controlled to some degree by means of leachate recirculation. A comprehensive study of such recirculation is in progress at the Georgia Institute of Technology under the direction of Dr. Fred Pohland. Also, in Sonoma County, California, the United States Environmental Protection Agency has instituted a field study to investigate leachate recirculation in six study landfill cells. No other investigations of leachate recirculation have been reported in the literature, although a third study on the topic has been initiated by the authors at the University of Louisville.

In the Georgia Tech study, four cylindrical vessels (each 36 in. in diameter and 14 ft high) were constructed and charged with refuse to simulate landfills; two were filled in the spring of 1971, and two were filled in the spring of 1972. Approximately 10 ft of compacted refuse weighing 1400 lb was placed in each test cylinder at a dry density of 535 lb/cu yd. At the initiation of the study, 250 gallons of tap water were added to the deposited refuse in each vessel to expedite the production of leachate. Operational characteristics of the four test cylinders were as follows:

Cell 1—Control cell in which no leachate recirculation, pH control, or raw sludge seeding was practiced.

Cell 2—Leachate recirculation cell in which no pH control or raw sludge seeding was practiced.

Cell 3—Leachate recirculation cell with pH control (between pH 6.8 and 7.4) but no raw sludge seed.

Cell 4—Leachate recirculation cell with pH control and raw sludge seed.

TABLE 5-2 Composition of Leachate from Control Landfill (Fill 1)

Time Since Leachate Production Began, Days	0	14	24	32	39	48	81	116	125	153	173	189	197	228	249
COD, mg/l	4,320	9,150	10,380	10,260	12,000	11,700	9,200	10,100	11,700	12,200	12,300	14,400	15,600	18,100	15,600
BOD$_5$, mg/l	2,500	5,000	9,200	6,330	11,000	8,200	8,800	9,600	8,700	11,100	9,200	12,000	9,300	13,400	12,600
TOC, mg/l	1,230	1,910	2,622	2,622	2,802	2,835	2,864	2,259	2,418	2,680	2,696	3,049	3,409	5,000	3,590
TSS, mg/l	125	34	59	61	47	213	270	640	550	292	470	360	175	85	175
VSS, mg/l	45	20	47	52	37.6	93	160	332	314	182	268	210	104	76	141
TS, mg/l	2,442	5,819	6,323	8,300	8,736	6,789	5,530	7,250	7,358	7,620	7,875	8,320	8,130	12,500	8,780
Total alkalinity, mg/l as CaCO$_3$	558	1,610	1,640	1,920	2,280	2,110	2,420	2,650	2,120	2,350	2,100	2,482	1,760	2,480	1,580
Total acidity, mg/l as CaCO$_3$	690	1,100	1,350	1,400	1,780	2,170	1,836	1,390	2,090	2,230	2,780	2,865	3,260	3,460	2,610
pH	5.2	5.6	5.3	5.3	5.3	5.3	5.7	5.3	5.2	5.3	5.1	5.2	5.1	5.1	5.2
Total hardness, mg/l as CaCO$_3$	450	1,400	1,850	1,810	1,940	1,754	1,410	1,429	1,694	2,232	2,354	2,306	2,449	5,555	3,463
Acetic acid, mg/l	500	2,111	2,360	2,664	3,666	3,268	2,789	3,285	2,590	3,280	3,440	3,393	3,550	5,160	3,754
Propionic acid, mg/l	369	1,595	1,834	2,038	2,313	2,108	1,875	2,625	2,110	2,290	2,190	2,400	2,214	2,840	1,742
Butyric acid, mg/l	110	965	1,075	1,050	1,280	1,164	1,000	1,203	1,424	1,195	1,215	1,350	1,750	1,830	1,770
Valeric acid, mg/l	Nil	425	575	625	535	612	643	893	656	708	652	730	801	1,000	705
Phosphate, mg/l PO$_4^{-3}$	26	3.0	5.0	7.8	2.8	2.9	3.3	4.2	3.4	2.8	1.7	1.6	1.5	1.3	1.5
Organic nitrogen, mg/l as N	56	47	61.4	62	75	48	40	177	64	6	20	12	43	107	116
Ammonia nitrogen, mg/l as N	56	150	167.6	187	185	192	148	103	130	260	214	218	264	117	52
Nitrate nitrogen, mg/l NO$_3^-$	13.3	32	89	84	115	15.0	—	9.5	12	—	—	—	—	—	—
Chloride, mg/l	322	385	109.8	105.1	97.9	340	—	170	240	210	208	312	308	180	300
Sulfate, mg/l SO$_4^{-2}$	84	126	108	81	156	17	2	7	1	16	—	—	—	—	—
Calcium, mg/l Ca	125	430	470	590	750	545	430	375	420	600	578	565	545	1,250	850
Magnesium, mg/l Mg	26	71.8	67	75	68	64	52	49	53	80	85	85	75	260	210
Manganese, mg/l Mn	3	10	5	6.2	8.8	8.5	10	7.5	10	16	14	15	16	18	19
Sodium, mg/l Na	63.8	125	132	132	143	150	180	118	135	155	154	155	148	160	140
Iron, mg/l Fe	9	21	70	30	95	65	60	155	230	200	300	290	420	185	250
Zinc, mg/l Zn	—	—	—	—	—	—	—	—	—	—	—	—	—	—	—
Total volatile acids, mg/l as acetic acid	865	4,310	4,925	5,399	6,721	6,133	5,370	6,750	5,655	6,370	6,420	6,693	7,000	9,300	6,785

TABLE (continued)

Time Since Leachate Production Began, Days	284	312	332	347	398	428	473	506	530	556	606	636	672	704
COD, mg/l	13,300	13,800	—	11,100	9,000	9,500	8,950	8,050	7,845	6,210	6,120	6,140	5,750	4,990
BOD_5, mg/l	9,560	8,800	—	7,750	5,300	6,500	6,050	6,600	4,800	3,835	4,300	4,200	4,300	3,350
TOC, mg/l	3,000	2,930	3,180	3,005	2,430	2,910	2,910	2,665	2,127	2,410	1,400	2,090	2,190	1,990
TSS, mg/l	605	610	308	880	1,243	800	680	800	540	1,170	1,010	510	750	750
VSS, mg/l	283	286	146	432	602	400	470	310	340	380	300	210	305	310
TS, mg/l	7,716	7,167	6,965	6,260	5,602	5,800	3,750	3,650	2,425	2,400	2,100	2,050	2,100	2,100
Total alkalinity, mg/l as $CaCO_3$	2,430	1,930	1,960	1,725	1,500	1,750	2,040	2,040	1,970	2,040	2,040	1,800	2,040	2,240
Total acidity, mg/l as $CaCO_3$	2,000	2,400	3,360	3,460	1,950	2,100	1,710	1,440	1,840	1,670	1,670	2,350	1,740	1,640
pH	5.2	5.3	5.3	5.25	5.33	5.60	5.68	5.90	5.95	6.10	6.00	6.10	6.20	6.30
Total hardness, mg/l as $CaCO_3$	2,424	2,299	1,622	1,326	1,576	1,840	1,580	1,310	1,190	1,170	1,160	1,840	790	750
Acetic acid, mg/l	3,640	2,830	2,275	2,210	1,000	2,410	2,520	2,220	2,750	2,920	2,910	1,750	1,750	1,550
Propionic acid, mg/l	1,640	1,580	1,380	1,330	720	1,100	1,520	1,260	720	400	410	1,200	1,100	1,150
Butyric acid, mg/l	1,800	1,740	1,540	1,460	970	940	500	704	714	90	40	410	400	400
Valeric acid, mg/l	750	768	590	560	855	710	395	428	420	70	30	395	395	200
Phosphate, mg/l PO_4^{-3}	0.9	1.1	.6	.40	.5	0.51	0.51	0.27	0.29	0.23	0.32	0.28	0.27	0.26
Organic nitrogen, mg/l as N	76	63	28	40	124	48	46	42	85	87	85	59	16	26
Ammonia nitrogen, mg/l as N	110	103	152	132	88	88	86	80	35	28	19	8	8	12
Nitrate nitrogen, mg/l NO_3^-	—	—	—	—	6.4	.09	.07	.07	.07	.06	.17	.04	.04	.15
Chloride, mg/l	280	295	124	137	143	150	130	164	164	200	134	134	85	110
Sulfate, mg/l SO_4^{-2}	—	—	—	—	—	—	—	—	—	—	—	—	—	—
Calcium, mg/l Ca	550	490	433	385	350	400	350	230	200	175	155	140	110	142
Magnesium, mg/l Mg	90	65	40	53	39	45	45	12	22	22	20	20	11	12
Manganese, mg/l Mn	12	12	19	11	10	15	6.5	7.5	3.5	4.5	4.5	3.5	4.5	2.5
Sodium, mg/l Na	85	140	103	110	130	130	145	130	140	170	275	235	235	210
Iron, mg/l Fe	370	440	190	70	292	240	280	295	280	270	270	250	245	240
Zinc, mg/l Zn	—	—	—	—	—	—	—	—	—	42.5	41	10	12	9
Total volatile acids, mg/l as acetic acid	6,460	5,745	4,795	4,615	2,745	3,950	4,570	3,915	3,420	3,300	3,820	3,670	3,240	3,090

Source: Chaw-Ming Mao, M. and Pohland, F. G., "Continuing Investigations on Leachate Stabilization With Leachate Recirculation, Neutralization, and Seeding," Special Research Report, Georgia Institute of Technology, September 1973.

TABLE 5-3 Composition of Leachate from Recirculation Landfill without pH Control or Sludge Addition (Fill 2)

Time Since Leachate Production Began, Days	0	10	18	24	31	39	48	58	67	96	111	126	140	161	189	197	219
COD, mg/l	4,280	9,288	8,870	9,080	8,111	7,700	8,140	9,580	10,400	10,025	10,500	10,500	10,350	8,890	5,810	4,270	3,550
BOD$_5$, mg/l	2,750	5,200	6,900	6,800	4,300	5,400	6,202	6,400	6,380	7,200	8,700	8,500	10,100	9,405	6,650	3,500	2,860
TOC, mg/l	2,130	1,120	2,260	2,040	2,394	1,818	2,665	2,000	2,675	2,798	1,990	1,979	1,952	1,542	1,280	1,067	914
TSS, mg/l	93	93	12	36.5	70.5	25	37.0	120	301	143	222	258	385	187	232	220	131
VSS, mg/l	22.5	13.6	9	27.5	45	18.8	16.9	70	161	78	158	142	188	118	156	116	76
TS, mg/l	2,349	4,329	4,552	5,023	5,400	4,728	4,941	5,250	5,440	5,980	5,830	6,918	6,106	5,336	4,090	3,987	3,240
Total alkalinity, mg/l as CaCO$_3$	302	700	865	1,080	1,200	1,370	1,525	1,438	1,035	1,900	2,350	1,640	1,670	1,640	1,550	1,342	1,115
Total acidity, mg/l as CaCO$_3$	554	1,900	1,540	1,350	1,000	1,390	1,265	1,530	1,765	1,798	1,730	1,830	1,700	1,630	500	333	240
pH	5.05	4.8	5.0	5.1	5.3	5.4	5.3	5.3	5.1	5.4	5.5	5.3	5.3	5.2	6.3	6.6	6.8
Total hardness, mg/l as CaCO$_3$	370	895	880	1,010	890	1,040	1,222	1,483	1,532	1,701	1,987	1,495	2,296	1,948	1,469	1,146	978
Acetic acid, mg/l	1,638	556	2,000	1,843	1,475	1,583	1,795	2,146	2,438	2,742	2,438	2,470	2,380	1,877	2,925	608	734
Propionic acid, mg/l	960	394	1,242	1,467	1,554	1,594	1,580	1,752	1,953	2,203	1,953	1,865	2,020	1,472	1,995	714	195
Butyric acid, mg/l	1,300	235	1,235	1,163	1,375	1,250	1,200	1,198	1,094	1,156	1,047	1,124	937	735	665	286	194
Valeric acid, mg/l	500	735	50	833	688	670	714	800	858	857	786	842	625	556	585	276	87
Phosphate, mg/l PO$_4^{-3}$	22	1.5	2.1	0.65	0.81	0.67	0.82	0.85	0.98	0.65	0.38	0.50	0.39	0.82	0.47	0.26	0.24
Organic nitrogen, mg/l as N	20	0	30	405	37.5	39.5	41	30	39	62	92	28	7	3	4	Nil	Nil
Ammonia nitrogen, mg/l as N	70	68	113.5	86.5	77.5	76.5	64	69	81	84	80	71	135	126	80	62	56
Nitrate nitrogen, mg/l NO$_3^-$	6.2	71.4	56.6	76.6	48	49	11.0	11.5	12.0	16.0	21.0	14.0	–	–	–	–	–
Chloride, mg/l	210	210	248	94.5	91	115	220	164	176	140	188	170	210	236	300	270	260
Sulfate, mg/l SO$_4^{-2}$	102	138	81	51	30	12	11	Nil	12	2	1	3	–	–	–	–	–
Calcium, mg/l Ca	60	315	350	435	420	430	420	415	440	500	550	385	600	475	400	340	290
Magnesium, mg/l Mg	16.5	59	53.5	62.5	56	56	50	50	53	55	62	44	70	60	50	45	40
Manganese, mg/l Mn	4	30	50	65	62	62	75	75	80	80	85	60	93	80	59	50	44
Sodium, mg/l Na	61.5	109	81.4	91.4	85	84	95	85	88	90	98	70	84	75	61	59	50
Iron, mg/l Fe	4.4	19.5	19	80	43	110	25	35	40	45	110	150	150	210	90	13	5
Total volatile acids, mg/l as acetic acid	8,605	1,465	3,875	4,370	4,080	4,120	4,315	4,855	5,270	5,815	5,195	5,245	5,025	3,895	5,340	1,545	1,075

Time Since Leachate Production Began, Days	228	249	284	284	312	332	366	398	428	473	506	530	556	606	636	672	704
COD, mg/l	2,970	2,840	2,580	1,950	1,280	1,050	1,110	800	870	490	225	258	192	113	56	84	70
BOD_5, mg/l	1,400	2,500	2,420	760	760	540	700	510	440	264	120	85	75	46	44	45	44
TOC, mg/l	710	565	500	308	256	480	475	545	510	515	375	325	310	325	520	345	250
TSS, mg/l	122	145	124	67	305	358	370	405	350	310	250	140	140	510	400	310	200
VSS, mg/l	74	87	56	37	18	41	69	72	50	100	90	110	80	280	250	110	70
TS, mg/l	2,792	2,370	2,510	1,848	1,627	1,784	2,038	—	2,100	2,800	2,000	820	720	950	900	850	700
Total alkalinity, mg/l as $CaCO_3$	952	980	925	738	692	800	780	800	800	840	840	780	760	620	840	880	840
Total acidity, mg/l as $CaCO_3$	180	166	133	84	80	152	200	250	250	250	250	230	240	260	110	180	140
pH	6.9	7.0	7.1	7.4	7.3	7.1	6.91	6.90	6.90	6.82	7.10	6.95	7.05	6.45	7.0	7.10	7.0
Total hardness, mg/l as $CaCO_3$	677	539	661	513	377	146	520	—	375	250	200	170	110	100	90	90	105
Acetic acid, mg/l	770	670	111	234	365	400	525	1,050	1,110	1,000	875	940	865	740	410	140	105
Propionic acid, mg/l	111	104	57	223	110	160	120	55	70	90	40	38	42	75	75	35	35
Butyric acid, mg/l	68	65	Nil	62	44	20	26	95	110	120	20	40	40	75	120	30	35
Valeric acid, mg/l	65	50	Nil	35	Nil	13	33	180	170	145	50	70	60	85	20	10	0
Phosphate, mg/l PO_4^{-3}	0.07	0.08	0.09	0.12	0.09	0.03	0.15	0.09	0.08	0.08	0.05	0.06	0.05	0.06	0.10	0.05	0.07
Organic nitrogen, mg/l as N	1	3	2	1	7	Nil	16	—	3	4	6.5	14	7	0	0	0	0
Ammonia nitrogen, mg/l as N	39	31	35	27	13	30	26	—	.18	15	3.5	0	0	0	0	0	0
Nitrate nitrogen, mg/l NO_3^-	—	—	—	—	—	—	—	—	.09	.04	.08	.06	.04	.05	.05	.05	.05
Chloride, mg/l	248	224	220	218	202	119	116	—	158	204	236	176	150	110	76	70	70
Sulfate, mg/l SO_4^{-2}	—	—	—	—	—	—	—	—	—	—	—	—	—	—	—	—	—
Calcium, mg/l Ca	190	145	175	135	82	115	136	—	40	25	27	27	11	11	9	9	9
Magnesium, mg/l Mg	40	38	40	35	38	32	34	—	30	14	13	12	11	11	10	11	10
Manganese, mg/l Mn	19	10	19	14	8	3	8	—	8	10	0	0	.4	.1	.2	0	10
Sodium, mg/l Na	60	55	60	55	75	52	63	—	60	70	40	60	100	120	120	120	120
Iron, mg/l Fe	1.4	1.9	14	4	1.2	5	14	—	0	0	40	0	0	0	0	0	0
Total volatile acids, mg/l as acetic acid	945	830	155	475	485	555	660	1,265	1,342	1,240	955	1,039	961	902	536	194	103

Source: Chaw-Ming Mao, M. and Pohland, F. G., "Continuing Investigations on Leachate Stabilization With Leachate Recirculation, Neutralization, and Seeding," Special Research Report, Georgia Institute of Technology, September 1973.

TABLE 5-4 Composition of Leachate from Recirculation Landfills with pH Control (Fill 3) and pH Control with Sludge Addition (Fill 4)

Parameter		2 5/3	8 5/9	17 5/18	24 5/25	31 6/1	38 6/8	45 6/15	52 6/22	58 6/28	68 7/8	73 7/13	80 7/20	87 7/27
Time Since Leachate Production Began	Days / Date													
COD, mg/l	A[a]	460	5,200	7,200	9,250	11,750	11,200	11,000	15,000	15,400	17,400	18,000	15,800	17,600
	B[b]	5,850	6,900	7,600	9,050	9,200	9,700	9,400	8,700	7,200	7,950	8,200	7,875	7,075
BOD$_5$, mg/l	A	195	3,350	5,600	7,900	9,200	8,500	8,000	7,600	10,300	12,100	11,200	12,300	14,650
	B	4,150	3,900	4,400	6,600	7,150	6,800	6,800	5,200	5,400	5,900	5,600	4,600	5,300
TOC, mg/l	A	332	2,030	2,720	2,860	3,655	3,820	3,440	4,000	4,430	4,330	4,800	4,500	4,925
	B	1,975	2,360	2,340	2,610	2,375	2,660	2,485	2,310	2,370	2,400	2,060	2,055	1,900
TSS, mg/l	A	—	146	210	355	441	558	364	814	768	1,225	1,101	690	463
	B	—	126	253	281	401	374	569	880	—	978	926	747	1,060
VSS, mg/l	A	—	100	72	111	146	205	85	270	280	393	342	192	151
	B	—	78	144	142	171	161	250	140	—	226	244	175	251
TS, mg/l	A	—	3,154	4,983	8,097	9,699	10,478	11,860	11,006	11,346	12,169	12,314	13,458	12,770
	B	—	3,896	4,745	5,206	6,219	6,811	7,756	5,678	6,012	6,135	6,534	6,912	6,387
Total alkalinity, mg/l CaCO$_3$	A	93	964	1,735	3,240	3,290	3,565	3,765	3,400	4,320	4,560	4,700	4,540	4,900
	B	1,500	1,870	2,530	2,830	2,710	2,660	3,220	2,740	2,940	2,780	2,540	4,360	3,150
Total acidity, mg/l CaCO$_3$	A	30	920	2,010	690	520	590	420	1,020	1,370	860	900	580	800
	B	325	485	830	860	930	630	835	550	640	500	560	550	400
pH	A	6.78	5.45	5.35	6.58	6.58	6.05	6.10	5.89	5.88	6.24	6.19	6.59	6.32
	B	6.61	6.52	6.28	6.50	6.32	6.34	6.30	6.81	6.69	6.61	6.55	6.88	7.00
Hardness, mg/l as CaCO$_3$	A	—	563	872	989	1,206	1,249	1,293	1,639	1,168	1,335	1,428	1,455	1,167
	B	—	537	790	863	997	1,043	1,405	1,055	642	847	898	1,057	492
Acetic acid, mg/l	A	44	1,000	1,875	2,150	2,300	2,910	2,950	3,140	3,950	4,000	2,400	2,530	2,200
	B	950	1,575	1,810	1,825	2,250	2,350	2,065	380	272	220	1,230	900	1,410
Propionic acid, mg/l	A	14	1,020	1,800	2,025	2,160	2,550	2,650	2,750	3,380	3,750	2,270	2,210	2,320
	B	440	1,140	1,460	1,235	1,275	1,360	2,600	2,260	2,620	3,580	2,970	2,430	2,650
Butyric acid, mg/l	A	13	350	800	850	1,075	1,275	1,425	1,500	1,770	2,000	1,495	1,475	1,350
	B	175	800	765	738	825	1,000	1,040	665	145	320	95	nil	50

Valeric acid, mg/l	A	13	88	295	375	475	610	725	855	1,220	1,970	1,790	1,820	1,670
	B	0	25	130	200	225	300	395	385	260	440	nil	100	100
Phosphate, mg/l PO$_4^{-3}$	A	0.27	–	1.47	0.27	0.50	0.45	0.25	0.34	0.31	0.22	–	0.20	0.28
	B	3.90	–	0.22	0.10	0.26	1.20	1.50	0.25	0.29	0.29	–	0.18	0.27
Organic nitrogen, mg/l as N	A	–	92	45	4	30	26	92	67	114	67	75	83	75
	B	–	107	119	119	126	130	66	42	32	47	55	50	46
Ammonia nitrogen, mg/l as N	A	–	172	270	318	320	324	335	339	376	400	400	400	400
	B	–	325	413	427	392	437	396	343	304	268	260	244	176
Nitrate nitrogen, mg/l as NO$_3^-$	A	–	–	3.1	2.7	4.0	3.3	2.2	3.1	3.3	2.7	3.1	0.4	0.6
	B	–	5.3	4.4	3.5	4.0	4.2	3.5	1.1	1.9	2.0	1.6	0.6	0.5
Chloride, mg/l Cl$^-$	A	–	191	186	243	257	–	286	250	238	272	276	286	268
	B	–	153	254	252	253	316	305	293	287	290	331	324	307
Calcium, mg/l Ca	A	–	136	246	290	335	366	382	440	305	325	340	365	300
	B	–	17	205	230	270	275	390	285	165	220	225	230	80
Magnesium, mg/l Mg	A	–	31	31	34	41	43	43	47	49	52	53	52	53
	B	–	19	38	40	44	46	55	50	53	63	67	67	36
Manganese, mg/l Mn	A	–	19	19	19	19	19	10	19	13	12	13	12	11
	B	–	10	19	19	19	19	15	19	6	8	14	14	23
Sodium, mg/l Na	A	–	118	294	1,210	1,410	1,880	1,600	1,100	1,590	1,400	1,600	1,600	2,300
	B	–	182	248	336	630	600	750	613	625	1,050	800	825	400
Iron, mg/l Fe	A	–	42	53	50	91	68	80	174	100	160	188	115	100
	B	–	29	48	49	59	73	99	56	nil	13	18	100	57
Potassium, mg/l K	A	–	–	–	535	595	710	550	530	570	600	590	605	563
	B	–	–	–	690	740	–	500	392	345	360	385	400	231
Zinc, mg/l Zn	A	–	–	–	1.3	1.0	1.3	5.0	4.3	7.5	20	22	12	17
	B	–	–	–	0.8	0.8	–	0.6	1.3	0.8	0.8	1.3	1.0	0.6
Total volatile acid mg/l as acetic acid	A	72	2,120	4,055	5,570	5,060	5,200	6,490	6,880	8,620	9,560	6,310	6,390	5,980
	B	1,425	3,060	3,587	3,450	3,975	4,305	5,105	2,890	2,645	3,600	3,695	2,930	3,575

(Continued)

TABLE 5-4 (Continued)

Time Since Leachate Production Began		Days 114	134	156	169	183	206	221	234	255	282	325	350	365	394
		Date 8/23	9/12	10/4	10/17	10/31	11/23	12/8	12/21	1/11	2/7	3/22	4/16	5/1	5/30
COD, mg/l	A[a]	17,710	16,650	16,510	14,000	13,200	14,500	13,000	11,800	7,100	5,500	2,480	1,450	950	780
	B[b]	1,860	950	850	840	745	560	560	490	403	376	350	340	290	270
BOD$_5$, mg/l	A	14,500	14,000	13,000	12,300	11,500	12,300	12,500	9,450	5,500	5,050	2,300	1,100	660	250
	B	1,400	860	500	367	232	220	130	125	44	62	66	85	90	88
TOC, mg/l	A	5,700	5,655	5,685	5,080	5,210	4,940	4,220	3,660	3,300	2,600	1,140	930	980	980
	B	1,650	815	745	660	610	540	610	570	275	250	347	325	450	470
TSS, mg/l	A	750	780	750	820	840	1,180	720	760	1,030	720	900	850	800	650
	B	470	490	480	510	610	420	350	330	280	310	260	310	340	410
VSS, mg/l	A	110	60	70	100	70	85	65	90	450	60	250	250	170	140
	B	140	140	130	120	130	160	110	90	120	140	150	120	120	120
TS, mg/l	A	12,000	10,500	8,500	8,500	7,800	7,000	6,440	5,610	4,380	3,440	2,435	2,710	2,210	1,460
	B	5,400	3,800	4,200	3,400	3,000	2,560	2,480	2,140	1,460	1,170	1,200	1,100	1,150	1,150
Total alkalinity, mg/l CaCO$_3$	A	4,450	4,400	4,560	4,280	4,840	4,870	4,880	5,400	5,800	6,010	6,180	5,840	5,760	5,420
	B	2,960	2,680	2,660	2,620	2,580	2,480	2,400	2,510	2,560	2,510	3,740	2,920	2,760	2,840
Total acidity, mg/l CaCO$_3$	A	890	1,010	1,090	1,070	1,240	1,340	1,390	1,310	800	810	310	260	240	240
	B	410	400	400	360	310	300	230	210	205	160	140	160	150	170
pH	A	6.45	6.60	6.70	6.65	6.65	6.75	7.40	7.40	7.45	7.50	7.40	7.20	7.20	7.10
	B	7.10	7.20	7.45	7.30	7.20	7.25	7.15	7.30	7.05	7.15	7.10	7.03	7.03	7.00
Hardness, mg/l as CaCO$_3$	A	750	540	260	210	200	190	180	180	160	160	140	140	140	140
	B	240	210	205	180	180	170	160	160	140	140	120	110	110	110
Acetic acid, mg/l	A	2,260	2,310	2,420	2,100	2,200	2,140	1,700	2,000	1,800	1,600	1,400	540	280	90
	B	1,160	1,120	1,000	640	310	210	110	120	100	85	80	78	100	75
Propionic acid, mg/l	A	5,780	5,350	5,100	3,620	2,420	1,540	890	680	540	540	640	110	75	40
	B	2,000	350	250	42	15	15	25	30	20	60	60	40	40	40
Butyric acid, mg/l	A	1,000	720	600	510	370	310	215	210	110	80	25	40	5	1
	B	50	50	40	25	12	0	0	5	20	0	10	0	2	0

Parameter															
Valeric acid, mg/l	A	1,420	1,300	1,200	640	540	400	210	200	80	0	5	0	0	0
	B	120	90	75	50	10	10	0	0	0	0	0	0	0	0
Phosphate, mg/l PO_4^{-3}	A	0.17	0.30	0.21	0.23	0.17	0.12	0.12	0.12	0.17	0.14	0.28	0.31	0.28	0.24
	B	0.42	0.37	0.22	0.30	0.31	0.24	0.23	0.37	0.21	0.17	0.06	0.09	0.11	0.14
Organic nitrogen, mg/l as N	A	48	48	32	96	254	157	133	131	70	30	26	16	50	44
	B	54	43	25	58	30	132	105	91	5	5	7	15	15	35
Ammonia nitrogen, mg/l as N	A	448	408	376	360	210	98	67	67	126	104	101	76	62	16
	B	216	224	192	176	197	154	105	56	49	3	1	25	40	65
Nitrate nitrogen, mg/l as NO_3^-	A	0.20	0.18	0.15	0.24	0.25	0.20	0.19	0.19	0.23	0.05	0.05	0.15	0.11	0.11
	B	0.23	0.17	0.17	0.15	0.17	0.13	0.12	0.13	0.14	64	64	9.6	12	12
Chloride, mg/l Cl^-	A	280	250	310	320	290	300	310	320	340	170	160	130	130	130
	B	330	300	380	380	350	350	340	360	340	240	180	130	110	130
Calcium, mg/l Ca	A	310	350	280	155	125	70	20	15	14	7	12	12	15	15
	B	80	2.3	2.4	2.6	5.5	5.5	11	12	14	9	8	13	12	12
Magnesium, mg/l Mg	A	55	24	12	12	31	31	34	34	36	36	20	18	15	12
	B	45	40	25.5	16	22	9	15	14	15	14	14	15	12	12
Manganese, mg/l Mn	A	13	11	4.5	3.1	2.7	0.4	0	0	0	0	0	0	0	0
	B	17	21	25	6.5	6.8	7.5	7.5	3.5	3.5	4.5	4.5	4.0	2.5	2.5
Sodium, mg/l Na	A	1,600	1,400	1,200	1,250	1,150	1,125	1,200	1,200	1,250	1,350	1,000	890	840	840
	B	560	520	500	490	470	470	480	490	500	470	490	500	290	290
Iron, mg/l Fe	A	25	120	150	110	75	75	40	20	21	18	15	15	22	22
	B	110	12	8	8	3	3	5	5	7	7	12	12	8	8
Potassium, mg/l K	A	550	550	510	515	500	495	480	310	310	310	300	300	300	300
	B	235	340	340	340	340	350	350	310	310	340	340	340	340	340
Zinc, mg/l Zn	A	1.0	15	10	0.95	0.05	0	0	0	0	0	0	0	0	0
	B		0	0.02	0	0	0	0	0	0	0	0	0	0	0
Total volatile acid, mg/l as acetic acid	A	8,480	7,920	7,686	5,760	4,740	3,600	2,690	2,415	2,360	2,035	1,940	640	350	125
	B	2,885	1,495	1,370	720	336	222	130	147	130	100	135	110	130	100

Source: Chaw-Ming, Mao, M. and Pohland, F. G., "Continuing Investigations on Leachate Stabilization With Leachate Recirculation, Neutralization, and Seeding," Special Research Report, Georgia Institute of Technology, September, 1973.

[a] Fill 4
[b] Fill 3

The leachate produced from these simulated fills was collected and sampled periodically. Twenty-four-hour leachate composite samples were secured from the three recirculating fills at 1 to 3-week intervals, whereas samples were obtained from the control cell only when a sufficient quantity of leachate was produced to yield a sample of 1 to 3 liters. Since these test cells were constructed outdoors, rainfall was not excluded from any of the fills.

Leachate samples from the four vessels were analyzed for various significant parameters. The results of these leachate tests are shown in Tables 5-2 through 5-4, reproduced from Pohland (Ref. 5-26 and 5-4).

The study results and general characteristics of the test cylinders may be summarized as follows:

1. Leachate recirculation produced a rapid stabilization of refuse because of the accelerated growth of an anaerobic biological population. Since biological stabilization of the organic fraction of the refuse proceeded at an optimal rate, BOD, COD, and TOC levels in the leachate from the recirculating fills were greatly reduced in a relatively short period of time.

2. Recirculation with leachate pH control resulted in more rapid biological stabilization when compared to leachate recirculation without pH control. Cell #3 (recirculation with pH control) produced a leachate BOD_5 of less than 100 in about 8 months, whereas Cell #2 (recirculation without pH control) produced a leachate BOD_5 of less than 100 in about 17 months.

3. Enhanced biological activity promoted by leachate recirculation resulted in a substantial reduction in some of the inorganic leachate constituents. Further reduction of some of these inorganics was achieved with pH control.

4. Leachate recirculation enhanced the surface settlement rate of the simulated landfills. Surface settlements of 30 percent of fill height were achieved in Cells #3 and #4 after approximately 13 months.

5. Recirculating leachate provided effective biological treatment, this treatment being more rapid and predictable than the degradation in a conventional sanitary landfill. Leachate recirculation reduced and controlled the total amount of pollutants entering the environment. Therefore, this process should be effective in improving landfill disposal and in returning land used for fill sites to other uses in shorter periods of time than that presently required for conventional fill operations.

Another pilot study investigating the effects of liquids on the stabilization of landfilled solid waste material has been underway since July, 1971 in Sonoma County, California under the direction of Donald B. Head (Santa Rosa Department of Public Works) with the consulting services of EMCON Associates (Ref. 5-34 and 5-35). The primary purpose of this EPA-funded project was to determine the effect on solid waste stabilization of applying water or septic tank pumpings to refuse cells or recirculating leachate through landfill cells. Five field-scale test cells were constructed and monitored, each cell having a different controlled moisture content and/or liquid character. The specific operating characteristics of each cell are outlined in Table 5-5. Significant study results obtained during the first 2 years of this investigation are presented in Figs. 5-2 through 5-21 and include:

Cell A—Control Test Cell. The relatively low leachate pH level coupled with a higher leachate volatile acid content implied that active anaerobic decomposition had been initiated. Methane had been measured in low concentrations indicating the presence of some methanogenic organisms, although conditions in this cell were far from optimum for these organisms. Phosphorus and nitrogen levels were lower than detected in other study leachates but still sufficient to sustain biological activity. Of the six heavy metals analyzed in the leachate, zinc, lead, and mercury were consistently found in quantities above or near the Drinking Water Standards (USPHS Standard: zinc $\leqslant$ 5 ppm, lead $\leqslant$ 50 ppb, and mercury $\leqslant$ 5 ppb).

Cell B—Field Capacity Test Cell. In general, anaerobic biodegradation occurred within the cell with little methanogenic organism activity. Correspondingly, pH levels were low, volatile acids were high, and methane production was

TABLE 5-5   Liquid Conditioning and Purpose of Sonoma County Test Cells

Cell Designation	Initial Liquid Conditioning	Liquid Used	Daily Liquid Application gal/day	Liquid Used	Purpose of Cell
A	None	None	None	None	Control cell
B	Field capacity[a]	Water	None	None	To determine the effect of high initial water content on refuse stabilization.
C	None	None	1000±	Water	To determine the effect of continuous water through flow on leachate character.
D	None	None	1000±	Recirculated water	To determine the effect of continuous leaching recirculation on leachate character.
E	Field capacity[a]	Septic tank pumpings	None	None	To determine the effect of high initial moisture content (using septic tank pumpings) on refuse stabilization.

[a]Field capacity is the condition when a sufficient quantity of fluid has been added to cause a significant volume of leachate to be produced from the cell.

Source: Sonoma County Dept. of Public Works and EMCON Associates, "Second Interim Annual Report," Vol. II, Solid Waste Disposal Demonstration Grant Project G-06-EC-00351, July, 1973.

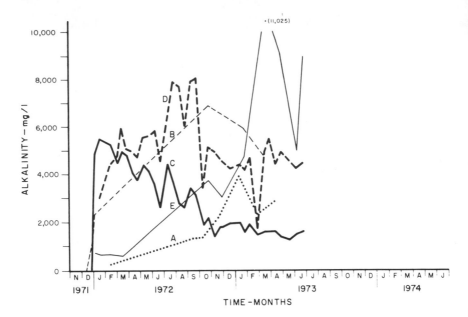

Fig. 5-2 Alkalinity of leachate in Sonoma County study. *Source:* Sonoma County Dept. of Public Works and EMCON Associates, "Second Interim Annual Report," Vol. II, Solid Waste Disposal Demonstration Grant Project G-06-EC-00351, July, 1973.

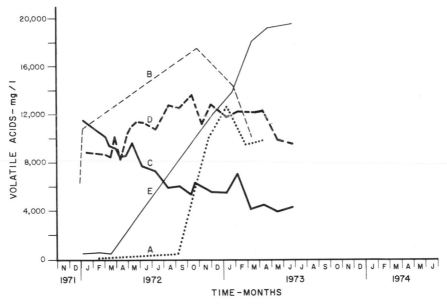

Fig. 5-3 Volatile acid concentration of leachate in Sonoma County study. *Source:* Sonoma County Dept. of Public Works and EMCON Associates, "Second Interim Annual Report," Vol. II, Solid Waste Disposal Demonstration Grant Project G-06-EC-00351, July, 1973.

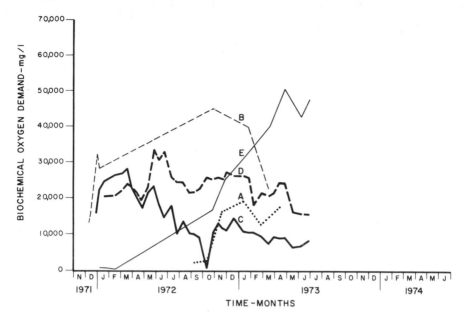

Fig. 5-4 Biochemical oxygen demand (BOD) of leachate in Sonoma County study. *Source:* Sonoma County Dept. of Public Works and EMCON Associates, "Second Interim Annual Report," Vol. II, Solid Waste Disposal Demonstration Grant Project G-06-EC-00351, July, 1973.

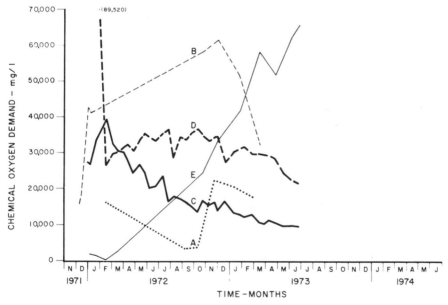

Fig. 5-5 Chemical oxygen demand (COD) of leachate in Sonoma County study. *Source:* Sonoma County Dept. of Public Works and EMCON Associates, "Second Interim Annual Report," Vol. II, Solid Waste Disposal Demonstration Grant Project G-06-EC-00351, July, 1973.

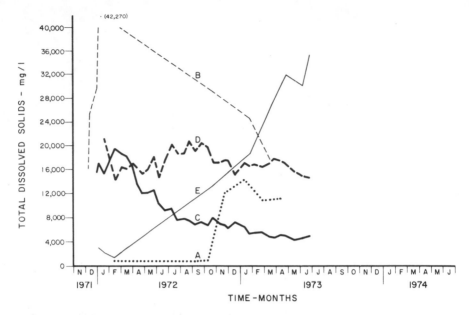

Fig. 5-6 Total dissolved solids content of leachate in Sonoma County study. *Source:* Sonoma County Dept. of Public Works and EMCON Associates, "Second Interim Annual Report," Vol. II, Solid Waste Disposal Demonstration Grant Project G-06-EC-00351, July, 1973.

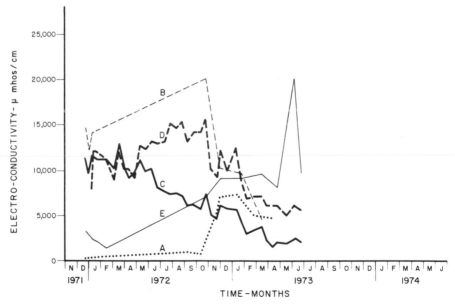

Fig. 5-7 Electroconductivity of leachate in Sonoma County study. *Source:* Sonoma County Dept. of Public Works and EMCON Associates, "Second Interim Annual Report," Vol. II, Solid Waste Disposal Demonstration Grant Project G-06-EC-00351, July, 1973.

250

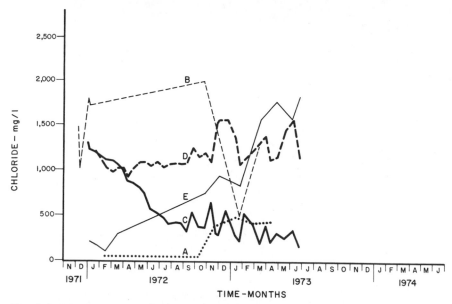

Fig. 5-8 Chloride concentration of leachate in Sonoma County study. *Source:*
Sonoma County Dept. of Public Works and EMCON Associates, "Second Interim Annual Report," Vol. II, Solid Waste Disposal Demonstration Grant Project G-06-EC-00351, July, 1973.

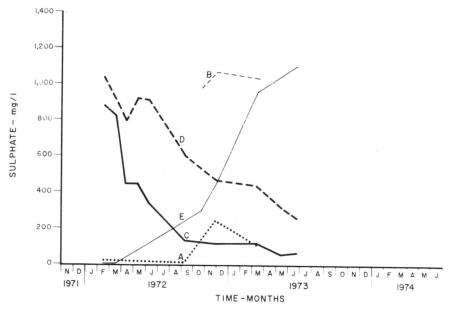

Fig. 5-9 Sulfate concentration of leachate in Sonoma County study. *Source:*
Sonoma County Dept. of Public Works and EMCON Associates, "Second Interim Annual Report," Vol. II, Solid Waste Disposal Demonstration Grant Project G-06-EC-00351, July, 1973.

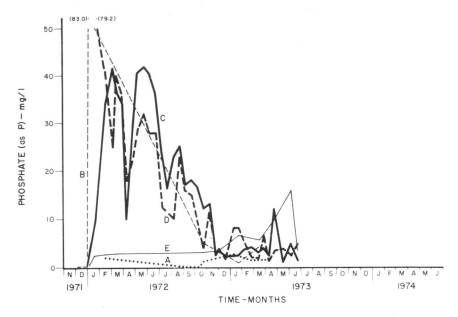

Fig. 5-10   Phosphate concentration of leachate in Sonoma County study. *Source:* Sonoma County Dept. of Public Works and EMCON Associates, "Second Interim Annual Report," Vol. II, Solid Waste Disposal Demonstration Grant Project G-06-EC-00351, July, 1973.

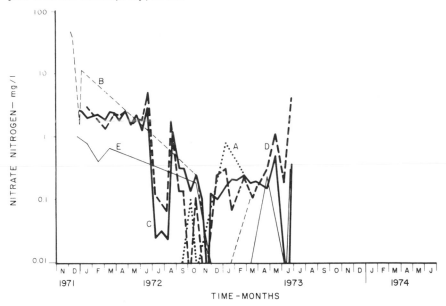

Fig. 5-11   Nitrate–N concentration of leachate in Sonoma County study. *Source:* Sonoma County Dept. of Public Works and EMCON Associates, "Second Interim Annual Report," Vol. II, Solid Waste Disposal Demonstration Grant Project G-06-EC-00351, July, 1973.

252

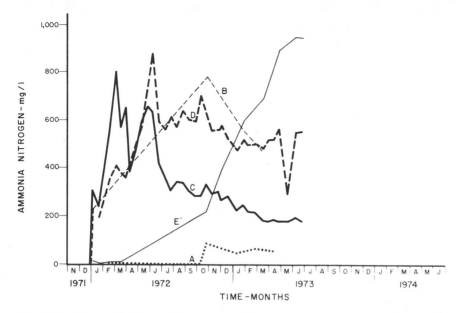

Fig. 5-12 Ammonia–N concentration of leachate in Sonoma County study. *Source:* Sonoma County Dept. of Public Works and EMCON Associates, "Second Interim Annual Report," Vol. II, Solid Waste Disposal Demonstration Grant Project G-06-EC-00351, July, 1973.

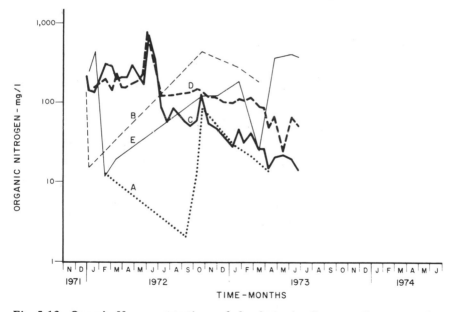

Fig. 5-13 Organic–N concentration of leachate in Sonoma County study. *Source:* Sonoma County Dept. of Public Works and EMCON Associates, "Second Interim Annual Report," Vol. II, Solid Waste Disposal Demonstration Grant Project G-06-EC-00351, July, 1973.

253

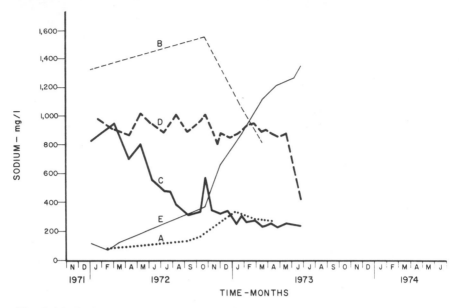

Fig. 5-14 Sodium concentration of leachate in Sonoma County study. *Source:* Sonoma County Dept. of Public Works and EMCON Associates, "Second Interim Annual Report," Vol. II, Solid Waste Disposal Demonstration Grant Project G-06-EC-00351, July, 1973.

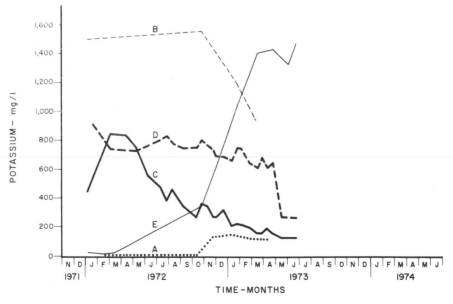

Fig. 5-15 Potassium concentration of leachate in Sonoma County study. *Source:* Sonoma County Dept. of Public Works and EMCON Associates, "Second Interim Annual Report," Vol. II, Solid Waste Disposal Demonstration Grant Project G-06-EC-00351, July, 1973.

254

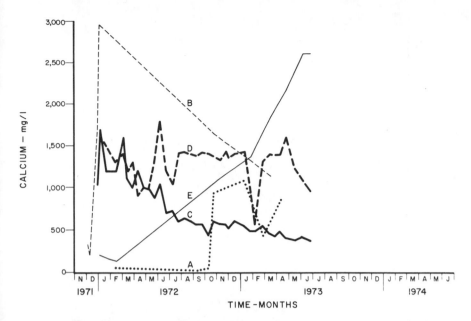

Fig. 5-16 Calcium concentration of leachate in Sonoma County study. *Source:* Sonoma County Dept. of Public Works and EMCON Associates, "Second Interim Annual Report," Vol. II, Solid Waste Disposal Demonstration Grant Project G-06-EC-00351, July, 1973.

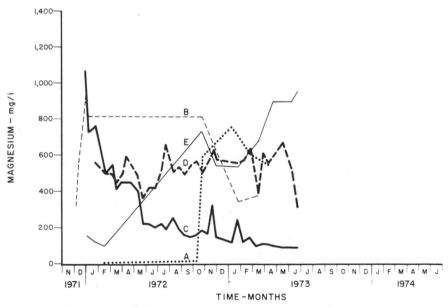

Fig. 5-17 Magnesium concentration of leachate in Sonoma County study. *Source:* Sonoma County Dept. of Public Works and EMCON Associates, "Second Interim Annual Report," Vol. II, Solid Waste Disposal Demonstration Grant Project G-06-EC-00351, July, 1973.

255

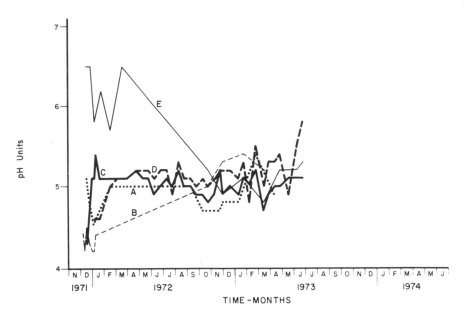

Fig. 5-18   pH of leachate in Sonoma County study. *Source:* Sonoma County
Dept. of Public Works and EMCON Associates, "Second Interim Annual Report,"
Vol. II, Solid Waste Disposal Demonstration Grant Project G-06-EC-00351,
July, 1973.

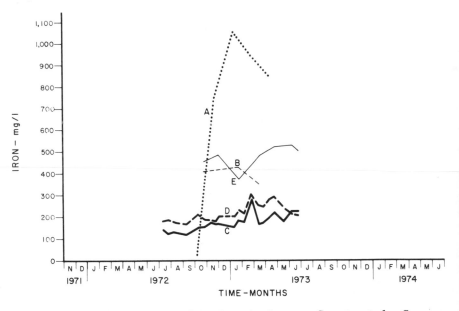

Fig. 5-19   Iron concentration of leachate in Sonoma County study. *Source:*
Sonoma County Dept. of Public Works and EMCON Associates,' "Second Interim Annual Report," Vol. II, Solid Waste Disposal Demonstration Grant Project G-06-EC-00351, July, 1973.

256

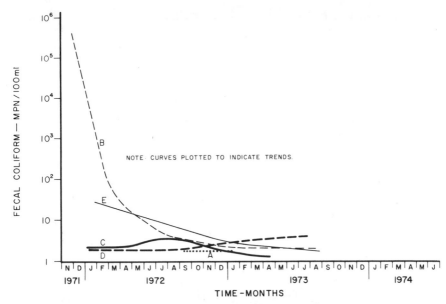

Fig. 5-20 Fecal coliform count of leachate in Sonoma County study. *Source:* Sonoma County Dept. of Public Works and EMCON Associates, "Second Interim Annual Report," Vol. II, Solid Waste Disposal Demonstration Grant Project G-06-EC-00351, July, 1973.

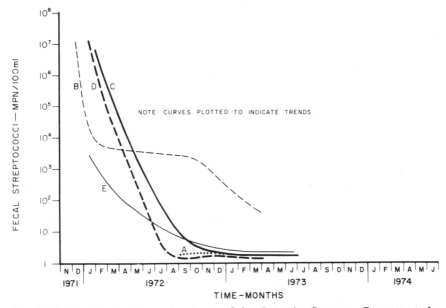

Fig. 5-21 Fecal streptococci count of leachate in Sonoma County study. *Source:* Sonoma County Dept. of Public Works and EMCON Associates, "Second Interim Annual Report," Vol. II, Solid Waste Disposal Demonstration Grant Project G-06-EC-00351, July, 1973.

extremely low. Trace metal data is similar to that developed for Cell A with zinc, lead, and mercury all at or near USPHS standards.

Cell C–Continuous Flow-Through Cell. All leachate parameters except pH have decreased with time demonstrating the general flushing action of the water added to the cell. Vigorous anaerobic biodegradation was indicated by BOD, COD, volatile acids, and gas composition data. Despite the low pH level, methane was produced at an increasing rate. An additional indicator of active biodegradation was the increasing percentage of ammonia-nitrogen and the decreasing percentage of organic-nitrogen. Although phosphorus and nitrogen levels indicated sufficient quantities existed for biological activity, the flushing action associated with this cell may eventually result in nutrient-limited conditions. The oxygen content of the water added to this cell may also eventually exert a detrimental effect on the methanogenic microorganisms which are strict anaerobes. Trace metal analysis indicated that all metals monitored were present in significant quantities except for cadmium. Zinc concentrations appeared to be decreasing with time indicating a flushing action. No other heavy metal trends were apparent. Additionally, the rate of settling of Cell C is more rapid than for Cells A, B, and E.

Cell D–Continuous Leachate Recycle Test Cell. Data suggested that active anaerobic biodegradation was occurring as indicated by high volatile acids and low pH levels. A sharp increase in methane production indicated methanogenic activity despite a pH level of approximately 5.0. The relatively large values of nitrogen and phosphorus forms suggested that these nutrients were not limiting. An accumulation of heavy metals in the leachate appeared to be occurring. The rate of settlement in Cell D was very similar to Cell C with approximately 3 or 4 in. (4 percent of the refuse depth) of settlement occurring in the 2-year study period.

Cell E–Biologically Seeded Test Cell. Data indicated that all leachate characteristics except pH were increasing in concentration with time. Initial biodegradation was accelerated due to seeding with septic tank pumpings (which is in agreement with Pohland's data). The nutrient levels present in the leachate indicated that nutrients will not limit biological activity. Leachate heavy metal content trends for cell E were very similar to Cells A and B.

Other study results indicated that the initial temperature response observed in the refuse cells was a quick rise during the first several weeks followed by a decrease to ambient air temperature in the upper layers. It was reported that the cell temperature was not affected by the absence or presence of a soil cover but was influenced by the temperature of the applied liquid when continuous liquid application was practiced.

The analysis of gases produced by the various cells indicated that the variation of carbon dioxide concentration with time is a reliable indicator of cell activity. Settlement measurements suggest that the initial settlement of the fill results from the densification of the refuse resulting from the daily addition of water; however, long-range settlements are the result of biological or chemical stabilization of the refuse. The amount of leachate retained or lost within the cells was greatest in Cell C and least in Cell D; leachate retention is believed to be greater in Cell C because of high evaporation losses caused by the use of less permeable material in the construction of the distribution system beneath the cell cover.

It should be noted that a substantially increased rate of waste stabilization is not currently being achieved at the Sonoma County project through the recirculation of landfill leachate. However, the test cells are not equipped to allow control of leachate pH characteristics. The elevation of leachate pH (from the low values associated with carbonic and volatile acid production) would greatly facilitate rapid biostabilization through an optimization of conditions necessary for the growth of methane-producing microorganisms. Also, the refuse placed in the study cells was not shredded as in Pohland's laboratory study. Such size reduction would greatly increase the surface area per unit weight of the refuse thereby serving to accelerate the rate of microbial stabilization as observed by Pohland.

5-2.3 Methodology—Landfill With Leachate Recirculation

The construction and operation of a landfill with leachate recirculation (LR landfill) closely parallels that of the conventional landfill with the exception that provisions must be made for the collection and recirculation of leachate and the monitoring of process parameters. For purposes of discussion, the operation of a LR landfill can be divided into three areas: basin construction, refuse placement and filling techniques, and monitoring.

LR Landfill Basin Construction. As in the case of conventional sanitary landfilling, it is essential that no groundwater contamination occur as a result of refuse deposition and degradation in conjunction with the LR landfill process. To prevent any leachate movement into an aquifer from the disposal basins, the bottom of the basins should be established below the permanent groundwater table and be provided with suitable collection sumps. The establishment of the basin bottoms below the water table and the pumping of all collected waters out of the collection sumps located in the basins creates a hydraulic regimen favoring flow of water toward the basins in general and the collection sumps in particular. Therefore, no hydraulic gradient exists for the flow of water away from the basins and out into adjacent aquifer layers. Even if the pumping system experiences temporary malfunction, the surrounding groundwater will initially invade the refuse disposal basins from which it can be easily removed. Flow will continue into the basins until the water level within the basins reaches the same level as the surrounding aquifer, thereby providing adequate time for pumping system repair.

The basin liner must be as impervious as practical to provide insurance against localized seepage of contaminants into the groundwater and to provide for leachate collection. Basin liners can be constructed of compacted soil in its natural state or mixed with a variety of soil additives including lime, pozzolana, or other soil cements. If lime is used, it will also serve as a buffering agent to assist in neutralizing leachate having low pH characteristics, thereby promoting an optimal microbial environment for solid waste degradation. In cohesionless soils or situations where the necessary degree of compaction is not practical, liners can be constructed of asphalt or polymer membranes.

Polymer membranes have not been used for solid waste disposal sites; however, they have been utilized to prevent seepage from reservoirs, ponds, pits, lagoons, and other operations requiring an impermeable liner. Therefore, the application of polymer membranes as impermeable liners for LR landfills seems appropriate. The materials used as membranes are high molecular weight synthetic organic polymers in the form of relatively thin, flexible, impermeable sheeting. The most frequently used polymer membranes are plasticized polyvinyl chloride (PVC), polyethylene, and butyl rubber (see Table 5-6).

If a polymer membrane is used for LR landfilling it must be covered for protection. The initial covering should consist of a fine-textured material which can be placed with a dragline, conveyor, or truck. Heavy equipment cannot move over the liners until they are protected with 6 to 12 inches of cover. Also, a side slope of at least 3:1 is necessary to assure stability of the cover material on the slope. Before placing the liner, all sharp rocks, stones, roots, and other sharp objects that might puncture the membrane should be removed or covered with a few inches of sand or other fine-textured soil (Ref. 5-33).

TABLE 5-6   Properties of Impermeable Membranes

	Vinyl	Butyl	Polyethylene
Tensile strength, psi	2,500	1,400	1,600
Elastic strain, percent	300	550	6
Puncture resistance	Highest	Good	Poor
Available gauges, mils	8–35	32–125	6–20
Available widths, ft	4–61	46–28	40–16
Method of joining	Heat, solvent or adhesive	Adhesive	Heat and adhesive
Weight, oz per sq yd per mil of thickness	1.0	1.0	0.8
Density, gm/cm^3	1.25	1.25	9.93
Cost	Medium	Highest	Lowest

Source: Staff, C. E., "Seepage Prevention with Impermeable Membranes," *Civil Engineering—ASCE*, February, 1967.

The type of liner to be used will depend on the hydrogeology of the site, the overall site design, materials available at the site, and economic considerations.

In a LR landfill, collection wells must be installed in each disposal basin to provide for leachate collection. A monitor well of sufficient depth to contain water throughout the year must also be provided in each disposal basin. The liner of each basin must be graded so that drainage will be directed toward the centerlines of the basin. The liner utilized must not only cover the bottom of the disposal basin but also must be extended upward around the basin sides to alleviate the potential for lateral leachate movement out of the cells into surrounding aquifers.

Perforated pipe laterals should be placed along the centerlines of the disposal basin in the LR landfill and covered with granular material such as coarse sand or gravel (see Fig. 5-22). These collection pipes should be connected to a vertical header pipe leading to the surface. To protect this header and the monitor well casing, and to provide access to the deposited refuse, an access tube should be positioned vertically in the center of the disposal basin. This access tube should be approximately 3 ft in diameter, fitted with sampling ports around its circumference for purposes of obtaining periodic samples of the degrading refuse, and covered with an unsealed manhole cover to reduce the possibility of hazardous conditions for operating personnel.

Vent pipes should be placed in each disposal basin extending from the surface to within several feet of the bottom granular drainage layer. Since each cell possesses an impervious liner, any escape of gas laterally from the cell will be

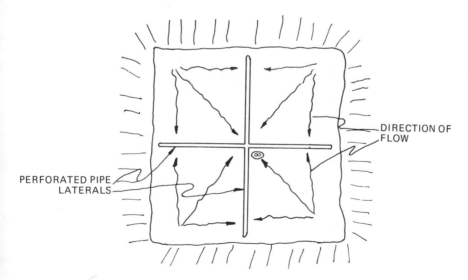

PERFORATED PIPE LATERALS

DIRECTION OF FLOW

Fig. 5-22   Arrangement of collection pipes in LR landfill basin bottom.

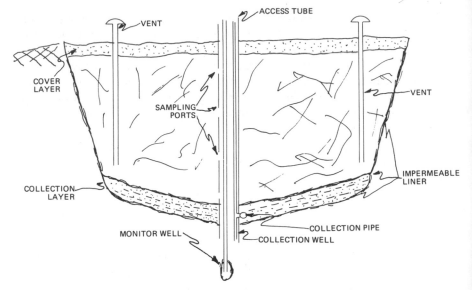

Fig. 5-23   Refuse cell design for a LR Landfill.

prevented. Therefore, vents installed vertically in the basins will collect and vent the gases not dissolved in the leachate to the atmosphere. Without such vent pipes, gravity drainage and vacuum pumping of collected leachate could be disrupted and dangerous gas accumulations could occur. It is important that the vent pipes do not extend into the granular layer, thereby avoiding oxygen contact with the recycled leachate. The introduction of oxygen in the leachate underdrain system might cause precipitation of chemical compounds (particularly ferric hydroxide) through the oxidation of various ions within the leachate. These precipitated compounds would significantly reduce the permeability of the granular underdrain layer.

Careful attention should be given to basin construction and design (see Fig. 5-23) since proper basin performance is essential to optimal LR landfill operation. The number of basins required for any particular landfill will depend upon the desired capacity of the landfill and the method of filling employed.

Refuse Placement and Filling Techniques. The disposal basins used in LR landfilling are filled according to standard methodology utilized in sanitary landfilling including compaction of refuse by tracked crawlers, daily cover, litter fences, etc. Lifts having an average thickness of 10 ft should be used, and the first 10-ft-thick layer of refuse should be placed over the entire cell before initiating the development of another lift. The major difference between conventional sanitary landfilling and LR landfilling results from the need for an irrigation and recirculation network between lifts in the LR landfilling process.

This system is then used to moisten the solid waste to elevate its moisture content to approximately 70 percent and for the subsequent recirculation of the leachate throughout the in-place refuse.

The collection and recirculation system for the leachate from the refuse cells must be designed to maintain the water table as close as possible to the bottom of the refuse cells. Water table control will insure that the potential within the landfill is always lower than that in the surrounding aquifer, and will facilitate the equal distribution of recirculated leachate throughout the in-place refuse. Discharge pipes on 20 ft centers over cells can be utilized to provide unhampered flow of recirculated leachate throughout the deposited solid waste. The discharge pipes should be installed to conform as nearly as possible with the face of the refuse and their locations should be staggered with respect to underlying collection pipes.

Because of the rapid decomposition of refuse, settlements as large as 25 percent of the total refuse depth can be expected during the first several years of leachate recirculation in any particular refuse cell. Therefore, the distribution and pumping systems must be placed with sufficiently flexible connections so that these piping systems will remain intact during the expected settlement period. A flexible piping system can be constructed through the utilization of irrigation pipes that are either connected to the manhole with flexible "accordion" joints or connected to an independent vertical header which can move downward with settling refuse. Also, since high temperatures will be generated as a result of the anaerobic decomposition of refuse, distribution system piping must be able to withstand temperatures of 160°F.

After the entire length of the fill area is covered with one refuse cell depth, placement of a second cell should commence at the original starting point and so forth until all cell areas have been filled. All appropriate vents and pipes must be extended for each subsequent refuse cell in accordance with the procedure discussed in the preceding paragraphs. Leachate pumping from the bottom collection system should be initiated during refuse placement to prevent any possible migration of leachate from the cell at any time.

In the unlikely event that the cell liner is ruptured, the pumping rate from the cell would be adjusted so as to insure a flow of the groundwater in the entire area into the cell, thereby preventing groundwater contamination until the liner can be satisfactorily repaired. Monitoring of groundwater quality will provide the information required to determine when such pumping may be necessary.

Monitoring of a LR Landfill. The basic philosophy behind the monitoring and testing of a LR landfill is to provide quantitative information concerning the following major considerations:

1. Analysis of the groundwater quality upstream and downstream of the landfill site before, during, and after proposed landfilling operations. Such data

will provide an overall assessment of initial groundwater quality conditions, thereby enabling a rational determination of whether or not landfill operations are responsible for any groundwater deterioration which may be noted during or after the landfill operation.

2. Determination of the degradation state of landfilled solid waste material throughout the proposed landfill operation. Such data will make possible a reasonable assessment of the solid waste decomposition rate, thus providing interested regulatory agencies basic information to show that their requests for accelerated solid waste degradation have been fulfilled.

3. Determination of chemical and biological parameters of recycled leachate during the proposed landfill operation. Such data will not only assist in an overall assessment of the solid waste decomposition state and the potential for leachate pollution of groundwater, but will also provide the municipal wastewater treatment facility with information required to formulate leachate treatment fees.

Initially, groundwater samples from the landfill site should be chemically analyzed before landfill operations begin, in order to determine:

1. Alkalinity
2. Aluminum
3. Arsenic
4. BOD_5
5. Boron
6. Bromide
7. Cadmium
8. Calcium
9. Chloride
10. Chromium $(^{+3}_{+6})$
11. Cobalt
12. COD
13. Copper
14. Cyanide
15. Hardness
16. Iron
17. Lead
18. Magnesium
19. Manganese
20. Mercury
21. Nickel
22. Nitrate
23. pH
24. Phenol
25. Phosphate
26. Potassium
27. Selenium
28. Sodium
29. Specific conductance
30. Sulfate
31. Total dissolved solids
32. Zinc

To determine groundwater quality during the landfilling period, monthly samples should be secured and analyzed for organic carbon content, pH, hardness conductivity, chloride, and specific heavy metals of interest. Such analyses will permit a constant check of the groundwater quality at the site, a consideration of paramount interest to concerned regulatory agencies.

Operational monitoring of the landfilled refuse and recycled leachate would encompass the remainder of the monitoring program. Associated parameters that should be monitored weekly, following landfill construction, include:

1. Volume reduction monitoring by direct observation of settlement.
2. Temperature profile study of the refuse cell by direct observation of a central thermometer connected to thermistors installed at various intervals throughout the cell depth.
3. Groundwater level below the cell by direct observation of the central study well.

The solid waste decomposition rate can also be estimated by analyzing typical wastes before and after landfilling for volatile solids and moisture content. Such analyses would be performed every 4 months throughout the monitoring period. The resulting data would define the ratio of organic material stabilized to the original organic content of the waste.

Additionally, the gases produced in the refuse cell should be monitored occasionally (every 4 to 6 months) to determine relative proportions of carbon dioxide (CO_2), oxygen (O_2), methane (CH_4), and carbon monoxide (CO).

Recycled leachate should be sampled at least monthly and analyzed to determine quality changes in organic content (BOD, COD, or TOC), pH, organic acids, total dissolved solids, hardness, phosphate, nitrogen, and one or more heavy metals. Such data collection permits an estimation of refuse stabilization, the potential for groundwater pollution, and the acceptability of leachate in biological wastewater treatment systems. Some general guidelines to be followed regarding the analysis of the leachate samples include (Ref. 5-34):

1. The analysis should seek accuracy instead of precision.
2. The high concentration of solutes and the changing nature of the leachate can interfere with analytical techniques and should be recognized as potential sources of error.
3. The high background color of the undiluted leachate and the complex nature of the solution indicate that colorimetric methods will not constitute a suitable analysis.
4. Calcium and magnesium cannot be analyzed by the normal EDTA titrimetric technique because the end point is masked by background color in undiluted leachate samples.
5. Because of color interference in undiluted leachate samples, chloride must be analyzed using a potentiometric titration technique utilizing an Ag/AgCl electrode.
6. The leachate will generally require dilution before analysis due to the high concentrations of most common constituents. Preliminary results from the Sonoma County project (Ref. 5-34) indicated that electroconductivity ratios may be of use in predicting dilution requirements for analyzing some parameters. In that project, good correlation was observed between values of electroconductivity and alkalinity, BOD, COD, and chloride (see Tables 5-7 and 5-8).

TABLE 5-7   Cell C Leachate Electroconductivity/Parameter Ratios

Parameter	Sample Date											Average Value
	1-18-72	2-15-72	3-2-72	3-14-72	3-28-72	4-11-72	4-25-72	5-9-72	5-23-72	6-6-72	6-20-72	
Alkalinity	2.01	2.10	2.25	2.53	2.11	2.22	2.53	2.53	2.38	2.78	3.08	2.41
BOD	0.45	0.42	0.37	0.44	0.44	0.44	0.55	0.52	0.41	0.56	0.54	0.46
Calcium	9.17	9.17	6.25	11.36	10.00	7.50	9.50	11.22	11.08	9.52	11.43	9.65
COD	0.33	0.28	0.31	0.41	0.33	0.32	0.39	0.41	0.40	0.49	0.39	0.37
Chloride	9.17	9.82	9.09	11.79	9.90	10.23	11.05	13.58	13.17	17.54	15.09	11.58
Magnesium	14.47	22.00	18.18	30.49	22.22	20.00	21.11	27.50	44.32	45.45	40.40	27.79
Potassium	–	–	11.83	–	–	10.75	–	14.67	–	17.86	–	13.78
Sodium	–	–	10.53	–	–	12.86	–	13.75	–	18.18	–	13.83
Sulfate	–	12.50	–	15.24	–	20.09	–	24.55	–	29.41	–	20.36
TDS	0.72	0.57	0.54	0.69	0.61	0.67	0.79	0.91	0.79	0.99	0.87	0.74

Source: Solid Waste Disposal Demonstration Grant Project G 06-EC-00351, Sonoma County Department of Public Works and EMCON Associates.

TABLE 5-8 Cell D Leachate Electroconductivity/Parameter Ratios

Parameter	Sample Date											Average Value
	1-18-72	2-15-72	3-2-72	3-14-72	3-28-72	4-11-72	4-25-72	5-9-72	5-23-72	6-6-72	6-20-72	
Alkalinity	3.93	2.47	1.88	2.02	1.98	2.02	1.91	2.27	2.18	2.24	2.83	2.34
BOD	0.59	0.53	0.41	0.50	0.44	0.46	0.45	0.54	0.36	0.42	0.39	0.46
Calcium	7.69	8.46	6.43	10.00	7.69	11.11	9.00	12.50	9.38	7.22	10.62	9.10
COD	0.13	0.42	0.30	0.40	0.31	0.31	0.29	0.37	0.34	0.38	0.38	0.33
Chloride	9.92	10.68	9.18	11.76	9.80	10.87	8.82	11.47	11.19	12.38	11.59	10.70
Magnesium	21.43	22.00	18.00	26.67	20.00	16.67	16.36	25.00	33.89	30.95	30.36	23.76
Potassium	13.19	–	12.16	–	–	13.75	–	17.19	–	17.10	–	14.68
Sodium	12.24	–	10.00	–	–	11.63	–	12.25	–	13.68	–	11.96
Sulfate	–	10.58	–	13.04	–	12.59	–	13.59	–	14.32	–	12.82
TDS	0.57	0.77	0.55	0.75	0.59	0.62	0.59	0.77	0.68	0.89	0.73	0.68

Source: Solid Waste Disposal Demonstration Grant Project G 06-EC-00351, Sonoma County Department of Public Works and EMCON Associates.

Verification of a stabilized solid waste decomposition state can be obtained through the interpretation of stabilization parameters. Stabilized conditions are realized when the decomposable organic solid waste material has been completely metabolized by both the organic acid-forming and methane-producing bacteria. Indicators of this stabilization include:

1. Solid waste volatile solids reduction of 25 to 50 percent.
2. Completion of cyclic increase and decrease of leachate organic acid content.
3. Completion of cyclic increase and decrease of leachate BOD_5.
4. Observation of extended methane gas production increase.

Following the completion of the landfilling operation, groundwater quality at the site should be monitored a minimum of once a year to determine whether the potential for delayed groundwater impairment may exist.

5-2.4 Method Parameters

The LR landfilling process accelerates the decomposition of solid waste material through the maintenance of optimal anaerobic conditions, thereby stabilizing the landfill more rapidly than conventional sanitary landfilling processes. Anaerobic decomposition of solid waste material in a LR landfill will result in:

1. The breakdown of complex organic solid waste material into simple compounds. If the process is continued long enough, an inoffensive end product will result.
2. The transformation of a portion of the organic solid waste material into liquids and gases, thereby reducing the overall solid waste volume. The volatile fraction of the solid waste material should be reduced 25 to 50 percent in dry weight.

The anaerobic decomposition process proceeds in two well-defined stages in the LR landfill method:

1. A period of acid production—this stage begins immediately after solid waste deposition wherein the more easily decomposable compounds are attacked. Facultative anaerobic bacteria will predominate during this stage and convert carbohydrates, proteins, and fats into organic acids and alcohols. Consequently, pH levels will decrease during this period if control is not practiced. The organic content (BOD_5) of the leachate from the fill will increase drastically during this period.
2. A period of gas production—this stage provides for the conversion of organic acids and alcohols by anaerobic methane-fermenting bacteria into methane and carbon dioxide. Consequently, pH levels will increase during this period and a concurrent decrease in leachate BOD_5 will be observed.

Important stabilization parameters to be monitored to determine optimum anaerobic decomposition conditions are directly related to these two stages of anaerobic decomposition and include:

1. Organic acid production—since various organic acids (acetic, propionic, butyric, and valeric) initially increase in the recirculated leachate during the period of acid production and then decrease substantially during the period of gas production, monitoring of leachate organic acid content will result in the typical data shown in Fig. 5-24.
2. Leachate BOD_5 concentration—since leachate organic acid content will increase and then decrease during the acid production and gas production stages respectively, a concurrent increase and decrease of leachate BOD_5 content should also be observed. This leachate BOD_5 increase is that which poses the potential groundwater pollution problem observed at many conventional landfill sites. In a conventional landfill this BOD_5 increase will occur over a long period of time. By contrast, in the LR landfill process, this leachate BOD_5 increase and decrease cycle will most likely be completed in several years. This rapid reduction in leachate BOD_5 content is the primary advantage of operating a LR landfill from an environmental impact point of view. A comparison of leachate BOD_5 variations in conventional and LR landfills is shown in Fig. 5-25.
3. Methane gas production—obviously another important parameter would be methane gas production at the site. An increase in methane gas would signal the beginning of the gas production stage.

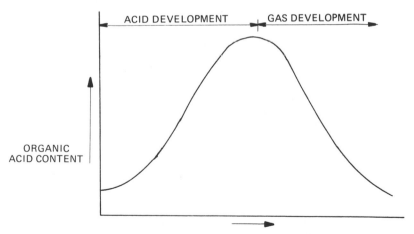

Fig. 5-24 Acid and gas stages in a LR Landfill.

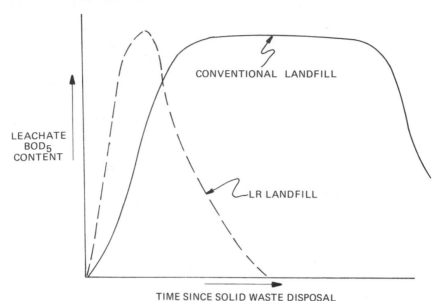

Fig. 5-25   Comparison of BOD Content in Conventional and LR landfills.

Monitoring these above-mentioned parameters will make it possible to determine if the anaerobic decomposition process is progressing at the proper rate in the refuse cache, and at the same time indicate what actions would be necessary to correct any discrepancies.

As mentioned previously, the parameters which affect the anaerobic decomposition of solid waste within the landfill include refuse composition, temperature, and moisture content. Refuse composition cannot be practically controlled at the landfill site; however, the temperature and moisture content of the in-place refuse can be adjusted through the recirculation of leachate.

5-2.5 Method By-Products

There are only two by-products from a properly operated LR landfill, recirculated leachate and gases emitted during anaerobic digestion. The recirculated leachate from a LR landfill should pose no problems regarding biological wastewater treatment. Values for an estimated range of the significant leachate parameters observed during Pohland's previously cited study (Ref. 5-26) include:

Biochemical Oxygen Demand (BOD_5)	200–750 mg/l
Total Suspended Solids (TSS)	100–300 mg/l
Total Dissolved Solids (TDS)	500–1200 mg/l
Hardness	200–400 mg/l

pH	7.2
Phosphate	0.1 mg/l
Nitrogen	20 mg/l

Corresponding parameters of a leachate derived from a conventional landfill without recirculation would be expected to be as follows:

Biochemical Oxygen Demand (BOD_5)	7500–10,000 mg/l
Total Suspended Solids (TSS)	400–600 mg/l
Total Dissolved Solids (TDS)	4000–6000 mg/l
Hardness	1400–4000 mg/l
pH	5.0–5.5
Phosphate	0.5–1.5 mg/l
Nitrogen	50–100 mg/l

Recent studies of leachate from conventional landfills have shown that such leachate could be added to domestic wastewater in an extended aeration-activated sludge plant at a level of at least 5 percent by volume without impairing effluent quality (Ref. 5-16). Since the characterizations given show that recirculated leachate from a LR landfill exhibits higher water quality characteristics than conventional landfill leachate, no problems should be encountered regarding leachate treatability in conventional biological treatment facilities.

An extension of previous leachate treatability studies at the University of Kentucky (Ref. 5-11) has demonstrated that the activated sludge process is an effective and efficient means of stabilizing sanitary landfill leachate. Optimal operational characteristics for typical leachate treatment during this study included a 10-day detention time corresponding to a mixed liquor volatile suspended solids concentration of at least 4400 mg/l in completely mixed, no-recycle systems. Over 97 percent COD removal efficiency was maintained in this investigation by activated sludge treatment alone, and over 99 percent COD removal efficiency was achieved with effluent polishing. Even when it is not economically feasible to construct an independent wastewater treatment facility at a landfill site, this recent study indicates that conventional landfill leachate (and certainly LR landfill leachate) should have no adverse effects on the performance of conventional activated sludge treatment processes.

The gases produced in conjunction with a LR landfill are identical to those generated during any anaerobic digestion process. These gases may be vented to the atmosphere according to EPA standards, or methane may be collected and utilized as a fuel source to generate power required to operate pumps and recirculate leachate.

In summary, the by-products from LR landfilling should not differ considerably from the by-products of conventional sanitary landfilling; however, the potential for groundwater pollution through leachate leakage is virtually eliminated during LR landfilling operations.

5-2.6 Method Requirements

The basic requirements for the operation of a LR landfill are quite similar to those for a conventional landfilling operation, i.e., weigh station, scales, litter fences, compaction equipment, etc. The major differences in requirements between the two methods results from the leachate recirculating system utilized in the LR landfilling process. This recirculation system would include an external water source, a pumping station, a holding lagoon, and a distribution network (refer to Fig. 5-26).

An external water source is necessary to increase the moisture content of the compacted refuse to 75 percent of the dry weight. The amount of water required for any particular LR landfill can be estimated using the following sample illustration:

Assume moisture content of the refuse at field capacity or slightly above, say 35 percent. Assume specific gravity of refuse = 1.00, and assume maximum dry density of refuse at approximately 26 lb/cu ft.

Volume, cu ft	Before	Weight, lbs	
0.41	Solids	26.00	$\dfrac{9.10}{26.00} = 0.35 = 35\%$
0.15	Water	9.10	Final % required = 75%
0.44	Gas	0.00	Add 75%–35% = 40%
1.00		35.10	

$$\text{Water required} = (26.0)(0.40) = 10.4 \text{ lb/cu ft}$$
$$= 33.66 \text{ gal/cu yd}$$

Therefore, approximately 34 gallons must be added per cubic yard of refuse to increase its moisture content from 35 to 75 percent.

The greatest economy can be realized if the water requirements are fulfilled by utilizing a suitable wastewater which would normally be discharged into a municipal wastewater treatment facility. Wastewater will not only provide a very inexpensive water source, but the nutrient-rich wastewater will also accelerate the

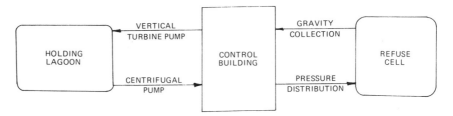

Fig. 5-26 Distribution system in a LR landfill.

anaerobic decomposition of refuse as shown by the Sonoma County project (Ref. 5-34). A concurrent benefit derived from wastewater utilization is the reduction of the load on the municipal wastewater treatment facility since:

1. About 20 percent of the wastewater will be "ultimately" disposed of during LR landfilling through evaporation and other losses.
2. The quality of the leachate discharged to the treatment facility will often be higher than that of the original wastewater.

If a suitable source of wastewater is not available then it will be necessary to purchase the required amount of water, thereby increasing the cost of LR land-filling operations.

The pumping station, or control building, represents a major capital cost difference between conventional sanitary landfilling and LR landfilling. The purpose of the pumping and control building is to collect leachate from the refuse cells and pump it either back into the cell or into the holding lagoon for treatment. The pumping station also represents an increase in operational costs over conventional landfills because of the increase in energy required to operate the pumping equipment. The pumping station should be of sufficient size to handle the anticipated daily flow of leachate from the refuse cells. The amount of daily flow expected can be calculated in the following manner:

$$q = \text{amount of leachate collected}$$

$$= 3.14 \, (P)(D^2 - d^2)/\left(2.3 \log\left(\frac{R}{r}\right)\right)$$

where

P = permeability coefficient (approximately 300 gpd/sq ft for refuse)
R = equivalent radius of landfill (ft)
r = radius of collection well (ft)
D = depth of zone in ft (well below static water table)
$d = D -$ well drawdown (ft)

For example, if

$R = 200$ ft, $r = 4$ ft, $D = 40$ ft, and $d = 10$ ft

then

$$q = 3.14 \, (300)(40^2 - 10^2)/\left(2.3 \log\left(\frac{200}{4}\right)\right)$$

$$= 221,000 \text{ gpd}$$

The holding lagoon should be utilized intermittently for leachate treatment and monitoring. However, under design operating conditions, leachate is recirculated directly into the landfill cells by means of a pipe bypass around the holding lagoon. The lagoon should be located above the statistical 100-year flood level and possess an impervious liner to prevent the possibility of groundwater contamination.

The final element in the leachate recirculation system is the pipe distribution network which is necessary for gas collection, leachate collection, and leachate recirculation. The piping utilized must be able to withstand major deformations as a result of large settlements, high temperatures ($160°F$), and corrosion from anaerobic decomposition products. The amount of piping needed will necessarily depend upon the size of the landfill; however, it must be arranged within the refuse (as outlined in the section on methodology) to assure proper operation of the LR landfill recirculation system.

A gas pump will also be required to pump gas from the gas probes. This pump may also function as a gas analyzer which registers the potential for a gas explosion hazard. It should be noted that the lighter volatile gases will accumulate at the top of the gas probe resulting in an excessively high initial reading of volatile organic gas production (Ref. 5-34).

5-3. SUMMARY—INNOVATIONS IN DISPOSAL

Both medium-temperature and high-temperature incineration and landfill with leachate recirculation appear to have advantages over conventional disposal methods. The medium-temperature and high-temperature incineration processes will optimally provide a volume reduction of about 98 percent and produce a totally inert residue which reportedly has several beneficial uses. The LR landfill process optimally alleviates groundwater pollution problems and returns the site to a wider range of land uses in a short period of time. However, since both processes are still in the developmental stage, many basic process problems exist. Only full-scale testing of these two innovations in disposal will conclusively determine their actual economic and operational feasibility in the field of solid waste management.

REFERENCES

5-1. Anderson, J. R. and Dornbush, J. N., "Influence of Sanitary Landfill on Groundwater Quality," *Journal of the American Water Works Association*, Vol. 59, No. 4, 1967, p. 457.

5-2. Calvert, C., "Contamination of Ground Water by Impounded Garbage Water," *Journal of the American Water Works Association*, Vol. 24, 1932, p. 266.

5-3. Carpenter, L. V. and Setter, L. R., "Some Notes on Sanitary Landfills," *American Journal on Public Health*, Vol. 30, 1940, p. 385.

5-4. Chaw-Ming Mao, M. and Pohland, F. G., "Continuing Investigations on Leachate Stabilization With Leachate Recirculation, Neutralization, and Seeding," Special Research Report, Georgia Institute of Technology, September, 1973.

5-5. Coe, J. J., "Effect of Solid Waste Disposal on Ground Water Quality," *Journal of the American Water Works Association*, Vol. 62, 1970, p. 12.

5-6. Davison, A. S., "The Effect of Tipped Domestic Refuse on Ground Water Quality," *Journal—Society for Water Treatment and Examination*, Vol. 18, 1969, p. 35.

5-7. Eliassen, R., "Decomposition of Landfills," *American Journal on Public Health*, Vol. 32, No. 9, 1942, p. 1029.

5-8. Environmental Protection Agency, "Demonstration of a High Temperature Vortex Incineration System," Demonstration Grant No. DO1-U1-00240.

5-9. "Evaluation of the Melt-Zit High Temperature Incinerator," USPHS Operation Test Report, Cincinnati, 1968.

5-10. Fife, J. A., "Solid Waste Disposal: Incineration or Pyrolysis," *Environmental Science and Technology*, Vol. 7, No. 4, April, 1973.

5-11. Foree, E. G. and Cook, E. N., "Aerobic Biological Stabilization of Sanitary Landfill Leachate," Office of Research and Engineering Services, University of Kentucky, September, 1972.

5-12. Fungaroli, A. A., "Hydrologic Considerations in Sanitary Landfill Design and Operation," Reprint Paper, National Industrial Solid Wastes Management Conference, March, 1970.

5-13. Fungaroli, A. A. and Steiner, R. L., "Laboratory Study of the Behavior of a Sanitary Landfill," *Journal—Water Pollution Control Federation*, Vol. 43, No. 2, February, 1971.

5-14. Hagerty, D. J. and Pavoni, J. L., "Geologic Aspects of Landfill Refuse Disposal," Presented at Southeast Section Meeting, GSA, Tuscaloosa, Alabama, March, 1972.

5-15. Hagerty, D. J., Pavoni, J. L. and Heer, J. E., Jr., *Solid Waste Management*, Van Nostrand Reinhold, New York, 1973.

5-16. Ham, R. K., "A Study of the Relative Attractiveness of Milled vs. Unmilled Refuse to Rats and Flies," Department of Civil Engineering, University of Wisconsin.

5-17. Harrington, W. M., "Sanitary Landfill Design Considerations to Protect Water Supplies," *Willing Water*, Vol. 15, April, 1971, p. 4.

5-18. Hershaft, A., "Solid Waste Treatment Technology," *Environmental Science and Technology*, Vol. 6, No. 5, May, 1972.

5-19. Hopkins, G. J. and Popalisky, J. R., "Influence of an Industrial Waste Landfill Operation on a Public Water Supply," *Journal—Water Pollution Control Federation*, Vol. 42, 1970, p. 431.

5-20. Hughes, G. M., Landon, R. A. and Favolden, R. N., "Summary of Findings on Solid Waste Disposal Sites," *Environmental Geology Notes*, 1971, p. 45.

5-21. Hughes, G. M., Landon, R. A. and Farvolden, R. N., "Hydrogeologic Data From Four Landfills in Northeastern Illinois," *Environmental Geology Notes*, 1969, p. 26.

5-22. Hughes, G. M., "Selection of Refuse Disposal Sites in Northeastern Illinois," *Environmental Geology Notes*, 1967, p. 17.

5-23. Lang, A., "Pollution of Ground Water By Chemicals," *Journal—American Water Works Association*, Vol. 33, 1941, p. 2075.

5-24. Mead, B. E. and Wilkie, W. G., "Leachate Prevention and Control from Sanitary Landfills," *Waste Age*, Vol. 3, 1972, p. 8.

5-25. Merz, R. C. and Stone, R., "Special Studies of a Sanitary Landfill," Final Summary Report, USPHS Publ. No. UI00r18-08, 1968.

5-26. Pohland, F. G., "Landfill Stabilization with Leachate Recycle," Annual Progress Report, EPA No. EP-00658, July, 1972.

5-27. Process Plants Corporation, Statement Issued in Hearings of the Senate Committee on Public Works on the Resource Recovery Act of 1969, March 4, 1970, Washington, p. 2204.

5-28. Qasim, S. R. and Burchinal, J. E., "Leaching from Simulated Landfills," *Journal—Water Pollution Control Federation*, Vol. 42, 1970, p. 371.

5-29. Qasim, S. R. and Burchinal, J. E., "Leaching Pollutants from Refuse Beds," *Journal—Sanitary Engineer Division, American Society of Civil Engineers,* Vol. SA1, 1970, p. 49.

5-30. Remson, J., Fungaroli, A. A. and Lawrence, A. W., "Water Movement in an Unsaturated Sanitary Landfill," *Journal—Sanitary Engineer Division, American Society of Civil Engineers*, Vol. SA2, 1968, p. 307.

5-31. "Report on the Investigation of Leaching of a Sanitary Landfill," State Water Pollution Control Board, Publication No. 10, Sacramento, California, 1954.

5-32. Salvato, J. A., Wilkie, W. G. and Mead, B. E., "Sanitary Landfill Leachate Prevention and Control," New York State Department of Health—Division of General Engineering and Rad. Health, 1971.

5-33. "Seepage Prevention with Impermeable Membranes," *Civil Engineering*, American Socity of Civil Engineers, February, 1967.

5-34. Sonoma County Department of Public Works and EMCON Associates, "First Interim Annual Report," Vol. I, Solid Waste Disposal Demonstration Grant Project G 06-EC-00351, July, 1972.

5-35. Sonoma County Department of Public Works and EMCON Associates, "Second Interim Annual Report," Vol. II, Solid Waste Disposal Demonstration Grant Project G-06-EC-00351, July, 1973.

6

Materials recovery

6-1. INTRODUCTION

A visitor to a sanitary landfill or to the storage area of a
municipal incinerator generally is appalled at the sight of so
much material being wasted. Municipal wastes amount to
approximately 200 million tons per year in the United States
alone. In addition to the municipal wastes, industrial solid
wastes which generally are collected and partially recycled
"in-house" amount to at least as much material as the municipal
refuse. Overshadowing wastes in both of these categories is the
amount of wastes generated each year in agriculture—over a
billion tons in the United States alone. Finally, the real giant
of waste producing activities is mineral extraction; almost
$1\frac{1}{2}$ billion tons of ore wastes are generated each year in the
continental United States.

When the finite limits of mineral resources are considered,
and when the above-mentioned amounts of solid wastes are
surveyed, the inevitable question arises "How much of these

wastes can be saved and recycled?" A great deal of the material in the solid waste stream is potentially recoverable, and an intensive search is under way for feasible and practicable materials recovery systems. To date, the success of materials recovery efforts has been extremely variable; outstanding successes in recycling certain materials have been accompanied by dismal failures in attempts to reuse other waste materials.

The materials which can be recovered from solid wastes (secondary materials) may be categorized into two general groupings: those materials which can be directly recycled, directly put back into use; and those materials which require considerable amounts of processing before they can be reused. An example of waste which may be directly recycled is the in-house glass waste produced during the manufacture of clear milk bottles. This glass waste, called cullet, may be collected and reinserted into the primary furnace in a glassmaking operation with no detrimental effects. On the other hand, a newspaper which travels to a consumer and thence into the solid waste stream must first be separated from the remainder of the refuse, then shredded and physically manipulated to separate individual fibers, processed in a deinking mill, and finally returned to a papermill for reuse. Obviously, it is much easier to recycle glass cullet than to recycle newspapers.

Because of the ease of reuse of in-house wastes mentioned in the previous paragraph, most of the secondary materials recovered today are retrieved as a result of direct recycling programs. In other words, industry recycles in-house wastes. Such industrial wastes are generally uncontaminated and are easily accessible. Thus, a recycling operation on in-house waste is easy and economical. The only obvious exception to this general statement is the type of waste produced in the textile industry, where the production of mixed-fiber materials complicates the reuse and recycling of fiber materials.

A second type of so-called "direct" recycling is being done today in the United States by citizens who voluntarily separate certain portions of the waste stream for recycling. For example, much of the current recycling of paper and textile wastes is the result of voluntary separation by the consumer population. Also, the aluminum and glass processing industries recently have mounted efforts to encourage the consumer to recycle aluminum and glass products, especially packaging products. These efforts have met with limited success.

Finally, another area where direct recycling is being practiced is the recovery and reprocessing of scrap metal. Every year in the United States, between 4 and 5 billion dollars worth of scrap metal is recycled and reused in metal processing. Comprehensive data on amounts and value of materials recycled in the United States are contained in Table 6-1. Further discussion of direct recycling of materials is contained in section 6-2.

Recovery efforts aimed at retrieving valuable materials from mixed municipal refuse have been relatively unsuccessful to date. Each year millions of tons of

TABLE 6-1   Consumption of Selected Recycled Materials[a]

Material	Consumption of Recycled Material (1000 Tons)	Value ($ Million)	Recycled Material as Percent of Total Consumption
Aluminum	1,056	553	23
Copper and copper base alloys	1,489	1,460	46
Ferrous	65,000	3,000	49
Lead	585	175	38
Nickel and nickel base alloys	42.1	209	29
Zinc	182	53	12
Precious metals (gold, silver and platinum)	79×10^6 Troy ounces	487	40
Paper	11,400	250	19
Textiles	1,400	84	27

[a]1969 figures with the exception of ferrous materials which are 1970 estimates.

Source: National Association of Recycling Industries, Inc., "A Study to Identify Opportunities for Increased Solid Waste Utilization," prepared by Battelle Memorial Institute, Vols. 1–9, New York, N.Y., 1972.

potentially recoverable materials are burned in incinerators and/or buried in sanitary landfills because no practicable method for separation and recovery of these materials has been developed. The bulk of this mixed refuse is paper. The composition of the municipal refuse stream, for example, has been discussed in previous chapters of this book dealing with incineration and composting. Generally, paper products amount to more than 50 percent by weight of the solid waste stream. On the other hand, valuable materials such as metals form much smaller fractions of the total amount of wastes. For example, metallic materials amount to about 9 percent by weight of the waste stream, glass accounts for approximately 9 percent by weight, and plastics amount to only 1 or 2 percent by weight of the total wastes generated. Virtually nonrecoverable food and garbage wastes amount to approximately 14 percent by weight of the solid waste stream. This mixed, heterogeneous character of solid waste is a major factor in the failure of recycling efforts aimed at recovering valuable materials from refuse. In addition to the expenditures of labor and money involved in separating desirable materials from the remainder of the waste stream, other factors also have militated against recycling efforts. For example, in many cases transportation costs for recycled waste materials are higher than the corresponding costs for virgin materials composed of the same substances. Additionally, a general lack of acceptance of recycled materials and the consequent lack of markets for those materials also has done much to precipitate the failures which occurred in recycling efforts. Sections 6-3 and 6-4 contain detailed discussions of the problems associated with recovering materials from mixed refuse.

Before proceeding any further in this discussion of materials recovery, it is pertinent to investigate the potential for recovery and reuse of the various materials and products found in the waste stream. In general, the potential for reuse of a particular product or material depends to a great extent upon the use for which the product was originally designed and manufactured. Materials may be in permanent use or in temporary use. Also, they may be used in such a way as to be nonrecoverable.

Any differentation between "permanent use" and "temporary use" will be arbitrary. An arbitrary dividing line of some practicality, however, can be based on a product life of approximately 7 to 10 years. In other words, a product in use for less than about 7 years may be considered to be in temporary use. Products in use in their original form for more than 10 years may be considered to be in permanent use. Examples of permanent-use products include structural components (wooden members, electrical wiring, structural metal members, concrete blocks, mass concrete, and pipes for plumbing), business and industrial machines, ships, trains, and hardbound books. On the other hand, examples for products in temporary use would include automobiles, magazines, newspapers, and packaging materials. The metal, plastic, and glass containers used in the packaging industry are in use for very short periods of time and therefore constitute a prime target of recycling efforts. Finally, some items are nonrecoverable: examples include facial and toilet tissues, plastic wrapping films, and protective coatings applied in packaging to various materials. Nonrecoverable products may amount to significant percentages of any given type of material which enters the waste stream. For example, approximately 12.5 percent of all paper products and 4 to 5 percent of all plastic products are nonrecoverable items.

Obviously, recycling efforts can best be focused on materials which fall in the temporary-use category.

6-2. DIRECT RECYCLING

Simple or direct recycling of waste materials may be arbitrarily divided into two categories of activity: in-house recovery of industrial wastes, and voluntary separation and recovery of municipal refuse components.

6-2.1 In-House Recovery

It is extremely difficult to obtain information about the in-house recovery of materials from any source other than industrial processors or associations of industrial processing groups. For this reason, the data which are presented in this section must be judged from the viewpoint of a prejudicial background; i.e., the amounts and percentages of recoveries listed in this section are those reported by members of the National Association of Secondary Materials Industries. There is little reason to believe that the data contained herein are in any way erroneous. Nevertheless, it is pertinent to point out the source of the information for the reader.

Metals. Recycling rates for in-house wastes in the metal processing industries are variable between 75 and 100 percent. There is little difficulty in reintroducing pieces from metal trimming or similar processing operations into furnaces. Most of the losses which do occur are a result of the manufacture of composite materials; wasted pieces of such materials are difficult to recycle because of the combinations of metals. An example is galvanized steel; obviously the steel and zinc metals in the composite product would be very difficult to separate. For this reason, the galvanized steel cannot be directly recycled but must be subjected to a chemical process.

Ferrous metal recycling rates are high. However, many ferrous metal products go into permanent use. Thus, the percentage of recycled ferrous materials in new products may never be predominant. For example, in the United States in 1970, 131 million tons of raw steel were produced. In producing this new steel, almost 25 million tons of in-house ferrous scrap were used (Ref. 6-31). Although this represents less than 20 percent of the new material production, the 25 million tons of scrap represent virtually 100 percent of the in-house wastes produced in

TABLE 6-2   Metals Recycling Rates

Material	1000 Tons of In-House Waste Recycled (1969)	Percent of In-House Waste (1969)
Aluminum	855	100
Copper	832	100
Lead	88	100
Zinc	141	68
Nickel	17.7	82
Stainless steel	219.5	100
Precious metals	(Troy ounces)	—
Gold	1,350,000	90
Silver	32,000,000	93
Platinum	540,000	90

Source: National Association of Recycling Industries, Inc., "A Study to Identify Opportunities for Increased Solid Waste Utilization," prepared by Battelle Memorial Institute, Vols. 1-9, New York, N.Y., 1972.

the steel processing operation. The recycling rate for in-house waste is high for many nonferrous metals also. (Table 6-2 provides detailed information on these metals.) Two exceptions (low recoveries of in-house wastes) are nickel and zinc. The low recycling rate for zinc is the result of inevitable losses during the production of brass and other losses from the manufacture of galvanized steel; much of the zinc which is used as coating for the galvanized steel is lost when the scrap steel is recycled. Likewise, most of the nickel losses may be attributed to the recycling of nickel steel alloys; complex separation methods are not justified for the minute amount of nickel compared to the amount of steel in those alloys.

Plastics. Most of the plastic wastes which are recycled are reprocessed as in-house wastes. If a plastic waste is produced after the material leaves the processing or manufacturing plant, it is virtually certain that the waste will not be recycled. For example, a certain amount of waste occurs during the processing of plastic film. That waste produced in-house has a good chance of being reused in a film form after it has been reprocessed. However, if the film is used to make a protective package for fresh produce, the film is almost certain to find its way into a sanitary landfill or a refuse incinerator.

If 1970, over 18 billion pounds of various types of plastic items were produced in the United States. Of this total production approximately 3.5 billion pounds were lost during manufacture and processing and ended as "prompt" industrial scrap. Almost 2½ billion pounds of this prompt industrial scrap were recycled by plastic processors. The remainder of the scrap entered the solid waste stream. The entrance of over 1 billion pounds of in-house plastic scrap into the solid waste stream presents a significant and important target for future recycling efforts (Ref. 6-17).

Wood. Because of sophisticated technological developments in wood processing, the wood industry produces a minimum amount of solid waste for disposal. Previously, lumber mill wastes such as sawdust, wood chips, and bark slabs were of little use and generally were eliminated by burning. At the present time, these mill wastes can be economically and efficiently converted into pulp for paper production (Ref. 6-10). Approximately one-fourth of all the paper pulp produced each year in the United States originates from the waste materials formerly discarded from wood processing operations. There is little room for improvement in the in-house recycling of wood wastes.

Glass. Only about 5 percent of all glass produced each year in the United States originates from recycled material. Almost all of the material which is recycled is generated from industrial sources; i.e., almost all of the recycled glass is in-house scrap. There are two major industrial sources for the cullet which is presently being recycled; bottle manufacturing, and the processing of sheet and plate glass. Most of the cullet originating from bottle recycling is in-house breakage and waste glass from breweries, dairies, soft drink plants, and other facilities where liquids are bottled. The sheet and plate glass wastes come from various industrial operations including the manufacture of windows, doors, and automobile window glass. Since waste glass is in reality a super-cooled liquid, it melts at much lower temperatures than do the virgin materials used in the manufacture of glass. For this reason, glass processors are quite ready to accept all usable cullet for recycling into manufactured products. However, there is little likelihood that any increase in the reuse of in-house scrap will occur since most of the glass waste from industrial processing is now being recycled. There is a significant market within the glass industry, however, for sorted clean waste glass from other sources (Ref. 6-1).

Textiles. Each year in the United States over 1 million tons of textile wastes are produced during cloth manufacturing operations. However, these in-house wastes amount to only about one-third of the total amount of textile wastes generated each year in the United States.

The textile industry is not nearly as successful in recycling in-house scrap as is the metal-processing industry; only about 30 percent of the in-house textile waste produced each year is recycled. Most of the recycling is done by producing wiping rags from the in-house wastes. However, some of the in-house waste is used in other forms also: fiber in high-grade paper, padding and batting, and fiber in roofing materials. Some of the in-house waste produced in textile mills in the United States is exported to countries where the resale value of such waste is higher than in this country. One reason for the higher value of textile waste in foreign lands is the relatively low price of labor for sorting and reprocessing of textile wastes there.

Paper. Almost 2 million tons of in-house paper and paperboard wastes are recycled each year in the United States. For example, in 1969, approximately 11.4 million tons of paper and board wastes were recycled, and of this total approximately 1.7 million tons were in-house wastes. These in-house wastes originate from trimming operations in printing houses, from cutting and trimming operations in envelope plants, from processing operations in card plants, and from overruns in the newspaper industry. Currently, approximately 90 percent of all in-house paper wastes are recovered and recycled. Recovery and reuse of such wastes is favored because of their accessibility and because of the lack of contamination of the in-house scrap. Contamination of paper waste in the municipal solid waste stream is one of the principal drawbacks to extensive recycling operations aimed at paper (Ref. 6-18).

Summary of In-house Recycling. In-house recovery of manufacturing wastes is practiced to a great extent at the present time in the United States. For metals, wood, glass, and paper, present reuse rates of in-house wastes are very high. However, large quantities of in-house plastic wastes are not reused; this situation could be changed.

6-2.2 Voluntary Separation

When any waste material is placed in the general municipal refuse stream, it is very difficult and expensive to separate that material from the other wastes and reclaim it. Consequently, voluntary separation of valuable materials by the consuming public is very advantageous. Voluntary separation of waste components for recycling has been attempted in many localities within the recent past. These efforts have met with varied success (Ref. 6-36).

Metals. A considerable amount of recycling of separated scrap metals is being done at the present time in the United States. For example, over 40 million tons of ferrous scrap are recycled each year and used in steel production. Additionally, over 400,000 tons of shredded tin cans are used each year in the copper processing industry; the tin cans are used in the western states in copper leaching operations (Ref. 6-4). The recovery of a large portion of these scrap metal items is based upon the collection of scrap metal by secondary materials dealers before the metal has become a part of the solid waste stream. In other words, scrap metal dealers have recovered metal before it becomes mixed with other forms of refuse in municipal collection operations. A typical example of a source for recovered scrap metal is the abandoned automobile. At times, the recovery of such scrap metal has been a thriving business; however, of late some uncertainty in markets and collection volumes has appeared.

In the 15 years from 1955 to 1970, the production of steel rose approximately 15 percent to over 130 million tons per year; during this 15-year period, the amounts of scrap ferrous metal purchased for recycling increased from approximately 37 million tons per year to just over 40 million tons per year (Ref. 6-31). The percentage increase in scrap purchases was approximately half the percentage increase in steel production. Also, no longer is scrap ferrous metal being exported from the United States to foreign lands since the foreign market for scrap ferrous metal has virtually disappeared. The domestic scrap market is in little better condition.

There are several factors which are responsible for the depression of the ferrous scrap market. The most important single factor has been a change in the production methods within the steel industry from the use of the open-hearth furnace to the use of the basic oxygen furnace. The basic oxygen furnace produces steel in a much shorter processing time period than does the open-hearth furnace. Because of this shorter processing time, the scrap metal content of the furnace charge must be considerably lower since the scrap metal (with a considerable content of iron oxides and other contaminants) requires a longer period of time to melt in the furnace. Therefore, the demand for scrap ferrous metal has decreased with the introduction of the basic oxygen furnace.

A second factor which has militated against the recycling of ferrous metal is the existence of freight rates which favor the transport of virgin materials as opposed to recycled items. Transportation rates set by the Interstate Commerce Commission in the United States have consistently favored the transport of virgin materials over transport of secondary materials. These high transport rates have driven down prices. Because of the low prices, metal users have discarded the scrap metal rather than save it for recycling.

A third and final factor which has detrimentally influenced the recycling rate for ferrous metals is the fragile metallurgy of steel production which requires pure scrap. In recent years, the use of different types of metal together in composite members has increased. The construction of an automobile, for example, involves the use of other nonmetal elements with ferrous metals. It is very difficult to separate the ferrous scrap from the remainder of the wastes in an automobile body. For this reason, scrap dealers have not attempted to segregate ferrous scrap from the remaining scrap in the automobile (Ref. 6-5). Consequently, the reusers of the ferrous scrap, the steel producers, have been reluctant to purchase scrap bundles containing other forms of scrap, both metallic and nonmetallic. Only a very small amount of extraneous material such as copper, tin, or other nonferrous metal is required to significantly affect the quality of steel produced according to current manufacturing methods. For example, the tin coating on steel cans causes difficulties in recycling and production of new steel; the tin in the new steel makes that steel quite brittle and undesirable for many uses. Additionally, the lead solder seam on the side of a tin can is detrimental to the furnaces now in use in the steel industry. Since the lead is denser than the ferrous material, the lead sinks to the bottom of the furnace and infiltrates the refractory material there (Ref. 6-31).

Because of the preceding reasons, the market price for ferrous scrap at the present time in the United States is quite low in comparison to prices for other metals. Tin cans, for example, are not being used in the recycling and reprocessing of steel, but presently are being used almost exclusively in the copper industry for leaching operations. Use in leaching operations can account for only a fraction of the total amount of tin cans produced in the United States each year. On the other hand, removing the tin coating from the steel body of the tin can is economically unfeasible.

There is some cause for optimism concerning the future recycling of ferrous scrap. The electric furnace being introduced into the steel industry can accommodate a charge consisting of all ferrous scrap. At the present time, less than 20 percent of the steel produced in the United States is processed in electric furnaces; however, this figure should increase in coming years (Ref. 6-7).

Environmental awareness among citizens in the United States may indirectly provide the impetus for increased recycling efforts by creating a demand that legislation reducing freight rates be instituted for ferrous scrap. If this were to occur, it is likely that the recycling of such scrap would increase dramatically. Also, a shift to the use of electric furnaces in steel production could occur under such circumstances (lower scrap costs).

In addition to more favorable freight rates, changes in technology may also affect the attractiveness of ferrous scrap. For example, the development of automated shredders for automobile bodies should increase the apparent value of the ferrous scrap by reducing processing costs necessarily incurred in separating the ferrous components in the automobile body from the other waste materials. Automobile shredders are an outgrowth of the tin can shredder which emerged during the World War II era, and they feature rather sophisticated combinations of hammermills, conveyors, special incinerators, and magnetic separation devices. The ferrous scrap processed in these automobile shredders is of much higher quality than that previously available.

Nonferrous Metals. Because of a smaller existing natural supply of ore, and because of difficulties in processing operations, nonferrous metals possess higher resale values than do ferrous metals. Table 6-3 shows amounts of various nonferrous metals recovered and recycled in the United States in 1969.

Conspicuous in this table is the low recycling rate for zinc. The apparent low rate for zinc is a result of two factors. A great deal of zinc is lost in sacrificial corrosion during field usage of galvanized steel products. Secondly, much zinc is recovered and reused after being made into brass. For this reason, those quantities of zinc do not appear in recycling estimates.

TABLE 6-3   Non-In-House Recovery of Nonferrous Metals

Material	1000 Tons Recycled	Percent of Non-In-House Scrap Recycled
Aluminum	201	15
Copper	657	40
Lead	497	38
Zinc	41	4
Nickel	24.5	29
Stainless steel	158.5	76
Precious metals	(Troy ounces)	—
Gold	450,000	64
Silver	43,000,000	65
Platinum	1,660,000	98

Source: National Association of Recycling Industries, Inc., "A Study to Identify Opportunities for Increased Solid Waste Utilization," prepared by Battelle Memorial Institute, Vols. 1–9, New York, N.Y., 1972.

A similar situation exists for nickel. Nickel is used extensively in producing steel alloys. Because the alloys consist predominantly of steel, the amounts of nickel recovered during recycling of steel are difficult to estimate. Consequently, the apparent amount of nickel being recycled is quite low and a rather low recycling rate appears in Table 6-3.

Perhaps the most disappointing statistic contained in Table 6-3 is the low recovery rate for scrap aluminum (about 20 percent for 1973.) In many ways, aluminum is easy to recycle and it yields a great return upon recycling (Refs. 6-33 and 6-34). For example, there is little or no problem with contamination of aluminum by other metals, in comparison to the contamination problems mentioned previously in connection with steel recycling. For instance, in the majority of cases, a composite waste item containing both aluminum and another metal may be used again to produce new aluminum with very little detrimental results. For many metals, the addition of those metals to the recycled aluminum improves the characteristics of the finished alloyed aluminum. An additional attractive factor in the recycling of aluminum is the low energy requirements for remelting recycled aluminum scrap in comparison to the energy requirements for conversion of bauxite ore into aluminum. Only about one-twentieth of the energy required to convert bauxite ore would be required to convert an equal amount (equal in yield of finished product) of scrap aluminum.

One of the reasons for the low present recovery rate of aluminum is the extensive use of aluminum in packaging, especially in throwaway packaging. A prime example of such throwaway packaging is the all-aluminum beverage can. Until very recently, aluminum cans were considered to be one-way items; i.e., the aluminum can was discarded on dumps or in municipal incinerators. However, in recent years some progress has been made in recycling aluminum cans (Ref. 6-34). The Reynolds Metals Company, in 1963, began a program to encourage recycling of aluminum cans. A resale price of 10 cents per pound was then paid for all aluminum turned in to the Reynolds Metals Company at the Can Reclamation Centers which have been located in all major metropolitan areas across the United States. The amounts of aluminum cans turned in at these centers have increased steadily in the last few years. For example, in 1970, 2000 tons (80 million cans) of aluminum were turned in; in 1971, approximately 19,000 tons (12 percent of the total number of cans manufactured) were recycled, and in 1972, over 28,000 tons of aluminum cans were recycled. This amounts to approximately 18 percent of the total number of cans produced in the United States in 1972. In addition to aluminum cans, the can reclamation centers have now expanded to take in all forms of clean aluminum scrap.

Plastics. At the present time, recycling of waste plastic is confined almost entirely to collection and reuse of in-house industrial waste. There are several reasons for this situation. First, much of the plastic which is used takes the form of packaging material, especially thin films and bubble packages. To reclaim this material is very difficult because of its fragile and flexible structural form and very low density. Thick, heavy containers made from plastic offer much more opportunity for recycling efforts than do thin packaging films. One type of container which has received some attention is the plastic milk bottle. In California, a recycling plan was introduced whereby plastic milk bottles were distributed and collected after use by a dairy. The empty plastic milk bottles were placed in a chipper which ground the bottles into small pieces. The small-sized plastic chips were then loaded and shipped to a plastics processor. This operation met with several difficulties. The plastic milk bottles were subject to contamination through contact with hydrocarbon derivatives. The use of a milk bottle to store, even for a very short period of time, petroleum products or similar chemicals effectively contaminated that particular bottle and rendered it unusable as a milk container. Moreover, one such contaminated bottle could ruin an entire shipment of recycled plastic chips as far as the waste processor and reuser was concerned. This difficulty has been surmounted through the development of a very sensitive gas sampling device which in a fraction of a second can be inserted into a milk bottle moving along a production line to procure and analyze a sample of the air contained within that bottle. The very rapid analysis will disclose the presence of contaminating hydrocarbon material within the pores of the plastic bottle; it may then be rejected from the processing assembly line. Through the use of this contaminant detector, "passed" milk bottles may be used over and over again safely. In the event that the bottle is damaged or in any other way rendered unfit for further service as a milk container, the plastic material may be recycled to another form.

An additional problem, however, in the recycling of plastics is the low density of the plastic wastes, such as packaging waste or the plastic chips produced from grinding returned milk bottles. Because of the low density of the material and because of the preferential freight rates, transport costs on a poundage basis are extremely high. This militates against the extensive recycling of plastic wastes.

In any event, even if plastics could be shipped economically to a processing plant, certain other difficulties arise in the reprocessing of the plastic wastes. Plastic wastes as they appear in the solid waste stream generally consist of a number of different formulations or combinations of hydrocarbon derivatives. There is difficulty in reblending such various derivatives in a recycling operation. There is some hope that this difficulty may be surmounted through the development of a compatibilizer, a chemical compound or catalyst which will allow the blending of various types of plastics together in a reforming or recycling operation.

The use of compatibilizers and sensing devices to reject contaminated waste plastics will not of necessity guarantee successful recycling of plastic wastes. Plastic wastes, as mentioned previously, occur in low-density flexible pieces which are very difficult to separate from a solid waste stream. Secondly, even discrete items such as plastic milk bottles may not be successfully recycled until and unless some reeducation of plastics users is attained. At the present time a consumer bias against recycled products exists, especially against recycled plastics with a suspicion of contamination. Contract specifications for materials used in manufacture presently call for virgin materials in most cases, and do not permit the use of recycled products. In some instances, reuse of plastics has been prohibited on the grounds of contamination and consequent health hazard. For example, for a time, public health agencies did not permit the use of plastic irrigation piping manufactured from recycled plastic products. This restriction was based upon the concept that previous use of the plastic component could have contaminated those plastics and presented a health hazard in the new use of the plastic pipe.

At the present time, the situation for recycling of plastic products does not appear favorable. In addition to the difficulties associated with plastics recycling which have been outlined above, there is some public feeling that a preferable method of resource recovery with respect to plastic waste is the combustion of those wastes and the recovery of the energy contained in the basic hydrocarbon materials composing the plastics. References 6-9 and 6-15 contain additional information concerning the reuse of plastics.

Wood. Currently in the United States, very little waste wood is being recycled in discrete form. Certainly the cellulose fibers of wood are being reclaimed (or will be reclaimed) in some of the recycling systems which have been developed to operate on mixed solid wastes. Most existing disposal systems do not reclaim wood items but rather dispose of the wood fibers as wastes, or at best reclaim the energy contained within the cellulose itself.

In relatively rare and isolated instances, intact structural units such as door frames, window frames, or interior furnishings are rescued from demolition operations and reused. In such cases, the wood is being salvaged for its antique aesthetic value. The reuse value inherent in most wood wastes is so low that separation and processing operations are not justified and the wood wastes are disposed of in combustion operations. A small portion of wood wastes in municipal refuse are recycled in pulping operations, but this amount is virtually insignificant compared to the amount of wood currently burned.

Glass. The major recycling effort applied to waste glass consists of the in-house reclamation previously described. However, some progress is being made in the voluntary separation of bottles, jars, and similar glass containers for reuse (Ref. 6-2). The wave of enthusiasm for resource recovery and environmental protection which is currently passing through the population of the United States has resulted in the organization of many volunteer glass recycling programs. The

Glass Container Manufacturers Institute has assisted environmentalist groups to set up a network of waste glass reclamation centers in more than 25 states. The recycling program based on such centers is expected to produce over 1 billion glass containers for reuse each year. In 1972, in one particular example, the Owens Illinois Company of Toledo, Ohio, purchased 3000 freight-car-loads of waste glass for over $1.6 million. This amounted to over 163 million pounds of glass bottles, jars, and other containers. However, even these large amounts of glass are only a small fraction to the total amount of glass containers produced each year.

The principal difficulty associated with glass recycling is the separation of glass wastes of different colors (Ref. 6-28). In the production of new clear glass, it is necessary that all waste glass materials be absolutely clear. The inclusion of only a small percentage of brown or green glass, for example, can ruin the color of an entire batch of new clear glass. The major advantage in the reuse of glass is the fact that considerably less energy is required to melt waste glass, as mentioned previously, than is required to obtain original melt from the raw materials of limestone and silica sand.

Since it is very difficult to obtain waste glass in segregated form (color separated), there has been considerable research devoted to finding uses for mixed color glass. The uses devised include: the inclusion of glass particles in an asphalt mix to produce glassphalt (Ref. 6-22), the reuse of glass bits in the production of mineral wool and foam insulations, and the use of waste glass as inert material in wall panels, tiles, and building blocks (Ref. 6-2). The economic profit and gain in such secondary uses for waste glass are not nearly as attractive as the return associated with the remanufacture of glass from waste particles.

Textiles. A little over 1 million tons of textile wastes are recycled each year in the United States exclusive of those wastes which are recycled in-house. The recycled textile wastes are used principally as industrial wiping rags. There is no existing major market for scrap textile materials other than as industrial wipers. Only about one-sixth of all the scrap textiles produced outside of industry are reused (Ref. 6-6). There are other uses for scrap textiles such as in the manufacture of high quality paper, padding and batting, and roofing materials, but these uses consume only very small percentages of the amount of waste textiles produced each year.

Paper. Efforts to voluntarily separate, collect, and recycle paper have drawn considerable attention in recent years, especially in the case of the recycling of newsprint. At the present time, only about 21 percent of the paper products made in the United States each year are recycled. A portion of the recycled paper is collected and reprocessed from nonindustrial sources. A very small portion of this nonindustrial production of waste paper originates from separation of corrugated board and newsprint from mixed municipal wastes. Generally such separation is a virtual necessity in such operations and the labor costs associated with the separation procedure are prohibitively high. Most of the paper, especially newsprint, which is recovered and recycled has been reclaimed before it is mixed with other refuse. Most of the reclaimed corrugated material and newsprint is put to use in the manufacture of inferior types of paper such as paperboard. Approximately one-fifth of all recycled paper fibers are used in construction materials. The remainder of all the recycled papers (approximately 15 percent) is used in the manufacture of recycled newsprint and various other paper items.

Reuse and recycling of paper is somewhat different in character from the reuse and recycling of other materials such as metals. Paper products are made up of sheets of interlocking wood fibers, and during recycling operations these fibers must be separated and recombined. During the separation operation the fibers are broken up and shortened in length. The shorter fibers produce changes in strength and stiffness of recycled paper; in other words, to produce a piece of corrugated boxboard from recycled paper which would equal in tensile strength a piece produced from virgin pulp fibers, the recycled piece must be made thicker. Because of uncertainty about strength of recycled papers, fibers from sources such as used newsprint eventually are used in paperboard and roofing manufacture because thickness and color are not crucially important in such reuse.

Recycling of newsprint is presently being undertaken in several urban areas in the United States. Some difficulty is encountered in all cases in the recycling of newsprint. There are several reasons for these difficulties. For example, only in major urban areas are sufficient amounts of used newsprint and used paper products available to furnish input for deinking and reprocessing mills. In small urban areas or in essentially rural areas, only small amounts of waste paper are generated, and the operation of a deinking mill with such low volumes of input waste paper is very uneconomical. For this reason, efforts to recycle newsprint must either be focused on major metropolitan areas or must be regional in character; i.e., in a regional scheme, newsprint and other waste papers would be collected in a large region and transported to one central location for recycling. The economics of transport of waste paper is such that freight rates are more favorable for transfer of virgin pulp materials than for transfer of recycled pulp. Such unfavorable freight rates militate against the regional approach to paper recycling.

Recycling of paper products, especially newsprint, is at a very crucial stage at the present time in the United States. On the one hand, increasing use of paper products and dwindling supplies of virgin timbers are combining to increase paper prices in an inflationary spiral. The stark reality of a depletion of forest supplies for pulp and lumber must be anticipated. On the other hand, the increasing amounts of paper in the solid waste stream have proportionately increased the heat energy of the wastes. For this reason (higher BTU values in the solid waste), considerable attention is now being given to the incineration of paper products in the solid waste stream to produce energy and to conserve irreplaceable fuels such as coal, oil, and natural gas (Ref. 6-8). The future of paper recycling efforts is extremely uncertain at the present time. A comprehensive report on paper recycling was prepared by Joseph E. Atcheson Associates, Inc., in 1971, for the paper industry.

6-2.3 Summary—Direct Recycling

Present Conditions. Currently, the major items being directly recycled consist of metals, paper, and, to a small degree, textiles. Only a very small portion of the plastics, glass, and wood materials produced each year are being recycled. At the present time, ferrous metals account for approximately 80 percent of the volume of scrap materials being recycled (approximately 48 percent of the dollar value in sales of secondary materials is attributed to ferrous scrap). Paper currently accounts for 14 percent of the volume of recycled materials and 4 percent of the dollar value of sales of recycled materials. Conversely, nonferrous metals amount to only approximately 4 percent of the volume of materials being directly recycled but they account for approximately 45 percent of the total value of sales of such recycled materials. Textiles account for only an insignificant volume of recycled material and only insignificant dollar values of sales (Ref. 6-6).

At the present time, the major portion of all of the direct recycling of materials accomplished in the United States each year is done by industry itself or by long-established scrap dealers. Of late, the so-called environmental movement has generated considerable interest in the recycling of materials and a number of voluntary recycling operations have sprung up all over the United States. Considerable doubt exists as to the permanency of such efforts and to their economic feasibility. Most of the volunteer efforts to date have been based upon a considerable profit being realized from the sale of aluminum scrap. This profit from sales of aluminum scrap tends to compensate for losses involved in the recycling of other materials by such volunteer groups.

The direct recycling of materials in the United States at the present time faces two major difficulties; uncertain markets and unfavorable freight rates. The market prices paid for recycled materials have characteristically fluctuated in a rather erratic fashion in past years. At the present time, in many cases it is now cheaper to use virgin ore than to purchase reclaimed secondary materials. The cost of separating, collecting, and shipping (at high freight rates) secondary materials necessarily increases the price required for such operations to achieve economic viability. The existence of habitual preference for virgin materials in the form of contract specifications prohibiting the use of recycled materials has also contributed to the present depressed market for recycled items.

Future Potential for Direct Recycling. The future of direct recycling of materials may appear to vary somewhat from material to material. For example, the future holds some promise for increased recycling of ferrous scrap because of changes in steel processing and increased efficiencies in recycling operations. The introduction of new casting processes in the steel industry will decrease the amount of steel scrap produced in-house and consequently increase the demand for ferrous scrap from external sources. The increased use of automobile shredders should produce increased demand and high prices for ferrous scrap materials (Ref. 6-5). In addition to these factors, the decreasing availability of virgin ores should contribute to greater recycling of all metal products.

Unfortunately, the optimistic outlook for ferrous metal recycling and recycling of other metals is not mirrored in predictions for recycling of other wastes such as glass, plastics, and paper. At the present time, unfavorable freight rates combined with a widespread bias toward virgin materials work against the increased recycling of these types of wastes. Moreover, paper and textiles constitute renewable materials and may represent sources of energy which are preferable to nonrenewable fossil fuels (this is discussed in Chapter 7). The heavy cost associated with collecting, separating, and transporting mixed glass wastes and the ready availability of the raw materials for glass production (limestone and silica sand), are unfavorable to the increased recycling of glass wastes. Recycling of plastic waste is very difficult to accomplish because of severe problems associated with separating plastics from other wastes. Discrete items or pieces of plastic may be recycled but difficulties arise in the combining of different types of plastic in a reforming operation. The development of compatibilizers which will facilitate the mixing of various types of plastics in recycling operations will create a favorable circumstance for the increased recycling of plastics; however, the convenience and ease with which plastics are disposed of in incineration operations does not favor direct recycling of plastic products. In all likelihood, plastics will be used as high heat content fuels in energy recovery operations.

Thus, there appears to be little likelihood of increased direct recycling of materials. In-house recovery of wastes at the present time is very efficient. The severe difficulties associated with increasing direct recycling of wastes from nonindustry sources have been enumerated. The most likely source of recycled materials will be the stream of combined and mixed solid wastes which are produced and collected in large urban areas. The next section of this chapter is devoted to a presentation and discussion of the separation of various components from mixed municipal solid wastes.

6-3. SEPARATION OF REFUSE COMPONENTS

Direct recycling of selected solid wastes presupposes separation of the desired materials in a rapid and easy manner; i.e., the voluntary separation of aluminum cans from household wastes for recycling at a can reclamation center is a basic ingredient in the success of aluminum can direct recycling. Likewise, recycling of used newspapers depends upon voluntary separation and has been most successful when collection and transport were voluntary. However, in many localities in the country today there exists no public program or official policy to manage and encourage direct recycling of recoverable materials. For many of the people in the United States there exists no background in voluntary separation, storage, and transport of waste materials to collection centers for their reprocessing. Because of this lack of facilities for direct recycling and the difficulty of separating many materials from the solid waste stream, there is a need for separation and recovery of components of the solid waste stream from that stream itself in a centralized processing facility. Because of the difficulties mentioned in effecting direct recycling, there is a demand for physical methods and processes which will facilitate the segregation of various materials for recovery from mixed solid wastes. This section is devoted to a discussion of the means, methods, and economics of separation.

6-3.1 Preparation for Separation

The truly mixed nature of solid wastes, their heterogeneous character, is perhaps the dominant characteristic of the waste materials produced in industrialized countries today. A brief survey of the materials which are collected in municipal packer trucks or a rapid examination of the refuse deposited in a sanitary landfill indicates that the materials which are thrown away by homeowners vary tremendously in size, shape, and physical-chemical characteristics. This heterogeneous nature creates considerable difficulties in the processing of wastes for resource recovery. Most of the equipment which is available at the present time for separation of wastes prior to actual material recovery has been developed for industrial processes in which the material to be treated is fairly homogeneous in size or physical characteristics. For example, the mechanical jigs which are quite commonly used in ore extraction processing are designed to separate materials of different specific gravity, but they operate on material pieces which fall within a rather narrow size range. Separation equipment, like the mechanical jig, has been designed to operate on relatively homogeneous material (whether by size or other characteristic), in order to make the separation unit operation itself most efficient. An overall savings in time and money has been achieved in the mineral processing industry, for example, by including a grinding and crushing process in the treatment of ore prior to the application of a mechanical jig for separation of the constituents of the given ore. Because of this type of design feature which is very common in separation equipment, it becomes necessary to consider an additional step in the processing of mixed solid wastes for materials recovery. This step which must take place prior to separation is size reduction.

Size reduction of solid waste has been mentioned as an auxiliary operation in the discussions of incineration, sanitary landfilling, and composting. The necessity for or benefit from size reduction for the conventional disposal processes is based upon the greater efficiency of the disposal process when it is used on a material of relatively constant size. For example, incineration of solid wastes is greatly facilitated (excess air requirements are reduced, grate sizes are reduced, combustion is improved, etc.), when the solid wastes are given prior treatment to reduce them to pieces of relatively constant size. Similar benefits in time and money saved have been achieved in the composting of solid wastes when the waste has been processed to yield a small, rather uniform size particle. Because of the efficiencies obtained in operating on a rather uniform size of material and because of the need for virtually constant input to most of the separation equipment available today, size reduction is a necessary part of materials recovery from solid wastes. Some resource recovery plants in Europe are being designed to operate without size reduction or other processing but such plants must be considered innovative and unproven (pers. comm., Dr. H. Hoff, Munich).

Size reduction equipment of various sorts has been mentioned briefly in prior chapters. However, some further mention with specific data on performance and unit costs of this equipment will be given here.

Size reduction may be achieved in a variety of equipment types, including crushers, cage disintegrators, cutters, chippers, shredders, pulverizers, shears, rasp mills, wet pulpers, disc mills, and hammermills. For the most part, current applications of equipment to size reduction of solid wastes have centered almost entirely on the application of hammermills. Some use of shredders, rasp mills, and wet pulpers has also been attempted with varying degrees of success.

While the types of equipment which may be used for size reduction are many, the actual process of size reduction is primarily the same in all types of equipment. The most efficient application of force to the material to reduce its size is that which causes the minimum expenditure of work energy. Work is defined as the product of force times the distance through which the force must act in order to accomplish the intended task. For material size reduction, tensile forces, compressive forces, or shearing forces could be employed. However, because of the mixed characteristics of the refuse pieces and particles in average municipal solid waste, the application of shear forces on the average will result in the minimum expenditure of energy, i.e., the minimum input of work in order to effect the desired separation. This is the reason for the predominant use of hammermills in the size reduction of mixed solid wastes.

Hammermills. As mentioned in the preceding paragraph, hammermills have been applied more generally to the size reduction of solid waste than any other type of size reduction equipment available today. Hammermills consist basically of single or multiple rotor axles with attached hammers. The hammers may be either fixed rigidly to the rotor shaft or mounted flexibly. Rotation of the rotor swings the hammer in an arc around the rotor axis and brings the hammer into contact with the material to be reduced (see Fig. 6-1). The necessary reaction to produce the breakage or shredding of the waste material is obtained by positioning fixed blocks or obstacles on the inner surface of a cylinder which forms the outside wall of the mill, or by positioning plates on the inside of the cylinder wall so that the material is caught between the swinging hammers and the wall with small tolerance. Many variations of this general principle are found in the hammermills which are in use today to process solid wastes. For example, in one variation of the basic device, moveable shredder teeth are mounted in the outside cylinder drum which encloses the rotor and swinging hammers. The outer hammers are positioned to pass between the moving hammers in the shredder device.

A useful variation of the standard hammermill is a mill in which the rotor is mounted horizontally and the hammers approach the input area for the wastes in an upswing manner, so that material which is not easily reduced is rejected by the hammers and is knocked back up through the input area into a reject chute. This

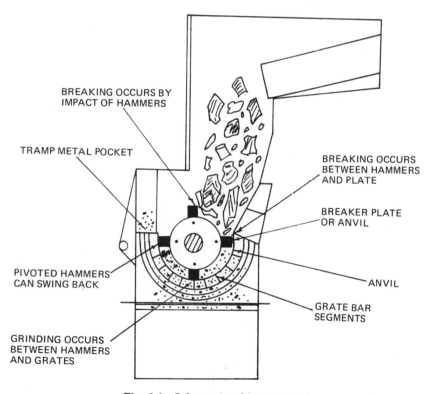

Fig. 6-1 Schematic of hammermill.

is shown in Fig. 6-1. A more sophisticated variation of this technique is the use of a vertical rotor and hammers which propel rejected material up and out of the mill along a spiral inclined surface into a receptable for rejected materials.

Because of the possible variety of hammers, clearances, and other features in hammermills, they are extremely flexible with regard to the materials which are input to be reduced in size. However, fluctuations in the size and type of material introduced into the hammermill require adjustments of the flow rate in order that the equipment is not overloaded and subjected to damage or excessive wear.

Wet Pulpers and Disc Mills. Another type of size reduction equipment which is becoming increasingly popular is the wet pulper which is a variation of an older type of apparatus, the disc mill. Disc mills are size reduction devices which consist of a single rotating disc and a fixed contact surface, or two counter-rotating high-speed discs. The discs may be modified to an irregular segmented shape as in some of the recent pulper designs but essentially the action of the apparatus remains unchanged. Material is introduced between the rotating and fixed surfaces and is subjected to many repetitions of impact from the rotating disc or the segmented wheel. It is reduced in size until it passes through openings in the fixed contact surface, or until it passes through the space between the outer rims of the counter-rotating discs in the case of a multiple disc mill. In the wet pulpers which are getting widespread attention, the wastes are mixed with water to produce a slurry (approximately 10 percent solids) and introduced into a pulper which consists of a segmented blade which rotates at a very high speed. The wastes are reduced rapidly to the desired size and passed out of the apparatus through openings in the bottom of the pulper for further processing. Materials which are not suitable for pulping are rejected ballistically by the rotating blades to the outer portions of the pulper drum where they may be collected separately. Recently much success in separation of municipal solid wastes has been reported by the Black Clawson Company through the use of wet pulping equipment.

Rasp Mills. Rasp mills, a third category of size reduction equipment, have been used quite extensively in conjunction with composting operations. Rasp mills and other similar size reduction equipment such as pulverizers operate in the following manner: a large rotor fitted on a vertical shaft rotates and carries heavy rasping arms around within the container drum of the mill, which may be as much as 20 to 25 ft in diameter. The rotor turns rather slowly in comparison to the speed of rotation of a hammermill, moving at 5 to 6 rpm. The swinging arms attached to the central rotor are quite heavy and they act to push input wastes around within the external housing. As the waste is pushed around, it passes over obstructions in the bottom plate of the mill which are called rasping pins and the size of the waste pieces is reduced (See Fig. 6-2). When the pieces have been reduced to approximately 2 in. in greatest dimension, they fall through holes in the bottom plate of the rasp mill and proceed for further

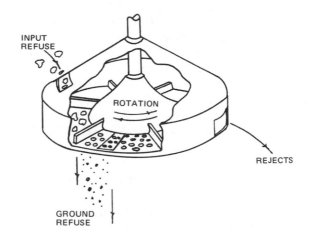

INPUT
REFUSE

ROTATION

INTERIOR IS
FLOORED WITH
ALTERNATE SECTIONS
WITH HOLES AND
RASPING PINS

REJECTS

GROUND
REFUSE

Fig. 6-2 Schematic of rasp mill (redrawn after USEPA).

processing. It is possible to reduce the size of input solid waste smaller than the 2 in. diameter just mentioned; however, the 2 in. size is the smallest piece size in most of the currently-operating rasp mills. As the rotor arms in the rasp mill pass around and around carrying the solid waste over the rasping pins and exit holes, resistant items gradually are forced to the outside of the mill by the centripetal action of the swinging arms. A reject chute or opening may be periodically opened to allow the exit of such resistant nonreducible items. Rasp mills have found very limited use in the United States; the only large scale use of the rasp mill to date has been in the solid waste demonstration composting plant funded by the Public Health Service and the Tennessee Valley Authority at Johnson City, Tennessee. This plant is discussed in detail in Chapter 2 on composting of solid wastes.

Shredders and Chippers. Various designs of shredders and chippers have been developed to operate on domestic solid wastes. One of the most common designs features the use of toothed wheels which rotate at different speeds and through which input solid wastes must pass. The toothed wheels act to penetrate, shear, and shred the encountered solid wastes. Shredders as such and as described herein are more generally applicable to the reduction of ductile materials or materials composed of elongated fibers. Therefore, this type of shredder is quite useful in the size reduction of the paper and fiber portion of municipal solid wastes.

Also of use in size reduction of solid waste are the yard waste chippers and shredders which are in quite general use in the United States at this time. Generally shredders of this type are employed by tree trimming and landscape design agencies who specialize in the disposal of yard wastes.

A special example or category of solid waste shredder is the automobile body shredder which is an outgrowth of the tin can shredders developed during World War II for the processing of scrap metal. These shredders are generally high-power input units and are rather costly in terms of capital expenditures. However, their usefulness in reducing large bulky automobile bodies to manageable pieces should not be underrated.

Other Types. Other miscellaneous types of size reduction devices also have been employed in the processing of solid wastes. These devices include single-blade shears which have been used to reduce bulky materials such as automobile bodies or timber wastes. Single-blade shears are of limited applicability for general solid wastes although their utility in scrap metal processing is considerable.

Various other pieces of equipment such as crushers, disintegrators, and drum pulverizers have been developed for the size reduction of boards and bulky materials. However, their applicability to the size reduction of solid wastes is questionable and limited.

Equipment Capabilities and Costs. In a comprehensive study of size reduction equipment suitable for solid waste processing, Battelle Memorial Institute personnel collected data on a large number of equipment types and process variations. These data are contained in the publication *Recovery and Utilization of Municipal Solid Waste*, distributed as report SW-10C by the Office of Solid Waste Management Programs of the Environmental Protection Agency. The following Tables 6-4 and 6-5 and Figs. 6-3, 6-4, 6-5, and 6-6 present a summary of the data collected by the Battelle research group. As is evident in these tables and figures, processing production rates are highly dependent upon available horsepower supplied to the size reduction equipment. It is also apparent that the costs associated with the reduction of solid wastes are directly dependent upon the desired product size after reduction. Conversely, as would be expected, unit production costs for reduction in size of the wastes are reduced as the production rate of the available equipment increases. As is apparent in the data presented, the cost estimates for reduction of municipal solid wastes vary considerably. The reason for this variation is felt to be caused by the wide disparity in equipment design, which in turn is the result of differences in materials for which the different reduction units were initially designed. In most cases the size reduction units currently applied to solid wastes were initially developed for size reduction of other materials.

The Battelle investigators concluded briefly that two trends are apparent in the developing size reduction equipment field. These trends were:

1. There is an increasing demand for greater and greater capabilities for size reduction equipment, for greater and greater tonnage rates for reduction of mixed wastes.

TABLE 6-4 Approximate Size Reduction Costs[a]

Capacity[b] (Tons per Hr)	Nominal Product Size (in.)	Costs (Dollars per Ton)					Variations from Packer-Truck Input
		Maintenance	Equivalent Annual Cost	Power	Labor	Total	
① [c] 17	6	0.22	0.12	0.45	0.25	1.04	Bulky waste included.
② 40	6	0.05	0.22	0.08	0.11	0.46	Wood only (nonbulky).
③ 10	6	0.27	0.19	0.19	0.41	1.06	12 percent prepicked.
④ 30	4	0.27	0.20	0.18	0.17	0.82	Some bulky extras; 4-in. product.
△ 9	2	0.06	0.33	0.09	0.48	0.96	30 percent durable bulky items rejected.
Ⓐ 10	6	0.18	0.18	0.15	0.43	0.94	
⑦B 7.2	4	0.25	0.25	0.21	0.61	1.32	10 percent reject of durable items.
⑧C 6.1	2	0.28	0.29	0.25	0.72	1.54	
⑨ 25	6	0.55	0.14	0.32	0.18	1.19	Bulky waste included.
⑩ 30	6	0.55	0.17	0.10	0.14	0.96	Harbor waste.
⑫ 50	6	1.10	0.18	0.40	0.09	1.77	Auto bodies.
⑭ 40	6	0.55	0.09	0.20	0.11	0.95	Bulky waste included.
[20] 12	1	0.11	0.08	0.13	0.36	0.68	Premilled input.
[21] 60	1	0.11	0.08	0.06	0.08	0.33	Premilled input.

[a] These costs are based on manufacturers' data modified by users' data and do not include site costs, crane costs, or other handling costs.
[b] Tons per year = average tons per hr × 8 × 286.
[c] See Fig. 6-3 for key to symbols.

Source: Drobny, N. L., Hull, H. E. and Testin, R. F., "Recovery and Utilization of Municipal Solid Waste," USPHS Publ. No. 1908, 1971.

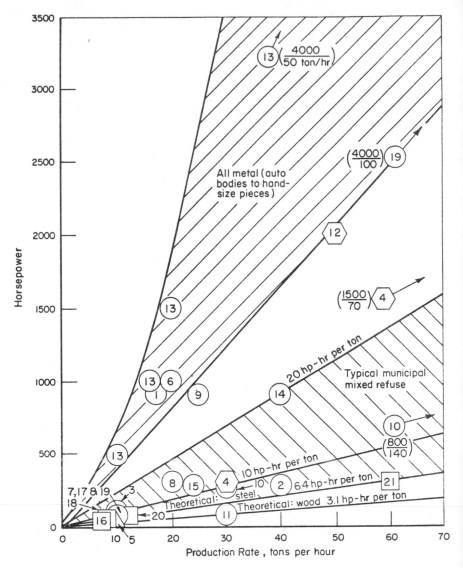

Fig. 6-3 Size reduction power requirements. *Source:* Drobny, N. L., Hull, H. E. and Testin, R. F., "Recovery and Utilization of Municipal Solid Waste," USPHS Publ. No. 1908, 1971.

2. The increasing diversity of the materials in the solid waste stream continues to cause difficulties in the design and operation of size reduction equipment.

Table 6-6 given below lists a summary of the basic types of size reduction equipment currently in use and also lists the potential application of this equip-

MANUFACTURERS' CODE NUMBERS

○ Hammermills
□ Drum Pulverizers and Wet Pulpers
△ Rasp Mill
⬡ Shredders

① American Pulverizer Company, St. Louis, Missouri.
② Buffalo Hammermill Corporation, Buffalo, New York.
③ Buhler Brothers (Swiss), Ontario, Canada.
④ Centriblast Corporation, Subdivision of Joy Manufacturing Company, Pittsburgh, Pennsylvania.
△ Dorr-Oliver, Netherlands.
⑥ Ferrox Corporation, Scott, Louisiana.
⑦ Heil Company (Gondard, French), Milwaukee, Wisconsin.
⑧ Gruendler, St. Louis, Missouri.
⑨ Hammermills, Inc., Cedar Rapids, Iowa.
⑩ Hazemag, U.S.A. (German), New York City.
⑪ Jeffrey Manufacturing Company, Columbus, Ohio.
⑫ Logemann Brothers, Fond du Lac, Wisconsin.
⑬ Newell Manufacturing Company, San Antonio, Texas.
⑭ Pennsylvania Crushers, Broomall, Pennsylvania.
⑮ Stedman Foundry and Machine, Aurora, Indiana.
⑯ John Thompson Company, England.
⑰ Tollemache, Ltd., London, England.
⑱ Vickers Company, England.
⑲ Williams Patent Crusher & Pulverizer Company, St. Louis, Missouri.
⑳ French Oil Mill Company, Piqua, Ohio.
㉑ Black Clawson, Middleton, Ohio.

Fig. 6-3 (*Continued*)

ment to size reduction of municipal solid wastes. The recent developments in size reduction of municipal wastes has indicated that there are possibilities for successful combination of a number of different types of size reduction equipment. Combinations of shears and various kinds of milling devices have been suggested for operation on ductile and nonductile materials because of the difference in response associated with differences in ductility.

TABLE 6-5 Adjusted Performance and Cost Data Size Reduction Equipment

Symbol	Horsepower	Capital Investment (Thousands of Dollars)	Adjusted Capacity (Tons per Hr)	Adjusted Power Requirements (Hp-Hr per Ton)[a]	Adjusted Unit Capital Investments (Dollars per Ton-per-Hr Capacity)
① b	800	35.0	17 × 1 = 17	47	2,060
②	300	8.5	40 × 0.43 = 17	19	500
③	200	33	10 × 0.65 = 6.5	31	5,080
④ {	350	80	30 × 1.39 = 42	8.3	1,900
	1,500	130	70 × 1.39 = 97	15.5	1,340
△	80	50	9 × 0.65 × 1.64 = 10	8.0	5,000
⑥	1,000	—	20 × 2.8 = 56	17.8	—
⑦	150	30	10 × 0.65 = 6.5	23	4,620
⑧	350	—	20 × 0.65 = 13	27	—
⑨	800	60	25 × 2.8 = 70	11.4	857
⑩	300	86	30 × 1 = 30	10	2,870
⑪ {	800	160	140 × 1 = 140	5.7	1,140
	150	17.5	30 × 0.43 = 13	11.5	1,310
⑫	2,000	150	50 × 2.8 = 140	14.3	1,070
	500	—	10 × 2.8 = 28	17.9	7,150
⑬ {	1,000	200	16 × 2.8 = 45	22	4,450
	1,500		20 × 2.8 = 56	27	3,570
{	4,000		50 × 2.8 = 140	29	1,430
⑭	800	59.5	40 × 1 = 40	20	1,490
⑮	300	14.2	24 × 0.43 = 10	30	1,420
⑯	50	—	7 × 0.65 × 1.64 = 7.5	6	—
⑰	150	53.0	10 × 1 = 10	15	5,300
⑱	60	—	8 × 0.65 × 1.64 = 8.5	7	—
⑲ {	150	10	10 × 0.65 = 6.5	23	1,540
{	4,000	—	100 × 2.8 = 280	14	—
⑳	150 + (12 × 15) = 330	15.0	12 × 2.38 = 28.56	11.8	536
㉑	350 + (60 × 15) = 1,250	69.5	60 × 2.38 = 143	8.8	485

[a] Primary size-reduction power added.
b See Fig. 6-3 for key to symbols.

Source: Drobny, N. L., Hull, H. E. and Testin, R. F., "Recovery and Utilization of Municipal Solid Waste," USPHS Publ. No. 1908, 1971.

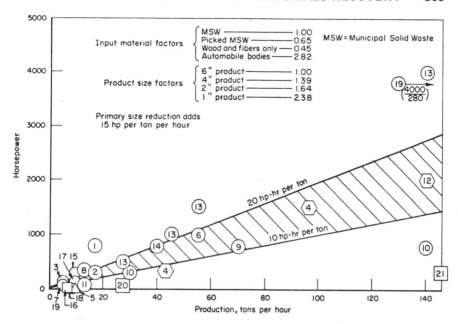

Fig. 6-4   Adjusted power requirements for size reduction. See Fig. 6-3 for key to symbols. *Source:* Drobny, N. L., Hull, H. E. and Testin, R. F., "Recovery and Utilization of Municipal Solid Waste," USPHS Publ. No. 1908, 1971.

Summary on Size Reduction. In summary, it appears that larger and larger units will be designed and manufactured for size reduction of municipal wastes. Additionally, further sophistication of individual pieces of equipment and improved combinations of various items of equipment should be expected in the near future. Size reduction is a necessary and vital step in the recovery of materials from the solid waste stream. As such, it is a necessary prelude to separation operations. Furthermore, size reduction may be justifiable in disposal operations also because of the greater efficiencies and effectiveness in both composting and incineration operations resulting from a continuous feed of virtually uniformly-sized pieces of solid wastes. Considerable effort should be devoted to research and development in size reduction equipment in the future because of the potential savings associated with improvement in this type of equipment (Ref. 6-13). Attention should be given in the future to the compilation of complete performance and cost records for existing types of equipment and size reduction devices which will be used in the future.

6-3.2   Separation Methods

The techniques currently used in the separation of solid waste constituents are based upon physical-chemical characteristics of the materials and generally fall

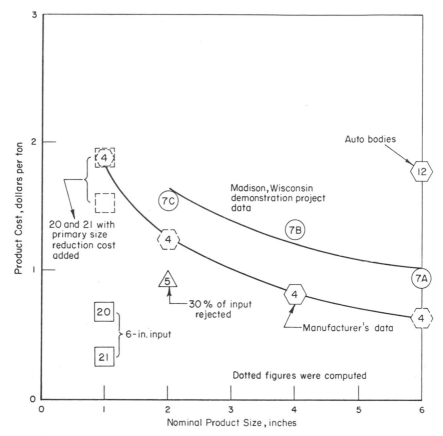

Fig. 6-5 Size reduction cost versus nominal product size. See Fig. 6-3 for key to symbols. *Source:* **Drobny, N. L., Hull, H. E. and Testin, R. F., "Recovery and Utilization of Municipal Solid Waste," USPHS Publ. No. 1908, 1971.**

into one of the categories discussed in the next few paragraphs. Separation of a desired material is achieved through identification of a given property of that material wherein the value or response of the given material differs considerably from the value or response of all other undesired materials. For example, ferrous metals are attracted by electromagnets and can be removed from a stream of wastes which do not respond to the presence of a magnet; in this case, electromagnetism is the property used in differentiating the ferrous metals from the remainder of the waste stream. The ability to separate the ferrous metals depends upon the difference in magnetic responses of those metals as opposed to the remainder of the waste stream. Therefore, the key to success in separation of materials is the selection of a given material characteristic which will be significantly different in value for the desired material as opposed to the remainder of the waste stream.

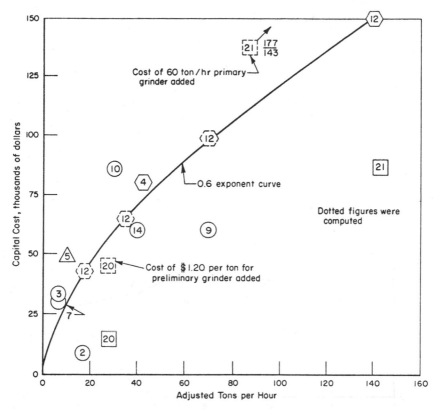

Fig. 6-6 Capital cost versus adjusted capacity for size reduction equipment. See Fig. 6-3 for key to symbols. *Source:* Drobny, N. L., Hull, H. E. and Testin, R. F., "Recovery and Utilization of Municipal Solid Waste, USPHS Publ. No. 1908, 1971.

Particle Size. One of the most widely used properties for materials separation is particle size. The simple characteristic of particle size appears to contradict the criterion given in the preceding paragraph for a property by which materials can be differentiated. Any given material may be reduced to any given particle size, so it would appear that differentiation of materials on the basis of particle size is not a logical or rational method of materials separation. However, physical properties such as hardness and ductility differ considerably from one constituent of the solid waste stream to another. Therefore, when a stream of wastes is subjected to a size reduction operation, the resultant sizes of particles of different constituents reflects their differences in hardness, resilience, ductility, etc. For practical purposes, therefore, resultant particle sizes after manipulation in a size reduction operation may reflect differences in the materials composing the various ranges of particle sizes. For example, waste glass

TABLE 6-6 Current Size Reduction Equipment and Potential Applications to Municipal Solid Waste

Basic Types	Variations	Potential Application to Municipal Solid Waste
Crushers	Impact	Direct application as a form of hammermill.
	Jaw, roll, and gyrating	As a primary or parallel operation on brittle or friable material.
Cage disintegrators	Multi-cage or single-cage	As a parallel operation on brittle or friable material.
Shears	Multi-blade or single-blade	As a primary operation on wood or ductile materials.
Shredders, cutters, and clippers	Pierce-and-tear type	Direct as hammermill with meshing shredding members, or parallel operation on paper and boxboard.
	Cutting type	Parallel on yard waste, paper, boxboard, wood, or plastics.
Rasp mills and drum pulverizers		Direct on moistened municipal solid waste; also as bulky item sorter for parallel line operations.
Disk mills	Single or multiple disk	Parallel operation on certain municipal solid waste fractions for special recovery treatment.
Wet pulpers	Single or multiple disk	Second operation on pulpable material.
Hammermills		Direct application or in tandem with other types.

Source: Drobny, N. L., Hull, H. E. and Testin, R. F., "Recovery and Utilization of Municipal Solid Waste," USPHS Publ. No. 1908, 1971.

may be successfully separated from a solid waste stream after wet pulverization on the basis of particle size alone because of the small size of the glass particles resulting from passage through the wet pulper.

Bulk Density. A second means by which materials may be separated is their specific gravity, or bulk density. The variation in force with which the earth attracts unit volumes of different materials may be used to separate those materials from each other. There are a variety of means for such separation which utilize the force of gravity in differing ways. Differences in bulk density of different types of materials may also be used in various inertial separators. The variation in apparent bulk density of materials is a result of differences in their actual mass per unit volume. The inertial forces mobilized in different materials when they are accelerated may be used successfully to separate those materials in a ballistic separator. Density is one of the most widely used properties for separation.

Electromagnetism. As mentioned previously in the case of ferrous metals, electromagnetism is a property of material which may be used to successfully separate various elements from each other. Magnetic separation of metals has received a considerable amount of attention and is widely applied in the waste separation industry because of the relative simplicity of a magnetic separator, consisting essentially of a permanent magnet or an electromagnet and a feed device which passes the waste stream near the magnetic source. Another reason for the wide application of magnetic properties to separate waste is that ferrous metals now are somewhat in demand (and formerly were in greater demand for reuse and recycling).

Electrical Conductivity. An additional means of separation has been suggested through the use of differences in electrical conductivity of various materials. Differences in conductivity can be used to separate materials by creating eddy currents within the pieces of the materials and actually causing repulsive forces between materials of different composition. This application of a material property for separation remains largely in the theoretical and experimental stages, and appears to hold little immediate promise for use in separation of solid wastes.

Color. The appearance of various constituents of solid waste has been used successfully for many years in hand-sorting operations. A more sophisticated means of separation based on material appearance is currently being developed based upon color differences of the material constituents (Ref. 6-28). Color separation is ideally suited for the segregation of materials such as glass, and for further separation of the various colors of waste glass. Separation on the basis of color has been accomplished for a number of years in the processing of industrial and gem diamonds, and an effort is now underway to modify the diamond separation procedures for use in the separation of various colors of waste glass. The separation of materials through color differences, however, requires rather sophisticated equipment and does not appear to be available for near-future use in operations on solid waste.

Signature Methods. Other means of separating collected wastes into individual components have been suggested at various times (Ref. 6-36). For example, the differences in resilience of materials has been used as a theoretical base for separation through obtaining a resilience "signature" from a sensor impact on an individual piece of material. Professor David Wilson of the Massachusetts Institute of Technology has been a leader in the development of separation concepts based upon material signatures. Resilience, infrared absorption spectra, and similar sophisticated properties of the collected materials have been suggested as a means of separation. At the present time this type of separation operation must be viewed as largely theoretical, and generally inapplicable to solid waste technology today. There is some evidence that this type of equipment will be developed in the future for use on waste materials. However, the instrumentation involved in such a process in any case would be expensive and would require considerable maintenance. Because of these requirements, the signature separation concept appears to have very limited present applicability to solid wastes and also holds little hope for practicable application in the future.

Chemical Separation. In addition to the physical means of separation described in the preceding paragraphs, chemical processing of wastes have been used to effectively separate constituents of the waste stream. The most obvious chemical means of separation is combustion. In solid waste incinerators as described in Chapter 3, noncombustible materials such as metals are easily separated from the combustible portion of the refuse. While this separation is an obvious operation, its importance should not be underrated. The utilization of metal in automobile bodies, for example, depends upon the effective separation of that metal from the combustible furnishings within the automobile frame. This has been most effectively accomplished through controlled combustion in special auto body incinerators. In addition to combustion, other chemical processes may be considered as effective means of separation. However, the heterogeneous nature of municipal solid waste generally prohibits the application of a specialized chemical process for separation. Combustion, with the attendant melting of various noncombustible constituents of the refuse, remains as the most widely applicable chemical separation process. Many separation applications now being proposed for resources recovery include prior combustion of the entire waste stream and then subsequent separation operations on the residue left after incineration (Refs. 6-19, 6-23, 6-27, 6-29, 6-30, and 6-32).

It is conceivable, and indeed quite possible, that other new methods and means of waste constituent separation will be evolved in the near future. The accelerated pace of technological change in modern society holds the promise of considerable development in such operations; however, present efforts in solid waste management cannot include proposed or possible operations but must be based upon what is immediately available. Therefore, the remainder of this discussion of separation will deal with methods and equipment which employ differences in properties listed in the preceding paragraphs as a means of attaining materials separation.

6-3.3 Methods and Equipment for Separation

Hand Separation. In the United States and in Western Europe today, the most widely applied means of separation of materials from the solid waste stream is hand picking and segregation of selected materials. Most of the hand picking of refuse is done in connection with composting operations, although some hand separation has been used successfully in conjunction with incineration. The range of materials which are selected by hand for salvage is somewhat limited, generally restricted to such things as corrugated board, large pieces of metal, glass bottles, and large pieces of plastic or rags. For the most part, hand sorting of refuse is not completely satisfactory and should find less acceptance in the future for a number of reasons.

First, the factor of human error is always present in the hand sorting of refuse. In some cases, a very high purity is desired or is necessary in recycled material in order to have a guaranteed resale price. Even an infrequent error in human hand sorting of refuse may destroy the acceptance and marketability of salvaged materials if subsequent manufacture using the resalvaged materials is prevented or is damaged because of the very occasional impurity which escapes the hand sorters.

A second factor which tends to limit the application of hand sorting is the limited ability of hand sorters to separate materials of various sizes; hand separation is virtually limited to materials of rather large particle size and is not applicable to materials in the size range of small gravel or sand-sized particles.

A third and perhaps preponderant reason that hand separation is not an attractive means of separation is that the rather low unit prices paid today for salvaged materials virtually preclude the use of a large work force of laborers, even if the laborers are being paid only a minimum wage. This statement is obviously more relevant to the situation in the United States than to the situation in Western Europe or in other foreign countries where labor is much less expensive than in the United States. For example, hand separation and sorting of refuse in Southeast Asia is an accepted practice not only in organized industrial or governmental operations but in independent operations by the entire populus. In the United States, on the other hand, hand sorting of refuse requires approximately $1\frac{1}{2}$ to $2\frac{1}{2}$ man-hours per ton of sorted refuse. This number of man-hours multiplied by the higher labor cost in this country militates against extensive use of hand sorting in resource recovery operations.

Because of the low potential for hand separation and sorting of materials for recycling, an effort has been made by the secondary materials industries in this country to develop mechanized means of sorting and separating materials from the solid waste stream. The major part of this effort has involved the adaptation, generally on a trial basis, of previously developed equipment for use with solid wastes. Most of the equipment which has been tried was initially developed for use in mineral or ore processing. Because of the differences between solid wastes and the materials for which the separation devices were initially designed, the devices have performed with varying degrees of success. The following paragraphs are devoted to a discussion of the devices which have been tried and which have been shown to have at least moderate applicability to waste processing.

Separation by Particle Size. The simplest means of separation of materials developed to date for use with solid wastes is separation on the basis of particle size. This means of separation is effective after size reduction operations. Considerable amounts of research have been done with regard to the sizes of particles which are produced in various materials after standard size reduction operations such as hammermilling. It has been found that, in general, the resultant particle size is dependent upon the characteristics of the material being milled. For example, in one study over 70 percent of the glass particles produced in a milling operation were smaller than $1/16$ in. in maximum dimension (Stirrup, F. L., *Public Cleansing Refuse Disposal*, Pergamon Press, Oxford, 1965). In this same study, approximately 60 percent of the metal can pieces produced after the milling operation were larger than 2 in. in least dimension. On the other hand, wood which had been subjected to the milling operation was reduced to pieces which ranged in size from subvisible particles to pieces of approximately 8 in. in smallest dimension. Almost 60 percent of the wood was reduced to pieces ranging in size from subvisible to $1^1/2$ in. in maximum dimension. Because of the differences in the sizes produced after any given milling operation, a combination of a milling operation and a size-classification separation technique may be made to selectively separate various materials from the waste stream. The devices which are commonly used to effect the size-based separation are vibrating screens or spiral classification units.

Vibrating screens are simple in operation and simple in explanation. Material is introduced to the top surface of the screen and only those sizes of particles which are smaller than the screen apertures pass through the vibrating screen and into a catchment container below the screen. The selection of aperture size for the screen, dimensions of the screen, use of dry or wet screening, and possible combinations of more than one size of screen all depend upon the product to be separated and the foregoing size-reduction operation. As a guide to the selection of screens and screening equipment, the Denver Equipment Company has compiled the material in Table 6-7. As is obvious in this table, the selection of vibrating screens in this manner is ideally suited for applications to the mineral processing industry. However, the table may be used successfully if the material to be separated is reduced to a given size in a previous milling operation, that size is known, and the reader correctly uses the data contained therein.

The use of vibrating screens for separation of solid wastes in the United States has been confined to experiences in composting plants. The Lone Star Organics plant at Houston, Texas used vibrating screens to size compost and separate materials from the waste stream, and a similar operation is used at Altoona, Pennsylvania. There is a scarcity of performance data on the use of vibrating screens for materials separation. However, certain generalizations may be made:

1. Dry screening appears to be much more favorable for use with solid waste than wet screening because of the problems associated with handling and

TABLE 6-7 Capacities for Vibrating Screens

Factor A

Size (in.) and Mesh Size of Clear Square Opening

Mineral (in.)	.0116	.0164	.0232	.0828	.046	.065	.098	.181	.185	
	48	35	28	20	14	10	8	6	4	
Sand	.143	.183	.226	.282	.36	.45	.57	.69	.73	.90
Stone dust	.120	.152	.188	.235	.30	.375	.475	.56	.595	.75
Coal dust	.691	.115	.142	.178	.226	.284	.86	.43	.45	.57

Size of Clear Square Opening (in.)

Mineral (in.)	¼	⅜	½	⅝	¾	⅞	1	1¼	1½	2	2½	3	4	5 (Use only in Single-Deck Screens)
Gravel	1.08	1.40	1.68	1.94	2.16	2.36	2.56	2.90	3.20	3.70	4.05	4.30	4.65	4.90
Crushed stone	.88	1.19	1.40	1.60	1.80	1.96	2.12	2.40	2.68	3.10	3.38	3.60	3.86	4.07
Coal	.68	.88	1.01	1.21	1.36	1.48	1.60	1.83	2.00	2.31	2.53	2.69	2.91	3.06

Factor B—Oversize

Determine or estimate percentage of oversize in feed to screen and use proper factor as given below. For example, if screen has 1-in. openings and 60 percent of feed to screen will go thru 1-in. openings, there is 40 percent of oversize and factor 0.95 would apply. Other percentages accordingly.

Factor C—Efficiency

Slight inaccuracies are seldom objectionable in screening aggregate and perfect separation (100 percent efficiency) is not consistent with economy. For finished products, 98 percent efficiency is the extreme practicable limit and 94 percent is usually satisfactory. Sixty to seventy-five percent efficiency is usually acceptable for scalping purposes.

Factor D—Undersize

Consider this factor carefully where sand or fine rock is present in feed. For example, if screen has ½-in. square openings and a large percentage of the feed is ¼ in. or less in size, such as sand or dust, determine percentage and use proper factor given below.

Factor E

If material is dry, use factor 1.00. If there is water in material or if water is sprayed on screen, use proper factor given below. Wet screening means the use of about 6 to 10 gpm of water per cu yd of material per hr.

Factor F

For single-deck screen, use factor 1.00. For a multiple-deck screen, be sure to use proper factor for each deck.

(Continued)

TABLE 6-7 (Continued)

Amount of Oversize (Percent)	Factor	Desired Efficiency (Percent)	Factor	Amount of Feed less than ½ the Size of Opening (Percent)	Factor	Wet Screening Size of Opening (in.)	Factor	Deck	Factor
10	1.06	60	2.70	10	0.55	¹⁄₃₂	1.25	Top	1.00
20	1.01	70	1.70	20	0.70	¹⁄₁₆	3.00	Second	0.90
30	0.98	75	1.55	30	0.80	⅛	3.50	Third	0.75
40	0.95	80	1.40	40	1.00	¼	3.50		
50	0.90	85	1.25	50	1.20	½	3.00		
60	0.86	90	1.10	60	1.40	⅝	2.50		
70	0.80	92	1.05	70	1.80	¾	1.75		
80	0.70	94	1.00	80	2.20	1	1.25		
85	0.64	96	0.95	90		Wet screening below 20 mesh not recommended.			
90	0.56	98	0.90	100	3.00				
92	0.50								
94	0.44								
96	0.35								
98	0.20								
100	0.00								

Source: Denver Equipment Company.

(1) To obtain capacity in tons per hr passing through screen cloth, multiply area of screen cloth by factors A, B, C, D, E, and F.
(2) To obtain total capacity of screen, add percentage of oversize to result of computation (1).
(3) To determine size of screen, obtain screen-cloth area needed by dividing capacity in tph less oversize in tph by factors A, B, C, D, E, and F.

$$CP = Area \times (A \times B \times C \times D \times E \times F)$$ in which: A, B, C, D, E, F are factors given above
$$TC = CP + Oversize$$
$$CP = \text{Capacity passing through screen (tph)}$$
$$TC = \text{Total capacity of screen (tph)}$$
$$Area = \frac{TC - Oversize}{A \times B \times C \times D \times E \times F}$$
$$Area = \text{Screen surface (ft}^2\text{)}$$

treating the water used in the operation (additionally, difficulties may arise in drying the separated materials).

2. It should be remembered that complete purity of the finished size-separated material has not been achieved because of the random distribution of material sizes after size reduction operations.

In other words, although most of the pieces of glass produced in a given milling operation may be of rather small size, a certain small percentage of glass pieces may be significantly larger than the majority of pieces. For this reason, large pieces of glass may occasionally appear with other materials which have been separated from the bulk of the glass on the basis of their larger particle sizes. Separation operations on the basis of size, therefore, must be considered to be rather general and approximate in nature.

The bulk of separation operations based upon particle size have been carried out through the use of vibrating screens. However, spiral classifiers have been used in the minerals processing industry for size separation. In the spiral classifier shown in Fig. 6-7, the material to be separated is mixed with a fluid and introduced into the classifier. The moveable spiral then agitates the mixture in the inclined tank. The fluid in the tank is removed by flow over a weir located at the end of the tank. Generally the smaller-size particles emerge over the overflow weir with the excess fluid. The larger-size particles settle more rapidly and pass to the bottom of the tank where the action of the revolving spiral passes them up

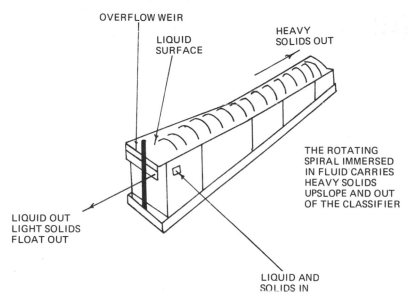

Fig. 6-7 Schematic of spiral classifier for separation by specific gravity (redrawn after Denver Equipment Company).

the inclined tank and out of the tank at the opposite end from the overflow weir. Generally, spiral classifiers are most applicable to separation of materials of virtually the same specific gravity but characteristically different sizes of particles. Therefore, the use of spiral classifiers on municipal solid waste would appear to be rather limited. The selection of a spiral classifier must be based upon the materials to be separated and the capacity desired for the unit. In the equipment manufactured to date only a rather narrow range of capacities has been achieved. For this reason, little applicability of spiral classifiers for separation of solid wastes is foreseen.

Materials Separation by Density. Various methods have been developed for the separation of materials on the basis of differences in material density or specific gravity. These methods include simple flotation, use of dense media, use of vibrating tables, use of jigs (principally in mineral processing operations), use of dry air stream classifiers, and use of inertial separators.

Flotation as a means of separation consists of mixing waste materials with a suitable medium and the subsequent separation of the materials on the basis of their attraction to or affinity for air passing upward through the medium. The details of the specific flotation operation, such as the selection of the medium to provide flotation, depend upon the characteristics of the material to be removed and separated from the remainder of the waste. The essential element in the operation is selection of the medium in conjunction with the characteristics of the waste to be separated so that the waste material to be separated has a greater attraction or affinity for the air bubbles passing up through the medium than does the medium for the material. In this way, as the air bubbles rise through the mixture of wastes and medium, the selected waste travels upward with the air bubbles to the surface and is collected through overflow over a weir or by racking and screening operations. As is obvious from the preceding description, flotation operations were initially developed for minerals processing. The technique appears to have some applicability for solid waste processing but certain difficulties are associated with the mixing of the heterogeneous wastes and the selected medium. Reactions between constituents in the waste and the medium may effectively eliminate certain media from consideration as vehicles for separation. Moreover, methods developed using flotation as a means of separation have been designed almost exclusively for use with small particle sizes. Therefore, the general heterogeneous solid waste stream may be a difficult material for the application of flotation separation techniques.

To date, the applications or trials for flotation as a separation technique have been very limited. The U.S. Bureau of Mines research team in College Park, Maryland has conducted a number of experiments using flotation techniques for separation of selected metals and other materials from the residue remaining after incineration of municipal solid wastes (Ref. 6-32). In the experimental work performed to date, a significant possibility for application of flotation techniques to separation of materials in incinerator residue has been demonstrated. However, as mentioned previously, little applicability to incoming mixed solid wastes is foreseen.

The specific gravity or density of waste constituents may also be used as a property for separation through the use of heavy media as supporting fluids. In other words, the specific gravity of the supporting medium may be selected so that only certain constituents of the solid waste stream float in the material or sink in the material. For example, very dense media may be used to support the majority of the materials in a solid waste stream allowing only selected high-density materials to sink and pass out of the separation device through the bottom of the medium container. Separation of wastes through the use of dense media has been contemplated for a number of years because of the widespread use of dense media separation in mineral processing industries. However, to date, very little application of dense media separation techniques to solid waste processing has been made. The U.S. Bureau of Mines has conducted experiments in College Park, Maryland using dense media separation techniques on incinerator residue but very little other work has been done in this area.

The basis for dense media separation is the creation of a fluid of such specific gravity that materials in a given specific gravity range either float to the surface and are recovered or sink to the bottom of a vessel containing the dense media and are recovered there. The media used for the separation operation may be truly dense media or may simply be a suspension of solid particles in water (or another liquid) to achieve in effect a dense medium. Since most of the use of heavy media has been in the mineral processing industry, rather high specific gravities have been desirable for the separation media. Materials such as magnetite or ferrosilicon with very high specific gravities have been mixed with water to provide a high-density separation medium. These two particular materials are also useful in creating the separation medium because they may be recovered from the waste-medium mixture through magnetic means. Thus, pieces of metal to be recovered from a solid waste stream could be successfully separated from the remainder of the wastes by mixture with a dense medium consisting of a suspension of magnetite in water. The dense metallic pieces would settle through the suspension and be collected at the bottom of the tank. The lighter portions of the solid waste would float on the surface of the suspension and could be gathered and collected with racks or similar skimming devices. Any magnetite mixed with the waste could be recovered through magnetic separation after the wastes had been collected from the surface of the suspension.

In addition to suspensions of solid particles in water, certain fluids may be used as dense media in separation operations. Typically, the fluids used are organic liquids—generally long-chain hydrocarbons and combinations of chlorinated and brominated hydrocarbons. These fluids have been used in the minerals extraction industry but their application to the processing of solid wastes would involve many disadvantages. For example, many of the more useful hydrocarbon liquids are hazardous in themselves since they produce dangerous fumes and also are injurious upon contact with the skin. Additionally, hydrocarbon compounds used as separation media could react with the heterogeneous constituents of municipal solid wastes and the resulting losses of the separation medium during processing could make the entire operation costly and inefficient.

Differences in the specific gravities of waste constituents and consequent differences in their inertial properties may also be used to effect separation of those constituents by means of vibrating tables. As developed in the mineral processing industry, these devices have been grouped into two categories, known respectively as "stoners" and "Wilfley tables." Stoners consist of vibrating tables which are inclined along their longitudinal axes with the feed of incoming refuse entering the'tables at their higher ends. The table is subjected to a transverse (perpendicular to the long axis) vibration consisting of a slow movement in one direction with a rapid reversal of movement back to the original position. The material on the vibrating table progresses, because of the inclination, from the input end to an outlet at the opposite end. Because of the transverse movement,

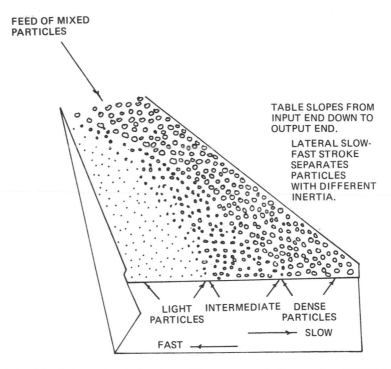

FEED OF MIXED
PARTICLES

TABLE SLOPES FROM
INPUT END DOWN TO
OUTPUT END.

LATERAL SLOW-
FAST STROKE
SEPARATES
PARTICLES
WITH DIFFERENT
INERTIA.

LIGHT   INTERMEDIATE   DENSE
PARTICLES   PARTICLES

→ SLOW

FAST ←

Fig. 6-8   Schematic of vibrating table separator (redrawn after USEPA).

materials of high density and therefore high inertial potential move on the table in the direction of the slow movement during the vibration. The heavy materials move with the slow initial movement and then during the rapid reversal of movement the heavy particles tend to remain in their new position and therefore congregate on one side of the vibrating table. A typical stoner is shown in Fig. 6-8. It is possible to effect approximate separation of materials of varying specific gravity through use of a stoner.

A Wilfley table is a device consisting of a vibrating table but, in contrast to the stoner previously described, the unit is positioned at an inclination transverse to its long dimension. A Wilfley table is shown in Fig. 6-9. Material is fed into the table at one end and the table is vibrated along its longitudinal axis. Located along the surface of the table are small ridges running longitudinally from one end of the table to the other end. Water is introduced into the table along one of its long sides and flows downslope from one long side to the opposite long side across the longitudinal ridges. Light or low-density materials are carried by the water over the longitudinal ridges and are collected on the lower long side of the table. Heavy materials tend to collect against the longitudinal ridges and are moved down the table away from the entrance end by the longitudinal vibrations. Therefore, heavy materials end up at the opposite end of the table from the input area and lighter materials are collected along the lower long side of the table.

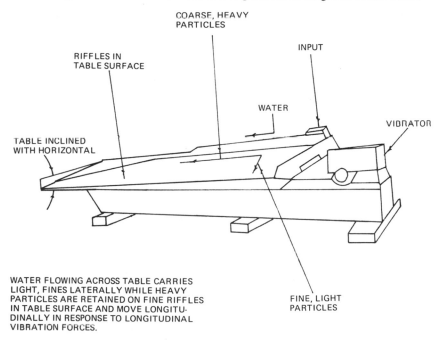

Fig. 6-9   Schematic of Wilfley Table (redrawn after Denver Equipment Company).

Both stoners and Wilfley tables are applicable to the separation of solid wastes which have been subjected to prior size reduction. However, the separation which is achieved is not a high purity operation, and therefore the application of vibrating tables to separation for recycling is somewhat limited.

Another kind of device which operates to separate materials on the basis of specific gravity is the jig. A jig is shown in Fig. 6-10. The action of the jig is to produce a fluctuation or pulse of water (or any other suspending fluid) through a collection of waste particles. As shown in the jig illustrated in Fig. 6-10, a plunger or piston may be used to cause a pulse in the water within the jig body. The water enters the waste compartment through a screen at the bottom and acts to propel the low-density materials to the top of the waste particle body. Denser materials flow down through the surging water stream and pass through openings in the screen and are collected at the lower port. Light or low density materials may be periodically removed from the top of the waste compartment.

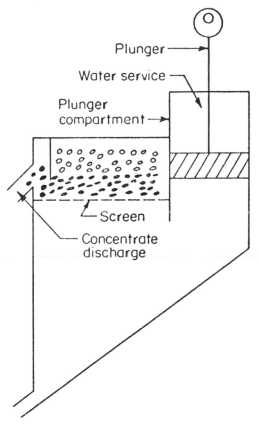

Fig. 6-10 Jig.

Jigs have been widely used, especially in series, in mineral processing operations, but to the present time no application of jigs to solid waste processing has been made. There exists a significant potential, however, for the use of jigs in the processing of incinerator residue, after an initial grinding operation on the raw residue produced from incineration.

Other devices which use the specific gravity of materials for separation include the Stanford zigzag air-classifier which has been developed by the Stanford Research Institute under support from the Environmental Protection Agency. This separator is shown schematically in Fig. 6-11. In essence, wastes are introduced into an upper port in the separator and tend to tumble down through the zigzag chute. The wastes thus are repeatedly entrained in an air stream which enters in the bottom of the zigzag chute and exits at the top of the chute. Light materials are carried by the air stream up and out into a receptacle. Heavy materials pass down through the air stream and are collected at the bottom of the zigzag chute. Thus, this classifier uses a combination of material density and the air resistance associated with particle size to effect separation. Its application to solid wastes processing would involve a prior size-reduction operation in order for the classification to be successful. Separation of any given material could be achieved by varying the speed of the air stream entering the bottom of the chute, since the supporting power of the air stream would vary with the stream velocity. The SRI zigzag classifier has been used successfully in trials to separate glass and mineral particles from mixed municipal refuse, with organics and aluminum particles

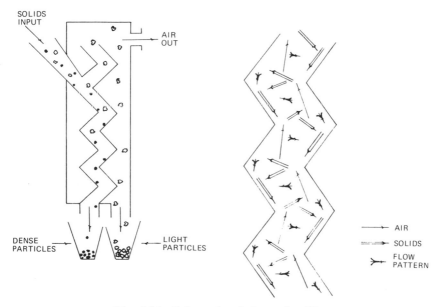

Fig. 6-11 Schematic of zigzag classifier.

carried out by the air stream from the zigzag chute and into an external collection chamber. Application of an air zigzag classifier to the waste processing facility of the city of St. Louis has been implemented as a means of reducing the wear associated with pneumatically feeding mixed refuse to the Union Electric Company boilers at the Meramec Power Station in the trial energy recovery operations conducted in St. Louis (see Chapter 7). Such dry air-classification of wastes may have great applicability to the processing of refuse for energy recovery. However, further separation operations would be required to separate aluminum particles from the organic fraction which is desirable for burning in utility furnaces.

The density or specific gravity of the material provides a direct measure of the force necessary to produce a given acceleration in volume of that material. This measure of force per unit volume necessary for a given acceleration is termed inertia. Differences in material specific gravity cause differences in the inertial properties of the different materials. Thus, if a group of pieces of material of identical size but varying composition all are propelled by a given force, the material with the greatest amount of inertia will travel the shortest distance. This statement may be examined from the viewpoint of practical experience, such as the experience of throwing light and heavy objects. The limits of force applied by a person's arm are finite, and the acceleration resulting for given materials varies with the weight of the piece of material. If a person were to impart the same force in throwing objects of the same size, the object which possessed the most inertia would fall closest to the person. On the other hand, if the person applied equal accelerations to all of the objects of the same size, the object with the greatest amount of inertia would travel farthest. The inertial properties of material have been used, in conjunction with the air resistance of the particles of the various materials, in proposed separation systems which have been given the general name "ballistic" separators. Other types of inertial separators have been developed which combine inertial properties of material with other properties such as resilience. Several inertial separators are shown in Fig. 6-12. The hammermill used in Madison, Wisconsin is purported by the manufacturer to function as a ballistic separator since the hammermill hammers swing upward and reject hard metallic or mineral particles up and out of the input chute. With the exception of the horizontal-rotor hammermills, however, no successful application of inertial separation devices has been made for solid wastes. Several unsuccessful trials of inertial separation equipment have been made. The masking effect of the various constituents in the solid wastes has prevented the effective separation of materials; i.e., there must be rather clear-cut differences in inertial or resilient properties of the material before inertial separators will be successful. Heterogeneous municipal solid wastes do not fulfill these requirements. Little hope for further application of inertial separators in solid waste processing can be held.

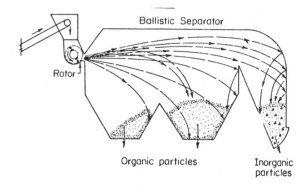

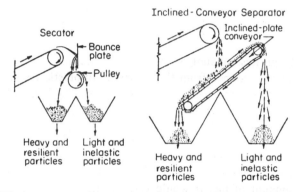

Fig. 6-12 Ballistic separators: separation by inertial properties. *Source:* Drobny, N. L., Hull, H. E. and Testin, R. F., "Recovery and Utilization of Municipal Solid Waste, USPHS Publ. No. 1908, 1971.

Magnetic Separation Devices. The effective transfer of metallic wastes in a junkyard through the use of an electromagnet suspended from a crane long ago suggested the use of magnetic separation devices for processing of ferrous metals in solid wastes. At the present time, two general types of magnetic separators are in use. In one instance, the magnetic separator consists of a fixed permanent magnet or electromagnet which is positioned above a point where the wastes pass on a continuous conveyor belt or similar transfer device. A permanent magnet will pull ferrous metals from the waste stream and the collected metals may be periodically removed from the surface of the magnet. An electromagnet may be intermittently turned off in order to allow removal of collected metal particles. This type of separation device is suitable for processing raw refuse; large pieces of ferrous metals may easily be removed from the waste stream by means of this type of separator.

The second type of magnetic separator consists of a drum-type device contain-

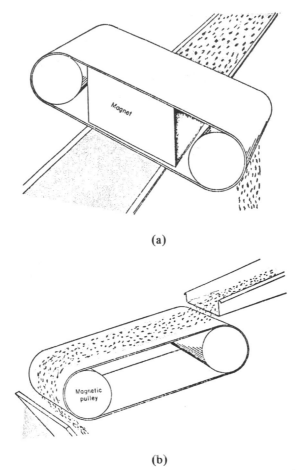

(a)

(b)

Fig. 6-13 (a) Suspended type permanent magnetic separator; (b) Pulley type permanent magnetic separator. *Source:* Drobny, N. L., Hull, H. E. and Testin, R. F., "Recovery and Utilization of Municipal Solid Waste," USPHS Publ. No. 1908, 1971.

ing permanent magnets or electromagnets over which a conveyor or a similar transfer mechanism carries the solid waste stream. The conveyor belt conforms to the rounded shape of the magnetic drum and the magnetic attractive forces pull the ferrous metal particles away from the falling stream of solid wastes. Both types of magnetic separators are shown in Fig. 6-13.

Magnetic separators will function efficiently whenever the ferrous metal particles in the waste stream have been successfully separated physically from other components of the waste stream; i.e., even though the ferrous metal pieces may be contained within a stream of solid wastes, they must be discrete particles separated from the nonmagnetic portions of the waste. For example, a piece of ferrous metal attached to a large chunk of wood or similar nonmagnetic material could easily continue in the solid waste stream because the attraction of a magnetic separator for the small piece of ferrous metal would not overcome the weight of the nonmagnetic wood plus the weight of the metal itself. For this reason, some sort of size reduction or grinding or pulverization operation should precede magnetic separation of ferrous metals. In some instances, the operation which has been used to separate ferrous metals from the other constituents has been combustion. For example, in incinerators ferrous metals are separated from organic portions of the wastes. The ferrous metals may be magnetically separated from the residue remaining after incineration.

Magnetic separation has little significant effect upon any other type of separation operation or on other waste processing operations. Additionally, magnetic separators have been in use for a considerable time in the scrap metal processing industry. For these reasons, the application of magnetic separators for the recovery of ferrous metals in a recycling operation may be achieved easily, rapidly, and inexpensively. Magnetic separators have been used extensively in the past and should find extensive use for metals recycling from solid wastes in the future.

Optical Separation Devices. As mentioned previously, color sorting of materials has been practiced for a considerable length of time in the minerals processing industry, particularly with respect to the color sorting of diamonds. Color sorting techniques have also been successfully applied in the food processing and agricultural industries. For example, sorting of kernels of corn on the basis of surface reflective properties (color) has been successfully accomplished. Therefore, considerable potential for color sorting of solid wastes materials seems to exist. Most of the optical equipment with possible applicability to the separation of solid wastes has been manufactured by the Sortex Company, Inc., located in Lowell, Michigan. Figure 6-14 shows a schematic diagram of a Sortex color separation unit. This type of color separation unit depends upon a rather uniform size of particle being fed into the feed hopper of the unit. The material comes from the feed hopper along a vibrating chute onto a grooved feeder belt and then into the separation chamber. Around the walls of the separation chamber are positioned four photocells and across from each cell is a colored background strip to which the photocell is sensitized. Light enters the separation chamber and is reflected from the colored background slides into the photocell. The light entering the separation chamber also is reflected from the particles of waste which fall through the chamber. When the reflection of light from the particle of waste matches the reflection of light from the colored background slide, the photocells produce a constant voltage. If a particle enters the chamber and is not the same color as the colored background strips, the light reflecting from the particle enters the photocells and produces a change in voltage across the cells. This

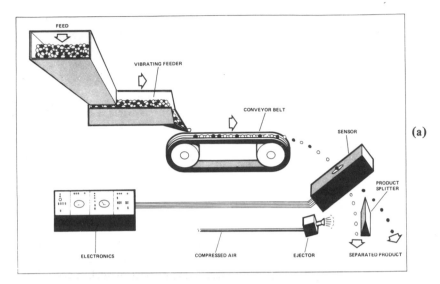

(a)

(b)

Fig. 6-14   (a) Schematic of Sortex separator (*courtesy* Sortex); (b) Sortex separator, model 962MB (*courtesy* Sortex).

change in voltage resulting from a different color particle (either lighter or darker in hue or brightness from the background strips) triggers a compressed air jet which blows the different-colored particle into a reject chute. Therefore, only particles of the desired color fall straight through the separation chamber and into the storage hopper.

At the present time, much interest and attention is being focused upon the application of a Sortex optical classifier unit at the Black Clawson wet pulping operation in Franklin, Ohio. At that plant, the Sortex unit will be used to separate various colors of glass fragments. Cost data for Sortex units are shown in Fig. 6-15.

Summary on Separation Techniques. A wide variety of equipment types and separation methods have been developed in manufacturing and mineral extraction processes. Very few separation devices or techniques, on the other hand, have been developed specifically for treatment of solid wastes. The application of the equipment used in other industries to the solid waste recovery industry has not met with complete success and has little prospect of such success. Modi-

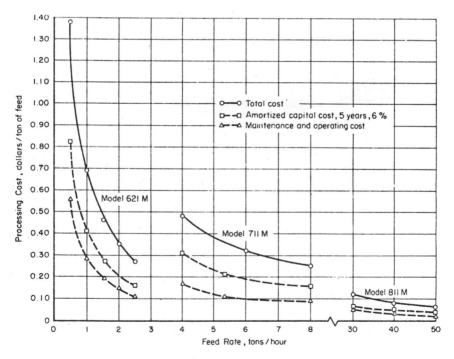

Fig. 6-15 Economics of Sortex. *Source:* Drobny, N. L., Hull, H. E. and Testin, R. F., "Recovery and Utilization of Municipal Solid Waste," USPHS Publ. No. 1908, 1971.

fication of existing equipment and development of new equipment are virtual necessities if materials recovery for the solid waste stream is to be achieved.

6-3.4 Economics of Separation

The economics of separation of materials from the solid waste stream has not received any significant amount of study and at this time the only economic data available for study are rather limited data concerning specific pieces of equipment and specialized separation techniques. No definite statements about separation economics may be made on the basis of this limited information. However, certain trends are obvious in the application of existing techniques and equipment to the separation of constituents of solid wastes. For example, magnetic separation is relatively inexpensive since the primary piece of equipment, a permanent magnet or electromagnet, is relatively inexpensive and the operating costs of the equipment, chiefly for electrical power for the magnet, are low. Magnetic separation therefore would appear to be a relatively inexpensive and effective means of materials separation; however, magnetic separation is limited to the retrieval of ferrous metals (high intensity magnetic fields may be used in the separation of colored glass from clear flint glass, but such application of magnetic techniques is very limited and is experimental in nature).

From the point of view of economics, hand separation or sorting of refuse is not a promising technique for materials recovery, at least not in the United States. The effectiveness of hand sorting of waste constituents is somewhat questionable from the point of view of purity and constancy; this human error factor is reflected economically in a lower resale price for recovered materials which have been separated from the solid waste stream through human agency.

There exists a wide variety of equipment types and separation methods which, in contrast to hand separation, are relatively inexpensive in terms of operating cost. However, much of this equipment has been designed and manufactured for use with wastes other than the heterogeneous waste from municipalities. For this reason, the effectiveness of this equipment is rather low. Stated in a different way, low cost mechanized separation of waste components has not been achieved. Future developments in the field of mechanized separation equipment may produce low cost, efficient, and effective means for separation of waste constituents. At the present time, however, the economics of separation are rather uncertain and the application of a given separation technique or method to a waste stream produced in any given municipality or locality must be evaluated individually. Disparate factors such as labor costs, power costs, capital costs for equipment, etc. must be considered in the design of a separation system for materials recovery from solid waste.

The preceding comments serve to summarize the economic situation with respect to materials separation. A review of the technical literature has produced little comprehensive data on equipment costs and method economics. However, some selected data on specific types of equipment have been presented in the earlier parts of this section. These data may serve to indicate preliminary costs and expenses associated with incorporating a given piece of equipment into a materials recovery system, but the application of these data to any specific design is discouraged.

6-3.5 Summary—Separation

In summary, it may be stated that a considerable potential exists for resource conservation through the separation and recovery of various constituents of the solid waste stream. However, methodology and equipment necessary for such separation have not been developed at the present time. Applications of existing equipment initially developed for processing of materials other than municipal wastes has not been successful. Some modification of existing equipment and the development of new methods and new apparatus has been attempted, but much work remains to be done in this area. The low value associated with the materials recovered from the solid waste stream militates against large investments in separation equipment or in research into separation methodology. However, the changing situation with respect to the market values for recovered materials (higher values through governmental incentives, changes in tax laws, and changes in public opinion) hold out promise that at some time in the near future considerable attention and study will be given to the development of better separation techniques. Development of systems of separation procedures will produce a synergistic effect on the overall materials recovery operation. In other words, combination of prior size-reduction operations with separation operations may make possible more effective and cheaper separation of the desired materials. On the other hand, lack of success to date in materials recovery efforts has somewhat discouraged further development in this field. Perhaps the most helpful sign to date with respect to materials recovery from solid waste has been the development of large systems for materials recovery from wastes. These systems will be discussed in the next section.

6-4. RECYCLING SYSTEMS

Within the last few years several systems for recycling materials contained in solid wastes have been proposed. The status of these systems at the present time ranges from simply a vague concept to an actual operating demonstration plant. Several of the systems proposed are simple and consist of combinations of two or more of the separation operations described in the preceding section. On the other hand, other systems have been designed to consist of a multitude of individual unit processes combined together to utilize to the fullest every component of the solid waste stream. In attempting to describe these systems, a first step is the classification of the type of operation. Generally, materials recovery and waste processing operations may be characterized in the following way: certain separation and recovery operations are intended for use on raw, untreated solid wastes before they are subjected to any detailed processing operation such as incineration—these systems will be termed "front end" systems. After an initial attempt at materials recovery from the wastes stream, almost all proposed recycling schemes contain a provision for the processing of the refuse which may entail composting, incineration, or pyrolysis of the wastes—this type of operation will be termed "processing." All of the processing operations described in the preceding phase produce some residue or "left over" materials which contain valuable portions of recoverable components—operations for recovery of these valuable components within residual matter will be termed "back end" systems. The basic reasoning behind the description of recycling systems in terms of "front end" and "back end" systems is simple. Certain unit processes such as mechanical separation of paper fibers, glass fragments, and bits of metals may be accomplished with moderate success on raw municipal refuse without detailed prior processing. Therefore, such operations may be grouped together in a "front end" systems of materials recovery. However, such operations generally will recover only a small portion of all the waste contained within the solid wastes stream. Typically, "front end" systems recover only about 15 to 20 percent of the volume of the incoming wastes. Therefore, the bulk of the wastes remain for disposal or for further recovery of resources, either as reused materials or as recovered energy. This is the reason for addition of further processing and "back end" systems of materials recovery. In general, more work has been done and more success has been achieved in operating on raw refuse before it is further processed; for example, mechanical separation of metals by magnetic means, glass particles separation by gravity, and paper fiber separation through wet pulping are relatively easy to achieve. Recovery of the value of the remaining waste material is considerably more difficult and more expensive. This is true because of the intimate mixing of organic and inorganic components of the remainder of the refuse. Typically, the value of the organic fraction may be recovered most easily through combustion and utilization of the developed energy.

Operations on the remaining portions of the solid waste stream after recovery of metals, paper, and glass have been described in Chapter 2, in the discussion dealing with composting operations. Additionally, processing of this portion of the wastes could be accomplished through incineration or through pyrolysis. The processing of wastes by high temperature and conventional incineration was discussed in Chapters 5 and 3, respectively. The recovery and utilization of the heat produced during incineration of wastes is discussed in Chapter 7. Pyrolysis, the combustion of material in the absence of oxygen, has been proposed as a processing method for recovery of energy contained in the organic fraction of solid waste and for recovery of various materials from that fraction (Refs. 6-23 and 6-30). Usually, pyrolysis operations are intended to produce hydrocarbon products such as oils, gases, or char and a residue from which valuable metals may be recovered.

All of the above-mentioned processing operations are intended to recover the worth of the wasted materials through their use as a soil conditioner in the case of composting, through utilization of recovered heat in the case of incineration, and through the production of available heat and useful by-products in the case of pyrolysis. However, all of these methods produce a residue which is essentially neither biodegradable nor combustible. The residue is separated from other components of the waste in incineration and pyrolysis operations but in composting this nondegradable residue remains within the compost on its way to a consumer. Therefore, there is ample justification for the development of separation and recovery systems designed to work on the residuals left after processing, whether the processing consists of composting or a combustion process.

In the following section, descriptions will be given of two subsystems developed by the U.S. Bureau of Mines which are typical of "front end" systems and "back end" systems. After these subsystems have been described, complete systems which incorporate "front end" recycling systems, processing operations, and "back end" materials recovery systems will be described.

6-4.1   Typical "Front End" and "Back End" Systems

The U.S. Bureau of Mines has conducted extensive research into the recovery of selected materials from both raw unprocessed solid waste and the residual material remaining after waste processing in incinerators (Refs. 6-19 and 6-32). This extensive experience with waste utilization has led to the development of a proposed dry separation method of materials recovery which is typical of the better kinds of "front end" systems devised by various persons and agencies. The "back end" system developed by the U.S. Bureau of Mines for separation and recovery of materials contained in incinerator residue is virtually unique and has not been duplicated or imitated in any commercial development. A significant need for further research and study exists in the area of resource recovery from previously processed solid waste.

The dry separation system developed for utilization with unburned solid waste is shown in Fig. 6-16. The initial step in this Bureau of Mines system is shredding of the raw refuse with immediate subsequent separation of ferrous metals through the use of a magnetic separator. The recovered metals are proposed for use in detinning operations which will render the purified metal usable as scrap in the production of steel and other ferrous products. The remainder of the solid waste travels to a horizontal air-classifier similar to the vibrating tables discussed in an

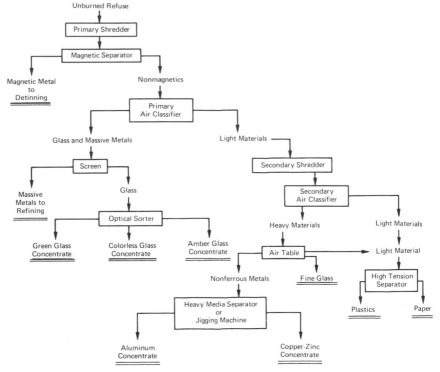

Fig. 6-16   USBM "front end" system.

earlier section. This air-classification separates the dense materials such as glass and heavy metals from the lighter fraction of the refuse. The collected glass and metals are subjected to screen separation which is feasible because of the very small average size of the glass particles in contrast to the moderate to large size of the metal particles in the waste ($1\frac{1}{2}$ in. or greater in least dimension). The obtained metals are then transported to a smelter. The glass particles are transferred for further separation in an optical separation device similar to the Sortex unit previously discussed. The light fraction of materials obtained from the primary air-classifier are subjected to a second shredding operation. After this second shredding/grinding operation, further air-classification, most probably in a series of cyclone separators, is used to separate the materials into a relatively heavier and obviously lighter grouping. The heavier materials are subjected to further separation on an air table as discussed previously, and divided into non-ferrous metals, glass, and light materials which are mixed with the light materials produced by the secondary air-classifier. The light materials coming from the air classifier are subjected to processing in a toothed shredder suitable for operations on ductile materials. This shredder effectively divides the light materials into a plastics fraction and a paper fiber fraction. The last step, shown in Fig. 6-16, is the separation of the nonferrous metals into an aluminum fraction and a copper-zinc fraction through the use of a dense media separator or a mineral jig.

The advantages inherent in the proposed U.S. Bureau of Mines system include no wet operations (thus, no water pollution, no vapor emissions, and no necessity for materials drying), the production of a large number of usable recycled materials, and the input requirements of only electrical power and a suitable medium for the dense media separator used in processing the nonferrous metals. The use of a dry separation methodology possesses considerable advantage over the use of water or a similar fluid in effecting the separation of materials.

The U.S. Bureau of Mines system for recovery of material from incinerator residue has been described in detail in other publications (Refs. 6-19 and 6-32). Many of the unit operations and unit processes used in operating on the incinerator residue are similar to the operations shown in Fig. 6-16 for the dry separation system utilizing raw refuse. There are some differences, however, including the use of water as a medium in screening operations. Water may be conveniently used in operating on the incinerator residue since the biodegradable portion of the refuse has been removed and thus no biological water pollution problem exists. Moreover, the volume of the materials in the incinerator residue amounts to approximately only 10 to 15 percent of the input waste stream; therefore, there is little necessity for drying out large volumes of organic materials which readily absorb water. The materials to be recovered from the incinerator residue consist of iron concentrates, pieces of aluminum, copper and zinc particles, and particles of clear flint glass and colored glass. Additionally, a carbonaceous ash is obtained. The initial separation of the residue from the carbonaceous ash is accomplished through the use of screens with water as a suspension medium.

The residue is then processed in a rod mill to reduce the particle size of the glass fraction. The glass particles, generally of the consistency of sand, are further processed after size separation through the use of a high-intensity electromagnet which separates the glass into clear flint and colored fractions. The remainder of the residue consists of nonferrous metals which are further separated into an aluminum fraction and a copper-zinc fraction through the use of a dense media separator such as that shown in Fig. 6-16. Table 6-8 below indicates the anticipated revenues, predicted by Bureau of Mines personnel, for sale of the recycled waste products. These values for recovered products should be compared with the estimated operating costs of the system of $13.30 per ton for a 400 ton per day plant (1971). Such a plant will also have a capital cost of approximately $2 million.

The preceding paragraphs have described two systems which characterize typical "front end" and "back end" systems for materials recovery from solid wastes. The following sections are devoted to a description of some of the

TABLE 6-8 Estimated Product Values (1971) U.S. Bureau of Mines Recovery System

Product	$ Value	Quantity (lb)/Ton Residue	$ Value Ton Residue
Ferrous metal	10/ton	610	3.05
Aluminum	0.12/lb	32	3.84
Copper–zinc	0.19/lb	24	4.56
Colorless glass	12/ton	552	3.31
Colored glass	5/ton	398	1.00
Total value			15.76

Source: Henn, J. J. and Peters, F. A., "Cost Evaluation of a Metal and Mineral Recovery Process for Treating Municipal Incinerator Residues," Dept. of the Interior, Information Circular 8533, 1971.

proprietary recycling systems which have been proposed and which characteristically include subsystems similar to those developed by the U.S. Bureau of Mines.

6-4.2 The Hercules, Inc. System

In 1970, the governor of the state of Delaware signed a contract with the Hercules Company of Wilmington to design what was termed "the most advanced solid waste reclamation plant ever devised" (Ref. 6-35). The design of this recycling plant was completed by the Hercules staff in 1972. The state of Delaware submitted the proposed design to the Environmental Protection Agency with a proposal for funds to meet the construction and operating costs of the plant during the first years of its use.

The system as designed by the Hercules engineers is shown in Fig. 6-17. It is designed to accommodate 500 tons of municipal solid waste per day and as much as 230 tons of wastewater treatment plant sludge per day (8 percent solids). Essentially, the design incorporates a "front end" system for reclamation of ferrous metals, a composting type of waste processing operation, a pyrolysis operation, and separation and recovery operations for the residue remaining after composting and after pyrolysis. The details of the procedures are not available from Hercules since they are considered to be proprietary in nature. The pretreatment operations will consist of a size-reduction operation accompanied by magnetic separation of ferrous metals. The material remaining after magnetic separation will pass into a composting plant where sewage sludge will be added to the solid waste. The compost produced in this plant will be subjected to a separation operation in which nondigestible organic materials and inorganic constituents will be removed. The finished product is termed "humus," and is proposed for use as a soil conditioner in agriculture or as a raw material in the manufacture of

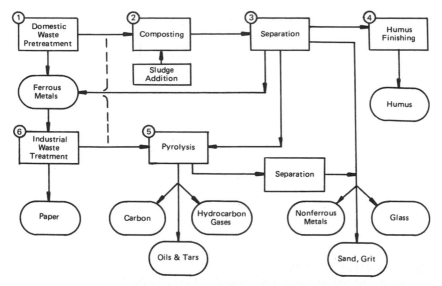

Fig. 6-17 Hercules system. *Source:* Delaware Reclamation Project, Hercules Corp. SWM Bulletin 2.

fiberboard. The nondegradable organic materials separated from the humus will be processed in a pyrolysis operation and converted to gases, oils, and carbonaceous char. Any residue remaining after pyrolysis will be separated and added to the inorganic portion of the residue removed from the humus and then subjected to further separation into glass, nonferrous metal, and sand fractions. An interesting feature of the Hercules plant is the inclusion of provisions for industrial waste treatment wherein the industrial waste will be subjected to magnetic separation, removal of paper fibers, and subsequent pyrolysis of the residue. Table 6-9 indicates the dollar values proposed to be received in the plant as designed by the Hercules staff. The table is based upon a 500 ton per day operation. It is significant to note that in this table, in contrast to the indications in the diagram showing the plant process flow, an unusable residue of approximately 21 percent of the input is proposed to be produced. This material which inevitably must be disposed of would create a loss in value associated with disposal costs. Therefore, in the table listed below, the notation "not applicable" following the category of material labeled "unused" is not accurate. A disposal cost or negative income of from $2 to $10 per ton should appear in the table at that point. However, the table is presented as it was originally compiled by Hercules, Inc.

The Hercules system described above is typical of a large number of reclamation systems proposed for use in recycling solid wastes at the present time. The probabilities of success or failure of such a system are exceedingly difficult to evaluate in early stages of design. Unfortunately, many of the recycling systems

TABLE 6-9   Recycling Rates and Expected Revenues:
Hercules Recycling Plant—500 Tons per Day

Material	Tons	Percent of Input	Expected Value per Ton
Humus	200	40.0	$10–15
Ferrous	34	6.8	$2–11
Nonferrous	11	2.2	$125–175
Carbon	8	1.6	$50–120
Paper	5	1.0	$10–20
Glass	24	4.8	Up to $10
Grit	14	2.8	Up to $2
Oils	15	3.0	NA
Hydrocarbon gas	85	17.0	NA
Unused	104	20.8	NA

Source: Hercules, Inc.

proposed for use in this country now exist only on paper. Other proposed systems can be examined for comparison. Another integrated system is the system proposed by the National Center for Resource Recovery.

6-4.3 The National Center for Resource Recovery (NCRR) System

The National Center for Resource Recovery is an agency funded by a number of users and manufacturers of packaging material (and therefore producers of significant amounts of solid waste). Examples of member companies are the Owens-Illinois Company, the Pepsi-Cola Company, and Continental Can Company. The purpose of the center is the investigation and sponsorship of means for advancing resource recovery principally through material recycling. Since the Center's creation, the focus of effort of NCRR personnel appears to have been the development of a so-called "network" system of plants for the recycling of solid waste constituents (personal communication, J. G. Abert, 1973). This recycling program is proposed to consist of twelve plants for recycling materials, each with an input capacity of 500 tons of waste per day. During 1973, the Center staff were conducting surveys and preliminary investigations to determine the most appropriate locations for the twelve plants. These locations obviously would be in metropolitan areas throughout the United States, but an effort is being made to obtain optimal locations with respect to transportation of recycled materials to nearby markets. In mid-1973, it was announced that New Orleans would be the site for the first plant. The basic concept of the network plants is the combination of a "front end" system of materials recovery with a processing operation which ideally is already in existence in a given locality. The type of "front end" system envisioned by the Center personnel is shown schematically in Fig. 6-18.

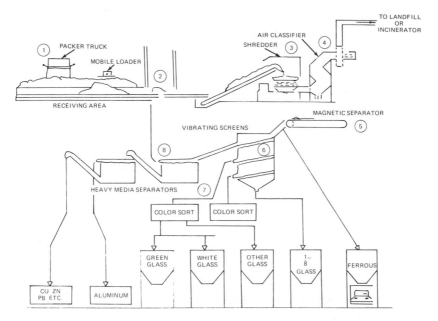

Fig. 6-18   NCRR "front end" system (redrawn after Abert).

As indicated in Fig. 6-18, about 80 percent of the incoming solid waste stream would pass from a "front end" separation system to a waste processing operation such as incineration, composting, or sanitary landfill. Particular emphasis has been placed on the combination of the "front end" system with an existing combustion operation, especially a combustion operation designed to produce power. Ideally, the waste would proceed through the "front end" materials recovery system, and after metals, glass, etc. had been removed in this system, the combustible remainder would be fed into an existing utility boiler for the production of electrical power as has been accomplished by the Union Electric Company in St. Louis, Missouri. The strategy associated with the type of recycling operation envisioned by the NCRR personnel is to obtain a local functionary or middleman (which in some instances could be an existing local utility or similar agency) who would undertake the building of the waste processing unit or who would accept the combustible material which would be produced from the initial materials recovery operation.

The NCRR network plant includes the best features of the materials reclamation systems developed by the Bureau of Mines and similar agencies over the last decade. Additionally, it has the advantage of being flexible with regard to the waste processing operation which is designed to accommodate the material remaining after initial materials recovery. Furthermore, the effort of the NCRR personnel is directed toward obtaining optimal locations for such facilities, where a high degree of success for recycling operations is probable.

As mentioned, many of the recycling systems proposed for use in the country today now exist only on paper. A few systems have progressed to the pilot or demonstration plant stage and therefore may be evaluated more thoroughly. One of the materials recovery systems which has reached the demonstration plant stage is the Black Clawson wet pulping and fiber recovery system. This system is discussed in detail in the next section.

6-4.4   The Black Clawson System

One of the materials recovery systems which has received considerable publicity is the wet pulping operation carried out by the Black Clawson Company under the proprietary names of Hydrasposal and Fiberclaim. The Black Clawson Company is a long-established supplier of paper processing equipment and the solid wastes recycling system now being promoted is based upon pulping methodology first developed in the paper industry. An existing 150 ton per day demonstration unit located in Franklin, Ohio was partially financed by the Environmental Protection Agency. Included in the pilot plant is a Hydrapulper for wet pulping of incoming solid wastes and a fiber reclamation system (Fiberclaim). Additionally, other materials are reclaimed from the pulped waste/water slurry.

The pilot plant was suggested to the City Council of Franklin by one of the Councilors who at the time was also a staff engineer for the Black Clawson Company. The city landfill was almost completely filled at the time and a new disposal facility was needed. A request for funds was made by the City Council to the Department of Health, Education, and Welfare for a disposal facility to demonstrate the wet pulping technique. The requested funds were granted and construction of the plant proceeded as scheduled (personal communication, J. Baxter, 1972). The plant was opened for operation in August, 1971.

Methodology. The pilot plant built in Franklin, Ohio (and proposed prototype plants) incorporate techniques and equipment developed in the Black Clawson Company test laboratory. The system is shown schematically in Fig. 6-19. In the initial year of operation of the pilot facility, the operating personnel were the same individuals who had developed and operated the trial plants in the Middletown, Ohio company laboratory.

Refuse is brought into the plant and deposited in an enclosed temporary storage area. From the storage area, the refuse is pushed onto a depressed conveyor belt and transferred to the Hydrapulper where it is mixed with water. The pulper reduces the size of the pieces of refuse until they pass out of the pulper through $3/4$-in. diameter openings. Pieces of metal or other durable materials larger than $3/4$ in. in diameter are rejected into a junk remover. The large pieces pass from the junk remover through a magnetic reclamation device where ferrous metals are retrieved. Nonferrous inorganics have been disposed of in a landfill.

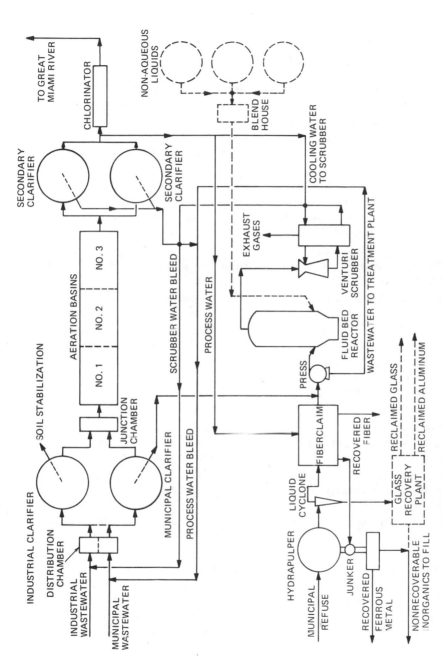

Fig. 6-19 Black-Clawson system at Franklin, Ohio (*courtesy* Black-Clawson Company).

The pulped waste materials pass in a slurry to a liquid cyclone in which glass and other dense materials are removed from the waste stream. The slurry passes to the Fiberclaim unit where long paper fibers are retrieved. These fibers amount to about 20 percent of the total weight of paper fibers in the input solid wastes. The remaining portion of the refuse, including 80 percent of the input paper and all of the plastic/organic material, is further dewatered in a screw press and then fed into a fluid-bed combustor. Combustion is purported to be 100 percent complete in this unit, with no residue. Exhaust gases from the reactor pass through a wet Venturi scrubber where they are cooled from the 1400-1600°F range of temperatures maintained in the fluid-bed reactor. The manufacturer has claimed that the reactor–scrubber combination eliminates all air pollution.

The materials removed from the liquid cyclone consist of glass, aluminum, and other inorganic particles. The Glass Container Manufacturers Institute has assisted the Black Clawson Company in developing a glass recovery and separation facility based on the Sortex process. Aluminum can be retrieved from the cyclone residue also, but the aluminum scrap is somewhat contaminated with other inorganic wastes.

The wastewater which is removed from the combustible wastes in the screw press is transferred to a wastewater treatment facility, a secondary treatment plant built by the Miami Valley Conservancy District in the case of the Franklin plant. The Franklin treatment plant incorporates an aerated lagoon process. The Black Clawson Company claims that the BOD_5 for the wastewater, which is approximately 6400 mg/l at the effluent port of the screw press, will be reduced to 35 mg/l at the effluent port of the wastewater treatment plant.

Critical Method Parameters. The factors which tend to limit the applicability of the Black Clawson System include: necessity for a market for recovered materials; requirement for utilities (electrical power and process water); necessity for air pollution control; and necessity for wastewater treatment. Since the Black Clawson system is designed to produce a salable end product—paper fibers—the success of the method, at least in the minds of the officials and citizens who are responsible for financing waste disposal, will depend on profitable sale of recovered fibers. Black Clawson engineers have indicated that recovered fibers will be in demand at prices of at least $20 per ton; the fibers obtained in the Franklin plant have been sold to a nearby roofing manufacturer at about that price.

The provision of utilities in the form of electrical power and process water is a serious requirement for the operation of the system. Horsepower demand is high (about 2 hp/input ton capacity) for the Hydrapulper units and as yet no provisions have been made for recovery of heat energy/electricity from the burning wastes in the fluid-bed reactor. Water requirements are also an important consideration since process water is lost in the wet scrubber and no provisions have been made for recycling such water.

The control of air pollutants from the fluid-bed reactor is a serious factor. The manufacturer has claimed that the fluid-bed process ensures complete combustion of wastes with no generation of unburned hydrocarbons or carbon monoxide. Nitrous oxides are not generated at the temperatures to be maintained in the reactor. Black Clawson engineers further claim that sulfur oxides and particulate matter will be removed from the exhaust gas stream in the wet scrubber. Wet Venturi scrubbers have not performed well in the past in such applications on conventional incinerators. The visible water vapor plume from the plant may constitute an aesthetic drawback for plants in urban areas.

The necessity for extensive wastewater treatment is perhaps the most serious consideration in the applicability of the Black Clawson method. The effluent Biochemical Oxygen Demand (BOD) of 6400 mg/l is high in comparison to other wastewaters such as domestic wastewater. The provision of a treatment facility for this effluent represents a necessary and costly adjunct to the waste disposal system.

Method Economics. The capital costs of the demonstration plant at Franklin, Ohio have been estimated at approximately $1,750,000, or about $11,700 per ton of rated processing capacity. The manufacturer has estimated that the processing costs will amount to about $6.00 per ton of input solid waste on the basis of revenues as listed below:

Paper fiber	$750/day (30 tons/day)
Ferrous metals	$ 96/day (12 tons/day)
Glass	$144/day (12 tons/day)

Other Considerations. In the Franklin, Ohio pilot plant, during the initial period of operation, difficulties were experienced in securing markets for reclaimed materials. Additionally, frothing problems occurred in the adjoining wastewater treatment plant.

Summary on Black Clawson. The Black Clawson Hydrapulper/Fiberclaim System has received much favorable attention because of its ability to process mixed municipal refuse at a supposedly low disposal cost. However, some doubt exists as to the guaranteed sale of all recovered materials. Serious technological requirements may include the control of air and water pollutants and the provision of electrical power and process water. Aesthetic problems (odors and steam plumes) may also exist for this system.

6-5. SUMMARY—MATERIALS RECOVERY

In conclusion, with respect to the utilization of solid waste materials, recycling operations hold great promise for resource conservation and recovery. However, there exists a significant need for further research in unit processes for separation and a further refinement of processing operations such as composting, incineration, and pyrolysis. The greatest factors of uncertainty with respect to recycling operations in the United States today are the fluctuating markets for secondary materials. Secondary-material markets are affected by changes in the seasons, fluctuations in the weather, changes in public opinion, rate changes for interstate and intrastate transport, and developments in processing equipment. Because of this dependency of markets upon a number of variable factors, the entire field of materials recycling is in a virtual state of flux at the present time. It is impossible to predict what the future will hold. However, some generalizations could be made: 1) it is highly likely that freight rates and tax laws will be changed to favor increased use of recycled materials; 2) public opinion is likely to favor an increased use of recycled materials; and 3) dwindling supplies of virgin materials should increase the unit value of materials recovered from the solid waste stream. However, serious doubt exists as to the proper choice between recycling materials, on the one hand, and recovering their value in energy through combustion on the other. Recovery of energy from solid waste is discussed in Chapter 7.

REFERENCES

6-1. Abrahams, J. H., "Utilization of Packaging Wastes," *Proc.*, Reuse and Recycle of Wastes, University of Rhode Island, Westport, Connecticut, 1971.

6-2. Abrahams, J. H., "Utilization of Waste Container Glass," *Waste Age*, Vol. 1 (4), 1970.

6-3. Abert, J. G., *Resource Recovery: A New Field For Technology Application*, National Center for Resource Recovery, Inc., 1972.

6-4. Alexander, J. H., "Banning the Can Won't Clean Up the Mess," statement before a State of Michigan Special House Committee studying disposable beverage containers, American Can Company, 1971.

6-5. "Automobile Scrapping Process and Needs for Maryland," HEW, Govt. Printing Office, 1970.

6-6. Battelle Memorial Institute, *A Study to Identify Opportunities for Increased Solid Waste Utilization*, Vols. 1-9, Columbus, Ohio, 1972.

6-7. Bennett, K. W., "The Scrap Industry's New Posture," *Iron Age*, December 24, 1970.

6-8. Bergin, T. J., Furlong, D. A. and Riley, B. T., "A Progress Report on the CPU-400 Project," Brochure issued by Bureau of Solid Waste Management, USPHS, 1970.

6-9. Burgess, K. L., "Reuse of Plastic Milk Container Materials," *Proc.*, Conference on Design of Consumer Containers for Reuse or Disposal, Batelle Memorial Institute, Columbus, Ohio, 1971.

6-10. Carr, W., "Value Recovery from Wood Fiber Refuse," *Proceedings of the Second Mineral Waste Utilization Symposium*, March 18–19, 1970.

6-11. Cheney, R., "Design Trends in Glass Containers," *Proc.*, Conference on Design of Consumer Containers for Reuse or Disposal, Columbus, Ohio, 1971.

6-12. Drobny, N. L., Hull, H. E. and Testin, R. F., "Recovery and Utilization of Municipal Solid Waste," USPHS Publ. No. 1908, Govt. Printing Office, 1971.

6-13. Duszynski, E. J., "A Case for Milling Refuse," *Pollution Engineering*, May/June, 1971.

6-14. Duszynski, E. J., "Madison's Newspaper Salvage Project," *TAPPI*, Vol. 54 (3), March, 1971.

6-15. Emich, K. H., "Returnable Plastic Milk Bottles," *Proc.*, Conference on Design of Consumer Containers for Reuse or Disposal, Columbus, Ohio, 1971.

6-16. *Engineering Services for Urban Forest Products Facility*, The Rust Engineering Company, Birmingham, Alabama, July, 1971.

6-17. Fong, G. P., Mack, A. C. and MacDonald, J. N., "Role of Plastics in Solid Waste—A Status Review," *Proc.*, New Directions in Solid Waste Processing, Framingham, Massachusetts, 1970.

6-18. Goetz, J. W., "The Utilization of Waste Paper," *Proc.*, New Directions in Solid Waste Processing, Framingham, Massachusetts, 1970.

6-19. Henn, J. J. and Peters, F. A., "Cost Evaluation of a Metal and Mineral Recovery Process for Treating Municipal Incinerator Residues," Dept. of the Interior, Information Circular 8533, 1971.

6-20. Hulbert, S. F., Fain, C. C., Cooper, M. M., Ballenger, D. T. and Jennings, C. W., "Improving Package Disposability," paper presented at the First National Conference on Packaging Wastes, San Francisco, 1969.

6-21. Hulbert, S. F., Fain, C. C. and Eitel, M. J., "Design and Evaluation of a Water Disposable Glass Packaging Container, Reports 1–6," Clemson University, Clemson, South Carolina, 1969–1972.

6-22. Malisch, W. R., Day, D. E. and Wixon, B. G., "Use of Salvaged Waste Glass in Bituminous Paving," paper presented at the Centennial Symposium, Technology for the Future to Control Industrial and Urban Wastes, University of Missouri—Rolla, 1971.

6-23. Mallan, G. M., "Preliminary Economic Analysis of the GR&D Pyrolysis Process for Municipal Solid Wastes," Garrett Research and Development Company, Inc., January, 1971.

6-24. Mallan, G. M. and Finney, C. S., "New Techniques in the Pyrolysis of Solid Wastes," paper presented at the 73rd National Meeting of the AIChE, Minneapolis, Minnesota, Aug. 27–30, 1972.

6-25. Midwest Research Institute, "The Role of Packaging in Solid Waste Management 1966 to 1976," USPHS Publ. No. 1855, 1969.

6-26. Midwest Research Institute, "The Role of Nonpackaging Paper in Solid Waste Management 1966 to 1976," USPHS Publ. No. 2040, 1971.

6-27. Ostrowski, E. J., *Recycling of Tin Free Steel Cans, Tin Cans and Scrap from Municipal Incinerator Residue*, National Steel Corp., May 26, 1971.

6-28. Palumbo, F. J., Stanczyk, M. H. and Sullivan, P. M., *Electronic Color Sorting of Glass From Urban Wastes*, Bureau of Mines Solid Waste Research Program, Technical Progress Report-45, October, 1971.

6-29. Park, W. R., Franklin, W. E. and Bendersky, D., "Economics of Resource Recovery from Mixed Municipal Wastes," presented at the 72nd National Meeting American Institute of Chemical Engineers, St. Louis, Missouri, May 24, 1972.

6-30. "Pyrolysis of Refuse Gains Ground," *Environmental Science and Technology*, April, 1971.

6-31. Story, W. S., "Ferrous Scrap Recycling and Steel Technology," *Proc.*, Conference on Design of Consumer Containers for Reuse or Disposal, Columbus, Ohio, 1971.

6-32. Sullivan, P. M. and Stanczyk, M. H., *Economics of Recycling Metals and Minerals from Urban Refuse*, Bureau of Mines Solid Waste Research Program, Technical Progress Report-33, April, 1971.

6-33. Testin, R. F., "Recycling of Used Aluminum Products," *Proc.*, Reuse and Recycle of Wastes, University of Rhode Island, Westport, Connecticut, 1971.

6-34. Testin, R. F., Bourcier, G. F. and Dale, K. H., "Recovery and Utilization of Aluminum from Solid Wastes," Reynolds Metals Company, Richmond, Virginia, 1971.

6-35. Varner, H. D., *A Systems Approach to Solid Waste Management: The Delaware Reclamation Plant, Proc.*, Mid-Atlantic States Section, Semiannual Technical Conference on "Solid Waste and Air Pollution," October 29, 1971.

6-36. Wilson, D. G., ed., *The Treatment and Management of Urban Solid Waste*, Westport, Connecticut, 1972.

6-37. Zinn, R. E., LaMantia, C. R. and Niessen, W. R., "Total Incinerator," 1970 National Incinerator Conference, The American Society of Mechanical Engineers, 1970.

7

Energy recovery

Two of the many challenges that must be faced by the United States during this decade, solid waste disposal and energy conservation, are not usually considered to be related. However, a great deal of research has recently been focused on energy recovery from solid waste.

The fact that an energy crisis has developed to an alarmingly real extent in this country is evident. The total energy demand in the United States has risen from 45 quadrillion BTU in 1960 to over 60 quadrillion BTU in 1970 and is expected to reach 90 quadrillion BTU by 1980. This increasing energy demand is expected to top the 200 quadrillion BTU mark by the end of this century. It is obvious that new sources of energy must be sought to curtail the relentless exhaustion of the world's available fuel.

One potential source of energy that has been neglected in this country is that which can be produced during the disposal of solid waste. Since typical solid waste consists primarily of paper, it possesses a relatively high BTU value. Several different solid waste processing systems have been developed during the last several years which produce energy in the form of electricity, steam, gas, or oil. These various processes, some of which are proprietary in nature, are referred to as energy recovery systems and include: the use of refuse as a supplementary fuel, waste heat recovery, gas turbine generation of electricity, pyrolysis, and methane production by anaerobic fermentation. This chapter analyzes each of these processes in depth and assesses their potential for energy recovery in the United States.

7-1. REFUSE AS A SUPPLEMENTARY FUEL

Most attempts to recover energy from municipal solid waste have focused upon the generation of steam. European incinerators equipped with waste heat boilers have utilized refuse as a source of fuel for many years. In most instances these facilities are quite sophisticated with all or part of the steam production used to generate electricity.

In the United States the use of solid waste as fuel is more of a rarity. The only notable incineration facilities in this country equipped to generate steam are the Chicago Southwest Incinerator, the Chicago Northwest Incinerator, and the Norfolk Naval Base Incinerator. In most instances, these installations utilize mixed municipal refuse as their primary fuel with no attempt to control refuse particle size or composition. Auxiliary fuel, such as oil or gas, is used in some units to assist in controlling steam pressure and temperature. Although existing facilities can produce steam for in-plant use or sale, it should be remembered that these plants are characterized by high capital, operational, and maintenance costs, low reliability, and markets for steam that are not readily available. As a matter of interest, it should be noted that the establishment of a continuing market for steam has been the major problem associated with the development of waste heat recovery incinerators in the United States.

The only established markets for power in this country are those associated with existing utility power plants. These large, efficient power-producing facilities are already integrated into a system for producing and distributing all the electricity possible to a rapidly expanding market. It would, therefore, seem apparent that the combustion of prepared refuse in existing power plant boilers presents the following advantages to both the community and utility in question:

1. Achieve maximum heat recovery from refuse.
2. Reduce air pollution emissions associated with solid waste disposal where the alternate means of disposal is incineration.
3. Conserve nonrenewable natural resources.
4. Reduce power production costs.

5. Possibly retain existing power plant boilers for longer life as base-loaded units.
6. Develop an environmentally acceptable method of solid waste disposal.

An important point to be noted regarding the use of refuse as a supplementary power fuel is that the process as presently conceived utilizes proven technology and commercially available processing equipment.

7-1.1 The St. Louis Project

The potential value of municipal solid waste as a fuel has been recognized for many years by cities, utilities, and their consulting engineers. Energy recovery in the form of steam from waste heat boilers had been proposed numerous times for the St. Louis, Missouri area, and consideration was given to the installation of such a system in the 1940s and then again around 1965. However, in both instances the idea was completely abandoned as not being practical because of the technical operating problems then associated with this type of system.

The basic concept of the existing St. Louis system was developed in late 1967 during an informal luncheon involving personnel of the Union Electric Company and Horner and Shifrin, Inc., Consulting Engineers. This luncheon centered around a discussion of the quantity and characteristics of the fuel fired to the utility's facilities, and the possibility of utilizing properly prepared solid waste as a supplementary fuel. It was agreed that the idea was worthy of further investigation. Consequently, the city of St. Louis (the only local agency which maintained control of the collection and disposal of large quantities of solid waste in the metropolitan area) was asked to express its interest with regard to investigating the potential of utilizing its solid waste as supplementary fuel in the Union Electric Company's facilities. The city indicated interest in the proposed study and applied for, and received, a partial grant-in-aid from the Environmental Protection Agency to study the process. Horner and Shifrin, Inc., was retained by the city to conduct this investigation in close cooperation with the Union Electric Company. The resulting comprehensive study concluded that although coal-fired boilers are not without operating problems of their own, these problems would not be significantly increased, if increased at all, by burning prepared refuse as a supplementary fuel.

At the completion of this study, the Union Electric Company displayed its interest in the project by offering the use of one, and eventually two, of its major boiler units for full-scale testing of the process, as well as the capital expenditures required to cover the cost of its portion of the prototype facilities required on its property. The city of St. Louis then submitted another application to the Environmental Protection Agency for a grant-in-aid to develop a prototype installation. This grant was approved on July 1, 1970, and operation of the study facility was initiated during April, 1972. Thus, for the first time in the United States, an investor-owned utility began to burn municipal solid waste as supplementary fuel for the direct production of electric power.

7-1.2 Methodology

Solid wastes currently being utilized as supplementary fuel in St. Louis consist of mixed municipal refuse generated in residential areas. No industrial, commercial, or bulky waste materials are being processed; however, certain selected industrial and commercial wastes may be processed at a later date.

The initial refuse preparation occurs at city-operated facilities (see Fig. 7-1). Refuse to be processed is discharged from packer trucks to the floor of the raw refuse receiving building. This refuse is then pushed to a receiving conveyor by front-end loaders. From the receiving conveyor, the refuse is transferred to an inclined belt conveyor, and then to a vibrating conveyor which feeds a hammer-mill. This hammermill is a conventional mill with a horizontal shaft, and possesses a hammer circle of approximately 60 in. and an interior rotor length of about 80 in. Power input to the hammermill is supplied by a direct-connected, 1250 hp, 900 rpm motor. All refuse entering the hammermill is reduced to a particle size of less than $1\frac{1}{2}$ in.

The shredded solid waste is then transferred by means of a vibrating conveyor which feeds an inclined belt conveyor leading to the prepared refuse storage bin. Magnetic metals are recovered from the solid waste stream at the head pulley of the inclined belt conveyor and discharged to trucks for shipment to ferrous metal buyers. The processed refuse is then conveyed from the storage bin to a stationary compactor for loading into self-unloading trucks for transport to the power plant facilities.

This initial preparation process was designed to be as simple as possible to minimize operational problems associated with refuse processing. Most of the operating problems that have occurred during the first several years of refuse preparation system operation have been of a mechanical nature, and are mainly due to the peculiarities of milled municipal refuse. The storage and handling of milled solid waste has required special consideration since this material has a tendency to compact under its own weight into a laminar, springy mass. Consequently, power requirements for equipment handling large volumes of shredded refuse have been constantly underestimated. However, operation of this process-ing facility has been quite simple and encouraging to date.

In the St. Louis prototype system, the processed solid waste is transported eighteen miles to the Union Electric Company's power plant. However, if the refuse processing facility was located near the power plant, it would be feasible to replace truck transport with pneumatic conveyance of the supplementary fuel directly to the boilers from the storage bin.

Figure 7-5 outlines the receiving and firing facilities at the Union Electric

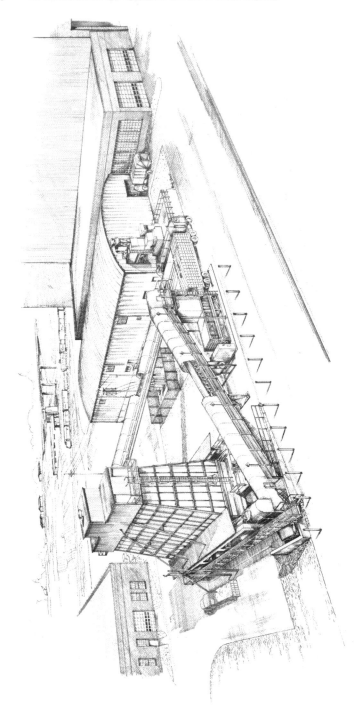

Fig. 7-1 Artist's rendering of the supplementary fuel processing plant in St. Louis, Missouri. *Source:* Horner & Shifrin Consulting Engineers, "Appraisal of Use of Solid Waste as Supplementary Fuel in Power Plant Boilers," St. Louis, Missouri, February, 1973.

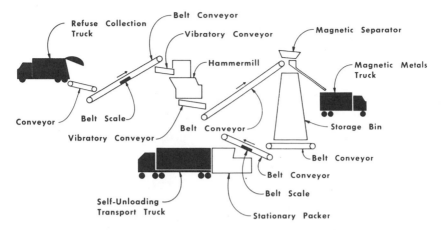

Fig. 7-2   Refuse processing facilities—St. Louis, Missouri. *Source:* Wisely, F. E., Klumb, D. L. and Sutterfield, G. W., "From Solid Waste to Energy," presented at the Conservation Education Association Conference, Murray State University, Murray, Kentucky, August 15, 1973.

Fig. 7-3   Shredded solid waste is transferred to the prepared refuse storage bin on an inclined belt conveyor at the refuse processing facility in St. Louis, Missouri (*courtesy Courier-Journal* and *Times*).

Fig. 7-4 Prepared refuse is transported to the Union Electric Company's power plant facilities in self-unloading trucks (*courtesy* Combustion Engineering, Inc.).

Company's power plant. The transported prepared refuse is discharged from self-unloading vehicles to a receiving bin which supplies a belt conveyor. The refuse is then discharged to a 12 in. diameter pneumatic feeder pipe and transported to a 8600 cu ft surge bin. A cyclone dust collector separates the refuse from the transport air. Embedded in the 40 ft diameter concrete bin floor slab are four drag chain conveyors each of which feeds refuse to its own pneumatic feed system for transport to one of the four refuse burners. Crude metering of the refuse feed to the boiler is accomplished by varying the speeds of the drive ring and conveyor drag chains. The supplemental fuel is fired independently of the normal boiler fuel.

The prepared refuse is burned in two 20 year-old tangentially-fired Combustion Engineering boilers located 700 ft from the storage bin. These boilers have a nominal rating of 125 megawatts each with a maximum gross output of approximately 142 megawatts. Although the steam systems of these boilers are not

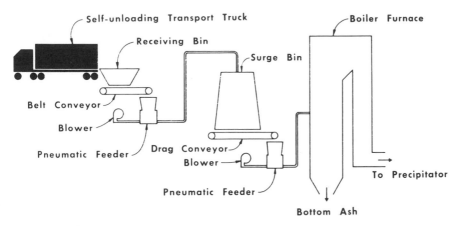

Fig. 7-5 Supplementary fuel receiving and firing facilities—St. Louis, Missouri. *Source:* Wisely, F. E., Klumb, D. L. and Sutterfield, G. W., "From Solid Waste to Energy," presented at the Conservation Education Association Conference, Murray State University, Murray, Kentucky, August 15, 1973.

Fig. 7-6 Prepared refuse receiving bin at the Union Electric Company's power plant in St. Louis, Missouri (*courtesy Courier-Journal* and *Times*).

"modern," the furnace and burner design is basically the same as newer units now being put into service. As shown in the boiler cross section (Fig. 7-7), four pulverizers each feed a level of four burners, one in each corner. Between each level of coal burners is located a pair of gas burners, one over the other. All burners are tiltable. The furnaces are about 28 ft by 38 ft in cross section, with a total inside height of 100 ft.

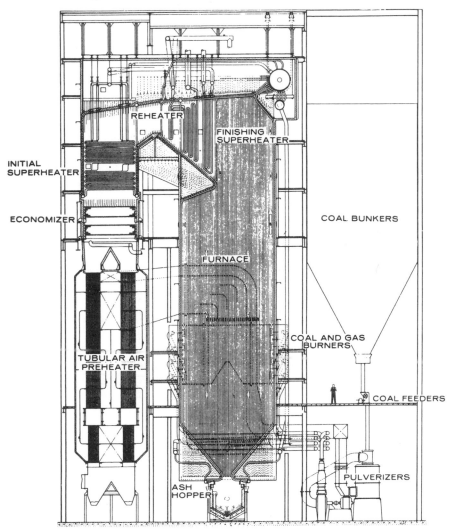

Fig. 7-7   Meramec Unit No. 1, Union Electric Company. *Source:* Wisely, F. E., Klumb, D. L. and Sutterfield, G. W., "From Solid Waste to Energy," presented at the Conservation Education Association Conference, Murray State University, Murray, Kentucky, August 15, 1973.

The Union Electric Company did not wish to mix the prepared refuse with coal in the feed system because of the possibility of spontaneous combustion in the pulverizers and the potential for disrupting the proper operation of the combustion control system. Therefore, space was provided for the installation of refuse burners by removing one of the gas nozzles in each corner of the furnace. The gas flow required to carry a full load of gas with four of the twenty-four nozzles removed was maintained by increasing the load gas header pressure to about 2 psig. The four resulting refuse burners are similar to 10 in. Combustion Engineering pulverized coal burners except that the refuse burner contains no grid work at the burner mouth. Boiler renovation costs for this project amounted to approximately $200,000.

Prepared refuse is fired through these burners at a rate equal to 10 to 15 percent, by heat value, of the full load fuel requirement of the unit. This 10 percent firing amounts to about 12.5 tons of refuse per hour or 300 tons per day. Since the city collects, processes, and delivers refuse to the power plant 5 days per week, no provisions have been made to supply refuse fuel to the power plant on weekends.

Since the prepared refuse is fired to the boilers at a constant rate, the rate of coal or gas firing may be varied to accommodate boiler heat requirement variations. If, for any reason, the boiler is suddenly taken out of service, an electrical interlock immediately stops refuse feeding. However, the pneumatic blowers are allowed to continue to function in order to clear the pipelines of any refuse remaining in them.

Since refuse firing was initiated in April, 1972, no discernible effects on the boiler furnace or convection passes have been noticed during operation by Union Electric Company personnel. In addition, no significant effect on the stack discharge has been observed. There has been no evidence of carry-over of unburned particles into the back passes of the boiler, nor any evidence of slagging due to the firing of refuse. It is important to note that the performance of the boilers when firing refuse and coal has been identical to the performance of the unit when fired only with coal. Data are not currently available regarding long-term corrosion effects of the refuse on boiler pressure parts. The low sulfur content of domestic refuse may tend to decrease corrosion potential, whereas the higher chlorine content may have an opposite effect.

All operational problems experienced to date are associated with the "heavy" fraction of the prepared refuse stream, i.e., nonmetallic metal, glass, wood, etc. These larger and denser particles have resulted in a milled refuse possessing highly abrasive qualities. This abrasiveness has produced excessive transport pipe wear and feed stoppages. Pipe wear has been concentrated at the pipe bends and elbows and appears to be a result of abrasion rather than impact. No significant straight pipe wear has been observed.

Another operational problem associated with the refuse "heavy" fraction has to do with bottom ash handling capability. Since the firing of refuse was initiated, the quantity of bottom ash has approximately doubled; however, no problems have been experienced with regard to handling this increased volume of bottom ash.

7-1.3   Critical and Limiting Process Parameters

The basic critical and limiting process parameters associated with the combustion of prepared refuse as a supplementary power fuel are essentially the same as those involved in the incineration process except for the physical and chemical refuse characteristics and the refuse "heavy" fraction problems.

Since refuse characteristics are of paramount importance to process performance, it is appropriate to briefly compare some of the more important characteristics of coal and refuse. Refuse characterization for the St. Louis project was performed by Research 900 laboratories of the Ralston-Purina Company. Since the chemical analysis methodology for refuse is currently in an elementary state, it is of interest to describe the techniques utilized by Research 900 laboratories for refuse sampling and preparation for chemical analysis.

The methodology utilized to develop a shredded, dry, uniformed texture sample required extensive method refinement. Initially, large-scale refuse samples were delivered by the St. Louis Refuse Commission to the Research 900 laboratories in 50–100 lb lots in plastic bags. The bags were then sprayed with a disinfectant at the receiving dock to effect disease control and transported to a weighing area where they were weighed and enclosed in an additional plastic bag. The sealed samples were then placed in an industrial batch oven and pasteurized. The contents of the samples were then emptied onto a dropcloth on the floor for subsampling. A classical quartering technique was utilized to accomplish the subsampling. Each quarter was then weighed, dried, and reweighed to determine its moisture content. The resulting samples were milled through a 4 mm diameter screen in a Fritz Mill followed by a 2 mm diameter screen in a Wiley Mill and then bottled for laboratory analysis.

Refuse samples prepared as above and samples of Union Electric Company's coal were then analyzed for various compositional properties using the American Society for Testing and Materials (ASTM) testing procedures. Table 7-1 illustrates the ranges of proximate and ultimate analyses of the two fuels. Major differences existed in moisture and carbon contents. The average refuse moisture content was about three times that of coal, whereas the average heating value of refuse (5000 BTU/lb) was somewhat less than half the BTU value of Illinois bituminous coal. The refuse sulfur content was found to be low, as expected; however, the chlorine content of refuse was found to be considerably higher than in washed coals. It should be noted, however, that the refuse chlorine level was of the same order of magnitude as that in run-of-mine coals.

A comparison of ash analyses for refuse and coal was also performed according

TABLE 7-1   Refuse and Coal Analyses Ranges of Composition (as received)

	Refuse[a] %	Coal[b] %
Proximate Analyses		
Moisture	19.69–31.33	6.20–10.23
Ash	9.43–26.83	9.73–10.83
Volatile	36.76–56.24	34.03–40.03
Fixed carbon	0.61–14.64	42.03–45.14
BTU per pound	4,171–5,501	11,258–11,931
Ultimate Analyses		
Moisture	19.69–31.33	6.20–10.23
Carbon	23.45–33.47	61.29–66.18
Hydrogen	3.38–4.72	4.49–5.58
Nitrogen	0.19–0.37	0.83–1.31
Chlorine	0.13–0.32	0.03–0.05
Sulfur	0.19–0.33	3.06–3.93
Ash	9.43–26.83	9.73–10.83
Oxygen	15.37–31.90	9.28–16.10

[a] From three samples of St. Louis refuse, with magnetic metals removed.
[b] From three samples of Union Electric Company coals.
Source: Wisely, F. E. and Hinchman, H. B., "Refuse as a Supplementary Fuel," *Proceedings of the Third Annual Environmental Engineering and Science Conference*, Louisville, Kentucky, March 5–6, 1973.

to ASTM test procedures and is shown in Table 7-2. As can be noted, both coal and refuse ash exhibited high silica levels; however, significant compositional differences were observed with regard to ferrous oxide, alumina, lime, and sodium oxide levels.

The ash fusion temperatures of refuse and coal were also measured and the results are presented in Table 7-3. As these data illustrate, the ash fusion temperatures of the two fuels are remarkably similar.

As previously mentioned, operation of the prototype facilities demonstrated that the following operational problems developed when only magnetic metal was removed from the solid waste stream prior to firing:

1. Excessive transport pipe abrasion.
2. Stoppage of pneumatic feeders due to the lodging of nonferrous metal particles between rotors and housings.
3. Significant increase in the quantity of bottom ash which included particles that possessed sufficient mass to preclude their combustion in suspension such as wood, dense plastics, rubber, and leather.

These heavy fraction operational problems resulted in plans to incorporate an air-density separator at the refuse processing facility. The installation of this air-

TABLE 7-2   Refuse and Coal Ash Analyses Ranges of Composition (as received)

Mineral Analyses (Ignition Basis)	Refuse[a] %	Coal[b] %
Phosphorus pentoxide	1.02–4.69	0.08–0.20
Silica	48.93–60.07	45.52–46.93
Ferric oxide	3.50–5.92	15.51–25.29
Alumina	5.02–13.72	16.54–18.53
Titania	0.74–1.60	0.81–1.01
Lime	7.54–18.19	2.13–6.31
Magnesia	1.14–1.91	0.80–0.92
Sulfur trioxide	1.84–12.54	1.41–6.28
Potassium oxide	1.57–2.70	1.70–1.78
Sodium oxide	3.62–5.95	0.30–0.62
Undetermined	0.08–0.69	0.39–5.25

[a]From three samples of St. Louis refuse, with magnetic metals removed.
[b]From three samples of Union Electric Company coals.

Source: Wisely, F. E., and Hinchman, H. B., "Refuse as a Supplementary Fuel," *Proceedings of the Third Annual Environmental Engineering and Science Conference*, Louisville, Kentucky, March 5–6, 1973.

classifier was completed in late 1973 at which time the schematic of the refuse processing facility was changed as shown in Fig. 7-8. Following installation of the density separator, it is anticipated that the operational problems experienced with unclassified waste will be either minimized or eliminated. Any pipe elbow

TABLE 7-3   Ash Fusion Temperature Ranges of Refuse and Coal

	Reducing °F	Oxidizing °F
Refuse[a]		
Initial deformation	1890–2070	2030–2100
Softening (H = W)	2190–2360	2260–2420
Softening (H = ¹/₂ W)	2210–2390	2290–2450
Fluid	2400–2560	2480–2700
Coal[b]		
Initial deformation	1940–2010	2020–2275
Softening (H = W)	1980–2200	2120–2455
Softening (H = ½ W)	2180–2220	2260–2470
Fluid	2250–2600	2390–2610

[a]From three samples of St. Louis refuse, with magnetic metals removed.
[b]From three samples of Union Electric Company coals.

Source: Wisely, F. E., and Hinchman, H. B., "Refuse as a Supplementary Fuel," *Proceedings of the Third Annual Environmental Engineering and Science Conference*, Louisville, Kentucky, March 5–6, 1973.

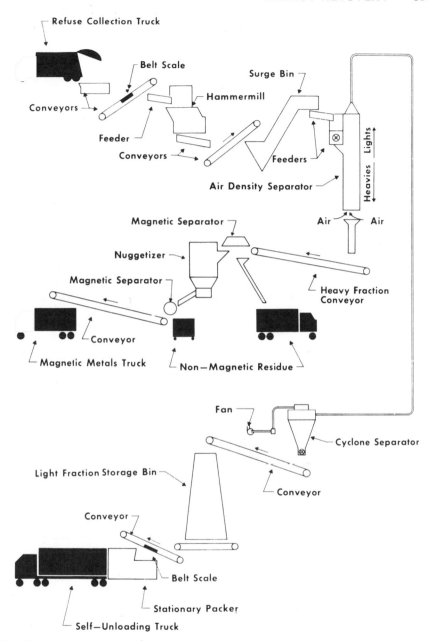

Fig. 7-8 Solid waste processing facilities with air density separation. *Source:* Wisely, F. E., Klumb, D. L. and Sutterfield, G. W., "From Solid Waste to Energy," presented at the Conservation Education Association Conference, Murray State University, Murray, Kentucky, August 15, 1973.

abrasion difficulties existing after air-classification is initiated will be addressed by special design of pneumatic piping.

7-1.4   Operating Characteristics

To date municipal refuse has been used as a supplementary fuel for power production only in the prototype project in St. Louis. Therefore, the premises presented in this brief summary regarding process operating characteristics are based solely on data developed during that project. The operational characteristics to be discussed include: burning rates, particle size, nonburnable and heavy materials, refuse heating value, refuse ash and moisture content, slagging, carryover, corrosion potential, refuse firing with oil, odors, air pollution effects, boiler efficiency, and cold weather operation.

Initially, it was proposed to operate the Union Electric Company's boiler at a nominal burning rate of prepared refuse of about 10 percent of the boiler heat requirement at full load. However, the prototype installation has operated consistently at a firing rate of approximately 15 percent of the boiler full-load heat requirement, and it appears that this rate could be increased to 20 percent without adverse effects.

The St. Louis installation has utilized milled refuse with a particle diameter of essentially 1½ in. A few refuse particles possessing a diameter of 4 to 5 in. have burned equally well in suspension; it is assumed that larger particles in this diameter range would not adversely affect boiler operation. However, the basic specifications for this type of process should limit the refuse particle diameter to 1½ in. or less to alleviate the potential for operational problems.

Most of the operational problems experienced at the prototype installation center around the presence of a "heavy" fraction in the prepared refuse which is fed to the boiler. Consequently, most of the noncombustible particles, and some of the burnable particles of sufficient mass to prevent their combustion in suspension (e.g., corn cobs) fall to the bottom ash hopper, and must be transported with the bottom ash to a disposal pond. The removal of this "heavy" refuse fraction is, therefore, warranted and should be accomplished by incorporating an air-density separator in the refuse processing operation. Air-classification will provide for the following operational improvements:

1. Decrease the boiler ash content.
2. Increase the prepared refuse fuel value.
3. Reduce the abrasive characteristics of the prepared refuse stream.
4. Reduce or eliminate mechanical problems associated with equipment jamming caused by heavy refuse particles.

The heating value of the prepared refuse utilized in the St. Louis project was somewhat variable depending upon its moisture content. Since this prepared refuse has a generally accepted heating value of 5000 BTU/lb as fired, it is logical to assume that this value will be a lower limit when considering classified refuse as fuel.

Raw refuse can be expected to yield an ash content of 25 to 30 percent. This percentage drops to 20 to 25 percent with magnetic metal removal, and to 10 to 15 percent with air-density separation. Although prepared refuse is fluffy in nature, a substantial portion of the refuse ash falls to the bottom ash hopper. Experience in the St. Louis prototype project indicates that the bottom refuse ash will contain small inert particles including glass but virtually no putrescible material. The physical quality of the flyash produced in this process is essentially the same as pulverized coal ash.

The moisture content of the fired prepared refuse may be expected to vary from 15 percent to about 35 percent, with an average in the 25 to 30 percent range.

There has been no evidence of slagging or carry-over problems in the St. Louis prototype operation. Evidently, refuse does not have a tendency to form more slag than Illinois bituminous coal. Likewise, operation has not indicated any carry-over of unburned materials into the back passes of the boiler by the gas stream.

Results of the corrosion potential testing program initiated in conjunction with the St. Louis project are not currently available (1974). The results of this testing program will indicate whether corrosion potential is greater when refuse is fired with coal than when coal is used alone. The removal of magnetic metals from the solid waste stream together with those elements bound to the metal should have a tendency to reduce the corrosion potential of raw refuse.

Currently, there is no reason why refuse cannot be fired in combination with oil instead of coal or gas. At an industrial installation at Rochester, New York, in-plant solid wastes are being fired in combination with oil and sewage sludge (Ref. 7-14). In solid waste–oil systems, air-classification becomes very important in order to minimize grate burning which may occur as a result of large particles not fully burned in suspension.

Odor problems associated with the process can be essentially eliminated if the solid waste delivered to the power plant is milled to a small enough particle size to allow dispersion of odor-producing materials. Additionally, closed pneumatic systems should be utilized for transporting and firing the prepared refuse. Although the St. Louis project does not include a completely enclosed materials handling system, there have been no significant odor problems.

Since the use of solid waste as a supplementary fuel in existing power plant boilers is only intended to replace a relatively small portion of the original fuel, any adverse air quality effects should be minimal. Since the average sulfur content of refuse is less than one-half that of No. 6 low-sulfur fuel oil, indications are that sulfur emissions should not be increased when firing refuse with fuel oil rather than oil fired alone. Actually, a reduction in sulfur emissions may result in the refuse-fuel oil system.

Although there is an absence of specific test data, it appears fairly certain that particulate emissions from a refuse-fuel oil system would be greater than those emanating from a fuel oil system, but less than those from a pulverized coal system (Ref. 7-35 and 7-14). It is estimated that the quantity of flyash carried to the precipitators in a system in which refuse replaced 15 percent of the fuel oil input (by heat value) would be only 20 to 25 percent of the flyash produced if pulverized coal was used alone (Ref. 7-14).

In refuse-fuel systems utilizing air-classification techniques for refuse preparation, it appears that both nitrogen oxide and chlorine emission levels would not be significantly higher than when power plants are fired with fuel only (Ref. 7-35).

Therefore, data presently available indicate that neither gaseous nor particulate emissions resulting from solid waste–fuel combination systems would differ significantly from emissions now being discharged from conventionally-fired power plants. It is also of interest to note that refuse-fuel system emissions would most likely be far less than those accruing from conventional incineration.

Throughout the St. Louis prototype project there has been no discernible effect upon boiler efficiency or power-producing capability when refuse–coal firing was practiced. Additionally, no adverse effects upon boiler operation were noted when prepared refuse was fired intermittently. Operating experience associated with the St. Louis project indicates that the handling and pneumatic firing of the prepared refuse was not adversely affected during cold weather conditions when ambient temperatures approached $0°F$.

It should be pointed out here that alternate means of refuse disposal must be maintained even after the installation of refuse-fuel systems since a substantial portion of the solid waste stream may not be recoverable or burnable. Likewise, the disposal of ash resulting from the use of refuse as a supplementary fuel must be provided. Therefore, it is mandatory that landfills or other economical means of solid waste disposal be maintained during refuse-fuel system operation. Likewise, this refuse disposal operation, as well as the refuse transportation system, must be flexible enough to prevent large accumulations of raw refuse during boiler outages or processing equipment breakdowns.

7-1.5 Economic Analysis

An economic assessment of the refuse-fuel system must include a consideration of the costs absorbed by both the public agencies and utilities involved. The major economic elements to be considered include:

1. Capital and operating costs associated with processing raw solid waste and delivering it to the utility in question. It is assumed that these costs would be absorbed by the public agency.
2. Costs required to install burning ports, pneumatic pipe lines, pneumatic

feeders, and possibly fuel storage facilities. These costs would normally be absorbed by the utility.

3. The value of materials recovered in the refuse processing operation. The most valuable recoverable materials would include ferrous and nonferrous metals. This income would be absorbed by the operator of the refuse processing operation.
4. Power plant operating cost allowance due to increased ash handling and other differential costs.
5. The equivalent fuel costs of the prepared refuse. Based on a heating value of 5000 BTU/lb, the prepared refuse would contain approximately 10 million BTU/ton.
6. Solid waste disposal costs by alternate means, i.e., sanitary landfill, incineration, etc.
7. The cost of landfilling materials which cannot be reclaimed or used as supplementary fuel.

From this breakdown it is apparent that costs and benefits may be absorbed by both the public agency and utility. Obviously, these economic considerations will vary from community to community, and therefore it is difficult to generalize on the process costs. However, wherever disposal and fuel costs are high, this process is almost certain to be economically attractive. In any situation, the financial arrangements must be negotiated between the public agency and the utility. Such negotiations would require an objective study of the situation which takes into account all of the economic elements on a mutually agreeable basis.

Horner-Shifrin, Inc., estimates that a processing facility (depicted in Fig. 7-9) located on the eastern seaboard processing 125 tons per hour of raw refuse, using two-stage milling with air-classification would cost about $5,300,000 to build (Ref. 7-39). This facility could be utilized to process approximately 1750 tons of refuse per day, 300 days per year, at a processing cost of $5 per ton of refuse, including amortization of the capital investment. This cost figure does not include savings to be realized from the sale of recoverable materials at the processing facility, costs associated with delivering the fuel to the utility, or the value of fuel replaced by the prepared refuse. These costs are comparable to those of a sanitary landfill when the landfill is in a location remote from the point of refuse generation (Ref. 7-39).

The estimated 1973 capital, operational, and maintenance costs for processing, transporting, and firing facilities involved in the St. Louis project are summarized in Table 7-4. These estimates are based upon the assumptions that the processing plant would include two-stage milling, that it would be less than twenty-five miles from the power plant, and that the supplementary fuel would be transported by truck-trailers.

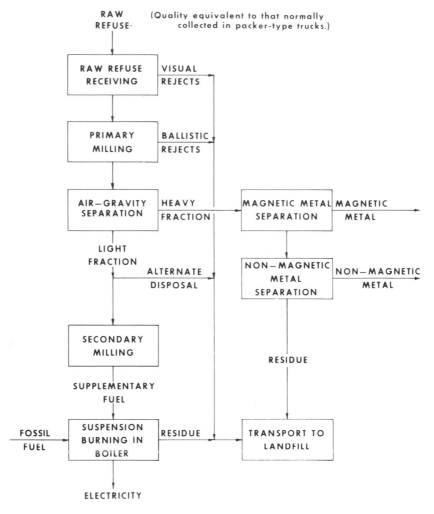

Fig. 7-9 Supplementary fuel production utilizing two stage milling and air classification. *Source:* Wisely, F. E., Klumb, D. L. and Sutterfield, G. W., "From Solid Waste to Energy," presented at the Conservation Education Association Conference, Murray State University, Murray, Kentucky, August 15, 1973.

7-1.6 Summary—Refuse as a Supplementary Fuel

From experience gained in the St. Louis prototype project, it appears that almost all fossil fuel-fired boilers possessing ash handling capability could be adapted for burning prepared refuse. Obviously, some boilers could be more easily adapted than others; however, Horner and Shifrin, Inc., estimates an

TABLE 7-4   Estimated 1973 Capital and Operating Cost Summary for St. Louis Project

	One Processing Unit[a]	Two Processing Units[a]
Raw Refuse Processed		
Tons/day	980	1,960
Tons/year, 5-day operation	254,800	509,600
Tons/year, 6-day operation	305,760	611,520
Supplementary Fuel Produced		
Tons/day	906	1,812
Tons/year, 5-day operation	235,560	471,120
Tons/year, 6-day operation	282,672	565,344
Estimated Capital Costs (1973)		
Processing facilities	$3,427,000	$6,432,000
Transporting facilities	579,000	832,000
Receiving facilities	232,000	456,000
Firing facilities	973,000	1,060,000
Total estimated capital cost	$5,211,000	$8,780,000
Estimated Annual Operation and Maintenance Costs (1973)		
Processing plant	$ 558,000	$ 975,000
Transport facilities	283,200	546,400
Receiving and firing facilties	55,700	75,000
Total estimated operation and maintenance costs		
−5 days/week operation	896,900	1,596,400
5 days/week operation	1,075,000	1,920,000
Amortization Costs (Annual)	$ 418,000	$ 704,000
Equivalent Unit Cost—5 Days/Week Operation		
Total unit cost—per ton of raw refuse	$5.16	$4.52
Total unit cost—per ton of supplementary fuel	5.59	4.89
Equivalent Unit Costs—6 Days/Week Operation		
Total unit cost—per ton of raw refuse	$4.89	$4.29
Total unit cost—per ton of supplementary fuel	5.29	4.64

[a]Two-shift operation.

Source: Wisely, F. E., Sutterfield, G. W. and Klumb, D. L., "St. Louis Power Plant to Burn City Refuse," *Civil Engineering*, Vol. 41, No. 1, January 1971, pp. 56-9. Reprinted with permission from the January, 1971 issue of *Civil Engineering* (ASCE official monthly publication of the American Society of Civil Engineers).

average boiler modification cost of about \$200,000 to \$400,000 per boiler. A typical boiler firing system could be constructed for approximately \$1 million (Ref. 7-14). At this point, it seems reasonable to assume that prepared refuse could be fired with gas, oil, or coal.

Even where process design stipulates a supplementary fuel firing rate of only 10 percent of the total fuel requirement, the capacity of existing suspension-fired boilers to consume prepared refuse is large enough to allow the process to become a principal means of solid waste disposal in major metropolitan areas. Table 7-5 lists existing power plants which have the potential capability of burning prepared refuse as a supplementary fuel. A 600 megawatt unit similar to the McDonough plant in Atlanta, Georgia could at full load consume approximately 1400 tons of prepared refuse per day at the 10 percent firing rate (Ref. 7-37).

Providing that negotiations between the utility and the solid waste supplier would result in a satisfactory financial arrangement, the process possesses the potential for extensive application. The use of prepared refuse as a supplementary fuel could provide an economic means of disposing of large quantities of solid waste in a manner which is environmentally acceptable while recovering energy by the direct production of electrical power. Concurrently, the implementation of this process would result in the conservation of irreplaceable natural resources and possibly help control air pollution levels associated with conventional power plants. These economic and environmental benefits could

TABLE 7-5   Potential Capability of Burning Prepared Refuse in Typical Steam-Electric Plants

Plant	Utility	Plant Capacity, mw[a]	Metro Area	Refuse Burning Capacity Tons/Year[b]
Dickerson	Potomac Elec. Power	586	Washington, D.C.	400,000
McDonough	Georgia Power	598	Atlanta, Ga.	410,000
Astoria	Con. Edison	1,550	New York, N.Y.	1,060,000
Portsmouth	Virginia Elec. Power	649	Norfolk, Va.	445,000
S. Oak Creek	Wisc. Elec. Power	1,170	Milwaukee, Wis.	800,000
Cromby	Philadelphia Elec.	417	Philadelphia, Pa.	285,000
Mitchell	W. Penn. Power	448	Pittsburgh, Pa.	305,000
Gannon	Tampa Elec.	900	Tampa, Fla.	615,000
Russell	Rochester G.&E.	260	Rochester, N.Y.	175,000
Cherokee	Pub. Serv. of Colo.	375	Denver, Colo.	255,000
Cane Run	Louisville Gas & Elec.	700	Louisville, Ky.	480,000
B. L. England	Atlantic City Elec.	300	Atlantic City, N.J.	205,000

[a] From Federal Power Commission data, 1968.
[b] Assuming 75 percent use factor—365 days/year, using 250 tons of refuse per day per 100 mw of plant capacity.

Source: Wisely, F. E., "Solid Waste and Electric Power Production," *Professional Engineer,* Vol. 41, No. 10, October 1971, pp. 40–41.

be derived through the cooperation of municipalities and utilities by implementing already proven technology with commercially available equipment.

7-2. WASTE HEAT UTILIZATION

With shortages of fossil fuels forecast for the late 1970s and early 1980s and with the failure of nuclear power to become widely available, increased interest has been generated in the use of solid wastes as fuel. In many communities refuse is being processed currently in incinerators wherein large quantities of heat are liberated and wasted either directly or indirectly to the atmosphere. Consequently, attention is being focused on ways and means of utilizing the heat liberated during incineration.

7-2.1 European Experience

Waste heat recovery has been practiced for a considerable length of time in Western Europe.

In Switzerland, the cities of Zurich and Berne have outstanding steam-producing incinerators. The Berne installation supplies 30,000 tons of steam each year to two hospitals, two factories, a dental institute, and another school (Ref. 7-33). The steam is produced at the rate of 1 ton of steam from each ton of refuse; over 5000 tons of coal are saved this way. The Zurich incinerator also is a successful waste heat recovery operation (Ref. 7-26).

The Issy-les-Moulineaux incinerator near Paris exemplifies many new French installations. In this plant approximately 450,000 tons of refuse are processed annually (Ref. 7-2). Steam produced from the refuse is used to generate electricity at the rate of over 90 million kilowatt-hour each year (Ref. 7-26).

Many cities in Germany and the Benelux countries are served by heat-recovery incinerators. In Dusseldorf, an income of over $700,000 per year has been realized through the production of electricity and sale of recovered metals from the 960 tons (nominal) of refuse burned each day (Ref. 7-28). Similar plants are in operation in Leverkusen and Hamburg (Ref. 7-26 and 7-28). In Rotterdam, an incinerator built in 1966 produces approximately 80 million kilowatt-hour of electricity from slightly less than 300,000 tons of refuse burned each year. Metal salvage and sale of residue generate additional income (Ref. 7-15).

Heat recovery is practiced to a somewhat more limited extent in Great Britain. However, in 1967, a 1600 ton per day plant was built in London to include steam production (Ref. 7-28). Revenue from steam sales and residue salvage operations amounts to about $1.5 million per year.

The success of these and other plants in Western Europe has created a trend there to include heat-recovery facilities in every large incinerator which is built.

A more detailed discussion of European solid waste management practices may be found in Chapter 8.

7-2.2 American Experience

Several incinerator heat-recovery operations have been conducted for a number of years in the United States. The history of these plants is a mixed story of success and failure.

All of the heat recovery units installed before the late 1960s consisted of boiler units or heat exchangers designed to obtain heat from exhaust gases produced in refractory-lined furnaces. The Providence, Rhode Island Field Point plant was equipped with two boilers in 1936 and two additional boilers in 1950. Electrical power produced at this facility is supplied to a local wastewater treatment plant. Deterioration of boiler tubes has been serious and hand-cleaning of deposits from tubes has been necessary (Ref. 7-6).

In the early 1940s two boiler units were installed behind the International Volund rotary kilns in the Mayson plant in Atlanta, Georgia; two additional boilers were added in the early 1950s. Some boiler-tube deterioration has been experienced and consequently, tube replacement was initiated after 15 years of operation. Hand-cleaning the boiler-tubes is necessary at this facility and the steam produced is used for heating nearby structures. Income from the sale of steam amounts to about $140,000 annually (1968) at this plant (Ref. 7-6 and 7-20).

In 1951, the Merrick incinerator was placed in operation in Hempstead, New York. A boiler unit was installed behind a secondary combustion chamber which serves four circular-hearth batch-fed furnaces. Electrical power is generated at a constant rate with over-production of steam being condensed. Hand-cleaning of tubes has been necessary and tube life has been about 8 years. A somewhat similar plant was built four years later in Miami, Florida to supply steam to a nearby hospital. Performance of the Miami plant has been comparable to the Merrick incinerator (Ref. 7-6). Shortly after the Miami plant went on-line, a plant was placed in operation in Oyster Bay, New York in which two boilers are supplied with hot gases from rectangular batch-fed furnaces. Internal corrosion in the boiler tubes has been severe and operation of the plant has been discontinued (Ref. 7-6 and 7-24).

In 1959, the South Bay plant began operating in Boston. This facility has three heat-recovery boilers fed by six rectangular batch-fed furnaces. The steam produced in the boilers (75,000 lb/hr at 250 psig) was intended originally for use in Boston City Hospital but this use was discontinued and the steam currently generated is used solely within the incinerator plant itself (Ref. 7-11).

In 1962, steam-producing heat-recovery units were installed in the Incinerator, Inc., plant in Chicago (Ref. 7-34). During the first five years of operation of the steam generators, the heat-recovery activities lost money because capital and operating costs experienced were higher than originally expected. With the negotiation of a contract for 1 million pounds of steam per day in 1968, the operation of this facility became profitable. The boilers at this plant are fed by two Volund rotary kiln furnaces which burn 250 ton per day each. Recovery of tin cans from the incinerator residue provides additional income (Ref. 7-7 and 7-34).

The successful operation of the Merrick incinerator in Hempstead led to the installation of a second heat-recovery incinerator in that city in 1965. Steam is produced in the newer facility's convection boiler units which are fed by continuous-feed furnaces. Electrical power is generated for use in the plant and additional steam is used to desalinize sea water. The steam production rate is approximately 150,000 lb/hr (Ref. 7-6 and 7-20). This unit is discussed in detail later in this chapter.

A new era of heat recovery began in the United States in 1967 with the installation of a water-tube-walled boiler in a new incinerator built at the Norfolk Navy Yard. This 180 ton per day plant produces approximately 50,000 lb of steam/hr at 275 psig. The use of a water-tube wall design permits operation at only 50 percent excess air with attendant savings in air pollution control costs. Revenue from this facility has been estimated at more than $40,000 per year (1967) (Ref. 7-20).

Other heat-recovery units have been installed elsewhere including the Montreal, Canada unit and the Northwest Incinerator in Chicago. The Chicago unit will be discussed in detail in a later section of this chapter. The Montreal unit has been described elsewhere (Ref. 7-30).

7-2.3 The Potential for Heat Recovery

The increasing development of heat-recovery systems is favored by two circumstances:

1. Solid wastes are "clean" high-energy fuels.
2. The energy content of solid wastes has been increasing for some years past and is likely to continue to increase (Ref. 7-19, 7-29, and 7-5).

Several studies have indicated that the heat content of municipal refuse is on the order of 5000 BTU/lb (10 million BTU/ton) or more. On the basis of this heat content value it is possible to estimate the amount of heat available for steam generation if the temperature of the exhaust gas from the incinerator is known. If this exhaust gas temperature is assumed to be approximately 600°F, then about 63 percent of the heat content is available for steam production (Ref. 7-4). Steam generation for this amount of available energy would be about 100,000 lb/hr in a 400 ton per day incinerator if the heat losses corresponded to the estimated reasonable values. The income from this steam would be on the order of $800 to $1200 per day, on the basis of 1973 steam prices. If the steam were used to generate electricity, approximately 6000 to 7000 kilowatt-hour could be produced for each 100,000 lb of steam.

The overall economic benefits of using refuse as fuel cannot be evaluated without consideration of the cost of refuse disposal in addition to the cost of preparing and transporting refuse to a steam-producing operation. Generally, 1 ton of refuse can be burned to produce about 8 to 10 million BTU (Ref. 7-9). If fossil fuel costs $0.30/million BTU, the replacement value of the refuse is about $2.40/ton. However, the cost of preparing, storing, and transporting refuse may easily exceed this value especially if transport distances are long. Nevertheless, the waste heat utilization operation may be profitable *in toto* when the cost of refuse disposal without heat recovery is considered as an expenditure.

Finally, since World War II, the paper and plastics content of municipal solid wastes has risen dramatically. Forecasts of increased packaging usage in the next decade indicate that the proportion of paper–plastics to total amounts of wastes will continue to increase. At the present time, the BTU content of unsorted solid wastes approaches 5000 BTU/lb. With increased utilization of combustible packaging materials, the heat content of the wastes may reach 7000 to 7500 BTU/lb. If the heavy inert materials such as glass and metals are removed through air separation, the heat value of the processed paper-rich wastes may exceed 10,000 BTU/lb.

The above-mentioned circumstances make waste heat utilization an increasingly attractive proposal.

7-2.4   Critical and Limiting Parameters

The utilization of heat wasted in incinerators is not without disadvantages and limitations. Some of these limitations are technical and some are socio-political. Technical problems include operational difficulties (slagging and deposition of material on boiler tubes, and both internal and external chemical attack on tubes) and economic problems (lack of markets for generated steam).

Almost all of the "add-on" convection boilers installed on refractory-lined furnaces in the United States have experienced difficulties with the deposition of material on boiler tubes. Steam generation capacity may be reduced by this occurrence; alternately, "hot spots" may develop at deposits on boiler tubes. In some circumstances, the deposits may be corrosive. To alleviate these difficulties, wider tube spacings, furnace wall cooling, or gas cooling to temperatures below about 2000°F by addition of excess air may be employed.

External tube corrosion may be avoided if temperatures in the metal tubes are maintained above 400°F. This corrosion is especially likely to occur during periods of low burning or when wet refuse is charged. "Dew-point" corrosion may also occur under such conditions (Ref. 7-6). Use of auxiliary furnace heat or circulated steam can remedy this situation by maintaining sufficiently high temperatures in the metal tubes.

Because refuse composition is geographically and seasonally variable, the heat content of refuse is equally variable. Fluctuations in moisture content and paper percentage in the refuse may lower its heat value and reduce the steaming capability of a boiler unit. To insure a constant output of steam or electrical power from a refuse-fired boiler, auxiliary fuel must be used to supplement wet or low heat refuse, or the steam supply must be rated on minimum production values with the overproduction of steam being condensed. Fluctuation in output has made "steam-from-refuse" a difficult concept to sell to utilities and similar consumers.

Other features of the process have caused different problems regarding the sale of generated steam. The economics of heat recovery in a refuse incinerator obviously are predicated on the sale of generated steam and electricity. Although the sale of such power may seem to be an easily accomplished objective in the highly industrialized society of the United States, the general experience in American incinerators has been that sale of waste heat is by no means guaranteed. In one municipal incinerator, heat-recovery boilers were to be installed aft of the secondary combustion chambers which were fed by four rotary kiln furnaces. The generated steam was to be sold to a large food processing plant nearby. However, when the advertising consultants for the food-processing facility learned of the proposed use of "steam-from-garbage" to prepare their clients' products, they advised against the operation because of the possible unfortunate consequences of public disclosure of the steam source to be used in food preparation. The incinerator managers were informed that the food processors were no longer interested in "garbage steam" and the boilers were not installed in the plant. As an additional example of this problem, the Chicago Northwest incinerator operated for at least three years with no customers for its generated steam; the steam was condensed in roof-top units at the Chicago plant during this period.

Finally, a conflict may arise in the socio-political sphere when a refuse disposal facility is proposed to include generation of power, a function frequently under the jurisdiction of another utility or agency. As stated concisely and well in a previous publication (Ref. 7-9), "We can generate power by burning refuse, and do so economically. But to put it into practice, the utilities will have to be aggressive and government at all levels must provide a favorable political and legal climate for expansion of utilities into this new phase of public service." Utility jurisdictions and authorities must be made compatible before heat-recovery operations are attempted.

7-2.5 Heat Recovery Systems

Various types of heat recovery systems are available for use on existing and planned refuse incinerators. The systems may be classified generally as "add-on" units designed to operate on exhaust gases from refractory-lined furnaces, integral water-tube wall units, and gas turbine electrical generator units. Gas turbine units are typified by the CPU-400 unit which will be discussed in the next section of this chapter.

Auxiliary "Add-on" Units. Heat recovery may be achieved by the addition of boiler or heat-exchanger units to refuse incinerators consisting of conventional refractory-lined furnaces. Typical of such an installation is the Chicago Southwest Incinerator which was constructed in 1963 for slightly less than $7 million. In this incinerator, refuse is burned in four 300 ton per day rotary kiln furnaces; gases are completely combusted in secondary combustion chambers. Hot gases pass through a damper from the secondary combustion chambers into four waste-heat boilers which each generate 50,000 lb/hr of steam. The generated steam is sold to consumers in the "stockyards" area of Chicago for an annual income of about $100,000. At the time of construction, this incinerator was one of the largest units on the North American continent (Ref. 7-1). Heat recovery in this type of unit is much less efficient than recovery in a water-tube wall unit. For example, in the Southwest Incinerator steam production per ton of refuse burned is only two-thirds of the per ton production in the Northwest Incinerator (water-tube wall) in that same city. However, heat-recovery boilers could be added to many existing refuse incinerators with proper engineering design. Even though the heat recovery efficiency in such units is lower than in water-tube wall units, the potential power production is great.

Water-tube Wall Units. The construction of the combustion chamber and boiler as an integral unit offers several distinct advantages over the "add-on" units previously described. Heat is exchanged radiantly with the water-tubes as well as by convection, and, therefore, the overall heat recovery efficiency is higher. In addition, refractory wear and deterioration are eliminated; such deterioration may be high in furnaces and secondary combustion chambers serving "add-on" units. Better heat removal also creates a greater reduction in waste gas volume and consequently results in savings for air pollution control.

Water-tube wall units also possess certain disadvantages such as the deposition and corrosion problems mentioned in the previous paragraphs.

The Northwest Incinerator discussed in the next section is a good example of a water-tube wall unit. Other units have been installed at Braintree, Massachusetts, Hamilton, Ontario, and Montreal, Quebec.

7-2.6 Process Case Histories

Waste heat utilization is currently being used or is planned for use in at least four areas of the United States at the present time (1974). Case histories will be presented on three of these installations including:

1. The Hempstead, New York dual-purpose facility.
2. The currently operating Northwest Incinerator in Chicago, Illinois.
3. The 1250 ton per day incinerator constructed in 1973-74 in Nashville, Tennessee.

Hempstead, New York Incinerator. The Hempstead, New York facility utilizes refuse for power production and also converts salt water into fresh water. The plant is located on a 200-acre site in the salt marshes of southern Long Island, within 400 ft of a salt water channel. The incinerator has a nominal capacity of 750 tons per day.

Feasibility studies for the Hempstead plant predicted that its installation would be economical if enough power were produced to satisfy both in-plant needs and future requirements of adjacent town park areas. The design of this facility was, therefore, developed to meet these criteria. In the Hempstead facility, steam is produced in two Combustion Engineering single-drum, controlled circulation boilers. Each boiler has a maximum continuous capacity of 85,000 lb/hr of steam. Steam is discharged at 462°F and 460 psig pressure. The steam is used to power two 1250 kilowatt Elliott turbines. Power generated is used in-house in the deaerator to heat condensate returning to the boilers and in the turbine drives for the boiler feed pumps. Standby and emergency equipment is provided by a diesel engine driving an electrical generator.

Fig. 7-10 Hempstead, New York incinerator.

Power is also used at this facility to desalt water because of the high consumption rate of fresh water for a plant of this type. Desalting operations became necessary because available potable water was too expensive in the quantity required, and the use of groundwater was impossible because of the danger of salt water intrusion. The desalting units serve rather like condensers in the steam cycle. Sale of desalinized water was not feasible at this plant because processing expenses made the cost of such water much higher than the cost of available public water.

Steam from turbines and overproduction steam reduced in temperature and pressure is used in submerged-tube evaporators to produce fresh water. Four desalinizers were installed, any three of which can consume the nominal total quantity of steam produced in the plant. Each evaporator has an equivalent condensing capacity of 46,250 lb/hr, or a fresh water production rate of 115,200 gallons per day.

Chicago Northwest Incinerator. The Northwest Incinerator of the City of Chicago is located at 700 North Kilbourne Street in the heart of the city. This incinerator is designed to process 1600 tons of refuse per day (30 percent of Chicago's refuse) in four furnaces each rated at 400 ton per day. However, the furnaces have been operated successfully at a feed rate of 475 ton per day.

Refuse is deposited from compactor trucks in the storage pit, which has sufficient capacity to accommodate refuse accumulations during periods of furnace maintenance, etc. Four grab overhead cranes each rated at 1.5 ton capacity move

Fig. 7-11 The Chicago Northwest Incinerator (*courtesy* Chicago Bureau of Sanitation).

the refuse into chutes which feed the combustion areas. Some problems have been experienced in the dumping floor–storage pit area because of accumulations of carbon monoxide from collector truck emissions. These problems should be alleviated by the proposed installation of extra fans above the dumping floor to carry exhaust gases to the outside air. No other difficulties have arisen in the dumping floor–storage pit area.

The refuse enters the main combustion area from the aforementioned chutes. Combustion occurs on closely-spaced grates which move in a reciprocating motion; the grates, manufactured by the IBW-Martin Company, are inclined downward from the chute-entrance and the motion of the grates tends to push ignited refuse back upslope and under the entering unlit refuse. The grates themselves are very heavy and have proven to be very durable. There is a very narrow opening between individual "teeth" in the grates. Underfire air is supplied by a complex system designed to create high upward velocities in the air moving between the closely-spaced grate teeth (Ref. 7-32). The overall effect is akin to the flotation achieved in a fluidized bed. The level of refuse on the grates and the residence time for the refuse is controlled by a clinker roller at the bottom of the grates. Siftings falling through the grates are carried beneath the grates downslope by a sequential air-jet arrangement into the residue removal system. Clinkers and ash also fall into the residue removal system.

The hot gases leave the grate area and travel upward into the water-wall boiler area where combustion is completed. Approximately 70 percent of the refuse heat value is recovered (Ref. 7-21). Flyash in the gas stream is precipitated as the gases travel through the Walther boilers; the heat in the gas stream is transferred to the water-tubes by radiation and the gases are cooled from approximately $2000°F$ to about $400°F$. In addition to the boiler proper, a heat economizer is used to preheat input water by using the hot gases. Steam is produced in the boilers at a nominal rate of 100,000 lb/hr in each boiler; actual production frequently exceeds 150,000 lb/hr in each boiler. The steam currently is condensed in units located on the roof of the structure.

The exhaust gases, cooled in the boiler and economizer, pass into electrostatic precipitators where particle collection efficiency has been measured at 97 percent or more. Difficulties have been experienced in the collection chutes of the precipitators because the flyash tends to clog and bridge in the narrowing chutes. To correct this problem, rappers have been installed to vibrate the wires and panels of the precipitator and the sidewalls of the chutes.

Residue from the quenching pit moves on a conveyor through a cylindrical screen separator where tin cans are removed. The remaining residue and flyash are removed in trucks for deposition is a former quarry. The separated metal is sent to western states where the tin cans are used in copper recovery operations. Cleaned gases are passed through the incinerator stack through an induced-draft fan.

Fig. 7-12 Overhead crane at the Chicago Northwest Incinerator (*courtesy* John J. Reinhardt).

Fig. 7-13 IBW Martin grates utilized at the Chicago Northwest Incinerator (*courtesy* John J. Reinhardt).

Some difficulties have been experienced in the residue conveyor system; flyash and grit have caused excessive wear to the rollers in the conveyors and the system has been replaced more than once.

At the present time steam produced in the plant is being used in the plant itself and in nearby facilities for cleaning refuse collection trucks and other equipment. Plans call for sale of steam from the Northwest Incinerator in 1974. Chicago's Refuse Commissioner (1973) had expressed confidence that a contract would soon be signed for installation of a steam pipeline to supply a nearby industrial park by mid-1974.

The operation at Kilbourne Street has encountered no major difficulties. Apprehension has been expressed by many individuals concerning corrosion of boiler tubes through contact with combustion gases produced through refuse burning. However, no evidence of corrosion problems has appeared at the Northwest Incinerator. At times, condenser capacities have been exceeded and the combustion chambers have been shut down. These problems have been minor.

Chicago officials are optimistic about the future of steam sales at the Kilbourne Street installation. This optimism may be due to the successful recovery of energy through steam production which has been achieved at the Chicago Southwest Incinerator. Since 1963, steam has been produced at that facility in a heat-recovery unit added to the basic refractory-wall furnace as described above. Sales of steam there have produced revenues of about $100,000 per year. The heat recovery system at the Southwest Incinerator is much less efficient than the water-wall boiler at the Kilbourne Street facility. However, similar recovery systems could be installed on existing incinerators in other parts of the country. In comparison to the Southwest Incinerator, the Kilbourne Street water-wall unit has potential for revenue amounting to some hundreds of thousands of dollars per year. The planned construction of a bulky waste shredder and a magnetic separator at the Northwest Incinerator should increase the utility and effectiveness of the facility.

Another unique feature of this installation is the degree of automation in the plant. Almost all plant operations are controlled from an air-conditioned room in the main building. Each furnace and its support systems are monitored via closed-circuit television. Manual controls are available, in case of emergency, for all operations of the plant. The estimated cost of the Chicago plant was $23 million.

Nashville, Tennessee Incinerator. The city of Nashville's proposed 1250 tons per day waste heat incinerator was scheduled to start operation sometime during the first quarter of 1974 (Ref. 7-36). The construction of this plant was initiated by two circumstances:

1. Environmental regulations passed for the state of Tennessee required that all land disposal of refuse must be by "sanitary landfill" (approved site with approved covering and backfill methods) no later than July, 1972.

2. Extensive urban renewal and downtown renovation in Nashville created a possible need for a central heating and cooling plant for the downtown area. After extensive study, the city officials decided that landfill sites would be difficult to obtain and that solid waste rather than fossil fuel could be used as the energy source for the heating and cooling plant.

The following plan was therefore developed. Refuse will be delivered to the plant located adjacent to the Cumberland River by the city at no cost under the terms of a 30-year contract. The steam generated by the waste heat recovery facility will be used for heating and cooling in twenty-seven adjacent buildings. Wet scrubbers will be used for air pollution control. On the basis of 1973 costs, the city of Nashville will save approximately $2 million per year in waste disposal costs by this system.

The heating and cooling plant and required transport pipe network is estimated to cost $18 million. Construction funds are to be raised by revenue bonds guaranteed by the clients buying the steam from the plant. Upon retirement of the bonds, ownership of the entire facility will revert to the city.

7-2.7 Summary—Waste Heat Utilization

Waste heat recovery incinerators have been used to some extent in the past in this country and will probably become more numerous as fossil fuel prices rise in the future. The success of these systems is primarily dependent upon long-term markets for the steam from the process. Presently the only facilities which have been consistently successful in selling steam, or are expected to be successful, are plants located in a central region with respect to prospective permanent clients. In the future, waste heat recovery systems should become more feasible economically because of predicted fuel shortages and the resultant public clamor for the development of a fossil fuel substitute. These systems will have the added economic advantage of eliminating solid wastes disposal charges to a large extent.

7-3. THE COMBUSTION POWER UNIT-400

The Combustion Power Company has under development a facility which is purported to be a pollution-free method of economically recovering both materials and energy from mixed municipal solid waste.

The CPU-400 system is designed to recover energy from solid waste as electrical power. To recover this energy, the waste is burned in a fluid-bed combustor. The hot gases from the combustor pass through a gas turbine to power an electrical generator. Additionally, the system includes provisions for materials recovery through separation and materials-processing subsystems. According to the estimates of the Combustion Power Company, a full-scale plant will be capable of producing approximately 5 percent of the electrical power needs of the urban area which generates the wastes processed in the plant. Forecasts are being made by the system designers of income from the process through the sale of electrical power and recovered materials.

A pilot plant was developed in Menlo Park, California in 1971–1972. The pilot plant was built to permit the testing of various prototype systems, and it is designed to process 80 tons of solid waste per day. From the 80 tons of solid waste processed in the pilot facility, approximately 1000 kilowatts of electrical power are generated. Additional testing of the pilot plant was scheduled for late 1973.

A full-size CPU-400 prototype is scheduled for operation late in 1974. This prototype will consist of three modules, each containing a fluid-bed combustor and a gas turbine unit. Each combustor will be capable of processing approximately 150 tons of refuse per day. The use of combustor-turbine modules will allow the construction of a plant of any desired size, from a minimum capability of 150 tons per day to 10 or 15 times that capability. The modules also are an attractive means of incorporating off-the-shelf equipment such as turbine units into the final system design. The full-size CPU-400 plant, as currently envisioned, will occupy less than 2 acres of land; this presents the possibility of the use of a number of such plants as refuse processing satellites around a large urban area. Such a satellite system could minimize expensive collection and haulage costs.

7-3.1 Basic Concept of the CPU-400 Unit

The basic stated philosophy of the manufacturers of the CPU-400 power generation unit is that the recovery of energy has been practiced for a number of years and that such recovery will become economically and environmentally more attractive with passing time. The electrical energy produced in the unit is considered to be a product with a ready market which will increase in magnitude over the coming years. Not only is the energy market available but the price is relatively stable. Moreover, as the paper content and plastic content of solid waste increase, the available BTU content and "power" content of the waste will also increase.

The design of the CPU-400 includes facilities for the recovery of ferrous metal, aluminum, nonferrous metals, and glass from solid wastes. The design of the combustion system virtually requires the removal of such constituents from the combustible portion of the wastes. However, the Combustion Power Company envisions the marketing and sale of all recovered materials whenever possible. The recycled ferrous metal, aluminum, and glass may find ready markets at the present time; all recovered aluminum certainly would be easily saleable. The income produced by the sale of recycled materials could serve to drastically reduce disposal costs if the economic forecasts of the system designers are correct. However, based upon current prices of recycled materials, the income associated with recycling noncombustibles would be rather minimal.

7-3.2   Methodology

Figure 7-14 shows the schematic diagram of the CPU-400 refuse disposal system. Packer trucks deliver solid waste to an enclosed receiving area where front-end loaders transfer the material onto conveyors. The conveyor belts carry the solid wastes into shredders, and the shredded solid wastes then enter a dry air-classifier. In the air-classification process, dense materials such as metals and bits of glass are separated from the remainder of the refuse. The dense particles then go to a material recovery module for further processing. The less dense materials such as the organic materials, shredded paper, pieces of plastic, and other like materials are carried in the air stream up and out of the air-classifier to a storage container. The storage container is specially designed to process shredded light refuse and to insure a continuous feed of combustible material to the fluid-bed combustors. The shredded waste then enters the combustors through high-pressure air lock feeders. In the combustors, inert sand-size particles are maintained in a fluid bed. These particles are stabilized in position by an upward flow of air through the combustor unit; the air originates in the compressor of the gas turbine unit. After the solid waste enters the combustor, it is supported on the fluid bed where combustion produces a relatively constant temperature of approximately 1500°F.

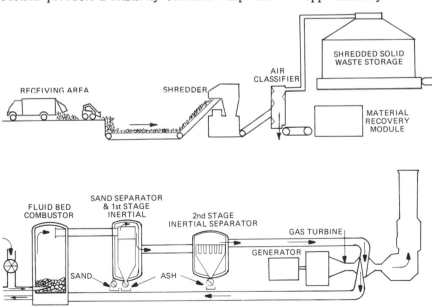

Fig. 7-14   Schematic diagram of CPU-400. *Source:* **Combustion Power Company.**

Designers of the system feel that the fluid-bed combustion process ensures rapid and complete combustion and minimizes the generation of air pollutants such as unburned hydrocarbons, nitrous oxides, sulfur oxides, and acidic gases. The hot gases passing from the fluid-bed combustor proceed through a series of inertial separators. Particulate matter is collected in these separators and the cleaned gases then pass through the turbine unit and out the exhaust stack.

The entire CPU-400 system is housed in two structures. A receiving building contains the shredding equipment, the air-classifier, and the heavy-media separation unit, and also provides storage room for approximately 100 tons of solid waste. The turbine building provides space for three modules, each containing a gas turbine, a fluid-bed combustor, and a particle collector unit. Also, in the turbine building, an area is set aside for storage and distribution of the shredded solid waste coming from the receiving building.

7-3.3 Critical Method Parameters

In general, the critical parameters for the success of this method of processing solid wastes are the same as the critical method parameters for conventional incineration operations. For example, the water content and percentage of combustible material of the incoming solid waste can have critical effects upon the incineration process in the fluid-bed combustors. If, for example, the percentage of combustible materials in the solid waste coming into the CPU-400 were to be drastically reduced, the entire unit would not operate efficiently. Such a change in the characteristics of the solid waste could possibly be brought about by a wide-scale paper recycling program.

An additional critical parameter for the operation of the equipment in the CPU-400 is a constant supply of clean hot gases from the fluid-bed combustor to the gas turbine unit. High purity of the hot gases passing through the turbine unit is crucial to the operation of the gas turbine/generator combination. The presence of particulate matter or corrosive gases in the gas stream passing through the turbine could cause excessive wear of the turbine components and could cause frequent costly repairs with much more expensive down-time. For this reason, air pollution control devices have been installed between the fluid-bed combustor and the gas turbine. However, the success of the fluid-bed combustor in preventing the production of air pollutants such as sulfur oxides has yet to be demonstrated. The presence of sulfur-bearing compounds or the presence of polyvinyl chloride in the input refuse could produce corrosive sulfur oxide or chlorine gases. Sulfur oxide combines with water vapor to produce sulfuric acid and could cause serious corrosion of metallic parts of the turbine. Polyvinyl chloride contained in the input refuse could produce hydrochloric acid in a similar fashion. For this reason, the air pollution control devices on the CPU-400 are one of the most critical parts of the system.

7-3.4 Method Requirements

The requirements for the facility include: the provision of utilities for the plant; equipment components necessary for operation of the plant; and trained personnel qualified to monitor and operate the various components of the CPU-400 system.

Since the CPU-400 is designed to produce electrical power, it is obvious that electrical power will not be necessary during the operation of the unit. However, electrical power must be used during facility start-up or maintenance of the plant equipment. On the other hand, sanitary facilities must be provided for the disposal plant.

A major requirement for this system is the group of ancillary equipment components necessary for successful operation. These components consist of shredders, air-classifiers, storage facilities, materials recovery facilities, particle collectors, and gas turbines.

The shredder used to reduce particle sizes in the incoming refuse is a very important part of the CPU-400 system. Each of the shredders in the prototype plant will be required to handle an average of 40 tons per hour of mixed municipal refuse; the peak input rates of refuse may be as much as 50 percent higher than this value. The emphasis in the CPU-400 system must be upon the reliability of the shredder to maintain a flow of small-size refuse particles to the air-classifier. At the present time, an Eidal SW-1000 shredder is being tested for use in the prototype CPU-400 plant.

A zigzag air-classifier will be used to separate the lighter combustible fraction of the refuse from the heavy fraction destined for the materials recovery system. At the present time (1974), an air-classifier manufactured by the Vancouver Company is being tested for use in the CPU-400 prototype system.

A special storage facility for the shredded refuse has been developed for use in the CPU-400. This refuse carousel is designed so that the refuse will enter via a pneumatic conveyor into the top of the storage compartment. Around the outside of the storage compartment is located a rotating pull ring. Fixed to this pull ring are four chains of sweep buckets. Each chain of buckets is fixed to the pull ring with the other end of the chain free to trail behind as it is pulled around the carousel by the pull ring. As the pull ring rotates, the bucket chains tend to drag toward the center of the bin, thereby causing the buckets to become filled from the outside of the refuse pile. The filled buckets then transfer the refuse onto the output conveyor to the fluid-bed combustor.

The materials recovery module presently being considered for use in the CPU-400 prototype consists of a magnetic separator for ferrous metal, a water-bath separator to remove plastics and organic material, a set of vibrating screens to separate glass from the remaining solid waste, and a heavy-media separation table to differentiate the remaining materials into aluminum and copper fractions. Details of the waste recovery module are not available at the time of this writing.

As mentioned previously, the most critical component of the entire CPU-400 system is the set of particle collectors now being tested (1974) at Menlo Park, California. This set of particle collectors includes an initial settling chamber designed to remove any particles of inert material from the fluid bed which may have escaped with the hot gases, and any particles of molten glass or aluminum which may have been entrained in the exit gas stream. From the initial settling chamber, the hot gases pass into a series of 6 in. diameter inertial separators and then into a set of $3\frac{1}{2}$ in. diameter inertial separators. The manufacturers claim that the present set of particle collectors will clean the exit gases sufficiently to guarantee at least 20,000 hr of turbine performance between maintenance periods. Additionally, the manufacturers claim that the gases escaping through the exhaust stack will contain less than 0.03 grains of particulate matter per standard cubic foot of exit gas. These claims have yet to be conclusively validated.

The gas turbine/electrical generator combination to be used in the CPU-400 represents an off-the-shelf item for inclusion into the basic system. Several models of turbine/generator combinations are available commercially. Some modifications of the commercial units are necessary but the modifications which are required are simply changes in the ductwork and manifolds to conduct the air from the compressor unit to the fluid-bed combustor and to convey the hot gases from the particle collectors to the turbine inlet. Such modifications should cause little difficulty in the prototype units.

The personnel requirements for operation of a prototype CPU-400 unit are not completely fixed at the present time (1974). However, the manufacturer estimates a labor force of at least 16 men working in four shifts to provide 24 hr per day operation of the plant. This corresponds to over $200,000 per year in direct labor costs. Previous experience with manufacturers' claims for labor requirements indicates that the labor costs estimates of the Combustion Power Company must be treated as preliminary estimates. In any case, the labor requirements for the full-size plant have not been completely determined since comprehensive tests on a prototype plant have not been completed to date. In all likelihood, the operation of the plant for the highly-mechanized system may be an appreciable factor in the overall economy of the method.

7-3.5   Method By-Products

As mentioned previously, included in the basic design of the CPU-400 is a material recovery system designed to recover waste metals, glass, and other materials. The by-products from this material recovery system are intended to be saleable secondary materials. However, the market for such materials at the present time in the United States is somewhat variable. Therefore, some consideration should be given to the possible necessity of disposal of portions of the materials recovered from the municipal refuse input into the CPU-400. However, increasing emphasis on resource recovery in the United States, together with possible changes in tax laws, freight rates, and governmental incentive buying should provide markets for all secondary materials recovered in the CPU-400 system.

In addition to the secondary materials recovered prior to combustion of the solid waste, the primary product of the unit will be electrical power. Although the manufacturers voice optimistic claims as to the marketability of the electrical power produced in this unit, experience in the United States to date indicates that there is no guarantee of a market for the variable power produced from solid waste. The fluctuations inherent in the incoming solid wastes will inevitably cause fluctuations in the power production rate of the unit. The design power generation capacity of the unit will correspond to the minimum rate produced by the most detrimental fluctuation in the characteristics of the incoming refuse. Therefore, a critical investigation of the marketability of the electrical power produced in such a unit is justified.

The third by-product generated in the CPU-400 unit will be an exhaust gas stream. In this respect, the CPU-400 is exactly similar to conventional incinerators which produce a stream of exhaust gases. The exhaust gases leaving the CPU-400 will be subjected to the same scrutiny by air pollution control agencies as the gases leaving conventional incinerators. Therefore, this waste gas stream may be a possible source of difficulty in the operation of the unit. Changes in air pollution codes and similar anti-pollution legislation may necessitate the inclusion of more sophisticated air pollution control equipment in the basic CPU-400 system. The increased capital and operating cost of such equipment may have a significant effect upon the system economics of the power unit.

7-3.6 Method Economics

To date, the economic analysis of the CPU-400 disposal system must be based upon the estimates of costs furnished by the manufacturer. All process economics presented must, therefore, be considered preliminary. Tables 7-6 and 7-7

TABLE 7-6 Materials Recovery Values: CPU-400

Material	% of Waste Min.	% of Waste Max.	Material Value $/Ton Min.	Material Value $/Ton Max.	Income $/Ton of Solid Waste Min.	Income $/Ton of Solid Waste Max.
Sand and ash	12.0	18.0	1	3	$0.12	$ 0.54
Glass (gravel)	3.0	8.0	2	10	0.06	0.80
Ferrous metal	4.0	7.0	30	70	1.20	4.90
Aluminum	0.1	0.5	300	400	0.30	2.00
Other metals	0.1	0.5	300	500	0.30	2.50
Potential total material income					$1.98	$10.74

Source: Combustion Power Company.

indicate the potential income associated with production of electricity and disposal of wastewater sludge. It should again be noted that all figures in these tables are manufacturers' estimates.

TABLE 7-7   Economic Data: CPU-400

CPU-400 Electricity Production

Basis

3 power modules burning 160 ton/day each of solid waste and 183 ton/day of sludge (0–10% solids) = 44,000 gal/day, 24 hr/day, 7 day/week, each generating 4000 kw of electricity, including steam turbine output.

"As received" solid waste includes solid waste burned plus 15% air-classifier fallout + 5% water loss on shedding.

"As received" solid waste = (160 ton/day) (3)/0.80 =   600 ton/day
At 85% utilization (310 day/year)   = 186,000 ton/year
Sludge consumption/year = (183/200) 186,000   = 170,000 ton/year

Costs

Capital cost $10.8 million
Annual capital cost at 8%, 20 years = $1,100,000
Capital cost/ton (186,000 ton/year)   = $ 5.91/ton

Operating costs
  Labor   =   600,000
    Payroll extras at 25%
  Maintenance   =   700,000
  Utilities   =   100,000
  Total annual operating costs = $1,400,000
  Unit operating cost at 186,000 ton/year   = $ 7.53/ton

                                                          Total cost = $13.44/ton

Income
Electrical
  8.93 × 10^7 kw-hr at $0.012/kw-hr = $1,071,000
    Unit electrical income at 186,000 ton/year   = $ 5.76/ton

    Material Income (mean value in Table 7-6 less
    capital and operating cost)   = $ 5.26/ton
Sludge disposal
  5300 ton/year dry solids at $50/ton = $265,000
  Unit sludge disposal credit at 186,000 ton/year   = $ 1.42/ton

  Assume 7#/cap/day solid waste and 0.20#/cap/day sewage
  solids then population generating 186,000 tons of solid
  waste will generate (0.20/7) (186,000) = 5300 tons of dry
  sewage sludge

  % solids = 5300/170,000 = 3.1% solids
                                                        Total unit income = $12.44/ton
                                                        Net disposal cost = $ 1.00/ton

Source: Combustion Power Company.

7-3.7 Summary—CPU-400

The CPU-400 system is designed to recover both saleable materials and energy from solid waste. The modular system concept and estimated process economics appear favorable; however, a complete evaluation of the process must await full-scale development and operation.

7-4. PYROLYSIS

The term "pyrolysis" refers to the act of decomposing organic compounds through the application of heat. More specifically, it is the process of destructive distillation carried out in an oxygen-free environment. Through pyrolysis, organic matter can be readily converted into gases, liquids, and inert char. The products of pyrolysis represent about 50 percent of the initial volume of the original matter, and they can be converted into energy to either sustain the process or produce excess power. For these reasons, and because solid waste has a high organic content, pyrolysis appears to be an attractive method of solid waste disposal. However, where pyrolysis has been applied in the past, such as in the production of charcoal and the recovery of methanol, acetic acid, and turpentine from wood, the feed material was homogeneous. Unfortunately, solid waste is not only heterogeneous, but may differ greatly from one batch to the next. Therefore, considerable research and pilot work is required before pyrolysis can be applied as a full-scale method of energy recovery from solid waste.

7-4.1 Literature Survey

In November, 1967, E. R. Kaiser and S. B. Friedman presented the results of a study conducted at New York University on "exploratory laboratory tests of destructive distillation of organic wastes, and the prospects for complete gasification of the organic matter" (Ref. 7-16). Although this study dealt with homogeneous samples of the type of matter found in solid waste and not with the heterogeneous combination which must ultimately be considered, it did provide an important step toward the application of pyrolysis to solid waste disposal.

The purpose of this study was to determine if the gases produced by the pyrolysis of organic matter of the type found in refuse could be used as a raw boiler fuel. This would allow the system to sustain itself without the addition of fuel. The matter used in this study was shredded and then dried by heating to 150°F. The types of material pyrolyzed and their chemical compositions are listed in Table 7-8. The average sample size was 5 grams.

A schematic diagram of the laboratory apparatus utilized in this study is shown in Fig. 7-15. Each sample was loosely packed in a 21 mm I.D. Vicor retort and heated from ambient temperature and pressure to 1500°F. Gases emitted during

TABLE 7-8   Composition of Refuse Pyrolyzed in Kaiser and Friedman Study

	Moisture	Volatile Matter	Fixed Carbon	Ash	BTU/lb	Ratio: VM / (VM + FC)
Low-vol. bituminous coal	3.61	17.41	74.84	4.14	14,587	0.189
High-vol. bituminous coal	3.56	37.18	56.55	2.71	14,220	0.397
Hardwood maple	0.0	76.1	19.6	4.3	8,190	0.795
Western Hemlock	0.0	74.2	23.6	2.2	8,620	0.759
White pine sawdust	7.0	78.76	14.10	0.14	8,260	0.848
Pine sawdust	0.0	79.4	20.1	0.5	9,130	0.798
Balsam spruce	3.67	77.75	15.52	3.06	9,196	0.835
Hardwood leaf mixture	9.97	66.92	19.29	3.82	7,984	0.776
Newspaper	5.97	81.12	11.48	1.43	7,974	0.877
Corrug. box paper	5.20	77.47	12.27	5.06	7,043	0.864
Brown paper	5.83	83.92	9.24	1.01	7,256	0.900
Magazine paper	4.11	66.39	7.03	22.47	5,254	0.905
Lawn grass	5.47	71.16	17.18	6.19	7,858	0.835
Citrus fruit waste	8.02	71.46	17.34	3.18	7,372	0.804
Vegetable food waste	5.43	73.31	16.37	4.89	7,820	0.818

Fats rubber, polyethylene plastic, and waste oils distill almost 100 percent of the ash-free organic matter on rapid heating away from air.

The cellulosic matter, VM + FC, is distilled 75 to 90 percent in the rapid heating of the ASTM crucible.

Source: Kaiser, E. R. and Friedman, S. B., "The Pyrolysis of Refuse Components," *Combustion*, May 1968, p. 31.

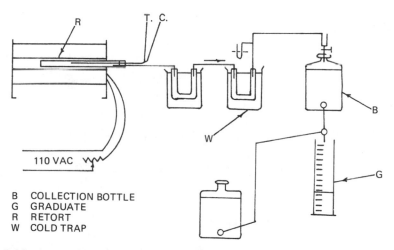

B   COLLECTION BOTTLE
G   GRADUATE
R   RETORT
W   COLD TRAP

Fig. 7-15 Apparatus utilized during Kaiser and Friedman's pyrolysis study. *Source:* Kaiser, E. R. and Friedman, S. B., "The Pyrolysis of Refuse Components," *Combustion*, May, 1968, p. 31.

pyrolysis were collected over saturated NaCl brine; liquids were condensed in U-tubes cooled by ice water.

Other than simply determining the amount of gas emitted during pyrolysis at 1500°F, Kaiser and Friedman also explored the effect of various rates of pyrolysis heating upon gas production. All tests of this type were conducted on shredded newspaper. The results of these studies indicated that the gas yield is highest at the high and low rates of heating, with lowest yields at intermediate rates. Therefore, even higher yields than those reported for the first phase of the study may be possible by flash heating.

Beyond the information provided by their study, Kaiser and Friedman suggested that the char produced by pyrolysis could be gasified through the addition of oxygen into the environment after pyrolysis. The heat produced from the "combustion" of the pyrolysis char would be ample to pyrolyze incoming raw refuse, thereby making the process self-sustaining.

This study left many questions unanswered concerning the pyrolytic destruction of heterogeneous solid waste. In an effort to answer these questions, a batch pyrolysis study utilizing shredded, separated, municipal refuse was undertaken by the Utilities Department of the City of San Diego, California under the direction of Donald A. Hoffman and Richard A. Fitz, in cooperation with San Diego State College (Ref. 7-13). The composition of the material pyrolyzed

TABLE 7-9   Composition of Refuse Pyrolyzed in San Diego Study

	% As Received	% Range during Survey
Combustibles	50.27	55.51–43.72
Incombustibles	49.73	56.28–46.33

	Combustibles, %			% on Dry Basis	
	As Received	Range	Moisture	With Metal and Glass	Without Metal and Glass
Constituent					
Paper	46.16	50.55–40.99	8.23	42.36	50.40
Yard trimmings	21.14	26.65–20.78	51.30	10.30	12.25
Wood	7.48	8.10–4.41	10.50	6.69	7.96
Rags	3.46	4.07–0.32	7.40	3.20	3.82
Rubber	4.73	4.76–3.60	9.74	4.27	5.08
Plastic	0.27	0.52–0.25	0.06	0.27	0.32
Garbage	0.81	0.92–0.77	57.80	0.34	0.40
Metal	7.64	9.68–6.30	0	7.64	0
Glass	8.31	10.38–7.67	0	8.31	0
Moisture	–		–	16.62	19.77
	100.00			100.00	100.00

Source: Hoffman, D. A. and Fitz, R. A., "Batch Retort Pyrolysis of Solid Municipal Wastes," Environmental Science and Technology, Vol. 2, No. 11, November 1968, p. 1023. Reprinted with permission from Environmental Science and Technology. Copyright by the American Chemical Society.

throughout this investigation is given in Table 7-9. A schematic of the apparatus utilized in the San Diego study is shown in Fig. 7-16. The retort was hand-charged with 0.37 lb of material, corresponding to a bulk density of 4.38 lb/cu ft. After charging, the retort was purged with helium, sealed, and placed into an electric furnace which had been raised to a predetermined temperature. The volatile portion of the matter escaped from the retort through perforated walls, the tar and heavier distillates condensed immediately and were collected in the tar trap, the lighter liquids were collected in receiving bottles submerged in acetone and dry ice, and the gases passed through a wet-gas meter and a sampling manifold and were collected in a receiving balloon. At the end of each run, the apparatus was purged with helium and disassembled to remove the char from the retort. When the process was completed, the various by-products were analyzed both quantitatively and qualitatively.

The results of this study confirmed that the products from the pyrolysis of organic solid waste are gases, pyroligneous acids and tars, and solid residue. All

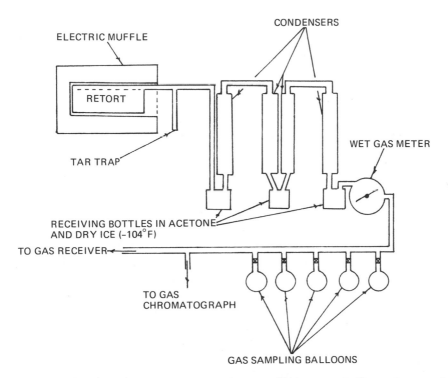

Fig. 7-16 Batch plant—San Diego pyrolysis study. *Source:* Hoffman, D. A. and Fitz, R. A., "Batch Retort Pyrolysis of Solid Municipal Wastes," *Environmental Science and Technology*, Vol. 2, No. 11, November, 1968, p. 1023. Reprinted with permission from *Environmental Science and Technology*. Copyright by the American Chemical Society.

of these constituents represent potential forms of energy, suggesting that once the process is started it could be self-sustaining.

In conclusion, this study showed that the process of pyrolyzing solid waste could be self-sustaining through the incineration of one or more of its by-products; the residual char is comparable to a semianthracite coal and thus constitutes an easily transportable fuel; the gases, liquids, and solids produced have a definite market value and could be salvaged; and the inert solids are sterile and represent a reduction in the original volume of the waste of over 50 percent. Although the results of this study seemed promising, further research was needed before pyrolysis could be incorporated for solid waste disposal on a pilot scale.

Another investigation utilizing larger solid waste samples was completed in August, 1970 for the U.S. Bureau of Mines by W. S. Sanner, C. Ortuglio, J. G. Walters, and D. E. Wolfson (Ref. 7-27). This study was concerned with three types of solid waste: 1) shredded and separated domestic waste, 2) processed waste containing plastic film, and 3) separated industrial waste shredded by two types of hammers—a Heil mill and a Gondard mill. Each of the refuse materials was analyzed before pyrolysis.

The plant utilized in this study is shown schematically in Fig. 7-17. The retort was considerably larger than the one used in the previously cited studies, measuring 18 in. in diameter and 26 in. deep. Corresponding batch weights for this retort were 80 lb for the municipal refuse, 54 lb for the processed municipal refuse, and 64 lb for the industrial refuse. After the retort was charged, it was sealed and lowered into the furnace, which had been preheated to $50°C$ above the desired temperature to allow for heat loss in charging. The retort was held at a constant temperature. The various by-products were separated by methods similiar to those used in previous studies indicated in the schematic diagram of the plant.

These tests demonstrated that a ton of municipal refuse could be converted into 154 to 424 lb of solid residue, 0.5 to 6 gallons of tar, 1 to 4 gallons of light oil, 97 to 133 gallons of liquor, 16 to 32 lb of ammonium sulfate, and 7380 to 18,058 cu ft of gas. One ton of industrial refuse consisting of paper, rags, and cardboard was converted to 618 to 838 lb of residue, 1.5 to 3 gallons of light oil, 68 to 75 gallons of liquor, 12 to 23 lb of ammonium sulfate, and 9270 to 14,065 cu ft of gas.

The residue from municipal refuse had the highest fuel value with a range of from 10 to 17 million BTU per ton. Again, it was shown that the energy from the gas was more than sufficient to provide the heat for the pyrolysis.

As part of a comprehensive study of solid waste management systems, J. Mc-Farland, et al. (Ref. 7-18) completed an investigation of pyrolysis for the National Environmental Research Center. This investigation included a laboratory study and a pilot plant operation phase. The objective of this study was to determine, design, and develop the equipment necessary for reliable and optimal

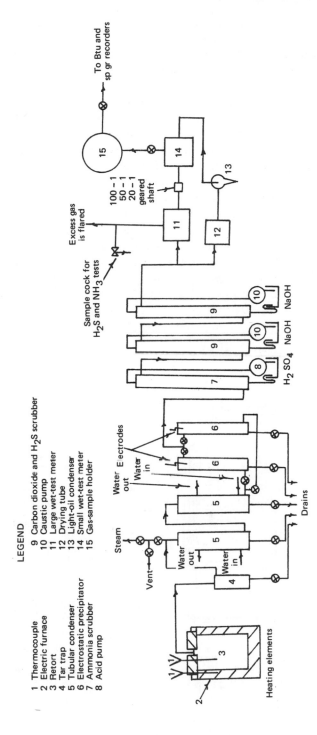

LEGEND

1 Thermocouple
2 Electric furnace
3 Retort
4 Tar trap
5 Tubular condenser
6 Electrostatic precipitator
7 Ammonia scrubber
8 Acid pump

9 Carbon dioxide and H_2S scrubber
10 Caustic pump
11 Large wet-test meter
12 Drying tube
13 Light-oil condenser
14 Small wet-test meter
15 Gas-sample holder

Fig. 7-17 U.S. Bureau of Mines laboratory pyrolysis apparatus. *Source*: Sanner, W. S., Ortuglio, C., Walters, J. G. and Wolfon, D. E., "Conversion of Municipal and Industrial Refuse into Useful Materials by Pyrolysis," U.S. Bureau of Mines, Report of Inv. No. 7428, August, 1970.

pyrolysis of municipal solid waste with a minimum amount of residue for ultimate disposal and no pollutant emission.

The refuse pyrolyzed in the laboratory phase was processed municipal refuse shredded to < ½ in. size. Wood was also pyrolyzed to provide a comparison of the results obtained with the refuse to results for a standard uniform lignocellulosic material. A schematic of the basic laboratory apparatus used in this investigation is given in Fig. 7-18. The reactor consisted of a 12 in. length of 3 in. diameter stainless steel pipe inserted into an insulated electrical furnace. The solid feed material was introduced into the reactor from the feed hopper through a 1 in. diameter stainless steel pipe as indicated by the schematic.

As pyrolysis began, gases and water vapor passed into the off-gas side arm and were directed into the condenser system. Uncondensed gases passing from the condensing system were collected in 21-liter gas collection bottles. During the run, feeding was interrupted to stop gas evolution as each of the sampling bottles became full. Records of temperature, pressure, and volume of each sample bottle were utilized to calculate the volume of gas collected in each bottle. The results of this study are thoroughly discussed in section 7-4.3 concerning critical

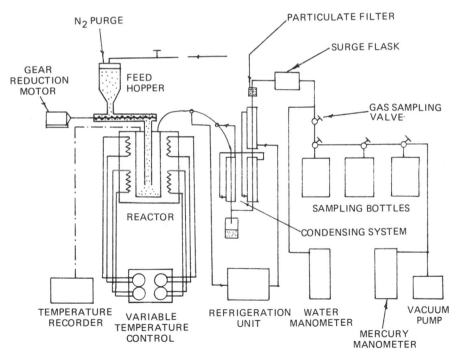

Fig. 7-18 NERC laboratory scale pyrolysis operation. *Source:* McFarland, J. M., et al., "Comprehensive Studies of Solid Waste Management," NERC Contract No. 2R01-EC-00260-01, SERL Report No. 72-3, EPA, May, 1972, p. 107.

parameters. Basically, these data correlated closely with the results of the Bureau of Mines study discussed above.

A 1973 study of solid waste pyrolysis of interest was conducted by Battelle Northwest under the direction of V. L. Hammond for the city of Kennewick, Washington (Ref. 7-12). The purpose of this investigation was to develop pilot plant data regarding the gasification process (pyrolysis with complete gasification, otherwise called "pyrolysis-incineration") which could be utilized in the design of a 100–200 ton per day capacity plant for the city of Kennewick.

In the gasification process, residue from the pyrolysis of solid waste as it progresses down through the reactor is reduced with an air-steam mixture to produce a low BTU content gas. This same process has been utilized by the coal industry to convert coal to natural gas. A schematic of the gasification pilot plant is shown in Fig. 7-19. Conclusions from the operation of this pilot facility may be summarized as follows:

1. Energy conversion by the process exceeded 80 percent.
2. Volume and weight reduction were on the order of those for incineration.

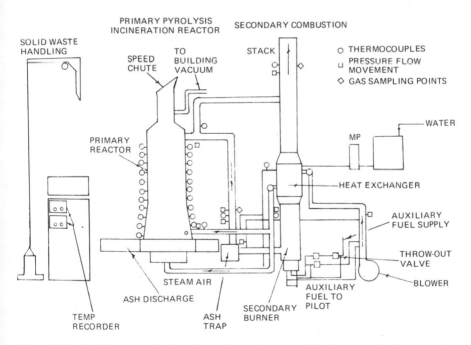

Fig. 7-19 Battelle pyrolysis pilot plant. *Source:* Hammond, V. L., et al., "Pyrolysis-Incineration Process for Solid Waste Disposal," Battelle Northwest, final report for the City of Kennewick, Washington, EPA 1-G06-EC-0032-1, December, 1972.

3. Gasification plants of capacity greater than 100 tons per day would be economically competitive with other methods of solid waste disposal.
4. The gas fuel from the process could be cleanly burned to produce steam or generate electricity.
5. Steam production appeared to be the most economical application of energy produced by gasification.

Further results of this study will be presented in subsequent sections discussing the Austin Method, which was developed from this pilot plant operation.

7-4.2 Methodology

The application of pyrolysis to solid waste disposal is a relatively new development. None of the pyrolysis methods discussed in this section have been operated on a full-scale basis; however, contracts for municipal application of at least three pyrolysis methods have been signed.

The five methods listed herein can be divided into two categories: high-temperature (slagging), and conventional pyrolysis. Conventional pyrolysis systems can be further divided into gas-producing and liquid fuel-producing systems.

The two high-temperature systems are actually little more than high-temperature incinerators. The difference lies in the fact that, in pyrolysis, the product gases are burned separately from the combustible waste in an afterburner; in incineration, these gases are consumed along with the waste. Therefore, pyrolysis generally provides for a more efficient recovery of heat produced through gas combustion. The two types of high-temperature pyrolysis systems include the Torrax system and the Urban Research and Development Corporation (URDC) system.

The Torrax system was developed by Torrax, Inc. to decompose solid waste into an inert residue by completely consuming combustibles and melting noncombustibles at high temperatures. Torrax, Inc. has operated a 75 ton per day pilot plant in Erie County, New York. A simple schematic of this system is shown in Fig. 7-20.

As-received refuse is charged into a vertical reactor through a seal. The particle size of the waste material is limited only by the capacity of the reactor. As the refuse descends through the reactor, it is dried, volatized, oxidized, and finally melted. Heat for the reaction is provided by a super blast heater.

Within the heater, ambient air is introduced into silicon carbide tubes where it is heated to approximately 2000°F by hot combustion gases surrounding the tubes. This heated, oxygen-rich air is then piped through tuyeres to the base of the vertical gasifier where it oxidizes the carbon in the pyrolysis residue to produce temperatures reaching 3000°F. This heat acts to produce a molten slag in the bottom of the reactor and to pyrolyze refuse charged into the top of the reactor. The molten material is collected in a quench tank to form a glasslike aggregate.

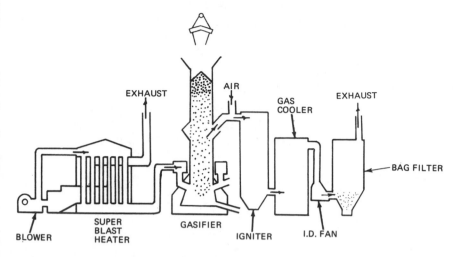

Fig. 7-20   Torrax Plant schematic diagram. *Source:* Stump, P. L., "Solid Waste Demonstration Projects," U. S. Environmental Protection Agency Publication SW-4p, 1972.

Gases produced by the pyrolysis of solid waste within the reactor are removed under negative pressure. In the process of reaching the exit vents at the top of the reactor, these hot gases pass from the pyrolysis section through the newly charged refuse column. This acts to dry the refuse and filter the gases. From the reactor, the gases pass to the secondary combustion chamber where complete combustion occurs, producing temperatures in excess of 2000°F. The hot gases from the afterburner may be utilized to produce steam or to continue the process. In the initial Torrax pilot plant, however, there was no provision for heat utilization and the hot gases were simply cooled, filtered, and vented to the atmosphere.

The URDC process is basically the same as the Torrax system. However, the URDC system was designed for self-sustained operation through the recycling of hot combustion gases. A schematic of the URDC process is shown in Fig. 7-21.

There are two conventional pyrolysis processes designed to optimize gas production: the *Landgard* and Austin methods.

The *Landgard* solid waste disposal system was developed by the Monsanto Research Corporation. The *Landgard* process consists of heating shredded waste in an oxygen-deficient environment at temperatures high enough to pyrolyze the combustibles into gases and solid residue. The gases are then reignited to provide heat to continue the process and steam to be marketed. Monsanto operated a 35 ton per day pilot plant in St. Louis from the fall of 1969 to the fall of 1971, and is presently (1974) constructing a 1000 ton per day plant in Baltimore, Maryland (see Fig. 7-22).

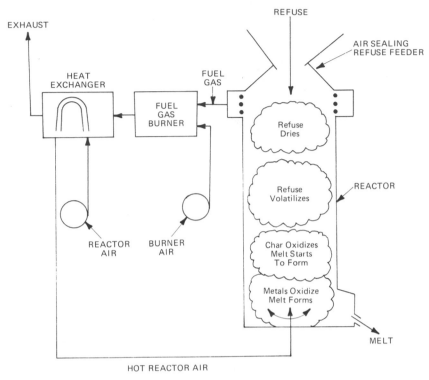

Fig. 7-21   URDC thermal processing system (approximate schematic).

The *Landgard* process begins by shredding the as-received refuse and storing it in a hopper from which it is metered into a refractory-lined rotary kiln (see Fig. 7-23). Direct-fire fuel is fed into the opposite end of the kiln. As the solid waste moves toward the burning fuel, it is exposed to increasingly high temperatures. The counterflow of hot gases allows for drying of the solid waste before "combustion" occurs. The kiln is designed to rotate every 5 to 20 minutes, depending upon the moisture content of the waste material.

When the modified pyrolysis process is completed, the hot residue is discharged to a quench tank from which it enters a flotation separator. After separation, the lighter residues are collected, dewatered, and landfilled. The metallics are separated from the heavier material which can then be sold as glasslike aggregate. The solid portion of the pyrolyzed refuse represents a reduction of about 75 percent in weight and 95 percent in volume from the original waste material (with metallics removed).

The gases generated by the process exit the kiln and are burned with air in a gas purifier. The high temperatures in the purifier destroy odors and consume combustible gases. Hot gases from the purifier pass through water-tube boilers to

Fig. 7-22   Artist's rendering of the *Landgard* 1000 ton per day pyrolysis plant in Baltimore, Maryland. *Source:* Buss, T. F., "The *Landgard* System for Resource Recovery and Solid Waste Disposal," presented at the Third Annual Environmental Engineering and Science Conference, University of Louisville, Louisville, Kentucky, March, 1973.

produce steam. From the boiler, the gases flow through a water spray scrubbing tower before entering an induced draft fan and being discharged to the atmosphere. The gases may be dehumidified and mixed with ambient air to eliminate plume formation if desired.

Another conventional pyrolysis process that emphasizes gas production is the Austin method. The Austin method was developed by the Austin Company through data obtained from the operation of the Battelle Northwest gasification pilot plant in Kennewick, Washington. A schematic of the Austin method is given in Fig. 7-24. Presently (1974), the Austin Company is under contract with the city of Kennewick to construct a 150 ton per day plant for solid waste disposal in that city.

In the Austin system, solid waste is shredded to < 4 in. size, passed through a magnetic separator, and then stored. From storage, the shredded waste is received by a rotary air lock which injects it into the reactor shown in Fig. 7-25. As the refuse enters the top of the reactor, it passes through the drying zone, the pyrolysis zone, and then settles in the ash bed. A multiscrew live bottom feeder removes the ash from the reactor and passes it to a crusher for reduction of clinkers. After crushing, the ash passes through another rotary air lock to a pan conveyor after which it is loaded onto trucks and is landfilled.

The steam and air mixture injected into the bottom of the reactor, at approximately 750°C, reacts with the char residue to produce the heat required to pyro-

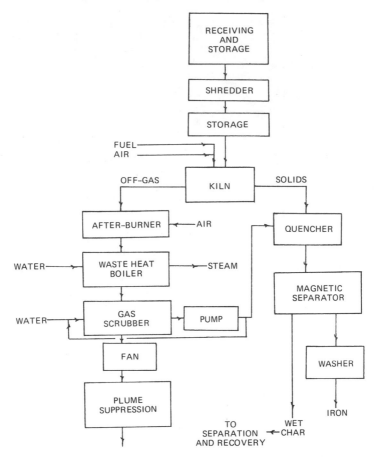

Fig. 7-23 Landgard Baltimore pyrolysis plant flow diagram. *Source:* Buss, T. F., "The Landgard System for Resource Recovery and Solid Waste Disposal," presented at the Third Annual Environmental Engineering and Science Conference, University of Louisville, Louisville, Kentucky, March, 1973

lyze incoming refuse. Half of the steam decomposes during this reaction while the other half passes through the reactor and heats the solid waste stream. Excess steam is then collected with the pyrolysis gases to be cooled and filtered. Water recovered from the condensers is piped to wastewater treatment facilities. The majority of the gas from the reactor is compressed and piped to a turbine for power generation. Steam to continue the process is produced in a waste heat boiler using the remaining pyrolysis gases and the heat from the turbine exhaust, as shown in the schematic (Fig. 7-24).

The final method to be discussed utilizes conventional pyrolysis to produce

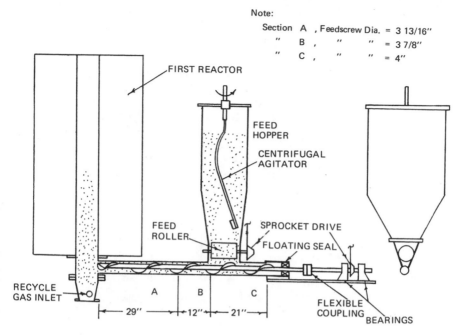

Note:

Section A , Feedscrew Dia. = 3 13/16"
" B ,   "   "   = 3 7/8"
" C ,   "   "   = 4"

FIRST REACTOR

FEED HOPPER

CENTRIFUGAL AGITATOR

FEED ROLLER

SPROCKET DRIVE

FLOATING SEAL

RECYCLE GAS INLET

A   B   C

FLEXIBLE COUPLING

BEARINGS

29"   12"   21"

Fig. 7-24 NERC pilot plant feed system. *Source:* McFarland, J. M., et al., "Comprehensive Studies of Solid Waste Management," NERC Contract No. 2R01-EC-00260-01, SERL Report No. 72–3, EPA, May, 1972, p. 107

liquid fuel oil. This method was developed by the Garrett Research and Development Company (GRDC), a subsidiary of the Occidental Petroleum Corporation.

The Garrett system employs a flash pyrolysis process designed to recover one barrel of synthetic fuel oil per ton of "as-received" refuse. This fuel is a low-sulfur oil with a heating value equivalent to 75 percent of No. 6 oil on a volumetric basis. Glass is also recovered by a froth-flotation method capable of 70 percent recovery at 99.7 percent purity. Although a full-scale plant utilizing the Garrett process has not been constructed, data provided by a 4 ton per day pilot plant were used by GRDC to project operating parameters for a 2000 ton per day plant.

A simplified flow diagram for the Garrett process is given in Fig. 7-27. For purposes of discussion, the process can be separated into two operations: feed preparation and pyrolysis of refuse.

Feed preparation is an integral part of the Garrett system. The Garrett preparation process is designed to provide an optimum material for fuel production through flash pyrolysis, while allowing for maximum recovery of recycleable

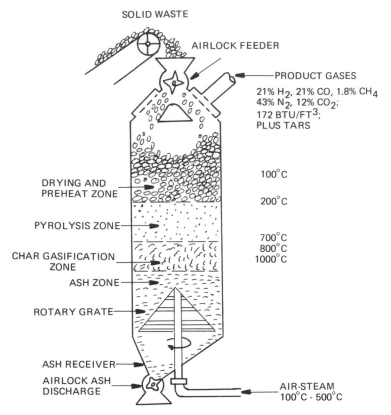

SOLID WASTE

AIRLOCK FEEDER

PRODUCT GASES

21% H_2, 21% CO, 1.8% CH_4
43% N_2, 12% CO_2;
172 BTU/FT^3;
PLUS TARS

DRYING AND PREHEAT ZONE

PYROLYSIS ZONE

CHAR GASIFICATION ZONE

ASH ZONE

ROTARY GRATE

ASH RECEIVER

AIRLOCK ASH DISCHARGE

$100°C$

$200°C$

$700°C$
$800°C$
$1000°C$

AIR-STEAM
$100°C$ - $500°C$

Fig. 7-25 Battelle gasification reactor. *Source:* Hammond, V. L., et al., "Pyrolysis-Incineration Process for Solid Waste Disposal," Battelle Northwest, final report for the city of Kennewick, Washington, EPA 1-G06-EC-0032-1, December, 1972.

glass and metals. The preparation incorporates the following operations: primary shredding to < 2 in. size; air-classification to remove most inorganics; drying to 3 percent moisture; screening to reduce inorganic content below 4 percent; recovery of glass and metal; and secondary shredding to < 24 mesh size. With the exception of glass recovery which is accomplished by a proprietary process developed by the GRDC (see Fig. 7-28), these operations may be performed using standard methods discussed elsewhere in this text.

Pyrolysis by the Garrett process is accomplished by rapidly heating the finely-shredded organic refuse to produce fuel oil, gas, and solid residue. Because of the proprietary nature of the system, details of the pyrolysis reactor are not available at this time.

In summary, several approaches have been applied to the pyrolysis of solid

Fig. 7-26 Garrett pilot scale pyrolysis reactor (*courtesy* Garrett Research and Development Company, Inc.).

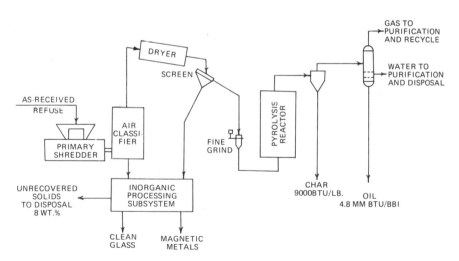

Fig. 7-27 Garrett pyrolysis process flow diagram. *Source:* Mallan, G. M. and Finney, C. S., "New Techniques in the Pyrolysis of Solid Wastes," presented at the 73rd National Meeting AIChE, Minneapolis, Minnesota, August 27–30, 1972.

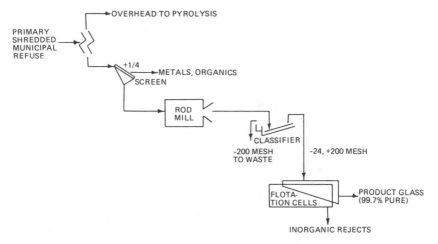

Fig. 7-28 Garrett glass recovery process. *Source:* Mallan, G. M. and Finney, C. S., "New Techniques in the Pyrolysis of Solid Wastes," presented at the 73rd National Meeting AIChE, Minneapolis, Minnesota, August 27–30, 1972.

waste. One approach, slagging, simply utilizes pyrolysis as a volume reduction process, similar to high-temperature incineration. This approach results in a minimum use of the waste as a "resource," and although it does an excellent job of volume reduction, its advantages over high-temperature incineration are questionable. The second approach, conventional pyrolysis, involves the utilization of the carbon content of solid waste to recover energy in the form of heat, oil, gas, and coal. Pyrolysis applied in the latter manner represents an efficient method by which solid waste is not only disposed but utilized as a resource.

In conventional pyrolysis, there are also differences in the approach taken. Most systems produce gas in a high-temperature, complete gasification process; other systems maximize oil production. In either approach, however, the value of the solid waste as a resource may be further increased if front-end separation is practiced to recover glass, plastics, metals, etc. Although some systems incorporate back-end separation, this eliminates the recovery of combustible items and increases the amount of bulk material present in the reactor, thereby reducing the efficiency of the entire system. The most efficient use of pyrolysis as a solid waste disposal system, therefore, would include a conventional system which maximizes energy recovery from the refuse and employs front-end separation to recover valuable resources from the waste material.

7-4.3 Critical Parameters

Laboratory studies and pilot plant operations have indicated that there are three basic critical parameters of solid waste pyrolysis: heating rate, temperature, and waste composition.

The effects of various heating rates were reported by Kaiser and Friedman (Ref. 7-16). The apparatus shown in Fig. 7-15 was charged with 4.37 to 5.20 grams of shredded newspaper, and the retort heated at various rates to 1500°F. Percentage changes in CO_2, CO, H_2, and CH_4, with changes in heating rate are given in Table 7-10; volume changes are shown in Fig. 7-29. Percentage changes for yields of gas, water, organic liquids, and char are presented in Table 7-11, and shown graphically in Fig. 7-30. These data show that gas yield was highest at the high and low rates of heating, char and water decreased with an increase in heating rate, and the heat of combustion of the gases was highest at the fastest rate of heating.

The effects of temperature on the pyrolysis process were studied by Hoffman and Fitz (Ref. 7-13) and McFarland, et al. (Ref. 7-18). Hoffman and Fitz pyrolyzed solid waste at 900, 1200, 1500, and 1700°F. Percentage yields of gases, organic liquids, and char at these temperatures are given in Table 7-12. Pyrolyzing temperatures were found to influence the quantity and quality of the gases produced (see Table 7-13) and their composition (see Table 7-14). Variation in pyrolyzing temperatures also affected the char composition (see Table 7-15). The difference in the pyrolysis products at various temperatures also affects the self-sustaining capacity of the system. Higher temperatures require more energy to sustain the process and produce by-products which contain less energy than those produced by low-temperature pyrolysis (see Fig. 7-31).

A detailed explanation of the effect of temperature on pyrolysis by-products was presented by McFarland, et al. (Ref. 7-18) in conjunction with the laboratory study they performed for the National Environmental Research Corporation. The temperature range studied in this investigation was from 500 to 1000°C (932 to 1832°F).

As in the previous studies, temperature was shown to influence the quantitative distribution of pyrolysis by-products and their composition. The product distribution and mass and energy balance for the six temperatures considered, cal-

TABLE 7-10   Effect of Pyrolysis Heating Rate on Gas Composition

Minutes to 1500°F	Volume Percent of Gas, Dry Basis								
	1	6	10	21	30	40	60	60	71
CO_2	15.01	19.16	23.11	25.1	24.7	25.7	22.9	21.2	24.01
CO	42.60	39.59	35.20	36.3	31.3	30.4	30.1	29.5	26.87
O_2	0.92	1.61	1.80	2.5	2.3	2.1	1.3	1.1	1.91
H_2	17.93	9.85	12.15	10.0	15.0	13.7	15.9	22.0	14.48
CH_4	17.54	21.70	19.95	20.1	20.1	19.9	21.5	20.8	22.18
N_2	6.00	8.09	7.79	6.0	6.6	8.2	8.3	5.4	10.55
BTU/cu ft	372	380	355	354	354	344	367	378	358
Gas vol., cu ft/ton	11,000	7500	6800	6170	6750	6560	7260	9170	8760

Source: Kaiser, E. R. and Friedman, S. B., "The Pyrolysis of Refuse Components," *Combustion*, May 1968, p. 31.

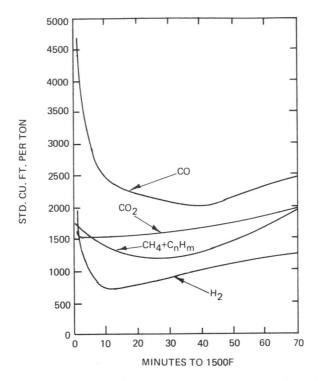

Fig. 7-29 Effect of pyrolysis heating rate on gas composition—Kaiser and Friedman study. *Source:* Kaiser, E. R. and Friedman, S. B., "The Pyrolysis of Refuse Components," *Combustion*, May, 1968, p. 31.

TABLE 7-11   Effect of Pyrolysis Heating Rate on By-Product Composition

Minutes to 1500°F	1	6	10	21	30	40	60	60	71
Gas	36.25	27.11	24.80	23.49	24.30	24.15	25.26	29.85	31.10
Water	24.08	27.35	27.41	28.23	27.93	27.13	33.23	30.73	28.28
CmHmOx liq.	19.14	25.55	25.70	26.23	24.48	24.75	12.00	9.93	10.67
Char C + S	19.10	18.56	20.66	20.63	21.86	22.54	28.08	28.06	28.52
Ash	1.43	1.43	1.43	1.43	1.43	1.43	1.43	1.43	1.43
BTU in cold gas per lb	–	–	–	–	–	–	–	–	–
Newspaper	2,045	1,425	1,207	1,092	1,192	1,129	1,330	1,730	1,568

Source: Kaiser, E. R. and Friedman, S. B., "The Pyrolysis of Refuse Components," *Combustion*, May 1968, p. 31.

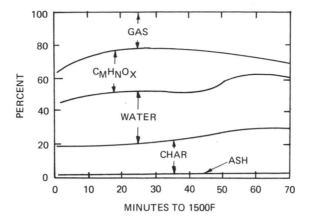

Fig. 7-30   Effect of pyrolysis heating rate on by-product composition—Kaiser and Friedman study. *Source:* Kaiser, E. R. and Friedman, S. B., "The Pyrolysis of Refuse Components," *Combustion*, May, 1968, p. 31.

TABLE 7-12   Effect of Pyrolysis Temperature on Product Yield

Temp °F	Gases	Pyroligneous Acids and Tars	Char	Mass Accounted For
900	12.33	61.08	24.71	98.12
1200	18.64	59.18	21.80	99.62
1500	23.69	59.67	17.24	100.59
1700	24.36	58.70	17.67	100.73

Source: Hoffman, D. A. and Fitz, R. A., "Batch Retort Pyrolysis of Solid Municipal Wastes," *Environmental Science and Technology,* Vol. 2, No. 11, November, 1968, p. 1023. Reprinted with permission from *Environmental Science and Technology.* Copyright by the American Chemical Society.

TABLE 7-13   Effect of Pyrolysis Temperature on Gas Yield per Pound of Refuse

Temp °F	Cu Ft	BTU per Cu Ft	BTU per Pound
900	1.90	300	569
1200	2.78	376	1045
1500	3.62	344	1245
1700	3.39	351	1190

Source: Hoffman, D. A. and Fitz, R. A., "Batch Retort Pyrolysis of Solid Municipal Wastes," *Environmental Science and Technology,* Vol. 2, No. 11, November, 1968, p. 1023. Reprinted with permission from *Environmental Science and Technology.* Copyright by the American Chemical Society.

TABLE 7-14   Effect of Pyrolysis Temperature on Gas Composition

Constituent	900°F	1200°F	1500°F	1700°F
H_2	5.56	16.58	28.55	32.48
CH_4	12.43	15.91	13.73	10.45
CO	33.50	30.49	34.12	35.25
CO_2	44.77	31.78	20.59	18.31
C_2H_4	0.45	2.18	2.24	2.43
C_2H_6	3.03	3.06	0.77	1.07
Accountability	99.74	100.00	100.00	99.99

Source: Hoffman, D. A. and Fitz, R. A., "Batch Retort Pyrolysis of Solid Municipal Wastes," *Environmental Science and Technology,* Vol. 2, No. 11, November, 1968, p. 1023. Reprinted with permission from *Environmental Science and Technology.* Copyright by the American Chemical Society.

culated on the basis of 100 grams total input, are given in Table 7-16. The effects of temperature on residue and organic condensate production, water consumption, total gas production, and gas composition are shown in Figs. 7-32 through 7-35. From this data, McFarland, et al. arrived at the following conclusions:

1. The carbonaceous residue decreased with increasing temperature due to the conversion of carbon to combustible gases at high temperatures by the reactions:

$$C + H_2O(g) \longrightarrow CO + H_2$$

$$C + CO_2 \longrightarrow 2CO$$

2. The amount of water present decreased with an increase in temperature due to the methane formation reaction:

$$CH_4 + H_2O(g) \longrightarrow CO + 3H_2$$

and to a lesser degree the "shift" reaction

$$CO + H_2O(g) \longrightarrow CO_2 + H_2$$

TABLE 7-15   Effect of Pyrolysis Temperature on the Proximate Analysis of Char

	Temperature, °F				Pennsylvania Anthracite
	900	1200	1500	1700	
Volatile matter, %	21.81	15.05	8.13	8.30	7.66
Fixed carbon, %	70.48	70.67	79.05	77.23	82.02
Ash, %	7.71	14.28	12.82	14.47	10.32
BTU per lb	12,120	12,280	11,540	11,400	13,880

Source: Hoffman, D. A. and Fitz, R. A., "Batch Retort Pyrolysis of Solid Municipal Wastes," *Environmental Science and Technology,* Vol. 2, No. 11, November, 1968, p. 1023. Reprinted with permission from *Environmental Science and Technology.* Copyright by the American Chemical Society.

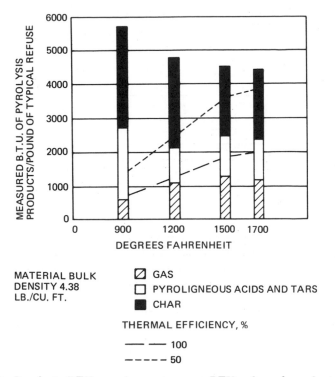

MATERIAL BULK
DENSITY 4.38
LB./CU. FT.

▨ GAS
☐ PYROLIGNEOUS ACIDS AND TARS
■ CHAR

THERMAL EFFICIENCY, %

— — 100
– – – – 50

Fig. 7-31 Pyrolysis BTU requirement versus BTU value of products. *Source:* Hoffman, D. A. and Fitz, R. A., "Batch Retort Pyrolysis of Solid Municipal Wastes," *Environmental Science and Technology*, Vol. 2, No. 11, November, 1968, p. 1023. Reprinted with permission from *Environmental Science and Technology*. Copyright by the American Chemical Society.

Some water was produced at lower pyrolysis temperatures by the dehydration reaction.

3. Organic condensate decreased with an increase in temperature due to cracking reactions which result in the formation of organic products of lower molecular weights. At higher temperatures, these products were converted to hydrogen, carbon monoxide, and methane. Except for benzene, organic products were absent from the uncondensed gas phases between 600 and 1000°C (1112 to 1832°F).

4. Gas production increased with temperature. Peak ethane, ethylene, and methane production occurred at 650, 800, and 900°C (1202, 1472, and 1652°F) respectively. Carbon dioxide production increased at a slow rate and equilibrated at 900°C; however, above 800°C (1472°F) a conversion of carbon dioxide to carbon monoxide occurred due to the high concentration

TABLE 7-16 Mass and Energy Balance for Laboratory Pyrolysis of Municipal Solid Waste

	500°C		600°C		700°C		800°C		900°C		1000°C	
	g	kcal	g	kcal	g	kcal	g	kcal	g	kcal	g	kcal
Input												
OD[a] solid waste	100.0	210	100.0	210	100.0	210	100.0	210	100.0	210	100.0	210
H_2O	39.7	0.0	38.2	0.0	45.2	0.0	32.8	0.0	45.3	0.0	52.9	0.0
Total input	139.7	210	138.2	210	145.2	210	132.8	210	145.3	210	152.9	210
Output												
H_2	0.12	4.1	0.50	17.1	0.94	32.1	1.24	42.4	2.74	93.4	2.96	101
CO	2.52	6.1	6.45	15.6	9.53	23.0	14.6	35.3	25.8	62.2	29.4	71.1
CO_2	11.4	0.0	14.5	0.0	20.2	0.0	20.7	0.0	24.2	0.0	24.2	0.0
CH_4	0.48	6.4	2.24	29.9	3.68	49.1	4.32	57.8	4.96	66.1	4.16	55.6
C_2H_4	0.28	3.4	1.40	16.8	2.52	30.2	3.08	37.0	1.68	20.2	0.28	3.4
C_2H_6	0.30	3.7	0.60	7.4	0.60	7.4	0.30	3.7	0.0	0.0	0.0	0.0
Total gases	15.1	23.7	25.7	86.8	37.5	141.8	44.2	176.2	59.4	241.9	61.0	231.1
Carbonaceous residue, ash free	14.0	110	11.5	90.0	12.9	101	9.1	71.2	5.7	44.5	3.6	28.3
Ash[b]	38.6	0.0	41.3	0.0	35.1	0.0	33.2	0.0	33.6	0.0	41.7	0.0
Organic condensate	6.3	73.4	5.0	58.3	3.2	37.4	1.9	22.2	2.0	23.3	2.0	23.3
H_2O in condensate	41.3	0.0	33.9	0.0	39.5	0.0	26.0	0.0	29.5	0.0	21.4	0.0
H_2O in gases	0.4	0.0	0.7	0.0	1.1	0.0	1.5	0.0	2.1	0.0	2.2	0.0
Total output	115.7	207	118.1	235	129.3	280	115.9	270	132.3	310	131.9	283
Recovery	82.9%	98.5%	85.4%	111.8%	89.0%	133.2%	87.3%	128.4%	91.1%	147.3%	86.2%	134.8%

[a]Oven dried.
[b]Determined on the basis of each carbonaceous residue.

Source: McFarland, J. M., et al., "Comprehensive Studies of Solid Waste Management," NERC Contract No. 2R01-EC-00250-01, SERL Report No. 72-3, EPA, May, 1972, p. 107.

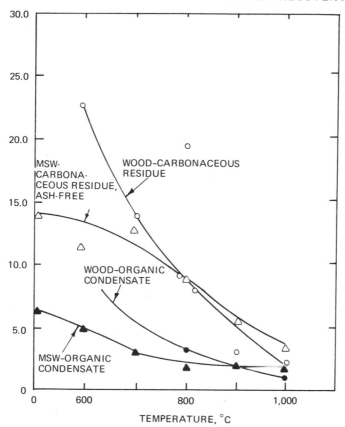

Fig. 7-32 Effect of pyrolysis temperature on carbonaceous residue and organic condensate. *Source:* McFarland, J. M., et al., "Comprehensive Studies of Solid Waste Management," NERC Contract No. 2R01-EC-00260-01, SERL Report No. 72-3, EPA, May, 1972, p. 107.

of hydrogen, low concentration of water, and negative temperature coefficient of the reaction:

$$CO + H_2O(g) \longrightarrow CO_2 + H_2$$

From the preceding discussion on temperature effects, an increase in pyrolysis temperature would be expected to:

1. Decrease ash content.
2. Decrease organic liquid content.
3. Decrease the total BTU content of the by-products.
4. Increase gas volume and heating value.
5. Increase water consumption.

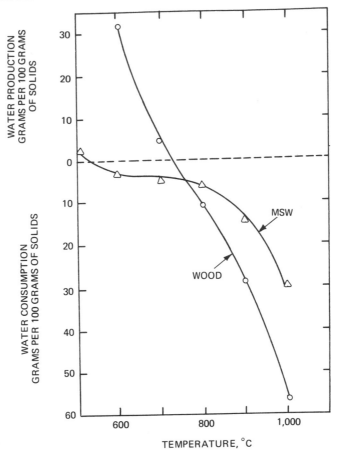

Fig. 7-33 Effect of pyrolysis temperature on water production and consumption. *Source:* McFarland, J. M., et al., "Comprehensive Studies of Solid Waste Management," NERC Contract No. 2R01-EC-00260-01, SERL Report No. 72–3, EPA, May, 1972, p. 107.

Waste composition and preparation methods also affect the composition of pyrolysis by-products. In the study conducted by W. S. Sanner (Ref. 7-27) for the Bureau of Mines, the yields of products from raw municipal waste, processed municipal waste containing plastic film, and industrial waste shredded by either a Heil mill or a Gondard mill (see Table 7-17) were reported for three pyrolysis temperatures (see Table 7-18). Chemical analyses were also given for the solid residue, gases, and tars produced in each situation (see Tables 7-19 through 7-21). Tar yields and a chromatographic analysis of light oils produced during pyrolysis

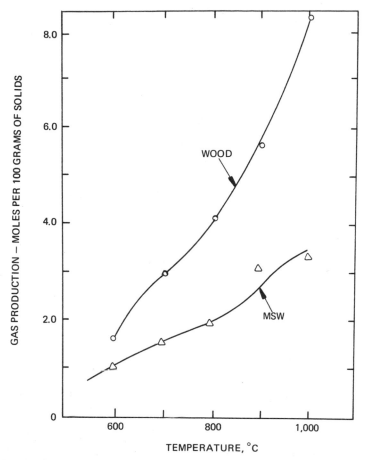

Fig. 7-34 Effect of pyrolysis temperature on gas production. *Source:* McFarland, J. M., et al., "Comprehensive Studies of Solid Waste Management," NERC Contract No. 2R01-EC-00260-01, SERL Report No. 72-3, EPA, May, 1972, p. 107.

are also given in Tables 7-22 and 7-23. The results from this study indicated that:

1. Raw municipal refuse pyrolyzed at 900°C (1652°F) underwent a 90 percent weight reduction, while industrial refuse was only reduced by 65 percent.
2. The residue from municipal refuse had the highest fuel value of the three materials studied.
3. In general, municipal refuse produced more ammonium sulfate, gas, tar, and liquor, and less solid residue and light oil than industrial waste.

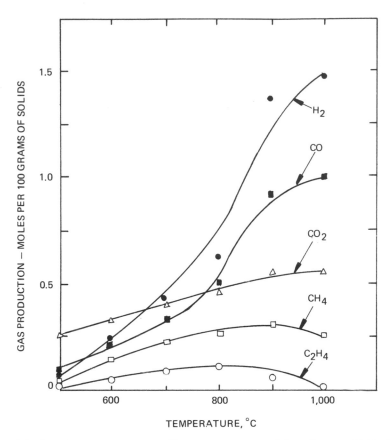

Fig. 7-35 Pyrolysis temperature effect on gas composition. *Source:* McFarland, J. M., et al., "Comprehensive Studies of Solid Waste Management," NERC Contract No. 2R01-ER-00260-01, DERL Report No. 72–3, EPA, May, 1972, p. 107.

7-4.4 By-Products

One of the major selling points of pyrolysis systems is their claim to pollution-free operation. Although the absence of full-scale pyrolysis operations has allowed this claim to stand uncontested, pilot plant operations indicate that undesirable by-products from the pyrolysis of solid waste can be kept to a minimum. It is even possible to eliminate plume formation.

Of the five systems discussed, only for the *Landgard* process has an analysis of stack gas been reported (see Table 7-24). However, since each of these systems employ a similar gas treatment process, those figures given for the *Landgard* process would probably be applicable to all of the pyrolysis operations discussed.

TABLE 7-17 Analysis of Refuse Pyrolyzed in Bureau of Mines Investigation

	Raw Municipal Refuse		Processed Municipal Refuse Containing Plastic Film		Heil Mill Industrial Refuse		Gondard Mill Industrial Refuse	
	As Received	Dry	As Received	Dry	As Received	Dry	As Received	Dry
Proximate, percent:								
Moisture	43.3	–	42.3	–	20.5	–	33.2	–
Volatile matter	43.0	76.3	44.3	76.8	40.3	49.7	33.5	50.6
Fixed carbon	6.7	11.7	5.6	9.7	9.9	12.3	4.6	7.0
Ash	7.0	12.0	7.8	13.5	29.3	38.0	28.7	42.4
Total	100.0	100.0	100.0	100.0	100.0	100.0	100.0	100.0
Ultimate, percent								
Hydrogen	8.2	6.0	7.6	5.0	6.0	4.6	6.2	3.8
Carbon	27.2	47.6	27.2	47.3	27.5	33.9	20.6	31.2
Nitrogen	.7	1.2	.8	1.4	.5	.7	.5	.7
Oxygen	56.8	32.9	56.5	32.6	36.4	22.4	43.9	21.8
Sulfur	.1	.3	.1	.2	.3	.4	.1	.1
Ash	7.0	12.0	7.8	13.5	29.3	38.0	28.7	42.4
Total	100.0	100.0	100.0	100.0	100.0	100.0	100.0	100.0
BTU per pound of refuse	4,827	8,546	5,310	9,180	4,570	5,645	3,415	5,155
Available BTU per ton of refuse, millions	9.654	17.092	10.620	18.360	9.140	11.290	6.830	10.310

Source: Sanner, W. S., Ortuglio, C., Walters, J. G. and Wolfon, D. E., "Conversion of Municipal and Industrial Refuse into Useful Materials by Pyrolysis," U.S. Bureau of Mines, Report of Inv. No. 7428, August, 1970.

TABLE 7-18 Effect of Pyrolysis Temperature on Product Yields in Bureau of Mines Investigation

Refuse	Pyrolysis Temp, °C	Yields, Weight-Percent of Refuse							Yields, per Ton of Refuse				
		Residue	Gas	Tar	Light Oil in Gas	Free Ammonia	Liquor	Total	Gas, Cubic Feet	Tar, Gallons	Light Oil in Gas, Gallons	Liquor, Gallons	Ammonium Sulfate, Pounds
Raw municipal	500–900	9.3	26.7	2.2	0.5	0.05	55.8	94.6	11,509	4.8	1.5	133.4	17.9
	750	11.5	23.7	1.2	.9	.03	55.0	92.3	9,628	2.6	2.5	131.6	23.7
	900	7.7	39.5	.2	–	.03	47.8	95.2	17,741	.5	–	113.9	25.1
Processed municipal containing plastic film	500–900	21.2	27.7	2.3	1.3	.05	40.6	93.2	11,545	5.6	3.7	96.7	16.2
	750	19.5	18.3	1.0	.9	.02	51.5	91.2	7,380	2.2	2.6	122.6	28.4
	900	19.1	40.1	.6	.2	.04	35.3	95.3	18,058	1.4	.6	97.4	31.5
Heil mill industrial	500–900	36.1	23.7	1.9	.5	.05	31.6	93.9	9,563	4.1	1.4	75.2	12.5
	750	37.5	22.8	.7	.9	.03	30.6	92.5	9,760	1.5	2.6	73.0	19.5
	900	38.8	29.4	.2	.6	.04	21.8	90.8	12,318	.5	1.6	51.1	21.7
Gondard mill industrial	500–900	41.9	21.8	.8	.6	.03	29.5	94.6	9,270	1.7	1.6	70.2	20.4
	750	31.4	25.5	.8	.8	.03	31.5	90.0	10,952	1.8	2.2	74.9	21.2
	900	30.9	31.5	.1	.5	.03	29.0	92.0	14,065	.02	1.4	68.5	22.9

Source: Sanner, W. S., Ortuglio, C., Walters, J. G. and Wolfon, D. E., "Conversion of Municipal and Industrial Refuse into Useful Materials by Pyrolysis," U.S. Bureau of Mines, Report of Inv. No. 7428, August, 1970.

TABLE 7-19 Effect of Pyrolysis Temperature on Solid Residue Composition in Bureau of Mines Investigation

| Refuse | Pyrolysis Temp, °C | Proximate, Percent | | | | Ultimate, Percent | | | | | Heating Value, BTU/lb | Heating Value, Million BTU/Ton |
		Moisture	Volatile Matter	Fixed Carbon	Ash	Hydrogen	Carbon	Nitrogen	Oxygen	Sulfur		
Raw municipal	500–900	2.6	4.4	29.6	66.0	0.4	32.4	0.5	0.5	0.2	5,020	10.040
	750	2.2	7.4	51.4	41.2	.8	54.9	1.1	1.8	.2	8,020	16.040
	900	1.0	4.7	31.7	63.6	.3	36.1	.5	.0	.2	5,260	10.520
Processed municipal containing plastic film	500–900	1.7	4.8	56.7	38.5	.6	57.7	.8	2.1	.3	8,800	17.700
	750	1.3	13.4	34.6	52.0	.8	41.9	.8	4.4	.1	6,080	12.160
	900	1.2	3.3	53.5	43.2	.5	53.4	.7	1.8	.4	8,090	16.180
Heil mill industrial	500–900	.9	2.6	15.2	82.2	.3	17.0	.1	.2	.2	2,520	5.040
	750	1.2	5.1	17.0	77.9	.5	19.4	.2	1.8	.2	2,900	5.800
	900	.1	2.5	12.9	84.6	.3	14.8	.2	.0	.2	2,180	4.360
Gondard mill industrial	500–900	.3	3.0	9.7	87.3	.2	11.8	.1	.4	.2	1,660	3.320
	750	1.0	3.6	16.6	79.8	.3	19.5	.2	.0	.2	2,680	5.360
	900	.2	6.4	16.2	77.4	.4	19.3	.3	2.4	.2	2,810	5.620

[a]Moisture on as-received basis; all other data on dry basis.

Source: Sanner, W. S., Ortuglio, C., Walters, J. G. and Wolfon, D. E., "Conversion of Municipal and Industrial Refuse into Useful Materials by Pyrolysis," U.S. Bureau of Mines, Report of Inv. No. 7428, August, 1970.

TABLE 7-20 Effect of Pyrolysis Temperature on Gas Composition in Bureau of Mines Investigation

Pyrolysis Temp. °C	Raw Municipal Refuse			Processed Municipal Refuse Containing Plastic Film			Heil Mill Industrial Refuse			Gondard Mill Industrial Refuse		
	500–900	750	900	500–900	750	900	500–900	750	900	500–900	750	900
Analysis, vol percent												
Hydrogen	45.47	30.86	51.91	44.86	25.27	42.41	45.50	47.89	49.12	47.45	49.63	51.14
Carbon monoxide	21.54	15.57	18.16	19.62	25.09	20.16	20.67	13.04	19.39	20.70	12.14	18.19
Methane	13.15	22.57	12.66	18.73	17.57	13.92	16.68	20.27	15.94	12.59	17.09	12.61
Ethane	1.30	2.05	0.14	2.08	2.01	0.25	1.27	1.50	0.37	1.49	1.25	0.32
Ethylene	4.67	7.56	4.68	4.54	10.36	7.89	2.77	3.97	3.86	3.39	5.30	4.42
Carbon dioxide	11.41	18.44	11.42	8.02	18.25	13.91	10.51	11.99	10.25	10.24	12.93	11.71
Propane	<0.01	<0.01	<0.01	<0.01	<0.01	1.17	<0.01	<0.01	<0.01	<0.01	<0.01	<0.01
Propylene	1.32	1.53	0.32	1.35	0.76	0.10	1.38	0.86	0.34	2.35	0.89	0.62
Isobutane	Trace	Trace	Trace	Trace	Trace	Trace	Trace	Trace	Trace	Trace	Trace	Trace
Butane	0.08	0.01	0.44	0.03	do.	0.11	0.05	0.01	<0.01	0.13	0.02	<0.01
1-Butene	0.16	0.15	Trace	0.11	do.	Trace	0.15	0.04	Trace	0.16	0.05	0.17
Isobutylene	0.17	0.15	do.	0.15	do.	do.	0.23	0.07	do.	0.25	0.12	Trace
trans-2-Butene	0.04	0.03	do.	0.08	do.	do.	0.12	0.02	do.	0.02	0.02	<0.01
cis-2-Butene	0.07	Trace	do.	0.03	do.	do.	0.07	<0.01	do.	<0.01	0.01	<0.01
Pentane	Trace	do.	do.	0.16	do.	do.	Trace	Trace	do.	Trace	Trace	Trace
Pentenes	0.60	0.87	0.20	0.17	0.53	0.07	0.55	0.29	0.63	0.70	0.51	0.39
Unidentified	<0.01	0.21	0.06	0.06	0.15	0.01	0.04	0.03	0.08	<0.01	0.03	0.36
BTU/cu ft of gas	473	563	447	536	570	511	478	518	498	471	502	447
Million BTU/ton of refuse pyrolyzed	5.473	5.421	7.930	6.188	4.207	9.228	4.571	5.056	6.134	4.366	5.498	6.067

Note: <0.01 = 1 part in 10^4. Trace = less than 1 part in 10^5.

Source: Sanner, W. S., Ortuglio, C., Walters, J. G. and Wolfon, D. E., "Conversion of Municipal and Industrial Refuse into Useful Materials by Pyrolysis," U.S. Bureau of Mines, Report of Inv. No. 7428, August, 1970.

TABLE 7-21 Effect of Pyrolysis Temperature on Tar Analysis in Bureau of Mines Investigation

Refuse	Specific Gravity at 15.6°C	Weight-Percent of Dry Tar		Boiling Range, Volume-Percent					Distillate, Volume-Percent of Dry Tar			Neutral Tar Oil, Volume-Percent		
		Anthracene	Naphthalene	0 to 170°C	170 to 235°C	235 to 270°C	270 to 350°C	Residue	Acids	Bases	Neutral Oil	Olefins	Aromatics	Paraffins and Naphthenes
Pyrolysis Temperature, 500–900°C														
Raw municipal	1.077	0.0	0.0	4.4	12.4	7.2	20.0	56.0	7.4	4.7	31.9	17.5	68.1	14.4
Processed municipal containing plastic film	.974	.0	.0	1.7	9.1	7.0	19.3	62.9	5.2	3.7	28.0	25.0	62.6	12.4
Heil mill industrial	1.111	.0	.0	3.3	11.1	5.4	10.2	70.0	5.9	3.7	20.4	23.0	68.5	8.5
Gondard mill industrial	1.093	.0	.0	3.7	14.1	7.4	16.5	58.3	7.0	5.3	29.1	21.3	67.7	11.0
Pyrolysis Temperature, 750°C														
Raw municipal	1.115	0.59	3.17	1.9	11.1	4.1	8.1	74.8	4.0	2.1	17.5	16.3	79.9	3.8
Processed municipal containing plastic film	1.101	Trace	Trace	4.2	13.2	7.5	12.0	63.1	4.9	6.3	25.7	23.5	69.6	6.9
Gondard mill industrial	1.099	.73	4.07	6.2	26.8	8.4	14.3	44.3	6.8	6.0	38.3	19.0	78.2	2.8

Source: Sanner, W. S., Ortuglio, C., Walters, J. G. and Wolfon, D. E., "Conversion of Municipal and Industrial Refuse into Useful Materials by Pyrolysis," U.S. Bureau of Mines, Report of Inv. No. 7428, August, 1970.

TABLE 7-22 Effect of Pyrolysis Temperature on Tar Components Yields in Bureau of Mines Investigation

Refuse	Tar	Acids	Bases	Gallons per Ton of Refuse					Pounds per Ton of Refuse	
				Neutral Oil	Residue	Olefins	Aromatics	Paraffins and Naphthenes	Anthracenes	Naphthalenes
Pyrolysis Temperature, 500–900°C										
Raw municipal	4.8	0.4	0.2	1.5	2.7	0.3	1.1	0.2	—	—
Processed municipal containing plastic film	5.6	.3	.2	1.6	3.5	.4	1.0	.2	—	—
Heil mill industrial	4.1	.2	.2	.8	2.9	.2	.6	.1	—	—
Gondard mill industrial	1.7	.1	.1	.5	1.0	.1	.3	.1	—	—
Pyrolysis Temperature, 750°C										
Raw municipal	2.6	0.1	0.1	0.5	2.0	0.1	0.4	0.02	0.14	0.76
Processed municipal containing plastic film	2.2	.1	.1	.6	1.4	.1	.4	.04	Trace	Trace
Gondard mill industrial	1.8	.12	.1	.7	.8	.1	.5	.02	.12	.67

Source: Sanner, W. S., Ortuglio, C., Walters, J. G. and Wolfon, D. E., "Conversion of Municipal and Industrial Refuse into Useful Materials by Pyrolysis," U.S. Bureau of Mines, Report of Inv. No. 7428, August, 1970.

TABLE 7-23 Chromatographic Analysis of Light Oils from Pyrolytic Processes in Bureau of Mines Investigation

Pyrolysis Temp. °C	Raw Municipal Refuse			Processed Municipal Refuse Containing Plastic Film			Heil Mill Industrial Refuse			Gondard Mill Industrial Refuse		
	500-900	750	900	500-900	750	900	500-900	750	900	500-900	750	900
Pre-benzene	25.04	3.70	0.80	28.08	0.74	0.79	14.25	11.05	0.45	22.88	5.33	3.38
Benzene	37.54	78.47	73.39	57.35	85.18	92.10	60.89	70.63	88.29	51.07	76.25	89.00
Toluene	23.76	14.06	12.25	10.31	11.54	3.82	15.96	14.19	6.40	15.93	14.16	5.10
Ethylbenzene	2.50	.31	.02	1.38	.14	<.01	2.23	.42	<.01	2.38	.30	<.01
m-,p-Xylene	2.39	.65	2.84	.71	.47	.67	1.43	.87	.86	1.42	1.09	.57
o-Xylene	1.01	.20	.81	.36	.20	.16	.59	.40	.12	.68	.35	.10
Unidentified	7.76	2.61	9.89	1.81	1.73	2.45	4.65	2.44	2.88	5.64	2.52	1.84

Source: Sanner, W. S., Ortuglio, C., Walters, J. G. and Wolfon, D. E., "Conversion of Municipal and Industrial Refuse into Useful Materials by Pyrolysis," U.S. Bureau of Mines, Report of Inv. No. 7428, August, 1970.

TABLE 7-24 *Landgard* Stack Gas Analysis

Nitrogen	78.7%
Carbon dioxide	13.8%
Water vapor	1.8%
Oxygen	5.7%
Hydrocarbons	<10 ppm
Sulfur dioxide	<150 ppm
Nitrogen oxides	<65 ppm
Chlorides	<25 ppm
Particulates (NAPCA method)	<0.02 grains/scf dry gas corr. to 12% CO_2

Source: Buss, T. F., "The *Landgard* System for Resource Recovery and Solid Waste Disposal," presented at the Third Annual Environmental Engineering and Science Conference, University of Louisville, Louisville, Kentucky, March, 1973.

Certainly any system put into operation should be capable of meeting current EPA air pollution standards of 0.08 grains/scf at 12 percent CO_2.

The desirable and possibly profitable by-products of pyrolysis (beyond recoverable inorganics) include gases, liquids, and solid residues (see Table 7-25). The percentage of each and their composition is controlled by the pyrolyzing method employed and is a function of the temperature, retention time, and waste composition. Laboratory studies, discussed previously, identified the by-products of solid waste pyrolysis and reported upon the effects of changing the parameters mentioned above upon by-product composition.

Of the five operational systems studied, for only one, Garrett, has information regarding by-product composition been reported. Both the Torrax and URDC

TABLE 7-25 Potential Value of Pyrolysis By-Products in the Garrett Process

	Composition (As Received) Wt %[a]	Commodity Value $/Ton	Estimated Recovery %	Potential Value $/Ton of Waste
Organic products	50–60			
Pyrolytic oil		12–15	40	2.40–3.60
Pyrolytic char		4–40[b]	30	0.60–7.20
Pyrolytic gas		4–5	20	0.40–0.60
Total potential value				3.40–11.40

[a] Based upon 25% moisture content.
[b] Potential value as activated carbon.

Source: Mallan, G. M. and Finney, C. S., "New Techniques in the Pyrolysis of Solid Wastes," presented at the 73rd National Meeting AIChE, Minneapolis, Minnesota, August 27–30, 1972.

systems developers report only that the by-products from their processes include an inert solid residue and secondary combustion gases which have been cooled and filtered. *Landgard* designers report that their system yields the by-products represented in Table 7-26.

The Garrett Research and Development Company has provided extensive data on the composition of its process by-products. A summation of these data is given in Table 7-27. The most important by-product of the Garrett system is the pyrolytic oil. This oil compares favorably with No. 6 fuel oil as shown in Table 7-28. The major drawback of pyrolytic oil is its high viscosity; however, tests have shown that this oil can be easily pumped at $160°F$. This pyrolytic oil can also be satisfactorily atomized for combustion with 50 psi steam when 10 gallons per hour of oil is delivered to the burner tip at 25 psi and $240°F$. Therefore, this oil could be utilized with only minor adaptations in an oil burning facility.

Another drawback of pyrolytic oil is its mildly acidic character. However, preliminary studies indicate that pyrolytic oil produced in the Garrett process and No. 6 oil are compatible and may be blended. If blending is proven to be practical, both the viscosity and the acidity of the pyrolytic oil may be reduced through blending.

The Garrett process also produces charcoal. Preliminary studies indicate that this product may be upgraded to activated charcoal, thus making it a valuable

TABLE 7-26 *Landgard* **Pyrolysis Process Residue Analysis**

(Wt % Dry Basis)			
Proximate Analysis		Ultimate Analysis	
Volatiles	5.5	Metal (Fe)	21.9
Fixed carbon	12.5	Glass + ash	60.1
Inerts	82.0		
		Carbon	14.5
		Sulfur	0.1
		Hydrogen	0.5
		Nitrogen	0.2
		Oxygen	2.7
Higher heating value = 2500 BTU/lb			
pH = 12.0			
Water soluble solids	2%		
Putrescibles	<0.1% (E/C anal. method)		

Source: Buss, T. F., "The *Landgard* System for Resource Recovery and Solid Waste Disposal," presented at the Third Annual Environmental Engineering and Science Conference, University of Louisville, Louisville, Kentucky, March, 1973.

TABLE 7-27   By-Products from the Garrett Pyrolysis Process

Char Fraction, 20 Wt %, Heating Value 9000 BTU/lb		
	48.8 wt %	Carbon
	3.9	Hydrogen
	1.1	Nitrogen
	0.3	Sulfur
	31.8	Ash
	0.2	Chlorine
	13.9	Oxygen (by difference)
Oil Fraction, 40 Wt %, Heating Value 4.8 mm BTU/bbl (10,500 BTU/lb)		
	57.5 wt %	Carbon
	7.6	Hydrogen
	0.9	Nitrogen
	0.1	Sulfur
	0.2	Ash
	0.3	Chlorine
	33.4	Oxygen (by difference)
Gas Fraction, 27 Wt %, Heating Value 550 BTU/cu ft		
	0.1 mol %	Water
	42.0	Carbon monoxide
	27.0	Carbon dioxide
	10.5	Hydrogen
	<0.1	Methyl chloride
	5.9	Methane
	4.5	Ethane
	8.9	C_3 to C_7 hydrocarbons
Water Fraction, 13 Wt %		
	Contains:	Acetaldehyde
		Acetone
		Formic Acid
		Furfural
		Methanol
		Methylfurfural
		Phenol
		Etc.

Source: Mallan, G. M. and Finney, C. S., "New Techniques in the Pyrolysis of Solid Wastes," presented at the 73rd National Meeting AIChE, Minneapolis, Minnesota, August 27–30, 1972.

commodity. In its present state, however, it may still be safely landfilled or burned as a low sulfur fuel.

When applied solely as a high-temperature process for volume reduction, pyrolysis has few advantages over high-temperature incineration. However, its potential for pollution-free, self-sustained operation may give it an advantage over incineration in the future. Conventional pyrolysis systems do not maximize

TABLE 7-28   Comparison of No. 6 Fuel Oil and Pyrolytic Oil

	No. 6	Pyrolytic Oil
Carbon, wt %	85.7	57.5
Hydrogen	10.5	7.6
Sulfur	0.7–3.5	0.1–0.3
Chlorine	–	0.3
Ash	<0.5	0.2–0.4
Nitrogen	} 2.0	0.9
Oxygen		33.4
BTU/pound	18,200	10,500
Specific gravity	0.98	1.30
Lb/gallon	8.18	10.85
BTU/gallon	148,840	113,910
Pour point °F	65–85	90
Flash point °F	150	133
Viscosity SSU at 190°F	340	3,150
Pumping temperature °F	115	160
Atomization temperature °F	220	240

Source: Mallan, G. M. and Finney, C. S., "New Techniques in the Pyrolysis of Solid Wastes," presented at the 73rd National Meeting AIChE, Minneapolis, Minnesota, August 27–30, 1972.

volume reduction. They do convert useless waste into a valuable resource, and as the supply of petroleum, natural gas, and coal dwindles, the pyrolysis of solid waste may provide an important source of power. However, the success of pyrolysis currently depends upon the marketability of its by-products.

7-4.5   Method Requirements

The operation of a pyrolysis system requires materials handling equipment, fuel generating equipment, power generating equipment, and air pollution control equipment. The requirements in each of these areas depend upon the type of system (slagging, conventional, oil-producing, gas-producing) employed.

An indication of the types of equipment required for each of the pyrolysis systems discussed was given in section 7-4.2, and a complete breakdown of the equipment requirements for the Austin method is presented in Appendix 7-1. Although equipment requirements are peculiar to each method, some general statements can be made concerning method requirements for pyrolysis systems:

1. Typical refuse handling equipment is required, to some extent, for each system and includes storage hoppers with a metered feed system, shredders, and resource recovery equipment. The degree of processing required is a function of the amount of materials recovery practiced and the type of end product sought from the system.

2. The amount of power generating equipment required depends upon the degree of energy recovery practiced, and is therefore greater for conventional systems.

7-4.6 Economic Analysis

The lack of full-scale pyrolysis operations has subsequently resulted in a lack of factual economic data on the various pyrolysis systems. However, most manufacturers have estimated costs and revenues for their respective systems, and that information is presented here.

The economics of basic *Landgard* systems having capacities of 250, 500, or 1000 ton per day are given in Table 7-29. Similar figures are presented in Table 7-30 for a *Landgard* operation employing steam generation and residue separation. Because of Monsanto's contract to build a 1000 ton per day in Baltimore, actual cost data for a 1000 ton per day plant are available (see Table 7-31). The actual cost exceeds Monsanto's estimate by $0.48 per ton due mainly to higher manpower, water, and maintenance costs and the addition of carbon char removal.

Based upon pilot plant operation, the Garrett Research and Development Corporation reports a disposal cost of $5.40 per ton for a 2000 ton per day capacity plant (see Table 7-32). However, based upon total sale of by-products, net revenue per ton is estimated at $5.70 (see Table 7-33). Therefore, if total sale of by-products is realized, it would be possible to operate the Garrett system at a profit. Even without total marketability of by-products, the Garrett system promises a very low net per ton disposal cost.

As part of the study of gasification through pyrolysis for the City of Kennewick, Washington, Battelle personnel conducted an in-depth economic analysis for five modes of operation: 1) disposal only without shredding, 2) dis-

TABLE 7-29 *Landgard* **Economics for Basic Pyrolysis Plant**

	1000	500	250
Plant capacity—tpd	1000	500	250
Sale price—$/million	9.7	5.8	3.7
Operating cost—$/feed ton			
Fuel	.89	.89	.89
Electric power	.99	1.08	.80
Operating manpower	.99	1.65	2.75
Water	.11	.11	.11
Maintenance	1.48	1.76	2.24
Miscellaneous	.33	.39	.48
Total operating cost	4.79	5.88	7.27

Source: Buss, T. F., "The *Landgard* System for Resource Recovery and Solid Waste Disposal," presented at the Third Annual Environmental Engineering and Science Conference, University of Louisville, Louisville, Kentucky, March, 1973.

TABLE 7-30 *Landgard* **Economics for Pyrolysis Plant Utilizing Steam Generation and Residue Separation**

Plant capacity—tpd	1000	500	250
Sale price—$/million	12.8	7.9	5.2
Operating cost—$/feed ton			
Fuel	.89	.89	.89
Electric power	1.06	1.18	.93
Operating manpower	.99	1.65	2.75
Water	.12	.12	.12
Maintenance	1.77	2.20	2.80
Miscellaneous	.41	.47	.52
Total operating cost	5.24	6.51	8.01

Source: Buss, T. F., "The *Landgard* System for Resource Recovery and Solid Waste Disposal," presented at the Third Annual Environmental Engineering and Science Conference, University of Louisville, Louisville, Kentucky, March, 1973.

posal only with shredding, 3) disposal with fuel generation for a steam boiler or industrial heating, 4) disposal with fuel generation for a gas-turbine generator, and 5) disposal and fuel generation for a gas turbine with heat recovery to generate steam for a steam-turbine generator (combined cycle) (Ref. 7-12). This

TABLE 7-31 *Landgard* **Economics for 1000 TPD Baltimore Pyrolysis Plant**

Plant capacity	1000 tpd
Sale price	$14,742,000
Operating cost $/feed ton	
Fuel	$ .89
Electricity	1.06
Manpower	1.02
Water and chemicals	.31
Maintenance	1.84
Miscellaneous	.42
Carbon char removal	.18
Total operating cost	5.72
Revenue-resource recovery	4.67
Steam	3.89
Iron	.44
Glassy agg.	.34
	4.67
Net operating cost	$1.05

Source: Buss, T. F., "The *Landgard* System for Resource Recovery and Solid Waste Disposal," presented at the Third Annual Environmental Engineering and Science Conference, University of Louisville, Louisville, Kentucky, March, 1973.

TABLE 7-32   Estimated Garrett Economics—2000 TPD Pyrolysis Plant

2000 TPD Solid Waste Pyrolysis Plant	
Total plant capital (excludes land)	$14,000,000
Working capital	400,000
Annual op. cost (amortization, 25 yrs at 6%)	3,800,000
Annual net revenue from sales	4,000,000
Annual profit without disposal charges	200,000
Cost per ton (350 days per year)	$5.40
Net revenue per ton (350 days per year)	$5.70

Source: Mallan, G. M. and Finney, C. S., "New Techniques in the Pyrolysis of Solid Wastes," presented at the 73rd National Meeting AIChE, Minneapolis, Minnesota, August 27–30, 1972.

analysis represents the most comprehensive study of pyrolysis economics to date, and although it is designed to apply to the gasification process, its general principles could also be applied to other pyrolysis systems.

The capital cost estimates were computed for the five modes mentioned, eight tonnage capacities, amortization periods of 15 and 25 years, and cost increases of 25 percent to 35 percent for engineering and contingencies. As might be expected, the capital costs were lowest for the first operational mode and highest for the second. A tabulation of the capital costs of a 150 ton per day plant for the five different modes is shown in Table 7-34, and indicates the items included in all capital cost estimates.

Operational costs include the direct disposal cost and the cost of amortization.

TABLE 7-33   Estimated Revenue from Garrett Pyrolysis Process—
2000 TPD Plant

Recovered Commodities	Wt % of As-Received Refuse (Approx.)	Percent Available for Disposal	Commodity Value $/Ton	Estimated Revenue $/Ton Refuse
Magnetic metals	7	95	20.00	1.33
Glass, mixed-color	8	70	14.00	0.78
Oil at 4.8 mm BTU/bbl	24	95	15.35	3.50
Char at 9000 BTU/lb	15	34	4.50	0.23
Gas at 600 BTU/cu ft	6	0	4.00	0.00
Water	33	34	-0.25	-0.03
Inorganic wastes	7	85	-1.50	-0.09
Net revenue per ton				5.72

Source: Mallan, G. M. and Finney, C. S., "New Techniques in the Pyrolysis of Solid Wastes," presented at the 73rd National Meeting AIChE, Minneapolis, Minnesota, August 27–30, 1972.

Direct disposal costs include labor and utilities as indicated by Tables 7-35 and 7-36. Insurance, taxes, maintenance, and materials were taken to be 7.4 percent of the yearly amortization cost. Total capital and operating costs for the forty situations considered are presented in Tables 7-37 through 7-41.

Beyond the operating expenses, potential revenue from the sale of by-products was also investigated. Hammond reported that electrical energy production costs, and therefore revenue value, would approach 12 to 16 mills/kilowatt-hour by 1976 and natural gas would sell for $1/million BTU. Based upon an energy conversion efficiency of 80 percent and an average heating value of 5000 BTU/lb of solid waste, these prices would yield revenues of $8.00 per ton of refuse for the sale of gas, $4.43 per ton of refuse for the sale of electricity (gas turbine), and $5.90 per ton of refuse for the sale of electricity (combined cycle). The increase in revenue from the more complex energy recovery systems will generally be offset by the increase in capital costs. However, the maximum utilization of pyrolysis by-products also supplies a new energy resource, the value of which may be limitless at a time when natural energy sources are becoming critical.

Actual capital cost data for an Austin method type of pyrolysis system, including a breakdown of individual cost items, are presented in Appendix 7-1. Many of the items presented in this itemized list would be typical of any operational pyrolysis unit.

7-4.7 Summary—Pyrolysis

The pyrolysis process may be utilized to convert the organic fraction of solid waste into gas, liquid, or char having a substantial energy value. Various pyrolysis systems are currently being studied by several companies in the United States on a pilot plant scale. Based on pilot plant data, some companies are claiming that their pyrolysis systems can produce a profit per ton of refuse processed through the sale of valuable end-products. Although the pyrolysis process appears to hold significant potential as an energy recovery system, a detailed evaluation of the process must await full-scale application. Operational data derived over the next several years from the first two full-scale pyrolysis plants, the 1000 ton per day *Landgard* plant in Baltimore and the 250 ton per day Garrett plant in San Diego, will most likely determine whether the pyrolysis process will become a major solid waste management technique in the future.

7-5. METHANE PRODUCTION BY ANAEROBIC FERMENTATION

Anaerobic digestion has been utilized for many years in wastewater treatment systems to reduce the amount of organic sludges that must undergo final

TABLE 7-34 Capital Cost Estimate for 150 TPD Gasification Plant—Battelle Northwest Study

Capital Cost Estimate for 150 Ton/Day Wet Basis Excluding Land, Engineering or Contingencies

	Scale Up Factors Ton/Day		Disposal Only No Shredding		Disposal Only With Shredding		Gas to Boiler		Gas to Turbine		Gas to Turbine Combined Cycle	
	50-150	150-1000	High	Low	High	Low	High	Low	High	Low	High	Low
Feed Preparation												
Truck Scales	0.2	0.5	$23,000	$15,000	$23,000	$15,000	$23,000	$15,000	$23,000	$15,000	$	$
Receiving conveyor and pit	0.1	0.6	75,000	65,000	65,000	50,000	65,000	50,000	65,000	50,000		
Shredder[a]	0.0	0.7			150,000	130,000	150,000	130,000	150,000	130,000		
Magnetic separator and conveyor	0.4	0.8	17,000	15,000	17,000	15,000	17,000	15,000	17,000	15,000		
Conveyor to storage[a]	0.1	0.6			70,000	55,000	70,000	55,000	70,000	35,000		
Shredded waste storage[a]	0.15	1.0			200,000	175,000	200,000	175,000	200,000	175,000		
Reject conveyor from shredder[a]	0.2	0.6			5,000	4,000	5,000	4,000	5,000	4,000		
Buildings—foundations (10,000 sq. ft)	0.7	0.7	120,000	100,000	160,000	120,000	160,000	120,000	160,000	120,000		
Instrumentation	0.5	0.6	15,000	10,000	15,000	10,000	15,000	10,000	15,000	10,000		
Power wiring	0.5	0.6	15,000	10,000	25,000	20,000	35,000	30,000	75,000	50,000		
Pyrolysis-Incineration Process Equipment												
Reactor feeder and hopper	0.7	1.0	25,000	12,500	25,000	12,500	25,000	12,500	25,000	12,500		
Reactor	0.7	1.0	60,000	50,000	60,000	50,000	60,000	50,000	60,000	50,000		
Reactor grate and residue handling	0.6	1.0	50,000	35,000	50,000	35,000	50,000	35,000	50,000	35,000		
Ash removal conveyor	0.6	1.0	10,000	5,000	10,000	5,000	10,000	5,000	10,000	5,000		
Gas condenser[b]	0.6	1.0	20,000	15,000	20,000	15,000	20,000	15,000	20,000	15,000		
Gas holder[b]	0.6	1.0					155,000	75,000	155,000	75,000		
Gas compressors[b]	0.6	0.6							127,000	100,000		
Gas burner	0.1	0.8	60,000	40,000	60,000	40,000	12,000	10,000	60,000	40,000		
Fans	0.3	1.0	9,000	6,000	9,000	6,000	9,000	6,000	9,000	6,000		
Cooling towers (not shown on flowsheet)[b]	0.5	0.6					15,000	12,000	42,000	30,000		

Note (Gas to Turbine Combined Cycle column): Same as gas turbine except as noted.

Item											
Cooling water pumped (shown on flowsheet)[b]											
Heat exchanger	0.4	0.9	25,000	15,000	25,000	15,000	25,000	15,000		
Valves and piping—ductwork sq ft	0.5	0.8	40,000	30,000	40,000	30,000	75,000	50,000	100,000	75,000
Building—foundations (3000 sq ft or 4000 sq ft)	0.3	0.8	50,000	35,000	50,000	35,000	68,000	48,000	68,000	48,000
Instrumentation	0.3	0.7	75,000	50,000	75,000	50,000	150,000	100,000	225,000	200,000
Power wiring	0.5	0.6	15,000	10,000	15,000	10,000	20,000	15,000	25,000	15,000
Power Generating Equipment										
Gas turbine generator[b,c]	1.0	0.85							620,000	500,000
Turbine waste heat recovery[b,c]	1.0	0.75							558,000 →	500,000 →
Cooling water pumps[b,c]	1.0	0.85							5,000	4,000
Supporting Equipment Fac. and Grounds										
Site preparation	0.6	0.5	25,000	10,000	25,000	10,000	25,000	10,000	25,000	10,000
Roads—parking	0.6	0.5	25,000	15,000	25,000	15,000	25,000	15,000	25,000	15,000
Landscaping—fencing—irrigation	0.6	0.5	45,000	30,000	45,000	30,000	45,000	30,000	45,000	30,000
Lights	0.6	0.5	15,000	10,000	15,000	10,000	15,000	10,000	15,000	10,000
Utilities	0.6	0.5	30,000	20,000	30,000	20,000	30,000	20,000	30,000	20,000
Fire and safety	0.6	0.5	30,000	20,000	30,000	20,000	30,000	20,000	30,000	20,000
Residue handling and mobile equipment	0.6	0.5	50,000	30,000	50,000	30,000	50,000	30,000	50,000	30,000
Office—shop—parts storage—lunch rooms—rest rooms	0.6	0.5	20,000	15,000	20,000	15,000	20,000	15,000	20,000	15,000
Tools	0.6	0.5	12,000	10,000	12,000	10,000	12,000	10,000	12,000	10,000
Furniture	0.6	0.5	4,000	3,000	4,000	3,000	4,000	3,000	4,000	3,000
Spare parts	0.6	0.5	12,000	10,000	12,000	10,000	12,000	10,000	12,000	10,000
Total			$972,000	$691,500	$1,437,000	$970,500	$1,702,000	$1,225,500	$2,674,000	$2,022,500
									$3,232,000	$2,522,000

[a] Not needed if shredding omitted.
[b] Not needed for disposal only.
[c] Not needed for gas generations.
Engineering 15% and contingency 10% for low estimate.
Engineering 20% and contingency 15% for high estimate.

Source: Hammond, V. L., et al., "Pyrolysis-Incineration Process for Solid Waste Disposal," Battelle Northwest, final report for the City of Kennewick, Washington, EPA 1-G06-EC-0032-1, December, 1972.

TABLE 7-35 Pyrolysis Plant Labor Distribution—Battelle Northwest Study

Plant Capacities, Tons/Day	Plant Manager	Plant Engineer	Clerk	Process Operator	Weigh-Master	Traffic Director	Plant Maintenance	Labor Cleanup	Total Number of Employees	Total Payroll/Yr
	$16,000	$15,000	$6,000	$12,000/yr	$7,500	$7,500	$12,000	$5,500		
50	1	1		8	1		2		12	$142,500
100	1	1		8	1		2		12	$142,500
150	1	1	1	8	1		2	1	14	$154,000
200	1	1	1	8	1	1	4	1	18	$178,000
400	1	1	1	12	1	1	4	1	22	$226,000
600	1	1	1	12	1	1	5	1	23	$238,000
800	1	1	1	16	1	1	5	1	27	$309,500
1,000	1	1	1	16	1	1	5	1	27	$309,500

Source: Hammond, V. L., et al., "Pyrolysis-Incineration Process for Solid Waste Disposal," Battelle Northwest, final report for the City of Kennewick, Washington, EPA 1-G06-EC-0032-1, December, 1972.

TABLE 7-36 Estimated Pyrolysis Electrical Power Use Costs/Year—Battelle Northwest Study

	Plant Capacities, Ton/Day							
	50	100	150	200	400	600	800	1000
Disposal only no shredding	$ 4,000	$ 4,500	$ 5,000	$ 9,500	$18,000	$25,000	$ 35,000	$ 45,000
Disposal only with shredding	12,000	14,000	16,000	23,000	45,000	68,000	85,000	105,000
Gas to fuel boiler	13,000	16,000	13,000	25,000	47,000	70,000	87,000	108,000
Gas to turbine generator	16,000	20,000	23,000	33,000	63,000	95,000	120,000	150,000
Gas and turbine combined cycle	16,000	20,000	23,000	33,000	63,000	95,000	120,000	150,000

Source: Hammond, V. L., et al., "Pyrolysis-Incineration Process for Solid Waste Disposal," Battelle Northwest, final report for the City of Kennewick, Washington, EPA 1-G06-EC-0032-1, December, 1972.

TABLE 7-37   Estimated Cost of Disposal only Without Shredding—Battelle
Northwest Study

Plant Capacity Ton/Day	Total Capital Cost		Operating Cost/Ton				Capital Cost/Ton of Plant Capital	
			High		Low			
	High	Low	25 Yr	15 Yr	25 Yr	15 Yr	High	Low
50	$ 792,391	$ 523,490	$14.72	$15.75	$12.43	$13.14	$15,847	$10,470
100	1,082,437	713,556	8.59	9.30	7.02	7.51	10,824	7,136
150	1,312,200	864,375	6.59	7.16	5.32	5.71	8,748	5,763
200	1,608,839	1,060,748	6.26	6.80	5.12	5.37	8,044	5,304
400	2,658,486	1,755,946	4.61	5.06	3.67	3.92	6,646	4,390
600	3,593,160	2,375,138	3.81	4.22	2.97	3.20	5,987	3,959
800	4,464,603	2,952,433	3.51	3.89	2.72	2.95	5,581	3,691
1000	5,293,906	3,501,769	3.18	3.54	2.44	2.65	5,294	3,501

Source: Hammond, V. L., et al., "Pyrolysis-Incineration Process for Solid Waste Disposal,"
Battelle Northwest, final report for the City of Kennewick, Washington, EPA 1-G06-EC-
0032-1, December, 1972.

disposal. This reduction in sludge quantity is accomplished by the conversion of
a portion of the solids to liquids and gases, and is accompanied by an increase in
the sludge's dewatering characteristics. The gas produced in this process is a
combination of methane and carbon dioxide possessing a high calorific value
and, therefore, has a high potential for reclamation. For example, the heat value
of the gas produced from the digestion of wastewater sludge varies from approximately 7000 to 9000 BTU/lb of organic matter destroyed, depending on the
nature of the digested solids (Ref: 7-22).

Anaerobic fermentation of complex organic material is usually considered to
occur in two distinct stages. In the first stage, facultative anaerobic acid pro-

TABLE 7-38   Estimated Cost for Disposal only With Shredding—Battelle
Northwest Study

Plant Capacity Ton/Day	Total Capital Cost		Operating Cost/Ton				Capital Cost/Ton of Plant Capital	
			High		Low			
	High	Low	25 Yr	15 Yr	25 Yr	15 Yr	High	Low
50	$1,199,682	$ 830,787	$18.51	$20.14	$15.44	$16.56	$23,993	$16,616
100	1,607,190	1,109,384	11.01	12.10	8.94	9.69	16,072	11,094
150	1,939,950	1,338,125	8.51	9.39	6.84	7.45	12,933	8,921
200	2,402,886	1,661,960	8.10	8.92	6.56	7.02	12,014	8,309
400	4,070,503	2,833,412	6.27	6.96	4.98	5.41	10,176	7,084
600	5,582,679	3,900,027	5.39	6.02	4.22	4.63	9,304	6,500
800	7,008,954	4,908,592	5.01	5.61	3.92	4.31	8,761	6,136
1000	8,377,828	8,878,312	4.64	5.20	3.59	3.97	8,378	5,878

Source: Hammond, V. L., et al., "Pyrolysis-Incineration Process for Solid Waste Disposal,"
Battelle Northwest, final report for the City of Kennewick, Washington, EPA 1-G06-EC-
0032-1, December, 1972.

TABLE 7-39   Estimated Cost of Steam Generation Using Gas to Fire Boiler—
Battelle Northwest Study

Plant Capacity Ton/Day	Total Capital Cost		Operating Cost/Ton				Capital Cost/Ton of Plant Capital	
			High		Low			
	High	Low	25 Yr	15 Yr	25 Yr	15 Yr	High	Low
50	$1,384,692	$ 931,508	$20.11	$21.99	$16.33	$17.58	$27,694	$18,630
100	1,888,991	1,262,376	12.24	13.52	9.63	10.48	13,890	12,624
150	2,297,700	1,531,875	9.55	10.58	7.42	8.11	15,318	10,213
200	2,861,210	1,907,352	9.08	10.05	7.10	7.64	14,306	9,537
400	4,911,103	3,271,725	7.16	7.99	5.45	5.95	12,278	8,179
600	6,788,494	4,519,695	6.24	7.01	4.66	5.14	11,314	7,533
800	8,570,507	5,703,210	5.83	6.56	4.34	4.80	10,713	7,129
1000	10,288,761	6,843,600	5.44	6.14	4.00	4.45	10,289	6,844

Source: Hammond, V. L., et al., "Pyrolysis-Incineration Process for Solid Waste Disposal,"
Battelle Northwest, final report for the City of Kennewick, Washington, EPA 1-G06-EC-
0032-1, December, 1972.

ducers convert complex carbohydrates, proteins, and fats into simple organic
acids and alcohols. In the second stage, anaerobic methane fermenters convert
these short-chain organic acids and alcohols into methane and carbon dioxide.
The relative proportions of these two gases are dependent upon the detention
time and temperature at which the process is operated.

Since solid waste is composed of essentially the same constituents that make
up wastewater sludge (carbohydrates, proteins, and fats), it would seem logical
that refuse could also successfully undergo anaerobic digestion. Several investigations have supported this conclusion by demonstrating that organic refuse,
especially garbage, can be anaerobically fermented provided that the proper

TABLE 7-40   Estimated Cost of Electrical Generation Without Heat
Scavenging—Battelle Northwest Study

Plant Capacity Ton/Day	Total Capital Cost		Operating Cost/Ton				Capital Cost/Ton of Plant Capital	
			High		Low			
	High	Low	25 Yr	15 Yr	25 Yr	15 Yr	High	Low
50	$1,967,248	$1,379,870	$25.14	$27.80	$20.24	$22.11	$39,344	$27,597
100	2,847,570	1,993,578	16.35	18.28	12.79	14.14	28,476	19,936
150	3,609,900	2,528,125	13.29	14.92	10.28	11.42	24,066	16,854
200	4,506,991	3,157,181	12.63	14.15	9.81	10.78	22,535	15,786
400	7,760,675	5,436,520	10.24	11.56	7.82	8.69	19,402	13,591
600	10,724,881	7,510,355	9.09	10.30	6.86	7.67	17,875	12,517
800	13,525,201	9,467,461	8.69	9.67	6.41	7.19	16,907	11,834
1000	16,214,173	11,345,160	8.03	9.13	6.00	6.74	16,214	11,345

Source: Hammond, V. L., et al., "Pyrolysis-Incineration Process for Solid Waste Disposal,"
Battelle Northwest, final report for the City of Kennewick, Washington, EPA 1-G06-EC-
0032-1, December, 1972.

TABLE 7-41 Estimated Cost of Electrical Generation Using the Combined Cycle—Battelle Northwest Study

Plant Capacity Ton/Day	Total Capital Cost		Operating Cost/Ton				Capital Cost/Ton of Plant Capital	
			High		Low			
	High	Low	25 Yr	15 Yr	25 Yr	15 Yr	High	Low
50	$ 2,218,348	$ 1,588,203	$27.23	$30.24	$21.98	$24.13	$44,367	$31,764
100	3,349,770	2,410,245	18.44	18.28	14.53	16.16	33,498	24,102
150	4,363,200	3,153,125	15.38	17.35	12.02	13.44	29,088	21,021
200	5,411,380	3,910,052	14.51	16.35	11.38	12.60	27,057	19,550
400	9,152,647	6,606,350	11.69	13.24	9.04	10.05	22,882	16,516
600	12,503,508	9,015,238	10.33	11.74	7.90	8.89	20,839	15,025
800	15,946,945	11,261,216	9.79	10.95	7.35	8.28	19,933	14,077
1000	18,614,095	13,396,522	9.03	10.29	6.85	7.74	18,614	13,397

Source: Hammond, V. L., et al., "Pyrolysis-Incineration Process for Solid Waste Disposal," Battelle Northwest, final report for the City of Kennewick, Washington, EPA 1-G06-EC-0032-1, December, 1972.

environmental conditions (temperature, pH, absence of oxygen, and nutrients) are maintained (Ref. 7-22).

7-5.1 Process Feasibility

The most definitive study of methane production by anaerobic fermentation was conducted in 1973 at the University of Illinois under a grant by the Environmental Protection Agency (Ref. 7-23). In this laboratory study, shredded residential solid waste having an average particle size of 1 in. was used as the substrate to be digested. Separation of the nondegradable portion of the refuse was not undertaken prior to shredding; however, mechanical difficulties (mixer jamming, etc.) caused by this inert fraction necessitated its removal prior to digestion. The compositional characteristics of the solid waste stream used in this study following separation are shown in Table 7-42.

Since the separated solid waste stream was found to be deficient in nitrogen and phosphorus, 3 to 6 percent by dry weight of raw wastewater sludge (primary plus waste activated) was added to each refuse digester. Eight completely mixed reactors were used in this investigation, each having an operating volume of 15 liters. The pH of the reactors was maintained above 6.6 with the addition of a base to insure an optimum environment for the acid producers and methane fermenters. Separated solid waste was added to the reactors on a daily basis.

TABLE 7-42   Composition of Solid Waste Utilized in the University of Illinois Fermentation Study

Determination	Number of Determinations	Average	Range
Volatile total solids	3	81.8 %	81.6–82.1
Lipids	3	6.2 %	5.98–6.36
Chemical oxygen demand	5	0.982 gm/gm Dry solids	0.943–1.021
Calorific value	2	7510 BTU/lb Dry solids	7410–7610
Carbon	2	44.1 %	43.68–44.50
Hydrogen	2	5.7 %	5.625–5.750
Nitrogen (CHN analyzer)	2	0 %	–
Nitrogen (wet Kjeldahl)	4	0.68%	0.666–0.700
Phosphorus	5	0 %	–
Total carbohydrates	5	52.75%	51.5–53.5
Cellulose	5	35.8 %	35.0–37.0

Source: Pfeffer, J. T., "Processing Organic Wastes by Anaerobic Fermentation," presented at the International Biomass Energy Conference, Winnipeg, Manitoba, May 13–15, 1973.

TABLE 7-43   Gas Production (cu ft/lb dry solid) in the University
of Illinois Fermentation Study

Temperature (°C)	Retention Time (Days)					
	4	8	10	15	20	30
35.0 (95.0°F)	1.40	2.30	2.58	2.86	2.95	3.05
40.0 (104.0°F)	2.33	2.98	3.20	3.60	3.85	4.10
41.0 (105.8°F)	2.49	3.09	3.25	3.71	3.95	4.20
42.1 (107.8°F)	2.59	3.14	3.33	3.74	3.95	4.20
43.0 (109.4°F)	2.58	3.09	3.26	3.67	3.82	4.09
45.0 (113.0°F)	2.08	2.55	2.70	3.05	3.14	3.36
48.0 (118.4°F)	2.64	3.14	3.33	3.57	3.82	4.06
53.0 (127.4°F)	3.33	3.88	4.07	4.29	4.51	4.70
56.0 (132.8°F)	3.74	4.20	4.35	4.54	4.75	4.88
60.0 (140.0°F)	4.23	4.49	4.57	4.79	4.93	4.96

Source: Pfeffer, J. T., "Processing Organic Wastes by Anaerobic Fermentation," presented at the International Biomass Energy Conference, Winnipeg, Manitoba, May 13–15, 1973.

The primary purpose of this study was to determine the amount of gas produced at varying process detention times and temperatures and the compositional variation of the gas produced at different digestion temperatures. Table 7-43 shows the amount of dry gas produced at various temperatures and detention times per pound of dry solids added. As expected, gas production increased with an increase in detention time and temperature from a low of 1.4 cu ft/lb of dry solids at 4 days holding time and 95°F, to 4.96 cu ft/lb of dry solids at 30 days holding time and 140°F. The compositional analyses of gas produced at 35°C and 60°C (95°F and 140°F) with varying detention times is

TABLE 7-44   Percentage of Methane and Carbon Dioxide in Gas Produced in the University of Illinois Fermentation Study

Reactor Number	35°C (95°F)			60°C (140°F)		
	Retention time (Days)	% CH_4	% CO_2	Retention time (Days)	% CH_4	% CO_2
1	4	69.7	30.3	3	53.5	46.5
2	4	69.8	30.2	4	55.0	45.0
3	6	64.3	35.7	6	51.9	48.1
4	8	58.6	41.4	8	50.2	49.8
5	10	57.2	42.8	10	52.3	47.7
6	15	53.8	46.2	15	53.5	46.5
7	20	53.4	46.6	20	49.1	50.9
8	30	53.8	46.2	30	53.7	46.3

Source: Pfeffer, J. T., "Processing Organic Wastes by Anaerobic Fermentation," presented at the International Biomass Energy Conference, Winnipeg, Manitoba, May 13–15, 1973.

shown in Table 7-44. Methane content was found to decrease with increasing temperature and detention time. Therefore, higher detention times and temperatures did result in increased gas production; however, the gas produced became more "contaminated" with carbon dioxide as these operational control values were increased.

To determine the weight reduction accomplished by the fermentation process, volatile solids destruction was also monitored at the various detention times and digestion temperatures studied (see Table 7-45). Since the volatile solids content gives an approximation of the amount of biodegradable material present in the refuse, and since approximately 70 percent of domestic solid waste is biodegradable, the amount of volatile solid reduction allows an estimate of the total amount of refuse that is destroyed during the process. The data presented show

TABLE 7-45   The Percent of Refuse Volatile Solids Destruction in the University of Illinois Fermentation Study

Temperature (°F)	Retention Time (Days)					
	4	8	10	15	20	30
95.0	17.4	28.6	32.1	36.6	36.7	38.0
104.0	29.0	37.1	39.8	44.8	47.9	51.0
105.8	31.0	38.5	40.5	46.2	49.2	52.3
107.8	32.2	39.1	41.5	46.6	49.2	52.3
109.4	32.1	38.5	40.6	45.7	47.6	50.9
113.0	25.9	31.7	33.6	38.0	39.1	41.8
118.4	32.9	39.1	41.5	44.4	47.6	50.6
127.4	41.5	48.3	50.7	53.4	56.2	58.5
132.8	46.6	52.3	54.2	56.5	59.1	60.8
140.0	52.7	55.9	56.9	59.6	61.4	61.8

Source: Pfeffer, J. T., "Processing Organic Wastes by Anaerobic Fermentation," presented at the International Biomass Energy Conference, Winnipeg, Manitoba, May 13–15, 1973.

volatile solids reduction increasing with temperature and detention time from a low of 17.4 percent to a high of 61.8 percent. Therefore, the maximum overall reduction in the weight of dry solids from the fermentation process would amount to about 0.70 X 0.62 or more than 42 percent. Assuming that the digester effluent could be dewatered, the fermentation process would reduce transportation costs to landfill sites as well as extending the life of these sites.

Since the economic viability of the process depends to a large extent upon the dewaterability characteristics of the reactor residue, a series of filter test-leaf experiments were conducted using Eimco 1/1 plain weave polyethylene test cloths. The data from these studies (see Table 7-46) indicate a high filter yield;

TABLE 7-46 Dewatering Characteristics of Reactor Residue in the University of Illinois Fermentation Study

Test Cloth Thread Count	Residue Slurry Concentration, %		
	3.5	6.8	9.3
	Filter Yield, lbs/sq ft-hr		
112 X 48	4.5	13.2	19.0
105 X 40	5.0	13.5	19.2
40 X 23	5.0	12.5	17.5
	Moisture Content of Cake, %		
112 X 48	75	73	75
105 X 40	75	77	75
40 X 23	77	72	76

Source: Pfeffer, J. T., "Processing Organic Wastes by Anaerobic Fermentation," presented at the International Biomass Energy Conference, Winnipeg, Manitoba, May 13–15, 1973.

however, lower cake moisture values would have to be achieved to reduce residue haul costs.

7-5.2 Process Economics

An in-depth economic analysis of the anaerobic fermentation process must include estimates of both the preparation system and the refuse digestion system costs. Fermentation process requirements necessitate size reduction and nonbiodegradable fraction removal prior to digestion. Obviously, this type of inert fraction separation could be achieved by utilizing various flow processes. Pfeffer has proposed the use of the unit processes outlined in Fig. 7-36 to accomplish the required refuse preparation (Ref. 7-22). In this system, mixed refuse would initially be shredded to a 1 in. particle diameter and then passed through a vibrating screen to remove glass, ash, and other nonmetals. The solid waste stream would then pass through a hydraulic separator in which metal removal would be achieved. The resultant refuse slurry would then be partially dewatered prior to introduction into the fermentation reactor. Following digestion, the reactor residue would be dewatered and transported to ultimate disposal sites.

Pfeffer has calculated both the preparation and digestion costs for this system based on a 100 ton per day facility operating at a feed slurry concentration of 6 to 20 percent solids. Shredding and separation costs were estimated to be $2.00 per dry ton of refuse processed. The cost of gas production by digestion for varying digestion temperatures and fed slurry concentrations at a detention time of 4 days is shown in Fig. 7-37. Gas production at 140°F is estimated to result in the

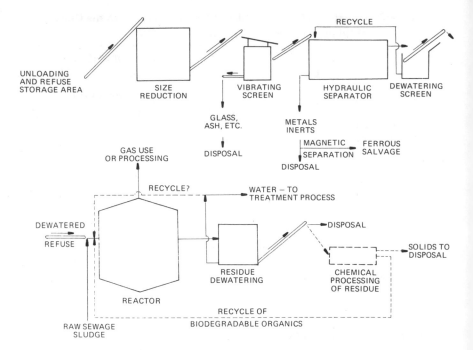

Fig. 7-36 Flow diagram for the anaerobic fermentation process. *Source:* Pfeffer, J. T., "Processing Organic Wastes by Anaerobic Fermentation," presented at the International Biomass Energy Conference, Winnipeg, Manitoba, May 13–15, 1973.

minimum process cost for a given feed slurry concentration. Apparently optimum process conditions occur at a feed slurry concentration of 20 percent and a digestion temperature of 140°F, resulting in a gas cost of 8 to 10 cents per 1000 cu ft of gas produced. Based on a gas sale price of $0.45 per 1000 cu ft, a return of $2.50 to $3.00 per ton of dry refuse could be expected. If the refuse preparation cost of $2.00 per ton is deducted, there is still a revenue of $0.50 to $1.00 per ton of dry refuse processed. If the as-received refuse is assumed to have a moisture content of 28 percent, this revenue figure is reduced to $0.36 to $0.72 per ton. Costs factors not included in the above estimate include:

1. Benefits attributable to disposal and transportation of reduced refuse volume following digestion.
2. Savings realized from disposing of wastewater solids in the fermentation process. A 100 ton per day plant would also process 2.9 tons of dry sludge solids. The cost credit for processing this tonnage of sludge is estimated at $0.44 to $1.74 per ton of refuse processed.

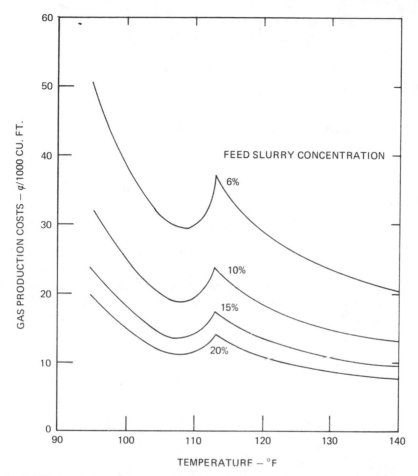

Fig. 7-37 Gas production cost for various digestion temperatures and feed slurry concentrations. *Source:* Pfeffer, J. T., "Processing Organic Wastes by Anaerobic Fermentation," presented at the International Biomass Energy Conference, Winnipeg, Manitoba, May 13–15, 1973.

3. Credit realized from salvaged materials.
4. Cost due to gas clean up prior to use.

7-5.3 Summary—Anaerobic Fermentation

Laboratory investigations have substantiated the feasibility of generating methane gas through the anaerobic fermentation of solid waste. Cost estimates for the process are very favorable. It has been estimated that the anaerobic fermentation process could satisfy approximately 11 percent of this country's 1970 demand for natural gas (2.0 quadrillion BTU per annum) if it was utilized to process the nation's entire solid waste stream (Ref. 7-22). Although the process does appear to hold promise, a comprehensive analysis of its overall applicability to solid waste management practices in the United States must await pilot and full-scale development.

7-6. SUMMARY—ENERGY RECOVERY

A wide variety of systems which produce energy while disposing of solid waste have been developed during the last several years. It seems apparent that the implementation of energy recovery systems in the near future is inevitable when the energy demands of this country and the world are considered.

The use of refuse as a supplementary fuel appears to hold a great deal of immediate potential for full-scale application. Longer range energy needs may be satisfied by pyrolysis or methane production processes with the concurrent possibility of revenue generation from these systems. However, the implementation of energy recovery processes in this country will require a realignment of solid waste management strategies and organization by city managers. Such a realignment is mandatory if this country is to recover one of its most valuable resources—energy.

REFERENCES

7-1. "America's Largest Incinerator," *Public Cleansing,* Vol. 56, No. 1, January, 1966.

7-2. Bender, R. J., "Incineration Plant-Plus," *Power,* Vol. 111, No. 1, January, 1967.

7-3. Buss, T. F., "The *Landgard* System for Resource Recovery and Solid Waste Disposal," *Proc.* The Third Annual Environmental Engineering and Science Conference, University of Louisville, Louisville, Kentucky, March, 1973.

7-4. Cohan, L. J., "Steam Generation From Solid Wastes," paper presented at Connecticut Clean Power Symposium, West Hartford, 1972.

7-5. "Combustible Rubbish Contents Favors Using Heat Recovery," *Refuse Removal Journal,* Vol. 10, November, 1967.

7-6. Day and Zimmerman, Engineers and Architects, "Special Studies for Incinerators; for the Government of the District of Columbia, Department of Sanitary Engineering," USPHS Publ. No. 1748, Govt. Printing Office, 1968.

7-7. DeBoer, J. G., "Incineration Should Be Allied With Landfill to Prolong the Use of Disposal Sites," *Refuse Removal Journal*, Vol. 11, No. 8, August, 1968.

7-8. DeMarco, Jack, "Advanced Techniques for Incineration of Municipal Solid Wastes," Open File Report (SW-38d.of) EPA, 1972.

7-9. Engdahl, R. B. and Hummell, J. D., "Power From Refuse," *The American City*, Vol. 83, No. 9, September, 1968.

7-10. Fife, J. A., "Solid Waste Disposal: Incineration or Pyrolysis?" *Environmental Science and Technology*, Vol. 7, No. 4, April, 1973, p. 308.

7-11. Flaherty, J. F., "Boston's Incinerator is a Steam Producer," *The American City*, Vol. 75, June, 1960.

7-12. Hammond, V. L., et al., "Pyrolysis-Incineration Process for Solid Waste Disposal," Battelle Northwest, final report for the City of Kennewich, Washington, EPA 1-G06-EC-0032-1, December, 1972.

7-13. Hoffman, D. A. and Fitz, R. A., "Batch Retort Pyrolysis of Solid Municipal Wastes," *Environmental Science and Technology*, Vol. 2, No. 11, November, 1968, p. 1023.

7-14. Horner & Shifrin, Consulting Engineers, "Appraisal of Use of Solid Waste as Supplementary Fuel in Power Plant Fuels," St. Louis, Missouri, February, 1973.

7-15. "Incinerator Uses Trash as Fuel in Rotterdam," *Refuse Removal Journal*, Vol. 10, November, 1967.

7-16. Kaiser, E. R. and Friedman, S. B., "The Pyrolysis of Refuse Components," *Combustion*, May, 1968, p. 31.

7-17. Mallan, G. M. and Finney, C. S., "New Techniques in the Pyrolysis of Solid Wastes," presented at the 73rd National Meeting AIChE, Minneapolis, Minnesota, August, 27–30, 1972.

7-18. McFarland, J. M., et al., "Comprehensive Studies of Solid Waste Management," NERC contract no. 2R01-EC-00260-01, SERL Report No. 72–3, EPA, May, 1972, p. 107.

7-19. Mullen, J. F. and Regan, J. W., "Energy Conversion of Refuse Advanced by Application of Basic Combustion Principles," paper presented at AIChE Symposium on Solid Waste Disposal, Atlantic City, 1971.

7-20. "Navy to Incinerate Rubbish For Power," *Refuse Removal Journal*, Vol. 10, April, 1967.

7-21. Pfeffer, J. T., "Processing Organic Wastes by Anaerobic Fermentation," presented at the International Biomass Energy Conference, Winnipeg, Manitoba, May 13–15, 1973.

7-22. Pfeffer, J. T., "Reclamation of Energy from Organic Refuse—Final Report, Grant No. EPA-R-800776," National Environmental Research Center, EPA, Cincinnati, Ohio, 1973.

7-23. Pikarsky, M. F., "Chicago's Northwest Incinerator," *Civil Engineering,* Vol. 41, No. 9, September, 1971.

7-24. Raisch, W., "Innovations in Refuse-Incinerator Design," paper presented at the ASME Annual Meeting, New York, December, 1957.

7-25. Regan, J. W., "Generating Steam From Prepared Refuse," paper presented at Fourth National Incinerator Conference, Cincinnati, 1970.

7-26. Rogus, C. A., "Refuse Collection and Disposal in Western Europe—Part IV: Refuse Disposal by Incineration," *Public Works,* Vol. 93, July, 1962.

7-27. Sanner, W. S., Ortuglio, C., Walters, J. G. and Wolfon, D. E., "Conversion of Municipal and Industrial Refuse into Useful Materials by Pyrolysis," U.S. Bureau of Mines, Report of Inv. No. 7428, August, 1970.

7-28. Sebastian, F., "Worldwide Rush to Incineration," *The American City,* Vol. 82, December, 1967.

7-29. Shequine, E. R., "Steam Generation From Incineration," *Public Works,* Vol. 95, August, 1964.

7-30. Spitzer, E. F., "Montreal's Combined Incinerator–Power Plant," *American City,* Vol. 85, No. 5, May, 1970.

7-31. Stephenson, J. W., "Some Recent Developments in Disposal of Solid Wastes by High-Temperature Combustion, Pyrolysis and Fluid Bed Reactor," for presentation before New York State Action for Clean Air Com., May 7, 1971.

7-32. "The Martin Stoker," *Public Cleansing,* Vol. 59, No. 2, February, 1969.

7-33. "The Swiss Don't Miss Incinerator Thrift," *The American City,* Vol. 75, February, 1960.

7-34. "Three Contractor Owned and Operated Incinerators Service Metropolitan Chicago," *Refuse Removal Journal,* Vol. 11, May, 1968.

7-35. Union Electric Company, Solid Waste Disposal Seminar, St. Louis, Missouri, October 26–27, 1972.

7-36. Wilson, M. J. and Gommill, A. M., "Putting Solid Waste to Work in Nashville," *Professional Engineer,* Vol. 41, No. 10, October, 1971.

7-37. Wisely, F. E., "Solid Waste and Electric Power Production," *Professional Engineer,* Vol. 41, No. 10, October, 1971, p. 40–41.

7-38. Wisely, F. E. and Hinchman, H. B., "Refuse as a Supplementary Fuel," Proc. The Third Annual Environmental Engineering and Science Conference, Louisville, Kentucky, March 5–6, 1973.

7-39. Wisely, F. E., Klumb, D. L. and Sutterfield, G. W., "From Solid Waste to Energy," presented at the Conservation Education Association Conference, Murray State University, Murray, Kentucky, August 15, 1973.

7-40. Wisely, F. E., Sutterfield, G. W. and Klumb, D. L., "St. Louis Power Plant to Burn City Refuse," *Civil Engineering,* Vol. 41, No. 1, January, 1971, p. 56–59.

7-41. Wisely, F. E., Sutterfield, G. W. and Klumb, D. L., "St. Louis Power Plant Burns City Refuse," *Civil Engineering,* Vol. 42, No. 12, December, 1972, p. 30–31.

APPENDIX 7-1

Preliminary Cost Estimate
Pyrolysis-Incineration Process
Kennewick, Washington*

Summary

Site development		$ 127,000
Pyrolysis building		243,000
Material handling equipment		614,000
Fuel generating equipment		588,000
Power generating equipment		621,000
Mobile equipment		30,000
Power wiring		96,000
Process piping		92,000
Instrumentation		239,000
	Subtotal	$2,650,000
Engineering and construction supervision		400,000
Contingencies		300,000
	Total	$3,350,000
Development of gas turbine		200,000
	Estimated Grand Total	$3,550,000

	Estimated Installed Cost
Site Development	
Site preparation	$ 14,900
Roads and parking	22,600
Fencing and gates	12,500
Landscaping and irrigation	30,000
Yard lighting	4,800
Sewers and water service	22,000
Fire mains and hydrants	20,200
Total	$127,000

*Source: Hammond, V. L., et al., "Pyrolysis-Incineration Process for Solid Waste Disposal," Battelle Northwest, final report for the City of Kennewick, Washington, EPA 1-G06-EC-0032-1, December, 1972.

		Estimated Installed Cost
Pyrolysis Building		
Foundations and floor slab		$ 30,000
Superstructure		110,000
Doors, partitions and finishes		25,000
Lighting		15,400
Plumbing and heating		33,600
Fire protection		29,000
	Total	$243,000

Material Handling Equipment		
Refuse receiving conveyor—48-in. apron conveyor, 40° rise to shredder, motor 25 hp, foundation		$ 64,700
Refuse shredder—Heil vertical shaft (15–18 ton/hr capacity) motor 200 hp, foundation		129,700
Reject conveyor from shredder—36-in. w × 12-ft l belt, motor 5 hp, foundation		4,800
Conveyor to storage—36-in. w × 332-ft l inclined belt, with intermediate transfer for magnet location, motor 15 hp, foundation		65,500
Self-cleaning magnet—42-in. w × 8-ft l pulley center, magnet 7½ kw with rectifier, motor 3 hp		12,600
Air blast fan—motor 3 hp		2,200
Ferrous metal conveyor to truck—30-in. w × 12-ft l belt, motor 3 hp, foundation		4,600
Refuse storage bin—Atlas #13TP3, 50,000 cu ft capacity with hydraulic sweep drive, motor 25 hp (sweep), foundation		
Bin out feed conveyor chain—with hydraulic drive, motor 7½ hp		195,400
Conveyor to reactor—36-in. w × 266-ft l belt, motor 15 hp, foundation		54,600
Weightometer—load cell to control feed rate to reactor		10,100
Live bottom waste hopper and feeder—15-ton capacity, motor 15 hp, foundation		42,500
Shredder monorail and hoist—10-ton capacity		$ 4,000
Truck scale—foundation		23,300
	Total	$614,000

	Estimated Installed Cost
Fuel Generation Equipment	
Reactor feeder—airlock feeder, water-cooled shaft and jacketed housing, motor 7½ hp	$ 12,600
Reactor—½-in. steel wall, 30-ft h X 10-ft id, foundation, refractory linings	55,400
Reactor ash discharger—mechanical feeder approx. 10-ft w X 14-ft o.a. l, motors two 5 hp with hydraulic drive	43,000
Ash collection conveyor—14-ft o.a. l, motor 3 hp	6,000
Ash crusher—(Dorr–Oliver), motor 3 hp	5,500
Ash feeder—rotary airlock with water-cooled shaft and jacket, motor 3 hp	8,300
Ash removal conveyor—apron-type pan 44-ft l, motor 2 hp, foundation	42,600
Gas condenser-cooler—1000 sf spiral counterflow, 48-in. od X 60-in h, gas blower to gas holder, motor 20 hp, foundation	27,000
Gas holder—40-ft id X 40-ft h dry seal double lift—piston type, foundation	$155,000
Compressor—gas to turbine (200 psi), motor 250 hp, foundation	127,300
Cooling tower—fan motors three at 15 hp, foundation	38,000
Pump cooling water to gas cooler—800 gpm, 100-ft tdh, motor 30 hp, foundation	3,000
Reactor air blower—motor 30 hp, foundation	8,800
Air and steam heater—(gas fired), foundation	53,000
Reactor monorail and hoist—5-ton capacity	2,500
Total	$588,000
Power Generating Equipment	
Gas turbine and generators—one 2500 kw, two 800 kw, waste heat recovery unit, foundation	$617,000
Pump cooling water to turbo-generator—500 gpm 150-ft tdh, motor 25 hp, foundation	4,000
Total	$621,000
Mobile Equipment	
Front end loader—Hough 50 series, 93 hp, 3½ cu yd bucket	
Total	$ 30,000

		Estimated Installed Cost
Process Piping		
Piping		$ 74,000
Ductwork and breeching		18,000
	Total	$ 92,000
Power Wiring		
Motor control centers		$ 22,700
Switchgear		45,200
Wiring		28,100
	Total	$ 96,000
Instrumentation		
Reactor—temperature, pressure, raw material input, ash output, gas output, central control panel		$129,600
Air heater—air, steam, air/steam mixture		20,100
Cooler condenser—temperature, process waste		13,400
Cooler tower—temperature, level, water quality		14,500
Atmosphere exhaust—stack gas		61,400
	Total	$239,000

8

European solid waste management

This chapter is devoted to the description and evaluation of solid waste management practices in Western Europe. General information on the quantities and composition of refuse generated in European countries will be given; collection and disposal practices will be described. Additionally, certain facilities will be described in some detail as examples of typical European disposal and resource recovery operations.

8-1. INTRODUCTION

In the following paragraphs, the results of a comprehensive study of European waste management practices are presented. This study was undertaken for several reasons. First of all, with greater population densities in the countries of Western Europe, it was believed that Western European countries have encountered problems of solid waste management before the same problems have arisen in the United States, be-

447

cause of the greater amount of available land for refuse disposal in the United States. Second, it was known that resource recovery, particularly heat recovery from refuse incineration, has been practiced for a considerable length of time on the European continent. Because of the increasing demand for energy in the United States, it was deemed advisable to investigate apparently successful energy recovery methods which have been employed in European waste disposal facilities. Thirdly, many ambiguous and vague statements have appeared in the professional literature from time to time concerning the success of materials recovery and energy recovery operations in Europe (e.g., composting, heat recovery, etc.). This study was undertaken in part, to elucidate the true situation with respect to resource recovery operations and waste disposal operations in the United Kingdom and several continental European nations. Finally, the benefits of an exchange of information with individuals interested in solid waste management in European countries were considered to be very significant. Consequently, the authors undertook a program of site visitations and conferences to inspect refuse management facilities in Western European nations and to confer with solid waste managers in those countries. The results of this effort are described in the following paragraphs.

8-2. GENERAL INFORMATION

Before describing the characteristics of the refuse generated in European countries and before discussing the collection and disposal operations undertaken in those countries, it is pertinent to describe the organizational framework in each country charged with the responsibility of solid waste management.

In the United Kingdom, the responsibility for waste management has devolved on municipal authorities, whether village, town, borough, or major metropolis. For the most part, collection throughout the entire country is accomplished by municipal agencies, with a minor portion of collection operations being carried out by private contractors under agreements with municipal agencies. In some instances, the disposal of refuse is being managed on a regional basis, but the regional concept of waste management generally has not been accepted and put into practice in Great Britain. Nevertheless, the first regional solid waste management authority in Europe was established in London under the auspices of the Greater London Council. In London, the disposal of refuse now is the responsibility of the Greater London Council, under the authority of the London Government Act of 1963. Prior to the passage of that legislation, ninety small local authorities had managed both refuse collection and disposal. Beginning in April of 1965, the disposal of refuse for the London area was allocated to the Greater London Council, while refuse collection responsibilities were allocated to the city of London and the remaining thirty-two London boroughs. After a transitional period wherein the transfer of functions was accomplished, the disposal service for London was completely transferred to the Greater London Council by early spring, 1967. Refuse is collected by the borough authorities in standard compactor-body collection trucks. Approximately one-half of the collected refuse is delivered directly to incinerators or land reclamation sites; the remainder of the refuse is deposited at transfer stations. There are over thirty refuse transfer stations in the Greater London area and the refuse is removed from these stations in highway trucks, railway cars, and barges on the Thames River. As soon as the refuse is deposited at the transfer stations or at a disposal facility, the refuse becomes the responsibility of the Greater London Council. In addition to the aforementioned collection of refuse, there is a public program for deposition of bulky wastes at the transfer and disposal facilities operated by the Council. The Council is required to provide sites for the deposition of bulky materials by the public, under provision of the Civic Amenities Act. Over 80,000 tons of bulky refuse are being deposited each year by citizens of London at the sites required under the provisions of the Civic Amenities Act. Included in this amount of bulky materials are motor vehicles, which yearly amount to over 25,000 abandoned or deposited vehicles. Approximately 40 percent of the vehicles are surrendered by citizens, and the balance is collected from the streets where they have been abandoned. The collected wastes are processed in a number of incinerators operated by the Council, or are transferred to land reclamation sites in abandoned quarries and chalk pits in Essex and Kent at a considerable distance from London.

The operation of the Greater London Council appears to have certain advantages in that economies of scale can be achieved in disposal operations. Additionally, the ability of a single agency to bargain for economic haulage rates for bulk transfer of refuse, for example, is a significant advantage of this type of centralized authority. However, the long-standing system of collection by local authorities seems to be an efficient operation and little advantage could be gained in centralizing the authority for refuse collection.

The administrative system for management of solid wastes in the United Kingdom is slated for revision in the immediate future (1974). According to pending legislation, functions of local government agencies will be redefined. According to the new system of functional arrangements, the boroughs throughout the United Kingdom will function as collection agencies and disposal management will be centralized in regional county authorities. All waste disposal sites in the United Kingdom will require a license, to be granted by a central environmental conservation agency. The requirement for a license would pertain to all private and public disposal sites. The license procedure effectively will establish limits on the types of wastes which could be brought to a given site for disposal; the license would not cover or restrict the actual operation of the disposal facility. At the present time in the United Kingdom, control of waste disposal operations is maintained through land use planning regulations. These regulations govern the approval of particular sites for disposal operations. This procedure is somewhat more cumbersome than the proposed system of control through management and restriction of wastes to be deposited or processed in any given facility.

In Switzerland, the responsibility for solid waste management has been allocated to individual municipalities and cantons. The Swiss federal government also exercises control and supervision of waste disposal operations through the offices of the Federal Water Pollution Control Administration. Although the basic responsibility for waste management has devolved on local governments in Switzerland, the federal government in some cases has granted subsidies to local governmental agenices in order to support innovative or experimental methods of waste disposal and processing. This support for demonstration projects is similar to the support which has been granted to innovative or experimental projects in the United States by the Office of Solid Wastes Management Programs of the United States Environmental Protection Agency. However, the major portion of waste management activities in Switzerland are carried out by local government, and because of the particular characteristics of the country, waste disposal through incinerators operated by urban governments has been a favored disposal technique. In the later sections of this discussion, several examples of the disposal facilities in Switzerland will be mentioned.

In the Federal Republic of Germany, solid waste disposal has been allocated to a number of agencies within the federal government. The major responsibility for waste disposal control is vested in the Federal Ministry of Public Health and its constitutent departments: the Institute for Water, Soil, and Air Hygiene in Berlin; the Central Office for Solid Waste Disposal (a cooperative organization of the federal government and individual state governments); and the Land Cooperative for Waste Management. In the private sector, the management of solid wastes is carried out by many organizations, including industrial corporations. The leading private organization for waste management is the Arbeitsgemeinschaft für Abfallbeseitigung. This organization is composed of experts from municipalities, industries, and agricultural associations and functions in advisory and consultative capacities for communities engaging in solid waste management. The actual responsibility for waste management devolves on local government bodies; however, the Central Office for Solid Waste Disposal is charged with the responsibility to provide guidelines and criteria for waste disposal and is responsible for optimizing refuse management in large regions through the development of comprehensive waste management plans. Additionally, at the state government level some regional authorities have been established to cope with the overall responsibilities of municipal government in providing services for the populace. For example, the Ruhr Regional Planning Authority operates in the state of North Rhine–Westphalia. This planning authority has a separate solid waste unit which is responsible for developing a comprehensive waste management plan for North Rhine-Westphalia. One of the facilities established by a regional planning authority in the Federal Republic of Germany will be discussed in a later section.

In France, the management of solid waste is centralized in five federal government ministries dealing with various aspects of waste management. The Public Health Council of France, the Departmental Governments, and the Regional Government of the District of Paris are also involved in waste management activities. In general, the federal ministries prepare the regulations and laws governing waste management, and the local governmental agencies implement the legislation and allocate monies and effort to the acutal management operation. The leading private organizations involved with waste management in France are the Association Generale des Hygienistes et Techniciens Municipaux and the Association des Ingenieurs des Villes de France. Industrial waste disposal is also conducted privately by a number of corporations and private contractors.

In Italy, the management of solid wastes has been neglected from the point of view of legislative control and regulation. The only existing law in Italy regulating solid waste disposal or refuse management is Public Law number 366 passed on March 26, 1941. This law is still in force at the time of this writing (1974), and calls for the recycling of all refuse; specifically, the destruction of refuse is forbidden. Article 22 of this law even calls for compulsory establishment of plants to salvage the contents of refuse for industrial and agricultural utilization. This law has not been implemented to any great degree. Legislation to replace this antiquated law has been introduced in various chambers of the Italian government. The proposed legislation is more realistic than the 1941 law in that the requirement for complete reutilization of refuse has been abandoned. However, this replacement legislation has not progressed through the parliamentary procedures necessary for enactment. Consequently, refuse management in Italy at the present time is regulated by the provisions of a law over 30 years old, which was passed during a period of economic uncertainty and wartime conditions. However, in 1971, the National Association of Sanitary Engineers of Italy established a commission to investigate the control of refuse disposal procedures. This commission, supported by the Polytechnic Institute of Milan, has promulgated a set of regulations for the organization and operation of various refuse disposal facilities. These regulations, which possess no enforcement authority, simply constitute a technical guide for refuse managers in Italy. It is anticipated that federal legislation governing the management of refuse will be forthcoming in Italy in the near future.

In summary, it appears that the overwhelming majority of solid waste management activities carried out in European countries are conducted by governmental agencies. A much smaller percentage of management operations (principally collection operations) are carried out by private contractors. Legislative control programs for waste management vary in character from quite comprehensive regulations and criteria, exemplified by German standards, to rather antiquated rules and regulations such as the waste disposal laws of Italy.

8-2.1 Refuse Characteristics

A number of previous studies have been published, describing the composition, generation rates, and particular management practices for refuse in Western Europe (Refs. 8-6, 8-7, 8-10, 8-13, 8-14, 8-15, and 8-19). Because these previous studies have been quite general in character or are now somewhat outdated, it is pertinent to describe the characteristics of the refuse generated in Western Europe. Table 8-1 lists approximate data on the composition and generation rates of solid wastes in six countries of Western Europe, with approximate corresponding data for the United States. There are some significant differences between the refuse composition in many of the European countries and the refuse composition in the United States. In general, the percentage of material of organic character is considerably higher in the European refuse; the relative amount of organic material appears to decrease with the relative industrial growth of any given country. In a similar way, the paper content of European refuse is not as high as the paper content of American refuse, but in the highly industrialized countries of Western Europe, the paper content is higher than in those countries which are dependent upon agriculture as a basis for the national

TABLE 8-1 Approximate Composition and Generation Rates of Solid Wastes, Western Europe

Component	UK	France	The Netherlands	Germany (BRD)	Switzerland	Italy	US
Garbage (organics)	27	22	21	15	20	25	12
Paper	38	34	25	28	45	20	50
Fines	11	20	20	28	20	25	7
Metal	9	8	3	7	5	3	9
Glass	9	8	10	9	5	7	9
Plastics	2.5	4	4	3	3	5	5
Miscellaneous	3.5	4	17	10	2	15	8
Average water content, %	25	35	25	35	35	30	25
Caloric content BTU/lb	4200	4000	3600	3600	4300	3000	5000
Generation, lb/cap/yr	700	600	455	770	550	465	1800

economy. Also, as is evident in Table 8-1, there is a considerably greater amount of dust and fine-sized materials in the European refuse. Additionally, the plastics content of European refuse is somewhat lower than American refuse but this plastic content is rapidly increasing. The average moisture content for European refuse is somewhat higher than the average moisture content for refuse in the United States. Of particular interest is the fact that the heat content, or caloric content, of the European refuse is considerably lower than that generated in the United States, principally because of relatively lower percentages of paper and plastics in the European refuse. Finally, the generation rate for domestic and commercial refuse in European countries is only a fraction of the generation rate for American refuse, as is evident in Table 8-1. Nevertheless, there appears to be a significant trend toward increasing generation rates in all of the countries listed in Table 8-1. At the present time, the volume of refuse generated per capita each year appears to be doubling every 18 to 20 years in the Federal Republic of Germany. Similar trends exist for generation rates in the United Kingdom, France, and Switzerland. Slightly smaller increases in generation rate appear to be occurring in the Netherlands and in Italy. The character of refuse in all of the European countries appears to be undergoing significant change. The relative amount of fine dust and ashes in the refuse has decreased in the last 100 years from over 80 percent to less than 20 percent in many nations. In that same time period the amount of paper has increased from an insignificant amount to as much as 45 percent of the total weight of refuse. Likewise, in the last 50 years the percentage of metals has doubled and the percentage of glass has more than tripled. Of course, plastics essentially are a postwar phenomenon and are rapidly increasing in use. The plastics content of the refuse in France, particularly, is rapidly increasing because of increased use of plastics in water and wine bottles. Plastic wastes are increasing rapidly in Italy also, but the use of plastic bottles for wine is forbidden by law in that country. Other plastic packaging seems to account for the increasing appearance of plastic in the solid waste stream. The relatively greater amounts of organics and fines in the European refuse indicates that composting may be more successful in Europe because of compositional differences in refuse between European nations and the United States. On the other hand, the smaller relative amounts of paper in the European refuse would indicate a lesser potential for heat recovery form European refuse as compared to American solid waste; the heat content of the refuse in Europe amounts to only 70 to 80 percent of the heat content of American wastes. However, because of the greater value of fuel supplies in many European nations as compared to the United States, heat recovery from waste incineration has been more widely practiced in European nations than it has been in the United States.

For further detailed information on European waste characteristics, the references previously mentioned and Refs. 8-22 and 8-25 contain much comprehensive descriptive material.

TABLE 8-2 Collection Characteristics, Western Europe

(a) Collection Characteristics
(Approximations based on total population served)

Country	Frequency of Collection, No. per Week	Separation		Point of Pickup		Collections	
		Percent Separate	Percent Combined	Percent Curb or Building Line	Percent Backyard or Building	Percent Municipal	Percent Contract
U.S.	10 percent: 3–6 47 percent: 2 43 percent: 1	50	50	50	50	90	10
England	1–5	90	10	10	90	95	5
France	10 percent: 7 40 percent: 6 50 percent: 3	5	95	100	–	100	–
Germany	100 percent: 2	–	100	–	100	100	–
Scotland	50 percent: 1 50 percent: 2	100	–	15	85	100	–
Sweden	50 percent: 1 50 percent: 2	–	100	–	100	15	85

(b) Refuse Containers

Country	Size, Gal Range	Size, Gal Median	"Hermetic" Collections	Ownership Percent Municipal	Ownership Percent Private	Maintenance Percent Municipal	Maintenance Percent Private	Use of Standard Size Fully Covered Containers, Percent	Empty Weight, Lb Range	Empty Weight, Lb Median
U.S.	10-50	25	–	5	95	5	95	50	7-25	17
England	10-25	20	–	20	80	20	80	80	–	25
France	5-20	15	25	10	90	10	90	100	–	15
Germany	20-30	25	30	80	20	80	20	100	25-50	50
Scotland	20-30	20	–	–	100	10	90	100	25-50	25
Sweden	20-65	35	35	50	50	50	50	100	25-60	55

(c) Collection Truck Data (Municipal and private)

Country	Population Served, per Truck Range	Population Served, per Truck Median	Truck Capacity, Cu Yd Range	Truck Capacity, Cu Yd Average	Type of Collection Body, Percent Noncompaction	Type of Collection Body, Percent Compaction	Type of Collection Body, Percent Dustless	Average Age, Yr	Average Purchase Cost, $	Capital Cost, $/1,000 Population Yr
U.S.	3,000-5,000	3,700	10-24	16	25	75	0	5	10,000	350
England	4,300-7,400	5,200	10-30	15	55	40	5	5	6,500	250
France	4,600-6,400	5,500	13-26	18	15	75	10	15	–	–
Germany	11,000-18,000	13,000	12-17	16	–	–	100	12	14,000	–
Scotland	5,200-7,200	6,000	7-25	18	65	25	10	5	8,000	260
Sweden	3,800-6,100	5,000	40-45	40	–	–	100	10	12,000	260
Switzerland	–	7,400	10-24	20	20	–	80	11	–	310

Sources: Rogus, C., "Refuse Collection and Disposal in Western Europe: Part I," *Public Works*, Vol. 93, No. 4, April, 1962. Rogus, C., "Refuse Collection and Disposal in Western Europe: Part II," *Public Works*, Vol. 93, No. 5, May, 1962.

8-2.2   European Collection Practices

Although this discussion has been primarily directed toward waste disposal methods and techniques for resource recovery, some mention may be made of European collection practices, since these practices have some influence on the planning and design of disposal facilities. This information may be useful to the practicing solid waste manager.

Table 8-2 contains comprehensive data collected by Casimir Rogus in a previous study of European solid waste management. The data contained in this table reflect the current collection situation in Europe, for the most part, with very few exceptions.

In almost all of the countries of Western Europe, with the possible exception of Scandinavian nations, the responsibility for refuse collection devolves upon municipal or governmental authorities. The refuse collection role of these authorities has been described in a previous section. In particular, in many municipal areas in Western Europe, the collection of refuse is done virtually on a daily basis. For example, in the city of Paris, refuse is collected each day of the year with the exception of holidays such as Christmas Day. As a general rule in France, in cities with population greater than 50,000, refuse is collected every day except Sunday. In smaller towns, the collection frequency varies with local needs. In Paris, the city government purchases compactor trucks and collection vehicles and then arranges for the actual collection to be done by private contractors who operate with the aforementioned trucks under franchise from the city government. Approximately 780 miles of collection routes along the streets of Paris are covered each day by approximately 800 trucks. Refuse collection begins when the collector crews meet at a central spot in one of the collection districts in the Paris area. The crew assembles and begin collection at precisely 6 A.M. All residents of Paris are supposed to place refuse for collection in covered containers no later than 5 A.M. each morning. Supposedly, all refuse will be collected by 8 A.M., but in the event of particular difficulties such as poor weather conditions, refuse collection may not be complete until approximately 9 A.M. Each crew makes only one trip along a collection route and then delivers the refuse to a transfer station or disposal facility. After the refuse collection is complete, the collection crew is employed for the remainder of the day in various municipal activities such as street cleaning. There is some indication that because of union pressure, collection on Sundays and holidays will be eliminated in the city of Paris in the near future.

A somewhat similar situation exists in the city of Rome, where approximately 800 collection trucks are used to collect municipal refuse. Each collector truck makes only one trip along a collection route and to a transfer or a disposal facility. However, the practice in Rome differs from that in Paris in that the collection crews are released after completion of the refuse collection; no further duties are assigned to the drivers and tippers after their 3 to 4 hour stint of refuse collection is completed.

In the Netherlands and in the Federal Republic of Germany, collection practices with regard to the allocation of men and trucks more closely resemble the conditions in the United States. For the most part, municipalities are responsible for collection of refuse, although some very small towns contract with private haulers to perform this service.

A wide variety of collection trucks and collection containers are used in European countries. Figure 8-1 shows some of the collection equipment typical of European practice. The use of standardized containers for refuse collection formerly was prevalent in many European nations. (see Table 8-2 (b)). However, in recent years, the use of plastic bags for refuse collection has increased considerably. In some European cities, the municipal collection agency even furnishes the plastic bags to the homeowner or commercial facility for the deposit of refuse. The standard refuse containers, until recent years, have been a type of fully-covered metal container, typified by the German "dustless" container which has a volume of 110 liters. In addition some smaller containers in the range of 35 to 50 liters in volume are used in Germany. Also, plastic bags are used to a certain extent. Finally, large containers for the deposit of refuse in institutions, shopping centers, or other points of refuse concentration are becoming more and more popular. The so-called dustless containers were developed during the time when most homes were heated by coal or wood. A very substantial and strong container was needed in order to support the weight of the ashes from the coal or wood fires. Moreover, a container was required to withstand the fires which occasionally resulted from placement of hot ashes in the container with other solid waste. Consequently, a heavy gauge, galvanized metal container was developed with a heavy flange around the bottom rim of the container. In order to prevent the scattering or blowing of dust and ashes from the containers, a heavy hinged lid was developed. These large metal containers obviously are quite heavy (the 110-liter container weighs over 40 lb empty) and when full of solid wastes the container is too heavy for a single man to lift. Therefore, the containers have been equipped with a thick metal flange on the bottom rim, as mentioned previously, so that the full containers can be rolled on the rim in the same way that heavy barrels are rolled. The refuse collector simply grasps a knoblike projection on the container (placed there for this purpose), and with his other hand rotates the container on the bottom edge. The refuse collector rolls the full container to the back of the collection truck. There a specialized tipping mechanism is brought into service. Lifting arms on the tipping mechanism engage two hinge points on the container and the container is lifted and rotated. A hoodlike device on the tipping mechanism engages the top

(a)

(b)

Fig. 8-1 (a) Refuse collection truck equipped with device to lift and unload collection containers, Rome; (b) Rome refuse collection truck, body open; (c) collection cart for refuse in Rome.

(c)

Fig. 8-1 (*Continued*)

of the container while it is being lifted and rotated so that the container is within the hood when the refuse is dumped into the truck; in this way, the so-called dustless dumping is achieved.

Because of the increasing use of oil and natural gas in space heating in Western Europe, the ash content of refuse has considerably decreased in recent years. Therefore, the need for heavy standardized metal containers has likewise decreased. Consequently, a trend has developed toward smaller and lighter collection containers, and, as mentioned previously, plastic bags are being used to an ever greater degree.

The number of tippers assigned to each collection truck in European countries varies considerably with local conditions and with the type of containers used in local collection efforts. Some typical data on collection practices are shown in Table 8-3, which shows some specific data for collection practices in the Federal Republic of Germany. An interesting feature of this table is the comparison in the collection efficiency, in terms of volume collected per tipper per shift, and the consequent collection cost for various types of containers used. The smaller, lighter containers appear to hold considerable advantage over the somewhat outdated 110-liter galvanized metal containers. However, the very large bulk containers which are used at points of waste concentration appear to hold considerable advantage in terms of collection efficiencies and consequent reductions in collection costs.

TABLE 8-3 Collection Practice in Federal Republic of Germany

(a)

Collection Area	Collection Time, Hr/Ton	Collection Crew	Collection Cost, $/Ton
Stuttgart, pop. 600,000	0.48	1 driver 4 tippers	21.00
Hemer, pop. 25,000	0.33	1 driver 2 tippers	10.50
Letmathe, pop. 27,000	0.38	1 driver 2 tippers	11.50
Average community costs pop. 10,000	0.36	1 driver 2 tippers	11.00
pop. 5,000	0.39		11.70
pop. 1,000	0.46		14.00

(b)

Container Size	35 Liters	50 Liters	110 Liters	1,100 Liters
Containers emptied per tipper per shift	700–1,000	600–900	100–170	40–60
Volume collected per tipper per shift	39 cu yd	48 cu yd	19.5 cu yd	7.15 cu yd
Waste density in container, lb/cu yd	420	420	340	305
Number of tippers per collection truck	2	2	4	2
Collection cost, $/ton	12.30	9.70	21.20	9.40

Source: Miethe, M., "Kosten der Abfallbeseitigung," 6th Fachlergang an der Universität Stuttgart fur Mull-u, Abfallbeseitigung, Herbst, 1973.

The principal effect of the collection practices in central Europe appears to be a normalizing influence on the amount of waste delivered to disposal facilities in those localities where refuse is collected each day. Also, because of the daily collection, adverse effects of the weather may be minimized. If refuse is placed outside shelter, and the refuse has accumulated for a period of one half-week or more, the possible consequences of rain or another form of percipitation in raising the refuse moisture content are amplified. If refuse is placed outside for collection each day, in the event of rain, a smaller amount of refuse is susceptible to wetting.

It is necessary to note, however, that local conditions may completely alter any normalizing effect of daily collection or similar practices. For example, in the city of Paris where refuse is collected daily, the Parisian habit of going on holiday in the month of August has a drastic effect on the total amount of refuse which is generated and collected in that city during the month of August. The refuse generation decreases to only 5 to 10 percent of the average generation rate for the remainder of the year. Consequently, many managers of disposal facilities in the city of Paris program maintenance activities and equipment downtime for the month of August when the service requirements for disposal of refuse are significantly lower than at any other time during the year. Many other examples of the influence of local conditions or customs could be cited.

Because of the tendency in some parts of Western Europe toward regional disposal facilities and localized collection systems, there has arisen a need for long-distance transport of wastes. In such systems, transfer stations wherein refuse is processed to obtain greater transport densities are virtual necessities.

8-2.3   Transfer Stations

Two facilities may be described as typical of the more modern and efficient refuse transfer stations which are operating in Western Europe at the present time. These two facilities are the Cringle Dock refuse transfer station in London and the Apeldoorn refuse transfer station in east central Holland.

The Cringle Dock transfer station is located on the Thames River at the foot of Cringle Street in Wandsworth near the Battersea Power Station. The transfer station incorporates refuse pulverizing equipment in order that the payload of transfer barges can be increased significantly. Also, the pulverized refuse is thought to be more suitable for land reclamation purposes in the chalk pits and old quarries in Essex and Kent where the refuse is deposited. The refuse is brought into the transfer station over a set of scales consisting of two units each 40 ft by 10 ft; these platforms have a load capacity of 40 tons. The scales are monitored through an electronic data collection unit and the vehicle identification code, tare weight, refuse type, and net weight of refuse are registered on paper tapes which are later processed through a centralized computer system operated by the Greater London Council. In order to determine the tare weights of collection vehicles which are not regular customers of the station, these vehicles are weighed in and weighed out. The vehicle entering the transfer station leaves the scale area and is directed to one of the eight tipping bays through a system of automatic traffic controls using indicator lights at the entrance to the building. After the collector truck has entered the building, a safety barrier is lowered at the appropriate tipping bay. This safety barrier is a common feature of tipping areas in refuse management facilities throughout Europe. When the collection vehicle leaves the tipping apron, the safety barrier automatically rises and a hydraulic ram pushes the deposited refuse into one of the main bunkers of the facility. There are two main bunkers, each 25 ft by 48 ft by 50 ft deep. The storage capacity in these bunkers is approximately 850 tons, more than the normal day's intake of approximately 800 tons. The refuse is transferred from these bunkers by two overhead traveling cranes. Each crane is equipped with an 8 cu yd clamshell grab with 9 tons capacity. The grabs deposit the refuse on vibrating conveyors. The feed conveyors are vibrated to reduce peaks in loading; the vibrating conveyors feed plate conveyors located at a lower level. The plate conveyors transfer the refuse into two Wakefield pulverizers, constructed by British Jeffrey-Diamond, Ltd. These pulverizers are heavy-duty devices capable of reducing bulky wastes such as old freezers or refrigerators as well as more standard domestic and commercial refuse. The design capacity of the pulverizers is 50 tons of refuse per hour per unit; the normal operational through-put for the units is approximately 33 tons per hour. The refuse emerging from the pulverizers consists of particles 80 percent of which measure less than $1\frac{3}{4}$ in. in any dimension. The remaining 20 percent (by weight) measures no more than 6 in. in any dimension. Explosion suppression equipment and water sprays are provided on each pulverizer. In the event of an explosion hazard, an inert gas under high pressure is released into the pulverizer units. Two elevating conveyor belts carry the pulverized refuse from the pulverizer under electromagnetic separators and over the central jetty in the dock building to the barge loading apparatus. The magnetic separators recover ferrous metals, which at the present time amount to approximately 6 percent by weight of the refuse input. The refuse is loaded into barges through a telescopic chute. Four barges of the type used for refuse transport on the Thames can be accommodated simultaneously in this facility. Additionally, provisions have been made to divert refuse from the barge loading area to bulk transfer vehicles for over-the-road transfer of refuse to disposal facilities. An interesting feature of this facility is the provision for dust control. Two dust extractor units serve the dumping bays and two additional units serve the two feed conveyor areas. These wet-type dust collectors principally are provided to improve working conditions. The barges used for refuse transfer can be loaded with approximately 180 tons of pulverized refuse; the same barges can hold only about 100 tons of raw refuse. The transfer station is operated on a normal $5\frac{1}{2}$ day week from 7 A.M. to 10 P.M., Monday through Friday and until noon on Saturday. A manager, scales clerk, and 8 men comprise the normal labor force on each of the two shifts. Also, a day-shift staff of 6 maintenance men includes a mechanic and an electrician. The total capital cost for this facility was approximately $5.8 million. At the present time, the cost for pulverization of the refuse is approximately $8.75 per ton. Figure 8-2 illustrates the barge loading facility at the Cringle Dock transfer station.

A somewhat different transfer station is operated at Apeldoorn in east central Holland by the refuse disposal agency Vuilafvoer Maatschappig (VAM). The transfer station at Apeldoorn is shown in Fig. 8-3. Refuse is brought to this transfer station by private citizens of the surrounding community or in municipal collection trucks. When a dumping vehicle approaches, the doors to the dumping area are opened and the vehicle enters onto a dumping apron. When the vehicle has entered the station, the doors to the dumping area close to prevent any dust or debris from leaving the plant. Whenever very dry conditions are

Fig. 8-2 Refuse, previously shredded, placed in barges at Cringle Dock transfer station, London.

encountered as a result of waste characteristics or as a result of climatic conditions, a water spray in the storage pit may be activated to aid in dust control. Refuse is dumped from collection vehicles into a storage pit at the facility. Bulky noncompostable items are placed in one portion of the transfer station and compostable refuse is placed in the other half. The principal purpose for this transfer station is the transfer of waste through the VAM system to compost plants located at a considerable distance from the generation area. The dumping floor is at the same level as the outside ground surface. The refuse is dumped into the deep storage pit at the facility and then lifted with grab buckets out of the storage pit and dropped into the top of a loading chute. The loading chute resembles an inverted truncated pyramid. The refuse drops through this chute and into waiting railroad cars of special design. The railroad cars are equipped with special gates along the top of the car which may be opened so that the car may be loaded from the top. When the railroad cars are full, they are moved out of the plant by means of cables connected to the cars (see Fig. 8-3(b)) and operated by equipment within the transfer station itself. Thus, when the railroad cars beneath the loading chutes are full, operators inside the transfer station may pull the full cars forward out of the plant and replace the full cars with empty cars beneath the refuse loading chutes. A special feature of the Apeldoorn transfer

(a)

(b)

Fig. 8-3 (*a*) Exterior Apeldoorn, Netherlands, transfer station; (*b*) refuse transport cars, VAM system.

station is the automatic operation of the transfer cranes and grabs. A compu-terized system enables an operator to set the controls on one of the cranes for completely automatic operation. The crane is systematically lowered into the different sections of the storage pit in a continuous fashion, loading the empty railroad cars, until the storage pit is completely empty. Load sensors in the grab device itself transmit appropriate signals to the control mechanism; when the grab is raised with no load, the sensor triggers a movement of the crane to an area of the storage pit not previously explored. After all portions of the storage pit have been explored, the crane system is automatically inactivated by the computerized control facility. The bulky wastes which are deposited in a sepa-rate section of the facility are transferred by the same type of crane, without the automatic control system. The railroad cars are arranged into a train and moved from the Apeldoorn area in the late afternoon and travel approximately 90 miles to a composting plant. At the composting plant, the cars are unloaded, bulky wastes are stockpiled for possible further processing, and compostable wastes are deposited in open windrows. The empty railroad cars are returned to the transfer station at Apeldoorn by early the next morning. The highly automated character of this transfer station is quite interesting; two operators maintain complete con-trol of the dumping operations, the transfer operations, and the movement of the railroad cars. Additionally, as the railroad cars leave the transfer station, a water spray is employed to wash each closed car. This water spray is shown in Fig. 8-3(b).

Numerous additional examples of sophisticated, highly automated transfer stations could be given here. This is one field of operation in which European practice may be significantly advanced over American practice.

8-3. DISPOSAL OF WASTES AND RESOURCE RECOVERY

Many ambiguous statements have been made in professional journals and similar literature concerning the widespread use of refuse incinerators for power genera-tion in Europe. Similarly, widespread use of composting as a refuse disposal method has been suggested. In reality, the use of composting as a refuse disposal method is severely limited in Western Europe. Additionally, refuse incineration with power generation is practiced only in certain localities, principally in large cities where a ready market exists for generated power. The predominant dis-posal method in Western Europe appears to be land disposal of refuse.

8-3.1 Disposal Methods in Western Europe

It is virtually impossible to determine exactly the fate of all the refuse generated in any given country. The reliability and comprehensive character of information on refuse management varies from one country to another in Western Europe.

However, an overwhelming impression is received (and substantiated by reliable comprehensive data in some cases) that land disposal of refuse is a predominant method of solid waste disposal in Western Europe. Table 8-4 shows percentages of waste disposed of in each type of system. The percentages shown in Table 8-4 pertain only to collected domestic and commercial wastes. A significant portion of all the waste generated in any given country is not collected. Estimates of the amount of total waste generated which are collected in each country are given in Table 8-4. Additionally, a distinction is made between controlled and uncontrolled landfill disposal of refuse. The term "controlled landfill" signifies the same requirements for landfilling as does sanitary landfill in the United States, with the exception of the provision of daily and intermediate cover. In some instances, controlled landfills in Europe are provided with underdrainage systems to collect leachate from the completed fill and prevent its migration into groundwater reservoirs. As is obvious in Table 8-4, landfill is the predominant method of waste disposal in almost all countries of Western Europe with the exception of Switzerland. The conditions in Switzerland favor incineration of refuse in that large concentrations of population exist in major urban areas, sites suitable for sanitary landfills are not readily available, and a large market exists for power produced from refuse incineration. Futhermore, over thirty-five incinerators are in operation in Switzerland at the present time and of these incinerators thirteen are equipped to produce power. Although the greater number of incinerators are not equipped to produce power, those which are capable of energy recovery service three times as much population as do the units without heat recovery facilities. In general in Switzerland, the incineration facilities located in all large urban areas are equipped with heat recovery facilities. Of the incinerators equipped with power generation facilities, only three serve areas with population of less than 120,000 persons. On the other hand, only two of the incinerators

TABLE 8-4   Waste Disposal Methods Used in Western Europe
(Estimated percentages of collected domestic and commercial wastes processed in each method)

Disposal Method	United Kingdom	France	The Netherlands	Federal Republic of Germany	Switzerland	Italy
Incineration	15	20	23	20	53	13
Composting	–	10	16	2	13	1
Landfill	85	70	61	78	34	86
Controlled	(60)	(7)	(12)	(16)	(4)	(6)
Uncontrolled	(25)	(63)	(49)	(62)	(30)	(80)
Estimated percent of total wastes collected for disposal	88	80	87	90	90	70

which are not equipped with heat recovery service populations of over 90,000 persons. The average size of the metropolitan area served by heat recovery incinerators is approximately 180,000 persons; the conventional incinerators without heat recovery serve metropolitan areas with an average population of 36,000 people. This situation in Switzerland is reflected also in other large metropolitan areas in other European nations. Most of the refuse incinerators in Europe are located in cities with populations of over 100,000, and most of those incinerators are equipped with some form of energy recovery system.

Another condition which favors the incineration of refuse and accompanying heat recovery for power generation is the greater relative value of fossil fuels in Europe as compared to the United States. As was shown in Table 8-1, the heat content of European refuse generally is considerably lower than the heat content of American wastes because of a lower percentage content of paper and plastics. Therefore, the situation with respect to incineration of refuse for heat recovery would appear to be less favorable in Europe than in the United States. However, the relative abundance of cheap fuels in the United States has militated against the use of refuse as a fuel for power production because other fuels of greater homogeneity have been available at low cost. In contrast, in Western Europe such fuels have not been in as great a supply as in the United States. Consequently, power production from wastes is favored in Europe.

With respect to composting, Table 8-4 indicates that composting is of little significance in the United Kingdom, Italy, and the Federal Republic of Germany. There are only one or two very small composting plants operating in Great Britain. Likewise, only a very small number are operating in Italy. In Germany, over twenty plants are operating at the present time, but only one plant, at Flensburg, is of significant size. Consequently, the twenty or more plants process only about 2 to 3 percent of the refuse collected in Germany. In France, there are more than forty composting plants in operation at the present time (1974). However, of these forty plants, only seventeen of the plants have operational capacities greater than 20 tons per day, and only five have capacities greater than 50 tons per day. The average cost of open windrow composting in France is approximately $1.60 per ton, and accelerated composting costs approximately $4.60 per ton. However, the costs for composting vary considerably with the size of the plant. Additionally, the relative location of the plant with respect to market areas in France has influence on the price of the composting. Costs for composting operations generally are considerably higher than the costs for landfilling, except where the compost can be marketed readily. Much of the compost in France is used in vineyards (approximately 60 percent). Also, about 17 percent of the compost produced in France is used in the cultivation of mushrooms. The amounts of compost used in these activities has increased dramatically in recent years. Most of the compost used in vineyards is transported to the Champagne and Bordeaux regions. A very minor part of the compost is used in conventional agriculture for the growing of food crops such as corn and wheat (about 8 percent). The remainder is used in specialized forms of horticulture and arboriculture.

In Switzerland, also, composting accounts for an appreciable amount of the total refuse disposal effort. However, most of the composting plants in Switzerland, which number more than sixteen at the present time (1974), are combined composting and incineration operations. In other words, about two-thirds of the composting plants in Switzerland are combined operations in which the organic fraction of the collected solid wastes is composted and large pieces of paper and plastic and other essentially noncompostable materials are incinerated, after which the incineration residue is landfilled. As in France, much of the compost produced in Switzerland is used in land reclamation and improvement. Compost has been found to be a suitable material, in Germany as well as in Switzerland, for use in vineyards and other cultivation along steep mountain slopes. The compost effectively prevents soil erosion in these areas. Some studies have shown that as much as 3 in. of soil have been eroded in the last 100 years from vineyard slopes along the banks of the Rhine River. The compost product appears to be quite useful in decreasing the amount of erosion which occurs.

Composting accounts for approximately one-sixth of all the refuse disposal in the Netherlands, and is more important in the Netherlands than in any other European country. However, at the present time (1974), only six composting plants are in operation in the Netherlands. The largest plants are located at Wijster and Mierlo and are operated by VAM. The VAM organization was founded in the middle 1920s, as a result of a decision of the government of the Hague to operate a processing facility to create compost from municipal solid waste and to sell it to farmers in the northern part of Holland. In the northern sectors of Holland, the top layers of soil consist of peat which has been removed for sale as a soil conditioner and as a fuel. The underlying parent material which is not suitable for raising crops has been exposed in wide areas of northern Holland. Therefore, it is highly beneficial to apply a layer of compost to the underlying parent material to improve the workability of the soil and to make soil nutrients available to plants. The VAM plant at Wijster will be described in detail in a later section. In addition to the two large VAM plants, four municipal composting plants are now in operation in the Netherlands: Arnhem, Baarn, Zaadam, and Haarlem Mer Meer. At Arnhem and Zaandam, Dorr-Oliver rasp mills are used in the compost plant and the compost is placed in windrows for curing. The Arnhem plant is scheduled to be closed in 1975 and the Zaandam plant is scheduled to be closed in about 1976. The two remaining small plants, at Baarn and at Haarlem Mer Meer, utilize the Dano process. They also use open windrows for final curing of the compost. The Wijster plant and the Mierlo plant utilize what is known as the Van Maanen system. In addition, at Mierlo, two rasp mills have been added to the plant to speed up the process. In general, all of the small municipal composting plants in the Netherlands are facing closure because of their incapacity to handle the large amounts of wastes being generated.

As mentioned previously, controlled tipping and landfill operations are similar to sanitary landfill operations in the United States with the exception that daily cover and intermediate cover of the refuse are not required in European practice. Although landfilling techniques are the predominant disposal operations in most of the Western European countries, only in Great Britain are these operations primarily controlled tipping operations. For example, in the Federal Republic of Germany, although landfilling accounts for approximately 78 percent of all the total refuse disposal effort, only about one-fifth of all the landfills are truly controlled landfills. However, at all landfills in Germany, a drainage system for collection and treatment of leachate must be provided in accordance with federal law. Drainage systems are also provided in the controlled landfill facilities in the United Kingdom. It is interesting to note that despite the copious publicity given to European heat recovery incineration and composting efforts, in most of the countries of Western Europe more than half of all the refuse disposal consists of uncontrolled land dumping.

The costs for waste disposal in the countries of Western Europe vary tremendously from one country to another and from one city to another within a given country. It is virtually impossible to give comprehensive data for comparison purposes between disposal costs in the United States and disposal costs in all of the countries of Western Europe. However, some isolated instances may be described and some data may be given on disposal costs. Table 8-5, for example, shows total disposal costs for solid waste in the Federal Republic of Germany at

TABLE 8-5 Total Disposal Costs in Federal Republic of Germany, $/ton

Suburban Community pop. 80,000	Cost of Collection	Cost of Transport	Cost of Disposal	Total
Composting			17.00	35.00
Controlled Fill	11.50	6.50	2.90	20.90
Incineration[a]			22.40	40.40
Suburban Community pop. 500,000				
Controlled Fill	11.50	8.00	2.50	22.00
Incineration[b]			8.00	27.50
Metropolis pop. 500,000				
Incineration[b]	20.00	3.60	8.00	31.60

[a]No heat recovery.
[b]Heat recovery, with attendant revenue.

Source: Miethe, M., "Kosten der Abfallbeseitigung," 6th Fachlergang an der Universität Stuttgart fur Mull-u, Abfallbeseitigung, Herbst, 1973.

three locations: a suburban community with a population of approximately 80,000 persons; an essentially suburban community with a total population of approximately 500,000 persons; and a densely populated metropolitan area with a population of approximately 500,000 persons. Some general data on costs include the fact that for the most part collection costs are approximately equal or slightly above costs for either composting or incineration. Collection costs are significantly higher than costs for landfilling. In the Federal Republic of Germany, landfill costs vary from about $2.00 per ton to as much as $8.00 per ton, with average costs in the range from $4.00 to $6.00 per ton. Composting costs in Germany vary from about $12.00 per ton to about $20.00 per ton. Incineration costs in that country vary from approximately $12.00 per ton to about $25.00 per ton, with significant reductions in net costs where sale of generated power is possible. In Switzerland, incineration costs vary from about $8.00 per ton to about $30.00 per ton with reductions in net cost for power sales. In Switzerland, collection costs average approximately $20.00 per ton with a range from about $18.00 per ton to about $25.00 per ton, and landfill costs amount to approximately $6.00 per ton. Some costs for composting operations in France have already been given. Incineration costs in France are very difficult to determine because of the complex situation governing the charging of fees and the sale of generated power through semi-public agencies and utilities. Landfill costs in France represent only a small fraction of composting costs. In the Netherlands, incineration costs range from approximately $12.00 per ton to about $18.00 per ton. Composting costs in that country range from approximately $9.00 per ton to about $10.00 per ton. Again, landfill costs amount to approximately one-third to one-fourth of composting costs.

In order to illustrate the techniques and particular characteristics of the disposal operations in Western European countries, some examples of various incineration, composting, and landfill facilities will be given in the following paragraphs.

8-3.2 Incinerator Plants

A number of incinerator plants can be described to illustrate the various techniques of waste incineration and power generation employed in different parts of Western Europe. The first incinerator which will be described is the facility at Edmonton, outside London in the United Kingdom.

As mentioned previously, the Greater London Council assumed responsibility for disposal operations in London in 1965. At that time, the disposal arrangements for the area of North London and the associated boroughs in that area were very inadequate. Consequently, it was decided to build a new incinerator and a site was chosen near a recreational development. This development is the Lee Valley Regional Park. To the west of the incinerator site, the development is primarily unattractive industrial facilities, and therefore the incinerator complex serves as a useful transition from the open space of the Lee Valley Park to the unattractive industrial developments. The incinerator building is 320 ft long and 100 ft high with a chimney 320 ft high. To minimize the impact of this complex of buildings on the aesthetics of the regional park, the main complex of buildings has been designed in a compact arrangement and the buildings are oriented at an angle to the Valley Park to present a minimum of building facade to the park. Additionally, the incinerator facilities have been covered with materials whose colors differentiate and define the various subelements of the incinerator structure and somewhat modify the impact of the facility. The careful landscape and design of the facility does much to mitigate the undesirable effects of having an incinerator plant near a greenbelt park.

The refuse collection vehicles enter the incinerator facility between the hours of 10 A.M. and 4 P.M. An average of 1800 tons of crude refuse is delivered to the plant each day in 600 to 700 separate truckloads. The collection vehicles enter a dual road system which allows continuous circulation around the main block of buildings. The road system has been built on two levels with ramps to elevate the delivery vehicles to the position of the tipping apron. As a vehicle enters the plant, it passes over an automated scale which has been fitted with electronic data collection apparatus. Each platform scale handles a vehicle every 40 seconds. The collection vehicles then proceed up a ramp to the enclosed tipping area. Ultrasonic detectors at each tipping bay indicate when a bay is occupied by a collection vehicle and a lighted number system directs drivers to an empty bay in one of the twenty-three unloading bays. One of these bays has been reserved for bulky refuse which is fed into a shearing machine before placement into a furnace bunker (see Fig. 3-17). Refuse is dumped from the collection vehicles into a shallow loading bay. Then after the collection vehicle has moved away from the tipping apron and the edge of the tipping bay, the refuse is discharged from the bay into the furnace bunkers. There are five large bunkers for domestic and commercial refuse with a small bunker, as mentioned previously, for bulky materials. The bunkers are 80 ft deep and have a total capacity of about 4300 tons. The normal daily input to the plant is approximately 1450 tons and the peak capacity is approximately 1800 tons. Refuse is transferred from the bunkers by two overhead grab cranes. A third crane is provided as a standby unit and for use when a fifth boiler at the plant is in service. Each crane has a feed rate of approximately 30 tons per hour. The refuse is transferred into furnace feed chutes and is pushed onto the furnace grates by hydraulic rams. To prevent escape of dust from the buildings, five fans (48,000 cu ft per minute capacity each) have been located on a floor above the cranes and are used to draw air through ducts from the tipping bays, the bunkers, and the furnace feed chutes. The extracted air is passed through wet dust collectors and is then returned to the atmosphere; the wet sludge from the dust collectors is fed back into the refuse bunkers. The refuse is fed into four boiler units; a fifth boiler unit is not normally in operation but is held in reserve as a standby unit. Each combustion unit is capable of consuming approximately 15 tons of refuse per hour. The combustion chambers are equipped with VKW stepped roller-type grates. These are the so-called "Dusseldorf" roller grates. In each chamber, an inclined series of seven parallel hollow rollers 5 ft in diameter and 11 ft 6 in. long carry the refuse through the combustion area. Each roller is controlled through an infinitely-variable gear arrangement to provide a speed range from 0.5 to 5 mph. The speed of rotation and the air feed through each grate is controlled individually to suit refuse conditions. The combustion temperature is maintained between 1700 and 1900°F. When the refuse has passed over the final roller, the residual material is passed through a quench bath onto conveyors which carry it to the residuals handling area. Exhaust gases from the boiler unit pass through electrostatic precipitators before discharge to the atmosphere. The precipitators reduce the dust content of the effluent gases to less than 0.05 grains per standard cubic foot. Residual materials coming from the combustion chambers consist primarily of ash and metals. The 6-ft-wide residue conveyors carry this material into a separate residuals building where they are sorted and stored for removal. The sorting operation consists of coarse screening, breaking, and magnetic separation. The ferrous metals which are extracted are baled and sold as scrap. The residual material after separation of the ferrous metals has found a ready market as road building materials. Oversize pieces of nonmetallic residual material are disposed of in a landfill. The flyash which is removed in the electrostatic precipitators also constitutes a residual material. The dry flyash is carried pneumatically from the precipitators to two bulk storage silos and water is added to the flyash to bring it to a slurry condition for compacting prior to disposal via road transport.

The combustion chambers in this incinerator are equipped with water walls and the exhaust gases from the combustion chamber pass through an economizer to aid in the generation of steam. Under normal conditions, generated steam is used to drive three 12.5 megawatt turbo-generators which create approximately 30 megawatts for export and sale. An additional turbo-generator is maintained as a standby unit. Additionally, a 2.5 megawatt generator is used to supply the electrical requirements of the incinerator facility itself. The main generators are direct-coupled units with single cylinder turbines. The generators are equipped with closed-circuit air cooling arrangements incorporating water coolers. Because the heat content of the solid waste is quite variable, a steam pressure governing system has been provided to maintain the pressure in the steam mains without a dependence on the heat output of the boilers. In other words, the boiler firing rate need not be controlled to maintain pressures in the steam lines. This arrangement is virtually a direct opposite to the standard arrangement in a fossil fuel power plant. A condenser is provided in the system to condense any steam which has been produced in excess of the turbines' ability to use the steam. Circulation water from the condensers is passed through induced-draft counter-flow timber towers which reduce its temperature by 15°F. The flow through the towers is 2 million gallons per hour. Makeup water for the closed circulating water system is drawn from the Deephams Wastewater Treatment Works effluent channel in the amount of approximately 1 million gallons per day. Electricity is generated at 11,000 volts. Power supplies for the Deephams Wastewater Treatment Works are supplied first and the surplus is transformed to 33,000 volts for sale to the Eastern Electricity Board.

In general, the Edmonton facility is very efficiently operated and very impressive. Some minor problems have developed in clogging of the grates and clogging of the exit chutes from the electrostatic precipitators. The clogging of the grates has resulted from the precipitation of a slag consisting primarily of light metals, in large part zinc. The zinc is thought to originate in storage batteries which are in ever-increasing use for portable appliances. Operating personnel at the incinerator facility expressed the opinion, however, that the precipitation of the zinc in a slag on the grates was preferable to the emission of some of the zinc into the atmosphere through entrained particulate matter or in the flyash from the electrostatic precipitators. The exit chutes from the precipitators experienced some clogging primarily due to sintering of the collected ash which was only periodically removed. A system for continual removal of the collected ash has eliminated the clogging in the exit chutes. No problems with traffic control and flow have arisen at the facility even though a maximum input of 110 collection trucks per hour has occurred. The vector control at the facility is excellent. No

Fig. 8-4 Dumping area, Edmonton Incinerator, UK.

odors were noted and no insects or other pests were seen after a thorough inspection of the facility. Ultraviolet insect controls are employed in addition to liquid insecticides and solid vaporizing insecticides. In summary, this incinerator operation appears to be very efficient and economical. Metals recovery amounts to approximately 4.5 percent of the input weight of refuse. A constant output of 30 megawatts of electrical power at 33,000 volts is supplied to the Eastern Electricity Board.

The Edmonton incinerator produces electricity and includes a facility for recovery of ferrous metals and residue. A second type of incineration activity wherein electricity and steam for space heating are produced is typified by the incinerator at Ivry, near Paris.

Since 1968, the disposal of refuse in the entire Paris district has been delegated to the city of Paris. The disposal region for the Paris district now covers four departments including the city of Paris and three suburban departments. The towns located in the three suburban departments have the option to join in the disposal effort conducted by the city of Paris or to develop an independent disposal system. In any event, collection of the refuse is performed by each town in an independent effort. Collection in the city of Paris itself has been described in a previous paragraph. In 1973, fifty-four suburban communities had joined the city of Paris in the disposal effort. These communities together with the city of Paris represented a population in excess of 5 million people. In addition to the refuse collected in this area, a certain amount of industrial refuse is also disposed of in the Paris disposal system. A total of 1,863,000 tons of refuse were disposed of in the Paris system during 1972. Of this amount, approximately 64,000 tons were contributed through private sources, principally industrial plants. The disposal practice in Paris is entrusted to a special branch of the Electricité de France called the Service de Traitement Industriel des Residus Urbains (TIRU). This semi-public utility manages waste disposal and by-products recycling under the supervision of a special division of the Prefecture de Paris. The city of Paris pays the operating expenses of TIRU and additionally pays the salary of the managing personnel. In turn, the city of Paris charges the member towns a fee for this service. In 1972, the fee was approximately $10.50 per ton.

The bulk of the disposal in Paris is carried out in incinerators such as the incinerator at Ivry. In these incinerators, steam is used for electrical energy production and also for space heating in Paris and its suburbs. The city of Paris altogether owns four plants specializing in waste disposal located around the city limits. Three of these plants, Saint-Ouen, Issy-les-Moulineaux, and Ivry are incinerators equipped for steam production. The remaining plant, Romainville, is equipped to produce rough compost by sifting and grinding of the material. Approximately 1,900,000 tons of refuse are delivered to these facilities in Paris each year. About 55 percent of the refuse originates in the city of Paris and only about 5 percent originates from industrial sources. The three incineration plants process approximately 1,700,000 tons each year; the remainder of the refuse, approximately 70,000 tons at Romainville and about 90,000 tons in various other operations, is processed in a different way. At Romainville, the refuse is ground and screened through vibrating screens and then shipped to vineyards in the Champagne region where it is cured for approximately 4 to 5 months and then used in the vineyards. The remainder of the refuse which is processed through a number of small operations is disposed of in controlled landfills at some distance from Paris. These landfills are becoming less and less utilized and function principally as auxiliaries for use in case of breakdown at one of the incineration plants. The three incinerators produce approximately 2 million tons of steam each year for space heating and produce approximately 200 million kilowatt-hours of electricity which is sold to Electricité de France. The Saint-Ouen plant which has been in use since 1955 is equipped with four rotary kiln furnaces and produces medium pressure steam for space heating. It is capable of processing approximately 400,000 tons of refuse each year but is not equipped for the production of electricity. The incinerator at Issy-les-Moulineaux has been in use since 1965. It is equipped with four furnaces with Martin grates and is capable of incinerating approximately 80 tons of refuse per hour. This incinerator produces high pressure steam which is used in space heating and in turbo-alternators for electricity production. The incineration capacity is approximately 600,000 tons per year. The plant at Ivry has been in use since 1969. It has two large furnaces equipped with Martin grate systems. These furnaces are equipped with water-wall boilers and are used to produce high pressure steam, in the same

Fig. 8-5   Storage pit, Ivry Incinerator, near Paris.

manner as is done at Issy-les-Moulineaux. The incineration capacity at Ivry is approximately 700,000 tons per year. The incinerator facility at Ivry is somewhat similar to that at Edmonton in that the flow of collection vehicles is designed to permit continuous circulation. Collection vehicles enter the facility, pass over automated platform scales, and proceed up a ramp to a completely enclosed dumping area. The refuse is discharged from the collection trucks into tipping bays which are similar to the bays at Edmonton. These bays are shaped in a triangular or quadrant fashion and the refuse is maintained in the bays until the collection truck has moved away from the front of the bay. At this time, a guard rail which had risen to prevent accidental movement of the truck into the bay lowers and the refuse falls from the dumping bay into the storage bunker of the facility. Twelve dumping bays feed into the single storage pit which has a capacity of approximately 10,000 cu yd of refuse. The collection vehicles leave the plant down a second spiral ramp and leave the facility grounds. Two overhead cranes with polyp buckets of 6 cubic meter capacity transfer the refuse from the storage pit to the feeding hoppers on the two combustion chambers. Additionally, facilities are provided so that in the event of a breakdown the refuse may be transferred to railroad cars or over-the-road trucks for transfer to landfills. The combustion chambers at Ivry are designed to process approximately 44 tons of refuse per hour at a heat content of 2500 kcal/kg, or approximately 55 tons of refuse per hour at a heat content of 2000 kcal/kg. The incinerator combustion chambers are equipped with Martin grates. These grates function in a reciprocal fashion but are unlike the conventional reciprocating grates employed in the United States. The Martin grate is composed of a series of sections. The sections move alternately to push the refuse uphill against the slope of the entire grate system and, therefore, push the refuse uphill against the force of gravity. The uphill action of the grate system creates turbulence in the fuel bed and equalizes the rate of combustion. Designers of the system claim that this upward action pushes burning material under newly charged refuse and accelerates the combustion process. Each individual section of the grate system is composed of large, highly-machined chrome alloy grate bars. These bars are designed with hollow fins on the bottom of each section so that underfire air circulates over the bottom of the bar before it passes upward between the bars into the fuel bed; in this way, the underfire air is used to cool the grate bars. The bars are set very close together and provide great resistance to the movement of the underfire air upward into the refuse; this close tolerance is designed to prevent the fall of siftings downward between the bars and also to provide high velocity in the air stream passing upward through the bars. The Martin grate system, the Dusseldorf system described in the section dealing with the Edmonton incinerator, and the Von Roll grate system are the three types of grates most often used in European incinerators. The Von Roll system will be described in a later paragraph in this section. At the bottom of the grate section, a roller or clinker bar regulates the depth of the refuse bed on the grate system. Residue falling from the end of the grate system falls into a quenching tank and is removed through two ash dischargers onto conveyor belts which remove it to a processing area outside the plant. The boiler system at Ivry consists of a primary chamber with water-cooled walls, two super heaters, and a tube-type economizer. Approximately 140 tons of steam are generated per hour at a design pressure of 96 atmospheres and at a super-heated temperature of 470°C. Approximately 275 tons of steam are produced each hour by the two boiler units together. The steam which is produced in the boiler system passes into a complex system of regulated steam lines to a turbo-alternator and to a space-heating system operated by a separate municipally-controlled agency. The system is equipped with a condenser so that the steam pressure in the system may be regulated. The turbo-alternator produces electricity at 5500 volts for distribution into the Electricité de France system. Additionally, transformers lower the voltage to 380 volts for general current distribution within the plant and within neighboring facilities. The turbo-alternator operates at 3000 rpm and has an output of 64 megawatts. Approximately 150 miles of steam pipes exist beneath Paris for distribution of steam from Ivry and the other two incinerator plants in the TIRU complex. These plants cannot supply the entire space heating needs of the city of Paris and a second company has been formed to create steam through the use of fossil fuels.

The ash which is removed from the residue hoppers via conveyor belts contains metallic particles and inert materials. The ferrous metals are salvaged from this material through the use of overhead magnetic separators. The inert materials are removed for use in road construction by over-the-road trucks. Flyash is collected from the effluent gas stream in electrostatic precipitators. The precipitators are equipped with pneumatic conveyors which empty the precipitator hoppers and discharge the flyash to two storage bunkers. The bunkers are emptied by two conditioning screws with a capacity of 33 tons per hour and the flyash is removed in trucks.

The operation of the incinerator facility and of all the facilities in the TIRU network is designed to be a nonprofit operation. Any expenses incurred over and above the income resulting from the sale of electricity or the sale of steam heat are balanced through the charging of disposal fees to municipal collection agencies and private individuals depositing refuse at the incinerator facilities. A significant portion of the operational budget at Ivry is accounted for in labor costs. The operational force for the incinerator amounts to approximately 150 persons. The actual operating personnel consists of three shifts of 8 workers. Because of holidays and sick days and other emergencies, a standby group of workers is added to the basic number of 24 workers to double the total operational staff to 48 workers. Maintenance personnel number approximately 50 persons, and administrative and housekeeping personnel account for another 50 individuals. Income for the incinerator comes principally from the sale of electricity and the sale of steam heat which is supplied to approximately 100,000 apartments and dwellings in the Paris area (from the Ivry incinerator). None of the flyash collected in the incinerator is sold and only about 50 percent of the grit and siftings are sold. The coarser portions of the residue are used in road covering operations.

As mentioned previously, during August many Parisians go on holiday and the refuse generation in Paris decreases drastically. Therefore, the power production capabilities of the TIRU network also decrease dramatically during August. This lapse is accommodated by the Electricité de France through the use of conventional fuels in operating power production plants. In actuality, the TIRU network generates only a small portion of the total amount of electricity which is needed in France each day. The TIRU power supply is used principally to satisfy peak loading demands on the entire system. Consequently, the operation of the TIRU network is geared to produce maximum income from the generation of either steam or electricity, depending upon the existing prices for steam heat and electrical power.

Only minor problems have been noted at the Ivry incinerator during its short tenure of operation. Corrosion of some of the pipes in the water-wall section of the boilers has been noted. This type of corrosion had been noted previously at the Issy-les-Moulineaux plant and has been described in detail in a previous publication (Ref. 8-3). This type of corrosion has been minimized through the operation of the facility such that temperatures are maintained above the dew point temperature. However, research is continuing into this problem. An additional problem has occurred in attempts to obtain a high degree of purity in the ferrous metal recovered at the incinerator. Although two-stage magnetic separation has been employed, a significant amount of nonmetallic materials appears in the recovered fraction of ferrous metals.

It is interesting to compare the facilities at Edmonton and Ivry with several other incinerator plants on the continent. For example, in the city of Dusseldorf, the municipal incinerator facility greatly resembles the Edmonton facility in that the Dusseldorf system of grates is used in the combustion chambers. The Dusseldorf incinerator plant was begun in 1963 and at present consists of a first stage of four boiler units; eventually the facility will house six boiler units. Three of the existing four units are used in continuous operation with one unit held in reserve. The capacity of each unit is 10 tons of refuse per hour. Twelve dumping bays feed the central storage bunker and overhead cranes with polyp buckets transfer the refuse to a hopper which feeds the Dusseldorf system grates. As in the previously described facilities, air is drawn into the facility through ducts in the dumping and storage areas and dust and odor problems are eliminated in this way. However, in contrast to the facility at Edmonton, the air drawn from the storage pit area is supplied to the combustion operation as underfire air. The boiler units and electrostatic precipitators at the Dusseldorf facility are quite similar to those operating at Edmonton and Ivry. However, the steam generated at Dusseldorf is not utilized at the plant itself but is delivered to a nearby power plant of the Municipal Works and used for the generation of electricity and supply of central heating in the city of Dusseldorf. Thus, this system is considerably different in that it supplies only steam which is utilized by another facility; however, both facilities are owned and operated by the city of Dusseldorf. The Dusseldorf incinerator differs from the previously-described operations also in the degree to which residue is processed after incineration. The residue is removed from the quenching tank and transported via conveyor systems to a special refuse treatment facility. There the material is classified and ferrous scrap is magnetically separated and baled for sale. Glass and ceramic particles are separated from the remainder of the fine ash by means of vibrating screens. The prepared ash is then sold for use in road construction. Additionally, at the Dusseldorf facility, a special shearing operation is provided to preprocess bulky waste. Collection vehicles enter the facility and deposit such bulky wastes in the shearing device. From this device the processed refuse is fed into the storage bunkers for incineration. As in the Edmonton and Ivry facilities, entry and logging of the refuse vehicles into the incinerator plant is highly automated. Collection vehicles are weighed upon their entry to the facility on a scale system equipped with electronic data processing equipment. The tare weight of all vehicles which are regular customers to the facility is stored in a centralized information bank and the vehicles need not be reweighed upon exit from the facility. As in the other incinerators previously described, electronic systems monitor the dumping operation at the tipping bays, and a lighted display indicates to the collection truck driver the proper dumping bay to approach with his particular load. In this way, 120 collection trucks per hour can be handled in this facility.

It is useful to compare the performance of various components of these systems. The Dusseldorf drum grates and the Martin grates can be compared by examining the performance of the incinerator at Stuttgart. This facility has one furnace equipped with Martin grates and one furnace equipped with Dusseldorf grates. The furnaces are routinely operated for 8 weeks and then stopped for maintenance and repair. This system has been in operation since approximately 1966, and indications of performance are that the Dusseldorf grates have greater mechanical reliability while the Martin grates entrain less particulate matter in the effluent gas stream and produce greater reduction of the incoming refuse (Ref. 8-10).

In addition to the refuse incineration facilities described in previous paragraphs which process principally domestic and commercial wastes, a number of facilities in Europe are designed for the processing of wastewater treatment sludge and industrial wastes. One facility designed for combined incineration of sewage sludge and domestic refuse is located near Zurich, Switzerland in the

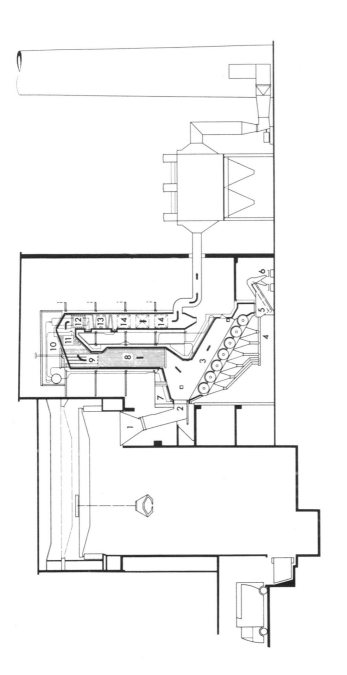

1 Refuse feed hopper
2 Refuse push feeder
3 "Düsseldorf System" roller grate
4 Humidifying worm conveyor
5 Wet type ash extractor
6 Ash belt conveyors
7 Roof light-off burner

8 Evaporator heating surfaces
9 Platen superheater
10 Nos. 1 and 2 spray attemperators
11 No. 2 primary superheater
12 No. 1 primary superheater
13 Evaporator coils
14 Continuous loop economizer

4 x 10 tons/hr of refuse throughput
4 x 16 tons/hr of superheated steam
150 kg./sq.cm.gauge design pressure
500 deg.C superheated steam temperature

Fig. 8-6   Cross section of Düsseldorf Incinerator (*courtesy* Vereinigte Kesselwerke AG).

Dübendorf area. This facility is designed for the incineration of wastewater treatment sludge in a multiple-hearth furnace designed according to the Nichols system. This system is shown in Fig. 8-7. The incinerator facility is located on a site adjacent to the wastewater treatment plant. The liquid raw sludge from the wastewater treatment plant is brought from that plant to the incinerator where it is mixed with domestic refuse. The domestic refuse is pretreated in the grinder to obtain rather uniform sizes of the waste particles. The mixed refuse and sludge is then incinerated in the multiple-layer furnace. Refuse which is rejected from the grinder is processed in an auxiliary furnace. The primary combustion gases produced in this auxiliary furnace are passed into the combustion chamber of the main furnace and are incinerated there. Additionally, this facility is equipped for incineration of used oils which are sprayed into the multiple-layer Nichols combustion chamber. This facility serves a population of approximately 70,000 and has been in operation since 1965. It has a capacity (combined refuse and sludge) of approximately 150 tons per day. Combustion gases are cooled through expansion and are passed through a wet scrubbing device before release to the atmosphere.

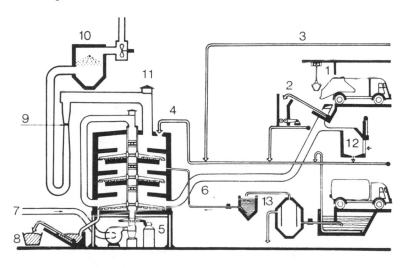

1—Refuse discharge	7—Cooling air
2—Grinder	8—Residue removal
3—Raw sludge	9—Flue gas cooling
4—Furnace input	10—Wet scrubber
5—Nichols oven	11—Emergency stack
6—Underfire air	12—Auxiliary furnace
13—Used oil input	

Fig. 8-7 Dübendorf Incinerator for refuse and sludge (*courtesy* Swiss Federal Institute EAWAG, Zurich).

In this facility, the combustion heat of the refuse and the waste oil is used to dry and then incinerate the sludge. The mixture of refuse and sludge discharged into the top of the combustion chamber is transferred from one level of the tower to the next lower level by means of large rotating arms. As the material travels from one level to another, the hot gases circulating through the tower dry the materials driving off moisture and then support combustion. Temperatures are controlled through the input of overfire air and the venting of gases through exhaust gas vents at the top of the combustion chamber. This system has been reported to have frequent mechanical breakdowns (Ref. 8-10).

Another interesting combined incineration facility has been completed recently near Rotterdam, in the Netherlands. This incinerator is owned by Rotterdam and twenty-three neighboring small communities. It is located in the Botlek area which is a highly industrialized zone on the seacoast of the Netherlands. Approximately 190,000 tons of domestic and household wastes are processed each year in the Rotterdam incinerator; also, about 440,000 tons of industrial waste and about 70,000 tons of chemical wastes are processed. The chemical wastes include chlorinated hydrocarbons, paints, acrylic nitriles, and oil sludges. The domestic wastes and nontoxic solid industrial wastes are processed in 6 furnaces equipped with Dusseldorf grates. Five furnaces are used at one time and a sixth furnace is kept in reserve. Each furnace is capable of incinerating approximately 16 tons of refuse per hour. This facility operates 24 hours a day 365 days per year. Each of these furnaces produces 50 tons of steam per hour at approximately 30 atmospheres pressure and at a temperature of 360°C. A computerized control facility regulates the steam circulation and mixing to ensure uniform temperatures and pressures in the outlet steam. One-third of the produced steam passes through a turbo-generator and is then condensed and returned to the boiler units. Two-thirds of the steam passes through two additional turbine units and exits to distilling plants. In these plants, contaminated and mineralized water, including seawater, is distilled. At the present time, three distilling plants are in operation, each producing 450 tons of demineralized water each hour. In the future it is planned to build two additional plants. The demineralized water is to be sold to industrial plants in the Botlek area for use as feed water in boilers, for chemical process water, etc. The generated electricity created in the turbine units will be sold through an electrical utility system. A minimum of 11 megawatts per hour are guaranteed by contract to this utility; the incinerator facility has a maximum producing capacity of 55 megawatts per hour. The furnaces used for combustion and incineration of domestic and nontoxic industrial wastes are very similar to the units at Dusseldorf and differ from those units only in small degree. Likewise, residue recovery facilities at Rotterdam are very similar to the Dusseldorf facilities. Flyash is recovered from the electrostatic precipitator at the plant and separated into two sizes in a centrifugal separator. Particles smaller than 75μ (microns) in equivalent diameter are sold for use as filler in asphalt.

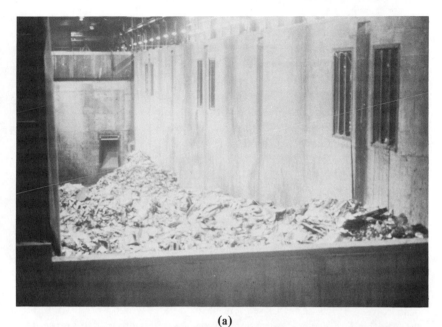

(a)

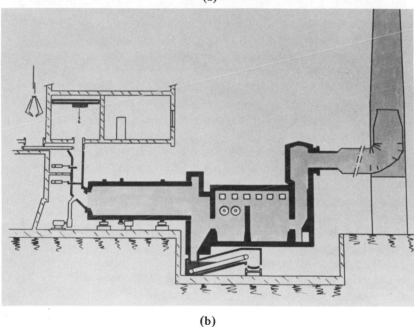

(b)

Fig. 8-8 (a) Storage pit at Rotterdam (Botlek) Incinerator; (b) cross section of chemical waste incinerator at Rotterdam.

Some problems have been encountered due to clogging of flyash in the collector bins in the precipitators when leaks developed in the water-tubes in the boiler portion of the incinerator. The damp flyash tends to bridge and clog in the collection bins. As in the Dusseldorf facility, a shredder is used to reduce the particle size of bulky wastes. This unit is manufactured by Hazemag in Germany.

Of particular interest is a special incinerator at the Rotterdam facility for the processing of chemical wastes. This facility consists of a rotating kiln primary combustion chamber followed by a secondary combustion chamber where liquid wastes are fed into the system and a third combustion chamber where gases produced in the preceding areas are completely oxidized. Approximately 30,000 tons of solid chemical wastes and 40,000 tons of liquid chemical wastes are brought to the facility each year from chemical plants. Some of the liquid wastes are contained in barrels and some wastes are contained in drums. The rotating kiln itself is approximately 4 meters in diameter and 10 meters long. In contrast to the rotating kiln units found in refuse incinerators elsewhere, this kiln is situated in a completely horizontal position. The weight of the refuse in the entrance chute to the kiln is the only force which impels the waste into the kiln and out of the exit end of the kiln. The loading mechanism first used for this kiln consisted of a series of sequential drop gates which were operated by a weight-sensitive leverage system; when 50 kg of waste were placed on the upper gate, it dropped this amount of wastes onto a lower gate. As the upper gate then moved upward to close, the lower gate dropped the charge of waste into the

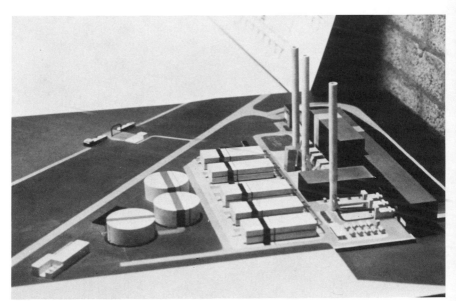

Fig. 8-9   Model, Rotterdam (Botlek) Incinerator.

entrance to the kiln combustion chamber. The wastes were transferred onto the upper gate by means of a conveyor supplied by an overhead crane from a 6000 ton storage pit. Soon after this unit was put into operation, a serious explosion resulted when flame backtracked through the sequential gates across the conveyor and into the storage pit. In 1974, a new system for charging solid chemical wastes was being installed in this incinerator. The new system was designed to include an elevator near the bottom of the entrance to the rotating kiln. An explosion-proof air lock was to be installed at this point. Liquid chemical wastes brought to the incinerator are fed into the second combustion chamber through two ports in each side of the combustion chamber walls. In the final combustion chamber, the combustion gases were to be completely oxidized. However, some problems arose during the early period of operation of this system with generation of deleterious effluent gases. An air pollution control system to prevent generation of hydrogen chloride and sulfur dioxide gases was to be installed to prevent such air pollution (1974). This industrial incinerator is being further modified so that industrial sludges may be pumped into the rotary kiln combustion chamber with the solid chemical wastes.

An operational feature of this incinerator plant which is of interest is the provision of extra capacity in each furnace. If two of the furnaces malfunction, the four remaining furnaces can process up to 20 tons of waste per hour and can be linked to one stack or the other at the facility in the event of a stack malfunction. This facility is highly automated and is operated at the present time by only 7 persons. Some minor problems have arisen in the transfer system for liquid wastes where nitrogen is used in a pneumatic transfer system. A new system of positive displacement pumps is being installed at the time of this writing (1974). Additionally, at this facility a dock is provided for transfer of wastes into the incinerator from barges and boats in the Saint Lawrence Harbor.

The fee system used at this facility is interesting. Chemical wastes which are easy to burn are processed for approximately $20.00 per ton. Average chemical wastes are burned for approximately $28.00 per ton and wastes which are difficult to burn are incinerated for $36.00 to $44.00 per ton. In contrast, domestic wastes are processed for $8.80 per ton. Solid industrial wastes are incinerated for approximately $11.00 per ton if a guaranteed amount of industrial wastes are provided by the supplier. If the amount of waste is not guaranteed at a certain amount per day, the waste disposal fee is increased to $17.00 per ton. Capital costs for this incinerator were approximately $88 million. Some special features of this facility include an automatic computerized control for the transfer cranes similar to the system used at Apeldoorn, which has been described in a previous section. Additionally, a foam spray fire prevention system has been installed in the refuse storage pits to prevent fires; the danger of fire is considered to be greater at this facility because the storage pit is approximately 70 ft deep and spontaneous combustion can occur in this large amount of refuse.

8-3.3   European Composting Facilities

The relative contribution of composting to the total solid waste disposal effort in Western Europe has been discussed in a previous section. In this section, several composting plants will be described to show the variety of composting operations which exist in Western Europe.

As mentioned previously, in the Netherlands the composting effort is handled in large part by the government-controlled company, VAM. VAM operates two large composting plants, at Wijster and at Mierlo. The Mierlo plant located in the Eindhoven region was built in 1955 and processes approximately 55,000 tons of compost each year. The Wijster plant was completed in 1931 and processes approximately 300,000 tons of refuse each year. It is one of the simplest and most successful composting systems in Europe. Therefore, it can be described here in greater detail. Refuse collected in municipalities serviced by VAM is brought to one of the sixteen transfer stations operated by this company in the Netherlands. One of these transfer stations, at Apeldoorn, was described in a previous section. At the transfer station, the refuse collectors must break open any plastic bags before they are deposited at the VAM facility. The transfer facilities are used to place the collected refuse in railroad cars. Although in the past road transport of wastes was done to a greater degree than at present, most of the transfer of waste is now accomplished by rail. Each railroad car contains two compartments with capacity of 50 cubic meters each. Approximately 25 to 30 tons of wastes are deposited in each rail car. Trains of up to thirty cars are composed at the transfer station and shipped to the compost facilities at Mierlo or Wijster. A train of twenty-five cars can be unloaded automatically in approximately 20 minutes at the Wijster facility. Each car is unloaded by an automatic system within the car that hydraulically tilts the two compartments in the car to dump wastes on either side of the car. When the trains reach Wijster, the cars are moved onto an overhead rail line and, during dumping, the wastes fall from the car into windrows approximately 30 ft high. The deposited wastes are turned over by overhead rail cranes every few months. The detention time for the compost is 6 to 8 months. Water is added to the deposited wastes at the time they are dumped from the railroad cars by means of water sprays located along the sides of the elevated rail line. Approximately 200 liters of water are added to each input ton of solid wastes. When the compost has been cured for 6 to 8 months in the open windrows, it is processed through roller grinders and then passed through a series of vibrating screens. The screening of the refuse produces two grades of compost, a super-fine grade which is used on sports fields and in surface layers of soil for landscaping purposes, and the second grade which is mixed in bulk with soil to produce a more fertile and arable growing medium. Some compost also is mixed with peat and fertilizers, and is bagged for sale to a rather small special agriculture market. Approximately 35 to 40 percent of the incoming solid waste is converted to compost. Approximately 50 percent of the input must be disposed of or recycled in another manner. The remaining 10 to 15 percent of the waste is transformed during the composting process into water vapor or carbon dioxide gas and enters the atmosphere. Formerly, coal was used for space heating in the Netherlands and the ash content of the refuse was much higher than at the present time when oil and natural gas are being used for space heating. In former times, approximately 70 to 80 percent of the solid wastes could be turned into compost. The residue which is obtained from the screening process at Wijster is further processed through magnetic separators to reclaim ferrous metals. The remainder of the material is deposited in a residue fill on-site at Wijster. During the 40 years of operation of the plant there, approximately 100 acres of the 500 acre site have been used for deposition of residue. At the present time, it is not economical to retrieve glass particles from the residue because current costs for recovery are approximately six times the current value of the recovered glass. Eighty to eighty-five percent of the finished compost is sold in bulk form, and is not separately packaged. The remainder that is packaged contains approximately equal amounts of super-fine compost (less than 5mm in particle diameter) and peat with artificial nutrients added. Some compost manufactured at the VAM installations has found use as supplemental animal feed. Newborn pigs in the Netherlands have been found to suffer from iron deficiencies. Iron can be obtained through the compost and the compost has been successfully used for pig feed. Additionally, VAM compost has been shipped to Switzerland for use as pig feed. The only problems which have arisen from the use of the compost as pig feed have been due primarily to small bits of glass in the compost. However, the zinc content of the pig feed has been rising in recent years and may preclude the use of this material for animal feed.

By the end of 1973, VAM was processing the refuse of approximately 2 million people living in eighty-six communities throughout the Netherlands. The direct cost of producing the compost at the VAM installations was approximately $4.20 per ton in 1973, and the selling price for the compost was approximately $8.80 per ton. The price for compost more than quadrupled in the last 15 years. Higher transportation rates have also increased the cost of the finished compost.

In contrast to the large composting plants operated by VAM in the Netherlands, four small municipal composting plants also are in operation in that country. The plant at Arnhem is typical of these small municipal plants. The composting plant at Arnhem consists of receiving facilities and two rasping mills which utilize Dorr-Oliver equipment. The incoming refuse is dumped into a receiving area and transferred from the receiving bay to conveyor belts which move the refuse into the plant proper. A man is stationed alongside each of the two transfer conveyors to break up plastic bags and remove as much of the plastic as possible from the moving stream of wastes. The refuse moves along the conveyor belt and beneath a magnetic separator where ferrous metals are removed.

Fig. 8-10   Turning refuse windrows at Wijster compost plant by overhead crane (*courtesy* VAM).

Separated ferrous metals are baled and placed on transfer trucks for sale. The remainder of the refuse falls into a rasping mill for size reduction. There are three rasping mills in this plant. Figure 8-11 shows a schematic of the Arnhem composting plant. The refuse passes out of the rasping mills in particles approximately 1 in. in greatest dimension. From the rasping mill the material is carried on a conveyor belt up an incline to a ballistic separator where dense and heavy items are removed. The rejects from the ballistic separator are mixed with the particles which were rejected from the sides of the rasping mill and removed as residue. The material which falls through the ballistic separator passes through a series of screens and then through two rotating crusher wheels for further size reduction. It is interesting to note that approximately 50 percent of the refuse

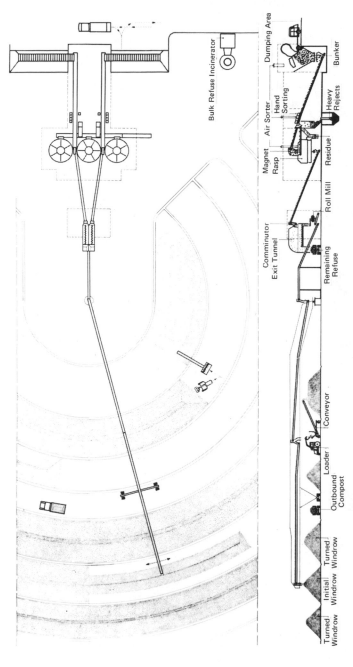

Fig. 8-11 Schematic, Arnhem, Netherlands compost plant (*courtesy* city of Arnhem).

collected in the Arnhem area consists of bulky items or industrial wastes which could not be composted and which are taken to a landfill for disposal. Of the material which comes through the composting plant, approximately 70 percent of the input is transformed into compost and the remainder is taken to a landfill for disposal. The plant processes approximately 100 tons of material per day. The material which has passed through the second series of rotating crusher wheels is conveyed by belt conveyor to the exterior of the plant and placed in open windrows for curing. It is turned periodically through the use of a front-end loader and portable conveyor. Cured compost is removed for sale. This plant was built in 1961 at a cost of approximately $1.2 million. It would now cost approximately $4 million to build. The compost produced from this plant is sold for about $4.80 per ton. Although ready markets exist for the compost produced from this plant, the plant is not capable of handling the amount of waste generated in the serviced area. Consequently, this facility will be replaced in 1975 or 1976 by an incinerator with a capital cost of approximately $14 million.

A somewhat similar but more modern plant for composting has been built at Meaux, about 30 miles outside the city limits of Paris. This facility serves an existing population of about 50,000 persons, but has been designed to accommodate the refuse from 120,000 persons. This extra capacity was built into the plant since the area around Meaux is thought to constitute an ideal commuter community for Paris and rapid growth has been forecast for this region. Collection trucks enter the facility at Meaux and pass across platform scales which are equipped with automatic data processing equipment. The trucks then proceed to the rear of the plant where refuse is deposited in a storage pit with a capacity of approximately 130 cu yd. The refuse is moved from the storage pit by a slowly moving metallic conveyor located at the bottom of the pit. The pit is equipped with duplicate conveyor systems which carry refuse to either side of the plant

Fig. 8-12   Exterior view, Arnhem compost plant.

where the refuse is inserted into Gondard hammermills. Each of the conveyor-hammermill combinations has a capacity to process 50 tons of refuse per day. The Gondard hammermills are vertical-axis, swinging-hammer type mills which include spiral ballistic separation attachments. Approximately 4 to 5 percent of the input solid waste is rejected from the hammermills via this separator system. After the material leaves the hammermill, it passes onto a dual vibrating screen; the upper screen has openings of approximately 40 mm diameter and the lower screen has openings of approximately 25 mm diameter. Only a small portion of the refuse is rejected after retention on the 40 mm screen. Any refuse which does not pass through the 25 mm diameter openings in the lower screen is reinserted into the storage pit and recycled through the grinding operation. The material which is rejected from the hammermill and from the 40 mm screen consists primarily of large pieces of plastic and paper. This material is conveyed to a central furnace which has a capacity of process approximately 20 tons of refuse per day. The material which passes through the 25 mm screen is subjected to further processing, including ballistic and magnetic separation. Ferrous metals are removed from the screened refuse and are sold to scrap dealers. The processed refuse from the 25 mm screen is removed by a truck to the storage area outside the plant. The refuse remains in the storage area of the plant for approximately 3 to 4 months and is turned periodically by means of a clamshell mobile crane approximately every 2 weeks until a finished cured compost product has been obtained.

The operating costs for this facility amount to around $1.30 per ton of which approximately $0.75 per ton corresponds to the expenses of operation for the hammermill. The total operational costs for the disposal facility are recovered through sale of the compost. Most of the compost is removed by truck to market areas within 30 to 40 miles of the plant. At the present time, the purchase price of the compost is approximately $3.75 per ton; this price covers the total operating cost and also pays a small fraction of the amortization of capital expenditures.

No problems have been encountered in the operation of the conveyors or hammermills or ballistic separators. However, some difficulties have been experienced with the furnace at this facility, since the refuse consists primarily of high heat content paper and plastics. The furnace was designed to operate on unprocessed raw refuse and the high heat release value of the paper and plastics in the rejects has produced considerable spalling of the refractories in the combustion chamber and has produced significant wear in the grates for the facility.

Another plant which combines incineration with composting was built in 1973 at Schaffhausen near the German–Swiss border. This facility serves a region of Switzerland in which more than 100,000 persons live. Provisions have been made at this plant for the processing of wastewater treatment sludge in combination with domestic and commercial refuse. Additionally, a small amount of industrial

Fig. 8-13 Meaux, France compost plant (near Paris) (*courtesy* John J. Reinhardt).

refuse (approximately 6500 tons per year) is processed in this facility. Furthermore, the system has been designed for the processing of used oils and water–oil emulsions. Approximately 1000 cubic meters of the waste oil materials are anticipated for processing each year at Schauffhausen.

Refuse is deposited at the facility in a storage bunker similar to that found in many incinerator plants. In a separate building near the end of the tipping area, transfer trucks deposit waste oil and emulsions. Transfer vehicles bring wastewater treatment sludge to this facility and deposit this material in a separate storage container. Some refuse is moved by overhead crane from the storage pit onto a chain conveyor and into a Bühler hammermill for processing in a compost system. The Bühler hammermill is a horizontal-axis grinder used for primary size reduction. The ground material is conveyed from the hammermill to a regulating chute (dosierrinne) which regulates the flow of the material into the remaining portion of the processing operation. After leaving the regulating chute, the refuse passes over a magnetic separator which retrieves ferrous metals. Then the material is inserted into a "Dano Bio-stabilisator," a long drum 25 meters in length and $3\frac{1}{2}$ meters in diameter in which the composting operation is initiated. In the drum, the sludge is mixed with the ground refuse and the mixture is tumbled from one end of the drum to the other by the rotary motion of the inclined drum. The travel time from one end of the drum to the exit end is regulated by the speed of rotation. The normal detention time for the mixture in the Dano drum at Schauffhausen is approximately 3 days. Process gases given off in the drum are vented through an exhaust fan and processed through an earth filter before release to the atmosphere. Oxygen for the composting process is supplied through a forced-draft fan at the exit end of the drum. When the processed mixture leaves the exit end of the drum, it drops into a vibrating screen separation device. Approximately 30 percent of the materials entering the vibrating screen are oversized paper and plastic materials which are retained on the screen surface. These materials are transferred by conveyor to a separate portion of the facility for incineration. The bulk of the material coming from the drum passes through the vibrating screen (approximately 70 percent passes) and is processed in a Bühler grinder which further reduces the size of the individual particles. From the fine grinder, the material is transferred to an outdoor storage area for curing. Some of the compost material is sold within a short time as so-called "fresh" compost for approximately $.80 per cubic meter. The remainder of the compost is allowed to cure for 2 to 4 months and is sold for approximately $2.00 per cubic meter. This facility is designed to process sludge and refuse in the ratio of 1 ton of sludge to 2 tons of refuse. Approximately 6500 tons of sludge will be received at this facility each year. The total amount of wastes brought to the facility each year will amount to about 44,000 tons. The excess amount of domestic and commercial refuse, the industrial refuse, and the rejects from the composting facility all will be processed in a Von Roll incinerator installation. The Von Roll combustion chambers are designed to operate on refuse with a heat content of 2500 kcal/kg. They are capable of processing approximately 65 tons of refuse per 24 hr. The grate system used at this facility is shown, in part, in Fig. 8-14. The segmented grates are inclined to the horizontal and are moved in a reciprocal motion by hydraulic cylinders which activate piston rods to move the grate sections. The overall motion is a slow progression under the force of gravity down the slope of the grate beds. Three sections of grates are installed in the Schauffhausen facility in each combustion chamber. The upper portions of the combustion chambers are equipped with water-tube walls. The principal purpose of these walls is not the generation of steam for the production of electricity or for use in space heating, but for the cooling of the exhaust gases from the combustion chambers. The exhaust gases are cleaned in electrostatic precipitators before being released to the atmosphere through a stack approximately 200 ft high. The steam produced in the water-wall chamber is condensed for the most part but a fraction is used for the production of electricity which is used for in-house power requirements.

Capital costs for this facility were approximately $4.5 million. The facility has not been in operation for a sufficiently long time to allow accurate determination of operating costs. However, these operating costs have been estimated at approximately $155,000, of which approximately $70,000 each year corresponds

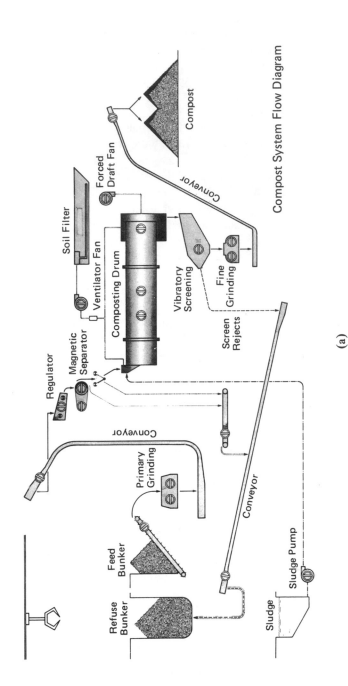

(a)

Fig. 8-14 (a) Compost system schematic, Schaffhausen, Switzerland plant; (b) Von Roll grates at Schaffhausen; (c) exterior of plant, Schaffhausen (*courtesy* Klaranlage-Verbandes, Schaffhausen).

(b)

(c)

Fig. 8-14 (*continued*)

to labor costs. Income from the sale of finished compost and from the charging of disposal fees for the processing of waste oil in the incinerator facility has been estimated at approximately $12,000 per year. None of the metal recovered at the facility during the first 6 months of its operation was sold because a buyer for this material could not be found.

This section on European composting plants can be concluded with a description of one of the largest Dano-equipped plants in the world, operated by Societa Laziale Imprese Appalti in Rome (SLIA). The SLIA plant is one of four refuse disposal plants which manage all of the refuse for the city of Rome. Until 1964, no formal disposal of refuse was accomplished in Rome. Since that time, the SLIA plant and three other plants have been charged with this disposal task. The three additional plants will be described in a later section of this discussion. The SLIA plant is located on the western boundary of the city of Rome and refuse is brought to the facility in standard collector trucks. The refuse is dumped into a storage pit and moved by polyp grab overhead cranes onto slowly moving metal conveyors. The conveyors pass the refuse on to large-grid screens where oversize pieces of material are immediately removed for incineration. Immediately after the first coarse screening of the material, magnetic separators are used to remove ferrous materials from the refuse. The refuse is then placed in one of ten Dano Bio-stabilisator drums. These drums are 27 meters long and $3^1/_2$ meters in diameter. Each drum is capable of processing approximately 55 tons of refuse per day; the refuse stays within the drum for a detention time of approximately 3 days before emerging at the downslope end of the Dano drum. The processed refuse emerges from the drum and is separated on a dual screen system; the upper screen has openings of 40 mm diameter and the lower screen has openings of 15 mm diameter. Any materials retained on the upper screen are removed for incineration. Materials retained on the lower screen are removed and cured for sale as "coarse" compost. The finer materials passing through the 15 mm diameter screen constitute "fine" compost and are stored for curing. The coarse compost product is distributed free of charge to farmers or is disposed of as necessary in landfill. The fine compost is sold for approximately $3.00 per ton. Sixty percent of the incoming solid wastes are processed in the composting operation and of this 60 percent, 27 percent emerges as fine compost and 33 percent emerges as coarse compost. The fine compost is used mostly in vineyards and in specialized agriculture. An interesting feature of this facility is the curing of the compost. The compost is placed in large sheds in layers to constitute stockpiles approximately 15 ft high at maximum height. Temperatures as high as 70°C (158°F) are achieved in the first several days after deposit of the compost. At no point in the entire process from the time of deposition until the time of sale of the finished compost is water added to the refuse in this facility. Moreover, the compost in the curing piles is not turned in any way at any time. The oversized pieces of refuse obtained during primary size separation and

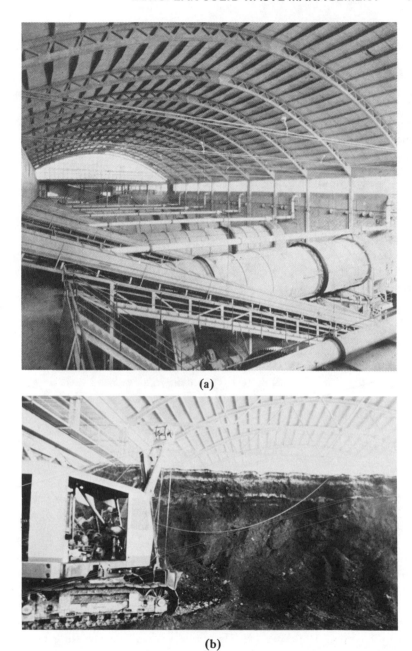

(a)

(b)

Fig. 8-15 (a) Dano Bio-stabilisators at SLIA plant, Rome (*courtesy* of Societa
Laziale Imprese Appalti, Rome); (b) curing pile, SLIA compost facility, Rome
(*courtesy* SLIA).

during secondary separation of the refuse emerging from the Dano drum are incinerated. The ash produced during incineration is processed to remove ferrous metals and then is separated into a fine fraction and a coarse fraction. The coarse fraction is deposited in landfills. The fine fraction is mixed in with the compost and is disposed of in that fashion. The incinerator at the facility is utilized to de-tin ferrous materials separated from the refuse. The de-tinned ferrous material is baled and sold to steel plants in the area for approximately $25.00 per ton (Ref. 8-11).

8-3.4 European Landfill Operations

As shown in Table 8-4, the majority of the landfills in Western Europe are not controlled operations. Therefore, no mention need be made of such inadequate facilities. As an example of a well-controlled landfill, the facility at Emscherbruch near Essen in the Federal Republic of Germany can be described. In the Area Development Plan set up by the Ruhr Regional Planning Authority, a regional greenbelt system was one of the principal goals established. The need for greenbelt to separate the big cities in the Ruhr and to provide recreational areas for the dense population of the area is very important. As part of the greenbelt development, the site of a former mine, the "Graf Bismarck," in the marshy areas along the Emscher River was selected as the site of a controlled landfill for the surrounding territory. As a result of the mining operations, a large slag heap was found on the site. The slag materials are being removed and used in road construction, and refuse is being used to replace the slag. The area of the landfill development is approximately 1 million square meters, and is estimated to have potential filling volume of approximately 25 million cu yd. The final height of the fill above original ground is scheduled to be approximately 230 ft. The site is ideally located near existing autobahns and expressways connecting major cities in the Ruhr area. Land acquisition and operation-initiation costs at this site amounted to more than $7 million. This landfill is extremely large, but well run.

Approximately 1000 vehicles transporting wastes enter and leave the Emscherbruch landfill each day. No liquid toxic or hazardous wastes are accepted at the site. The landfill is completely enclosed with fencing to prevent promiscuous dumping. Vector control at the site, with regard to rats and mice, consists of the distribution of rat poison in ceramic pipes on the slopes of the finished landfill cells. The ceramic pipes are placed in single lengths in a horizontal position on the slopes of the finished cells; the pipes are approximately 1 meter long and 15 cm in diameter and are placed approximately 30 meters apart near the bottom of the fill slopes. The pipes are filled with a special powder-form rat poison which has a time-delay effect. This poison is innocuous to human beings and to other animals. The operation at Emscherbruch has a number of other interesting features. All leachate and surface drainage at the site is collected and fully treated before release into natural waters. The leachate and surface drainage flow into the Emscher River, which is completely channelized and which empties into a waste-water treatment facility. Furthermore, in several locations, leachate is being collected from various experimental plots and the collected leachate is being analyzed. In addition to experimental work carried out by governmental agencies, a number of chemical firms which deposit wastes at the site participate in the monitoring activity by furnishing chemical analyses of the collected leachate and help support the costs of the experimental program. In addition to leachate collection, temperature measurement is performed in the experimental plots at several different depths through the deposited refuse. The temperatures are monitored daily and the data are collected and sent to a central governmental clearing house in Berlin for collation and analysis (Ref. 8-26). The mode of filling at this site is somewhat different than American practice. Collection trucks enter the facility and pass over an automated platform scale equipped with electronic data collection apparatus. Regular customers of the facility have a "credit card" arrangement for identifying their trucks and receiving appropriate billings. The trucks proceed on paved roads to the dumping area. The trucks drive to within 10 meters of the operating face and deposit their loads there. Tracked bulldozers and special wheeled compactor dozers then push the refuse forward and pass over the refuse to compact it. The refuse is pushed forward and down over the working slope by these vehicles. Very little compaction is actually done on the sloping working face. The refuse is placed in lifts of 2 meters thickness. At the dumping face, approximately five to six vehicles are present at all times depositing refuse along the dumping face (which can accommodate approximately ten vehicles). Two workers at the dumping face direct incoming truck traffic. Altogether, 20 to 25 persons are employed daily at this site; the number of workers varies with weather conditions and with traffic estimates. Five bulldozers are used near the working face. Approximately 1 million tons of refuse are placed in the fill each year. Thirty-five percent of the incoming refuse is domestic refuse and about 7 percent of the refuse is industrial solid wastes. A large portion of the remainder is demolition wastes from nearby cities. The large percentage of demolition wastes deposited at this site reflects the philosophy of the regional planning agency. The overall aim of that planning agency is the deposition of refuse in an organized fashion so that refuse which is combustible can be processed in incinerators for the production of heat and electricity. Refuse which is compostable is assigned to compost plants wherever possible. All residue from these operations and all materials which are neither combustible nor compostable are placed in landfills. Complete details on this facility are given in Ref. 8-26.

Landfill is practiced to a great extent in Italy, perhaps to a greater extent than in any other European country. Therefore, some information on Italian landfills may be given here. A recent publication (Ref. 8-2) contains some interesting

(a)

(b)

Fig. 8-16 (*a*) Rodent control device (poison in pipe) Emscherbruch landfill, Federal Republic of Germany; (*b*) dumping area at Emscherbruch.

facts concerning the costs of operation and initiation for landfills in specific Italian cities. For example, installation costs for a controlled landfill at Genoa amounted to approximately $0.20 per ton, and operating costs at that facility amounted to approximately $0.60 per ton. Combined costs for installation and operation of a controlled landfill at Brescia amounted to approximately $1.00 per ton in 1973. At Florence, the total installation and operating costs for a landfill there amounted to approximately $2.70 per ton in 1973. An interesting comparison can be made for the city of Milan where a central incinerator is used to process approximately one-fifth of the generated refuse and the remainder is deposited in two sanitary landfills. The costs of incineration at the central facility amount to approximately $7.95 per ton while the costs for a landfill at Gerenvano outside Milan amounted to approximately $2.97 per ton (about $0.56 per ton operating expenses and approximately $2.41 per ton transport costs). For a landfill outside Milan at Corneliano-Bertario (approximately 15 miles from the city of Milan), total disposal costs amounted to approximately $3.53 per ton, of which approximately $0.46 per ton were operating expenditures. It appears that the operating costs for landfilling in Italy are considerably lower than similar costs in the United States and other European nations. The cost for transportation and deposition of refuse in landfills in Italy appears to be quite competitive with the costs of central incineration. One reason for this favorable situation with respect to landfilling is the monopolization of electrical power supply by a governmental agency. This monopoly militates against the sale of generated electricity by refuse incineration facilities.

8-4. INNOVATIVE EUROPEAN PRACTICES

Several innovative practices are becoming somewhat widespread in Western Europe. Others are still in trial stages. A widespread innovation is the use of grinding as a preparatory step in landfilling. For example, approximately eighty grinding facilities now exist in France. A large number of these (more than forty-five) are used as prior operations before sanitary landfilling. Gondard, Tollemache, Hazemag, Bühler, and Clero apparatus are used in these operations. Grinding in European practice has been described in a previous publication (Ref. 8-10). As landfilling space becomes more valuable in the United States with increasing urban sprawl, it is anticipated that grinding of refuse will be practiced prior to landfilling at American sites.

Another innovation becoming more widespread in Europe is the installation of pneumatic collection systems. These pneumatic systems appear to have significant advantage for installation in new multi-family dwellings or in new institutional facilities. Their application to existing facilities is not as advantageous. These facilities have been described in previous publications (Ref. 8-10).

Pyrolysis, which is receiving increased attention in the United States, has not been adopted in any large-scale plants in Europe to date (1974). An experimental facility capable of processing 5 tons of refuse per day has been built in Copenhagen, Denmark. Also, small experimental units have been constructed in governmental laboratories in Great Britain. Recently, plans for construction of a large pyrolysis plant in Munich were announced. However, developments in pyrolysis in European systems have not been extensive.

Fig. 8-17 Pneumatic collection system at Olympic Village, Munich, exterior chute.

(a)

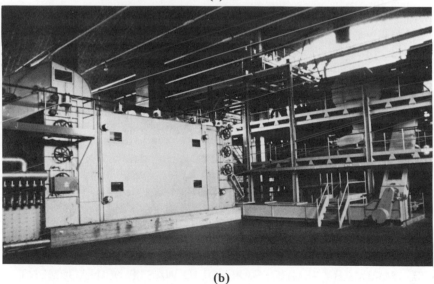

(b)

Fig. 8-18   (a) Paper pulper SARR plant, Rome (*courtesy* SARR); (b) equipment
for processing refuse into animal feed, SARR plant, Rome.

An interesting and somewhat innovative system of refuse disposal has been developed, however, in the city of Rome in three plants which were mentioned previously in connection with the SLIA composting-incineration plant in that city. These facilities have been described in detail in Ref. 8-11. The plant operated by the Societa Agricola Recupero Residui (SARR) may be described as typical of these three plants. In this plant, refuse is deposited by collector trucks into a shallow storage pit. An automated grab which resembles a large mechanical arm pulls the refuse from the floor of the storage pit onto a conveyor and the conveyor carries this material into the plant proper. Special patented machines within the plant open and empty the plastic sacks in which the refuse was collected. Utilizing various separation techniques, ferrous materials are separated and a significant percentage of rather clean paper particles are obtained from the general waste stream. The remainder of the refuse is divided into a stream of coarse particles, a stream of medium-sized particles, and a stream of fine-size particles. The ferrous material is separated and taken to a rotating furnace, $2\frac{1}{2}$ meters in diameter and 12 meters long, which is supplied with a liquid fuel burner. In the furnace the metallic portion of the refuse (approximately 2.5 percent of the input) is de-tinned; the processed ferrous scrap is then baled and transported to nearby steel mills. The paper selected from the refuse stream constitutes about 70 percent of the paper input to the plant. This paper is temporarily baled and stockpiled. After temporary storage, the paper is pulped and reprocessed for recycling. The coarse fraction of the refuse which has passed through the special processing machine is fed into the incineration section of the facility. About 45 to 50 percent of the input to the plant is incinerated. This facility is equipped with three furnaces, two of 100 ton per day capacity and one of 60 ton per day capacity. The medium fraction of the refuse is processed in a series of special devices for use as animal food. After sterilization in autoclaves, this material is subjected to a series of grinding and sorting operations. The steam for the autoclaves and the power for the further processing of this fraction is obtained from incineration of the coarse fraction. The finest fraction of the refuse is processed in a Dano composting drum similar to those drums previously described.

This operation appears to constitute a high degree of materials and energy recovery from the input of solid waste. An even higher degree of recovery may be attained in the near future, since operators of this facility are contemplating the installation of an optical separator for recovery of glass particles from the refuse stream.

8.5. SUMMARY—EUROPEAN SOLID WASTE MANAGEMENT

This examination of European practice is not intended to be comprehensive. However, a number of observations may be made concerning European practice

Fig. 8-19 Interior, lobby, Edmonton Incinerator, UK, showing emphasis on facility aesthetics.

in general. These observations may be of use and value for solid waste managers in other countries.

The approach to refuse management in most European countries is very professional, with the officials charged with collection and disposal responsibilities being drawn from civil service organizations. Political ramifications in the staffing of waste management positions in some countries of Western Europe appear to be significantly absent. There is a definite emphasis on fulfillment of a public service in the management of wastes in European countries. There does not appear to be as great an emphasis on environmental protection among European waste managers as among their counterparts in the United States. In all instances observed by the authors in their travels throughout Western Europe, the refuse collection and disposal service appeared to be reliable and efficient.

It should be noted that the composition and quantities of European refuse are significantly different than the composition and quantities generated in the United States. European refuse in general contains less paper and more organic matter than American refuse. However, increasing amounts of paper and plastics are appearing in European refuse. Additionally, the heat content of European

Fig. 8-20   Analytical laboratory, Rotterdam (Botlek) Incinerator, showing professional approach to refuse analysis.

refuse is less than that of American refuse, but European refuse is more valuable as a substitute fuel because of the greater value of fossil fuels in Europe as compared to the United States. An entirely different set of economic conditions exist in European refuse management practice as compared to practice in the United States. Nevertheless, many of the same problems that plague European managers irritate and aggravate their American counterparts. Decreasing availability of sites for landfills and increasing costs for labor and equipment appear to be problems common to Western Europe and the United States. Organizational solutions for the waste management problem in Europe are quite different than the organizations created for waste management in the United States.

The most successful European operations appear to be very simple. Successful operations in Europe also appear to incorporate a measure of built-in flexibility so that changes in waste characteristics, secondary materials markets, or energy values may be accommodated within existing systems. European successes in energy recovery through incineration of refuse seem to indicate a large potential for similar success in the United States. Finally, an overall analysis of European practice indicates that an optimum system for solid waste management in any given locality may rely on various modes of disposal such as incineration, composting, and landfill.

REFERENCES

8-1. Billig, G. and Tenaille, G., "Le Marche du Compost en France," *Techniques et Sciences Municipales*, Vol. 68, No. 8–9, August–September, 1973.

8-2. Bonomo, L. and Comolli, P. C., "La Decharge Controlee en Italie," *Techniques et Sciences Municipales*, Vol. 68, No. 8–9, August–September, 1973.

8-3. Defeche, J., "Corrosions Produced by the Incineration of Domestic Refuse," *Proc.*, 4th Int. Cong. IRGR, Basel, June 2–5, 1969.

8-4. Greater London Council, *London's Refuse*, GLC Information Centre, County Hall, London, SE1, 1969.

8-5. Greater London Council, *Annual Report*, Department of Public Health Engineering, 10 Great George Street, London, 1973.

8-6. Hart, S. A., *Solid Wastes Management: Composting European Activity and American Potential*, DHEW, USPHS, Cincinnati, 1968.

8-7. Hart, S. A., *Solid Wastes Management in Germany*, DHEW, USPHS, Cincinnati, 1968.

8-8. Horstmann, O., "Biespiel: Das neue Kompostwerk Heidelberg," 6th Fachlehrgang an der Universität Stuttgart fur Mull-u, Abfallbeseitigung, Herbst, 1973.

8-9. Jager, D., "Art, Menge und Zusammen-setzung der Abfall," 6th Fachlehrgang an der Universität Stuttgart fur Mull-u, Abfallbeseitigung, Herbst, 1973.

8-10. Jensen, M. E., *Observations of Continental European Solid Waste Management Practices*, USPHS Publ. No. 1880, Govt. Printing Office, 1969.

8-11. Mensurati, E., "Disposal of Solid Refuse and Production of Paper Pulp: Rome Plants," *Proc.*, Separate, EUCEPA Conference, Rome, May 7–12, 1973.

8-12. Miethe, M., "Kosten der Abfallbeseitigung," 6th Fachlergang an der Universität Stuttgart fur Mull-u, Abfallbeseitigung, Herbst, 1973.

8-13. Rogus, C., "Refuse Collection and Disposal in Western Europe: Part I," *Public Works*, Vol. 93, No. 4, April, 1962.

8-14. Rogus, C., "Refuse Collection and Disposal in Western Europe: Part II," *Public Works*, Vol. 93, No. 5, May, 1962.

8-15. Rogus, C., "Refuse Collection and Disposal in Western Europe: Part III," *Public Works*, Vol. 93, No. 6, June, 1962.

8-16. Tabasaran, O., "Wiederverwendung der Abfallstoffe," 6th Fachlehrgang an der Universität Stuttgart fur Mull-u, Abfallbeseitigung, Herbst, 1973.

8-17. Tietjen, C., "Utilization of Composted Domestic Refuse," *Proc.*, 4th Int. Cong. IRGR, Basel, June 2–5, 1969.

8-18. van der Kooi, I., "Refuse Disposal in the Rotterdam Industrial Area," *Haus der Technik-Vertrags veroffentlichungen*, No. 155, Rotterdam, 1972.

8-19. Wiley, J. S., "Some Specialized Equipment Used in European Compost Systems," *Compost Science*, Spring, 1963.

8-20. "Evacuation et Traitement des Ordures Menageres," Circulaire du 14 avril 1962, *Journal Officiel de la Republique Francaise*, Paris, 1963.

8-21. "Het'einprodukt' van onze welvaart hoeft geen probleem te zijn," Brochure, VAM (NV Vuilafvoer Maatschappij), Amsterdam, 1972.

8-22. "Programm 'Umweltgestaltung-Umwelt-schutz'," Sonderdruck, *Mull-und Abfallbeseitigung*, Handbuch, Erich Schmidt Verlag, Berlin, 1972.

8-23. "Refuse Organization in Paris and Suburbs," Information Circular Electricité de France, Paris, 1972.

8-24. *Solid Wastes Management in Switzerland*, Swiss Federal Institute for Water Resources and Water Pollution Control, Dübendorf, 1972.

8-25. "World Survey Finds Less Organic Matter," *Solid Waste Management/ Refuse Removal Journal*, Vol. 10, No. 26, 1967.

8-26. "Die Zentraldeponie Emscherbruch in Gelsenkirchen," Sonderdruck, *Mull-und Abfallbeseitigung*, Handbuch, Erich Schmidt Verlag, Berlin, 1972.

9

Summary and procedures for selection of short-term and long-term solid waste management techniques

The preceding chapters have attempted to present comprehensive information and data describing currently available techniques of solid waste disposal and resource recovery. Topical areas discussed have included conventional and innovative methods of disposal, materials and energy recovery systems, and European techniques utilized in solid waste management. It is appropriate at this point to summarize the information presented in the preceding chapters and to provide a framework from which this information may be utilized to rationally select both short-term and long-term solid waste management systems.

9-1. CONVENTIONAL DISPOSAL METHODS

Simple disposal of solid waste can be defined as the final deposition of refuse without any attempt to gain returns. Currently, there are three conventional methods for disposing of solid waste: sanitary landfilling, incineration, and composting. At present, sanitary landfill is the simplest, most economical, and most popular method, with incineration a distant second. Technically, composting does not fit the simple disposal definition since this process does produce a usable end product, humus. However, since there is little demand for humus in the United States, composting practice in this country is insignificant. Nevertheless, composting is discussed herein because of its relative popularity as a solid waste management technique in other parts of the world.

9-1.1 Sanitary Landfill

The landfilling of solid waste has been practiced since the beginning of civilization. Originally, landfilling consisted of dumping refuse in an open fill and allowing it to decompose. By the turn of the 20th century, the nuisances of odor and litter and the presence of disease vectors associated with these open dumps spurred far-reaching revisions in landfilling practices. For the first time municipalities began burying their wastes. The term "sanitary landfill" was first utilized to describe an operation in Fresno, California in the 1930s. Today a sanitary landfill must satisfy the following criteria:

1. Daily cover of the solid waste with earth.
2. No open burning of waste materials.
3. No deterioration of surface or groundwater quality at the site.

Sanitary landfilling is the most widely used and most economical refuse disposal method known today. A sanitary landfill operation requires no expensive physical plant, has low labor costs, involves standard construction equipment, and is the only final disposal method in present use. However, unfavorable public opinion resulting from experiences with improperly operated fills, decreasing land availability (especially in urban areas), and the potential for surface and groundwater pollution from a landfill must be regarded as serious obstacles to sanitary landfilling.

There are three filling techniques utilized in connection with sanitary landfilling: the trench, area, and ramp methods. In the trench method, a long, relatively narrow trench is excavated and the solid waste is placed in the trench in layers, compacted, and covered with fill obtained from the original excavation.

In the area method, the refuse is dumped onto open ground, spread into uniform layers, compacted, and then covered with earth obtained at some other site. The ramp (or progressive slope) method is a hybrid of the trench and ramp methods. Here, an existing slope is further excavated at the toe, the refuse is dumped, spread, and compacted on the working face, and then it is covered with fill from the original excavation.

Operational parameters largely determine the method of operation and the site selection for a sanitary landfill. Significant operational parameters include: the geology and topography of the site, characteristics of the cover material, characteristics of the refuse, hydrology of the site, and climatic conditions at the site. Topography and geology determine the best filling technique to be used. The hydrologic conditions at the landfill site will determine the potential for water quality deterioration. The cover material for a sanitary landfill operation should be easily workable and moderately cohesive. Ideally this cover material should be obtained either at or in close proximity to the landfill site. Surface and groundwaters must not be allowed to come into contact with the decomposing refuse if pollution of local water supplies is to be avoided. Rain, wind, and temperature are significant operational parameters: precautions must be taken to provide adequate drainage, fencing must be installed to prevent the blowing of litter, prolonged cold temperatures can make excavation difficult, and extremely wet weather will reduce cover material workability and hinder refuse compaction.

The primary components of a sanitary landfill are the site itself, the equipment used, the personnel required, and the site structures. The landfill site should be within reasonable haul distance and provide ready access for refuse collectors. The equipment at a sanitary landfill is used for refuse handling, earthmoving, or placement and compaction. The number of personnel required for a landfill depends mostly on the size of the operation. From 4 to 6 laborers can deposit 100 cu yd of refuse per day. In addition to the foreman or supervisor, equipment operators, gateman, etc., an overall project supervisor, secretarial assistants, auditors, bookkeepers, and equipment operators may also be needed. The site structures should provide shelter and sanitary facilities for the operating personnel and storage for the equipment. Adequate fencing (to keep out intruders and keep in blowing refuse), sufficient lighting, and scales should be installed. Fire protection must also be provided either at the plant site or by local fire districts.

Auxiliary operations associated with a sanitary landfill operation may include shredding, grinding, or volume reduction of refuse. Shredding and grinding refuse results in the intimate mixing of both the decomposable and nondecomposable components of the waste. This mixing decreases the incidence of disease vectors by destroying insect eggs and larvae present in the refuse, and by making the refuse unsuitable to rats and other animals for forage and harborage. Shredding and grinding of the refuse also reduces odors, diminishes the incidence of blowing litter, and allows easier handling and compacting of the refuse.

Volume reduction prior to disposal in a landfill may be accomplished in various manners: in home compactors, in refuse compactor trucks, at transfer stations, and occasionally by bailing at the landfill site. Volume reduction increases the efficiency of collection systems and extends the life of landfill sites.

The limiting parameters for sanitary landfills are land availability, cover material availability, gas and leachate production, nuisances created by improper operation, and public antipathy.

Land availability is the most critical limiting parameter. Ever-increasing urbanization has made both land and cover material required for sanitary landfill operation rarities. Increasing transportation costs coupled with the scarcity of suitable land and cover materials make sanitary landfilling an unlikely long-range solution to the total municipal solid waste problem.

Leachate is a potentially serious pollutant for ground and surface waters, and must either be prevented from forming or collected and treated. The hazardous or noisome gases created during the decomposition of refuse must also be collected and vented to the open air or burned.

An improperly run landfill is often a public eyesore. Nuisances such as blowing litter, odors, open fires, and disease vectors have generated extensive public opposition to sanitary landfills. Public education programs coupled with model landfill operations are usually required to gain public acceptance.

Consideration of the future use for completed landfill sites has seldom been incorporated into landfill design. Completed fills evolve into either recreational sites, agricultural lands, or commercial or industrial parks because of a lack of adequate land use planning. The use of completed fills for agricultural activities is rare because of the detrimental effects of the decomposing wastes on plants. It would seem that proper planning could significantly contribute to the logical land use of completed landfill sites.

Sanitary landfilling is an advantageous method of solid waste disposal since there is no large capital outlay for a physical plant, labor costs for the process are not extremely high, the operation does not produce a residue or by-product, and the equipment utilized in the landfill process is standard construction industry equipment.

The disadvantages associated with sanitary landfilling include: the production of water, air, and land pollution through poor operating procedures, the development of public disfavor because of the association with poorly maintained sites, the continuing reduction in the amount of land available for this method of solid waste disposal, and loss of potentially valuable materials and energy.

9-1.2 Incineration

Incineration is a combustion process that reduces solid, liquid, or combustible wastes to carbon dioxide, other gases, and a relatively inert residue. An accompanying landfill is required to dispose of the incinerator residue.

Incineration is probably the second oldest method of solid waste disposal known to man. The earliest form of incineration most likely originated when early man discovered he could burn waste material rather than dump it outside his immediate living area. This process disposed of solid waste and also provided warmth.

The first municipal incinerator is reported to have been constructed in Nottingham, England in 1874. A government facility constructed on Governor's Island, New York Harbor in 1885 is reported as being the first American incinerator. The use of incineration as a solid waste disposal method in the United States increased drastically during the first several decades of this century, and in 1920 there were more than two hundred plants in this country.

The operational parameters of an incinerator include waste characteristics, combustion time, turbulence, combustion temperature, air supply, auxiliary fuel, and effluent production. The moisture content of the waste determines the amount of extra heat that may be needed to complete the combustion. The combustion time is regulated by adjusting the waste travel rate through the combustion chambers. Turbulence is required in an incinerator to provide thorough mixing of the waste with air. Without turbulence, combustion would not be complete. Baffles, constrictions, and tumbling of waste from the grates are means of providing turbulence.

One of the most important factors in the combustion process is temperature. Temperatures in the combustion chamber of a conventional incinerator range from 1600 to 2000°F. The completion of the combustion process and control of the temperature in the incinerator both require air. Air is added to the incinerator by natural draft or fans. Openings in the furnace area below the grates supply "underfire" air and in the upper part of the furnace ports are provided for "overfire" air. Underfire air supplies oxygen for the combustion process, while overfire air provides necessary turbulence and temperature control.

Auxiliary fuels are desirable for furnace warm-up, primary combustion promotion when the moisture content of the waste is high, secondary combustion to control odor and smoke due to incomplete combustion, and supplemental heat for recovery units when the solid waste is deficient in heat value. Gas or oil usually serves as the auxiliary fuel.

Incineration of solid waste produces waste gases such as nitrogen dioxide, nitric oxide, sulfur dioxide, sulfur trioxide, and carbon monoxide that are air pollutants and also contain entrained particulates. Incinerators must be provided with air pollution control devices so that the discharge of these undesirable gases and particles can be controlled. The incineration process also generates wastewater from residue quench water and air pollution control systems which must be treated prior to disposal.

Poor design and operation of incinerators in the past have led to adverse public opinion. Public acceptance of an incinerator can be enhanced by the selection of an acceptable site—a location within economical haul distances from the waste generation area where no other activities are disrupted, no traffic impediments are created, and residential influx around the site is prevented.

The basic components of an incinerator include: scales, a storage pit and tipping area, cranes, charging hoppers, furnaces, grates, a residue removal system, and air and water pollution control equipment.

Scales are used to weigh incoming solid waste, residue, and salvable materials. Improved operations, assistance in management control, planning facilitation, and equitable fee assessments are some of the things that can be accomplished when accurate and meaningful weight records are employed in an incineration operation.

The storage pit should hold at least several days' collection volume and should be enclosed to exclude weather and confine odors, dust, and noise. The tipping area should be large enough to permit rapid unloading and should have sufficient clearance to allow the unloading of the largest vehicle in the collection system.

Cranes fitted with grab buckets or grapples are used to convey the refuse from the storage pit to the furnace charging hopper. They are usually powered by electric motors and are best controlled by operators in an overhead cab.

Usually, each furnace cell is provided with its own charging hopper. Most charging hoppers have the shape of an inverted, truncated pyramid; they are generally constructed of steel and are sometimes lined with concrete. The most commonly used furnaces for refuse incineration are the rectangular furnace, the multicell rectangular furnace, the vertical circular furnace, and the rotary kiln furnace.

Grate systems in an incinerator are used for drying the waste, igniting the waste, tumbling the waste to create turbulence, and burning the waste.

Air and water pollution control equipment used in incinerators is designed to prevent or control detrimental liquid and gaseous emissions. The incinerator designer must consider what type of emissions will be encountered and then select the appropriate control device.

Important process requirements for incineration include power, water, land, and personnel. Particular utility requirements associated with an incinerator encompass auxiliary fuel, sewers, potable water, electricity, and communications.

The incineration process can have both desirable and undesirable by-products. The desirable by-products include salvable materials such as metals, glass, and heat for power production. Undesirable by-products are odors, dust, litter, wastewater, flyash, noxious gases, and the residue end product.

In summary, incineration is still a popular mode of solid waste disposal in the United States because of several process advantages. The land requirements associated with the process are small, the operation of an incinerator is not dependent upon weather conditions, the haul distances from solid waste generation areas can be reduced because of incinerator location in urban industrial areas, significant volume reduction is accomplished, landfill requirements are reduced, and a stable odor-free residue is produced.

The disadvantages of incineration include: high construction and operating costs, the production of air and water pollutants, the requirement of highly skilled employees to insure proper incinerator operation, the need for continuing maintenance of the plant, and the necessity for the disposal of noncombustible residue.

The use of simple incinerators as a method of solid waste disposal will probably decrease in the near future because of increasingly stringent air pollution regulations. These regulations will increase the cost of air pollution control equipment associated with the incineration process, thereby increasing the capital costs of the total incineration system. Such an increase in capital costs would probably cause the economics of incineration to be uninviting to most municipalities selecting a new method of solid waste disposal. However, heat recovery incineration for energy conservation may be more successful and more appealing to municipal officials.

9-1.3 Composting

Composting is the biochemical degradation of organic wastes. The end product of composting is a humuslike material that can be used primarily as a soil conditioner. The first applications of composting occurred in Oriental countries. Vegetable matter and animal manures were placed in piles and allowed to decompose. The first systemized composting processes were developed by Sir Albert Howard and Becarri.

Since the 1920s a great deal of information has been accumulated concerning the European practice of refuse composting. This information is not directly applicable to the solid waste disposal situation in this country, however, because of the significant compositional differences between American and European refuse and differing demands for soil conditioners.

Since composting entails the degradation of organic compounds by naturally-occurring microbes, a variety of microorganisms are involved in, and are necessary to, the composting process. The key to commercial composting is to provide an environment in which these microorganisms can perform most efficiently, thereby reducing the time required for stabilization. The most important parameters of the composting environment are temperature, the amount of oxygen available, the nutrient content of the waste, the moisture content of the waste, and the pH of the waste.

There are five basic steps involved in all composting operations: preparation, digestion, curing, finishing, and storage or disposal. Differences between various composting processes arise from the method of digestion and/or the amount of preparation and finishing required.

There are five basic parameters that limit composting as a viable method of solid waste disposal, including: the physical and chemical composition of the refuse, land availability, the size of the community served, the need for secondary disposal, and the existence of a compost market.

Composting depends upon the existence of certain microbes and nutrients in the matter being composted. The necessary microbes exist in sufficient quantity in all domestic refuse while the necessary nutrients, when not already present, can be easily supplied through the use of wastewater sludge or other appropriate additives.

Increasingly, composting plants are depending upon the sale of recycled materials as a major source of income. However, recycling requires extensive sorting and cleaning. Adequate nearby markets for the recycled items must be available in order to make separation of recycleable materials a profitable operation.

Available land for windrow composting plants may be difficult to find in large metropolitan areas. The appearance of windrowed refuse will most likely be objectionable to people living in the area. Mechanical systems, however, can be operated in an enclosed area occupying about 8 to 10 acres and need not create objections if properly managed.

The size of the service community influences the operation of a composting plant. The largest plant constructed in the United States to date has had a capacity of 300 tons per day, and could serve a population of only 90,000. Manufacturers of existing compost systems have shown no desire to build any larger than the 300 ton per day system, and at least one has indicated 200 tons per day as the "breakpoint." The basis for this limitation apparently is related to compost market size, off-shelf equipment availability, and physical constraints.

If composting is chosen as a primary disposal method, a secondary disposal process, usually landfill, must be provided for the disposal of noncompostable, nonrecycled wastes and unsold compost. Approximately 20 to 30 percent by net weight of municipal refuse consists of noncompostable matter which must be landfilled.

The single most important limiting factor is the need for a viable market for compost. Without a market for the finished compost, composting becomes very uneconomical when compared to other refuse disposal systems. No viable large scale compost market has developed in the United States to date.

Composting by-products fall into two categories: those by-products which are the result of improper management, and those which are a natural result of the process. Improper management by-products include odors, disease vectors, noise from comminution equipment, and pathogens in compost. Odors and disease vectors (flies, rats, etc.) usually appear when the refuse is allowed to remain too long in the storage area. Any composting operation will produce noise from comminution equipment. However, isolation of noise sources and proper use of insulating materials will alleviate this problem. Pathogenic bacteria will not be a problem if the temperature is kept at sufficiently high levels (above 140°F). To date, there have been no reports of any worker or handler of compost having been infected. Proper by-products of a composting operation include the 20 to 30 percent of municipal solid waste that is noncompostable and the humuslike compost itself. If the compost cannot be sold, it will have to be landfilled. With the cost of composting at $10–$20 per ton, the additional cost of landfilling would make a composting operation prohibitive for most areas of the country.

A composting plant has three physical requirements: land, power, and labor. There are two basic land requirements for a compost plant: the plant site and an area for secondary disposal. Power needs are a function of the degree of mechanization and capacity of the plant. Labor requirements are directly related to plant capacity. As a general guide, a modern 200 ton per day plant will require 20–25 employees. Generally speaking, windrow systems require more land and labor, while mechanical systems need more power and a smaller but more highly skilled work force.

Equipment necessary for the operation of a composting plant includes: scales, tipping facilities, and conveyors; sorting equipment; comminution equipment; and finishing equipment.

In summary, the composting process has apparently been doomed to failure in the United States for two reasons: the biological decay of American refuse is harder to attain because of its high cellulose content, and synthetic fertilizers are cheaper and easier to use than most of the products derived from the composting process. However, with the rising costs of landfilling and incineration, composting may become an economical method of size reduction and resource recovery prior to landfilling in the future.

9-2. INNOVATIVE DISPOSAL METHODS

The great majority of solid waste disposal today is accomplished by conventional methods, primarily sanitary landfilling. However, with land for solid waste disposal rapidly becoming scarce in urban areas, and with the depletion of natural resources, a search is on for new methods for disposition of solid waste. Proposals have included both new methods of disposal and the utilization of solid waste as a raw material. Two innovations in disposal, medium-temperature and high-temperature incineration and landfilling with leachate recirculation, are among the methods currently being proposed as superior alternatives to conventional disposal methods.

9-2.1 Medium-Temperature and High-Temperature Incineration

A new development in incineration, the so-called "high-temperature incineration" process, consists of combustion and fusion of wastes at temperatures between 3000 and 3300°F, which is 1200 to 1500°F higher than combustion temperatures in conventional incinerators. The residue from this process is approximately 2 percent of the initial refuse volume and is an inert slag. Flyash collected by air pollution control devices amounts to about another 2 percent of the original volume.

The principal objectives of high-temperature incineration include: maximum volume reduction of solid wastes; complete combustion producing a solidified slag that is sterile, free of putrescible matter, compact, dense, and strong; complete oxidation of the gaseous products of incineration with discharge to the atmosphere after adequate treatment for air pollution control; possible salvage and reuse of the residue; and a minimum use of water, with minimum pollution of groundwater and streams.

Fusion of the noncombustible residue can be accomplished at temperatures in excess of 2600 to 2800°F, with the actual temperature dependent upon the composition of the refuse. However, to insure adequate fluidity of the slag, temperatures approaching 3000°F should be maintained. At the high temperatures necessary to cause total incineration, deposits of clinker and slag form on the furnace walls and linings. These deposits can cause serious damage by falling and jamming grates, and causing the refractory to wear, melt, and cave in.

Air pollution from a high-temperature incineration system can be controlled, but costs may be higher than for conventional incineration since a high-temperature system may produce higher particulate loadings and more nitrogen oxide emissions.

The term "high-temperature incineration" has also been applied to several units which operate at temperatures only slightly above the operating temperatures of conventional incinerators. These medium-temperature incinerators (such as the General Electric Vortex incinerator in Shelbyville, Indiana) operate well below the 3000°F range of slagging incinerators and may not accomplish volume reductions equivalent to high-temperature incineration.

The critical and limiting process parameters of high-temperature incineration are the same as for conventional incineration.

The components of a high-temperature incinerator include refuse conveyors, furnaces, and residue removal systems very similar to those in conventional incinerators. However, in a high-temperature process furnaces are sometimes classified as gasifiers because they convert the refuse into gas and a liquid slag.

The by-products of high-temperature incineration are gases and solidified slags. The gases are processed in air pollution control devices. The slags are either disposed of or recovered. In one case the slag has been spun to form a "wool" which may have use as an insulator. Other uses for the slag have been suggested, including the manufacture of foam insulation and various types of building materials.

Preliminary results of high-temperature incineration pilot plant studies indicate that various advantages can be claimed for the process. These advantages include: a residue free of putrescible matter, maximum density in accompanying landfill operations, a minimal pollution of groundwater and streams, the potential for use or salvage of residue, and a minimal water use.

High-temperature incineration pilot plant operational problems due to the compositional variability of the solid waste stream coupled with the process' large fuel consumption detract from its attractiveness as a future means of solid waste disposal. However, if operational difficulties can be overcome, the process may possess full-scale applicability.

9-2.2 Landfill with Leachate Recirculation

This innovative disposal technique is basically a sanitary landfill in which the leachate is recirculated through the deposited refuse with ultimate treatment and disposal of the collected leachate after waste stabilization. Leachate recirculation in a landfill results in rapid stabilization of the organic fraction of the deposited refuse because of the accelerated growth of an anaerobic biological population.

During the leachate recirculation landfill process, the moisture content of the solid waste is increased from 25–30 to 65–70 percent so that anaerobic microbial activity can be maximized. The moisture content of the refuse may be increased by adding water to the cell concurrently with the placement of the solid waste in the cell. However, water added in excess of the field capacity of the refuse will drain from the cell, and subsequently, leachate recirculation is required to maintain an elevated refuse moisture content in the cell.

The basic differences between the construction and operation of a conventional sanitary landfill and a leachate recirculation landfill are the collection, recirculation, and treatment of leachate and the monitoring of process parameters such as biochemical oxygen demand, total organic carbon, chemical oxygen demand, total dissolved solids, total solids, alkalinity, acidity, pH, hardness, volatile acids, phosphorus, nitrogen, chloride, sulfate, calcium, magnesium, manganese, sodium, iron, and moisture content.

The basic requirements for a leachate recirculation landfill operation are similar to those for a conventional landfill operation with the exception of the leachate recirculation system utilized in the recirculation process. This system includes an external water source, a pumping station, a holding lagoon, and a distribution network.

There are only two by-products from a properly operated leachate recirculation landfill: recirculated leachate, and gases emitted during anaerobic digestion. Recent treatability studies concerning leachate from conventional landfills have shown that such leachate could be adequately treated in an extended aeration activated sludge plant. Since leachate derived from a leachate recirculation landfill has been shown to exhibit higher water quality characteristics than conventional landfill leachate, no problems are anticipated in treating leachate recirculation landfill leachate in conventional biological wastewater treatment facilities.

The gases generated in a leachate recirculation landfill may be vented to the atmosphere according to Environmental Protection Agency standards, or methane may be collected and utilized as a fuel source to generate power.

In summary, the leachate recirculation landfill process appears to be a technically feasible method of solid waste disposal which alleviates the potential for groundwater pollution usually associated with landfilling operations and returns the site to a usable land area in a relatively short period of time. The results of more pilot plant and full-scale operations will ultimately determine the future of leachate recirculation landfilling as a viable disposal method.

9-3. MATERIALS RECOVERY

A great deal of the material in the solid waste stream is potentially recoverable, and solid waste management personnel are seeking feasible and practicable materials recovery systems. To date, the success of materials recovery efforts has been extremely variable; outstanding success in recycling certain materials has been accompanied by dismal failures in attempts to reuse other waste materials.

The secondary materials which have the potential of being recovered and recycled from the solid waste stream may be classified into two general categories: those materials amenable to direct recycling, and those materials which require processing before reuse.

9-3.1 Direct Recycling

Most materials recovered today are retrieved through direct recycling programs. Currently, most direct recycling in the United States is done by industry itself or by long-established scrap dealers. Ferrous metals account for approximately 80 percent of scrap materials being recycled and 48 percent of the dollar value of secondary materials. Paper accounts for 14 percent of the volume and about 4 percent of the sales value. Conversely, nonferrous metals contribute only around 4 percent of the volume but nearly 45 percent of the sales value. Textiles account for only an insignificant amount of the volume and sales. Direct recycling of secondary materials faces two major problems: uncertain markets and unfavorable freight rates. The future for recycling metals, both ferrous and nonferrous, looks optimistic; however, the future for recycling other wastes such as glass, plastics, and paper looks very uncertain.

9-3.2 Separation of Refuse Components

In many localities there exists no public program to manage and encourage direct recycling of recoverable materials. Because of the lack of facilities for direct recycling, the lack of a public demand and support for direct recycling, and the difficulty of separating many materials from the solid waste stream, there is a need for separation and recovery of components of the solid waste stream from that stream itself in a centralized processing facility.

The heterogeneous nature of solid wastes is perhaps the most dominant characteristic of the waste materials produced in this country today. This creates considerable difficulties in resource recovery. Most of the equipment available at the present time for separation of wastes has been designed to operate on relatively homogeneous material (whether by size or another characteristic). Because of this, it becomes necessary to consider an additional step in the processing of mixed solid wastes for materials recovery, i.e., size reduction.

For the most part, current applications of equipment to size reduction of solid wastes have centered almost entirely on the application of hammermills. Some use of shredders, rasp mills, and wet pulpers has also been attempted with varying degrees of success.

The techniques used in separation of materials in solid wastes today are all based upon physical-chemical characteristics of the materials which generally fall into one of the following categories: particle size, bulk density (specific gravity), electromagnetism, electrical conductivity, color, signature methods (i.e., resiliency), and chemical separation (most commonly combustion). The key to success in separation of materials is the selection of a given material characteristic which will be significantly different in value for the desired material as opposed to the remainder of the waste stream.

The primary methods of separating recoverable materials from mixed municipal refuse include hand separation and separation by particle size, density, electromagnetism, and optical properties.

Most separation equipment and separation methods have been developed in manufacturing and mineral extraction processes. Few if any have been developed specifically for the treatment of solid waste. The application of this equipment to solid waste has not met with complete success and has little prospect of such success. Modification of existing equipment and development of new equipment must be accomplished if adequate materials recovery is to be achieved.

A considerable potential exists for resource conservation through the separation and recovery of various constituents of the solid waste stream. However, methodology and equipment necessary for such separation have not been developed at the present time. Perhaps the most hopeful sign to date with respect to materials recovery from solid waste has been the theoretical development of large systems for materials recovery from wastes.

9-3.3 Materials Recycling Systems

Certain materials recovery operations are intended for use on raw, untreated solid wastes and are therefore termed "front end" systems. All solid waste disposal operations except sanitary landfilling produce some residue which may contain recoverable components. Processes for recovery of these valuable components from this residue are termed "back end" systems. Typically, front end systems recover only about 15 to 20 percent of the incoming wastes. Therefore, the bulk of the wastes remain for further recovery of resources, either as reused materials or as recovered energy. In general, more work has been done and more success has been achieved in operating on raw refuse than in back end systems for processed refuse. Back end recovery of material is considerably more difficult and more expensive because of the intimate mixing of organic and inorganic components of the remainder of the refuse.

After extensive research, U.S. Bureau of Mines personnel proposed a dry separation method of materials recovery for a front end system. Unit processes included in this system are shredding, magnetic separation, air classification, screening, and dense media separation. The advantages of this system include no wet operations, the production of a large number of usable recycled materials, and the input requirements of only electrical power and a suitable medium for the dense media separator.

A back end system also developed by the Bureau of Mines is intended for the recovery of materials from incinerator residue. Many of the unit operations and unit processes used on the incinerator residue are similar to the operations for the dry front end system. There are some differences, however, including the use of water as a medium in screening operations. The materials recovered from the residue consist of iron concentrates, pieces of aluminum, copper and zinc particles, and particles of clear flint glass and colored glass. Additionally, a carbonaceous ash is obtained. The operating costs of this system are estimated to be $13.30 per ton for a 400 ton per day plant (1973). Such a plant would have a capital cost of approximately $2 million.

Several other materials recovery systems have been developed by private corporations and include the Hercules, Inc., System, the National Center for Resource Recovery Network, and the Black Clawson Process.

The Hercules System is designed to process 500 tons of mixed municipal refuse per day. Following size reduction, ferrous metals are removed from the waste stream by magnetic separation. The remaining refuse is composted, after which the organic humus is separated from the noncompostables. The nondegradable organic material is then separated from the inorganic fraction and pyrolyzed to form oils, gases, and a carbonaceous char. Process residue then undergoes separation into nonferrous metal, glass, and sand fractions.

Presently (1974), this system is only in the construction stage and has not been operated at the full-scale level. Technically, the system is feasible and could be very attractive as a future solid waste disposal method. However, more study, laboratory testing, pilot plant operations, and full-scale operations are needed to fully evaluate the system.

The National Center for Resource Recovery (NCRR) derives its support from packaging material users and manufacturers. NCRR is proposing the development of a system of twelve plants throughout the United States for recycling materials, each with an input capacity of 500 tons per day. The basic concept of the plants is the combination of a front end system of materials recovery with a processing operation (incineration, landfill, etc.) which ideally is already in existence in a given locality. NCRR's front end system consists of size reduction followed by air-classification. The resulting heavy fraction then undergoes magnetic separation, screening, optical separation, and dense media separation designed to recover ferrous metals, nonferrous metals, and glass. The light fraction derived from the air-classification unit feeds the processing operation.

Particular emphasis has been placed on the combination of the front end system with an existing combustion operation designed to produce power. Ideally, the waste would proceed through the front end materials recovery system and the combustible remainder would be fed into an existing utility boiler for the production of electrical power as has been accomplished by the Union Electric Power Company in St. Louis, Missouri. This NCRR system is in existence on paper only; however, it includes the best features of the materials reclamation systems developed over the last decade. Additionally, it has the advantage of being flexible with regard to the waste processing operation which is designed to handle the material remaining after initial materials recovery.

The Black Clawson Hydrasposal–Fiberclaim pilot plant facility is located in Franklin, Ohio and consists of a wet pulping operation similar to that used in the paper industry. In the 150 ton per day plant, refuse is conveyed from a storage area to the Hydrapulper by means of a depressed conveyor or belt.

In the Hydrapulper the refuse is mixed with water and reduced to a size small enough to pass through $3/4$ in. diameter openings. Metals and other materials larger than these openings are collected in a junk remover. Ferrous metals are then removed magnetically from the junk remover residue for potential recycling. Nonferrous metals are currently landfilled, but plans are underway to install nonferrous metal recovery facilities (1974).

The waste-slurry from the Hydrapulper is passed through a liquid cyclone where glass and other dense materials are removed. These materials possess a potential for recycling but are also currently landfilled (1974).

The remaining slurry enters the Fiberclaim unit where the longer paper fibers are retrieved and sold to a nearby roofing company for use in the production of roofing material. The long fibers account for approximately 20 percent by weight of the total paper from the incoming waste. The remaining 80 percent of paper and other materials are dewatered in a screw press and combusted in a fluid bed incinerator. Combustion is claimed to be 100 percent complete in this reactor, with no residue. Exhaust gases then pass through a wet Venturi scrubber.

Several disadvantages of this wet-pulping operation may be noted. This process produces a relatively large wastewater stream. This wastewater is high in organic pollutant levels and the treatment of this wastewater stream requires both expensive and sophisticated treatment systems. Total process economics do not appear to be favorable. These disadvantages will have to be resolved before this system can be termed successful.

In summary, materials recovery operations may help to enhance resource conservation and recovery efforts. However, further research focused on the optimization of separation unit processes is needed. The entire field of materials recovery is currently in a state of flux because of the fluctuating markets for secondary materials. Materials recovery will probably increase in the future since freight rates and tax benefits will probably be changed to favor recycled materials, public opinion is likely to favor an increased usage of recycled materials, and dwindling supplies of virgin materials should increase the unit value of materials recovered from the solid waste stream.

9-4. ENERGY RECOVERY

Today, the United States is facing an "energy crisis." Fossil fuel supplies have not met demands and new energy sources are being sought. Solid waste is now being investigated for its potential as an energy source.

The heat values for mixed municipal refuse range from 3000 to 8000 BTU/lb. New methods to exploit energy found in solid waste are being developed. Solid waste is being used as a supplementary fuel with coal in the generation of electricity, as a heat source in the production of steam, as a direct source of power for the generation of electricity, and as a natural gas source.

9-4.1 Refuse as a Supplementary Fuel

In 1967, engineers from the Union Electric Power Company (UEPC) in St. Louis, and engineers from Horner and Shifrin, Inc., Consulting Engineers discussed the possible use of municipal solid waste as a supplementary fuel source. The city of St. Louis showed an interest in the experiment and applied for and received a grant from the Environmental Protection Agency to cover part of the cost. In April, 1972, full-scale operations began.

Before delivering the refuse to the Union Electric Company plant, the city of St. Louis sanitation technicians process the refuse in the city's own facility. The refuse is initially fed into a hammermill after which it is processed through a magnetic separator which removes the ferrous metals. The refuse is then conveyed to a stationary compactor for loading into self-unloading trucks, for transport to the power plant. The city of St. Louis is currently (1974) installing an air-classification system for use in the refuse processing facility. After the air-classifier installation is completed, the refuse will go from the magnetic separator to the air-classifier. The combined system will then remove metal, glass, oversize pieces of wood, plastic, etc., which represent about 20 percent of the collected refuse. The remaining 80 percent of the refuse will be light, easily burnable particles.

At the power plant, a pneumatic feeder transports the refuse to a surge bin. The surge bin meters and divides the refuse into four streams for feeding into the pneumatic transport system. The refuse is burned in two 20 year-old Combustion Engineering boilers which have a nominal rating of 125 megawatts each. Refuse is fired at a rate equal to 10 to 15 percent by heat value of the full-load fuel requirement of the unit. As a rule-of-thumb measure, for a 10 percent BTU ratio of refuse to coal a 100 megawatt unit will consume 250 tons of waste as supplementary fuel per day. The city of St. Louis refuse processing facility was designed to handle 300 tons per day of raw refuse and had capital costs of $2.5 million. Capital costs for the UEPC facility were $600,000. Estimates have set capital costs at $5.3 million and operating costs at approximately $5 per ton of refuse for a 1750 ton per day refuse processing facility.

The critical and limiting process parameters for this process are the same as those involved in the incineration process except that the degree of separation of glass and metals affects the operation of such a facility. Obviously, the fuel value per ton of the refuse is increased when glass and metals are removed. The removal of glass and metals will also prolong the life of the waste transporting pipes. To date, there have been no significant adverse effects on the boilers using the prepared refuse as a supplementary fuel.

The most troublesome problems associated with this system have been accelerated pipe wear caused by glass and small metal pieces striking the bends of feeder pipes, and blockage of pneumatic feeders by the larger and denser particles. The installation of the air-classification device should help to alleviate both problems. Other, less serious problems have been the incomplete burning of soggy organic materials such as corn cobs and orange peels, and some slight mechanical difficulties with the refuse grinding operation.

Advantages claimed for this system include:

1. A reduction in coal consumption, projected at about 400 tons per day for the St. Louis installation, with a resultant minimum savings of $8000 (1974) per day (low-sulfur coal at $20.00 per ton).
2. A reduction in SO_x production because of the low sulfur content of the refuse and apparent precipitative action of lime-rich solid wastes (precipitation of $CaSO_4$ appears to be occurring within the combustion chamber).
3. The potential for recycling glass and metals.

The use of refuse as a supplementary fuel must be considered one of the most promising of the energy recovery processes developed to date. The decrease in low-sulfur coal resources and the passage of more stringent air pollution control regulations will no doubt make this process even more attractive to cities in the future. Accordingly, major cities such as Chicago, New York, and Toronto are now contemplating the installation of such a system. Also, the St. Louis system is being expanded to serve a much larger area.

9-4.2 Waste Heat Recovery

The recovery of waste heat from an incinerator in the form of steam has previously been a predominantly European application of solid waste disposal. Waste heat recovery in the United States is a rather recent development, but is being more widely used each year. The principal reasons for the increased popularity of waste heat recovery systems include:

1. The heat value of the solid waste generated in the United States is increasing.
2. Water-wall furnaces used in heat recovery incinerators generally require less maintenance than that required for a comparable conventional refractory furnace.
3. Air pollution control is easier to maintain because less combustion air is required in the furnaces, and because the exhaust gases are cooled, and thus their volumes reduced, in the waste heat boilers.

The critical and limiting parameters of a waste heat recovery and utilization system are much the same as those for a conventional incinerator. The primary difference in critical parameters between a conventional incineration system and a waste heat recovery system is the marketability of the steam or hot water produced by the waste heat system. For economic reasons, the plant should be located as near as possible to prospective steam purchasers. A waste heat system is severely limited as a feasible means of solid waste disposal if the steam is not purchased.

The simplest method of recovering waste heat from existing or new incinerators is a boiler immediately following the combustion area of the incinerator. The major problem incurred with this system is that many refractories cannot withstand the high temperatures necessary for the production of steam.

The refractory problems can be alleviated by the construction of an integral boiler in the combustion chamber of the incinerator. The principal problems with this system are corrosion and pitting of the boiler tubes. Maintaining the water temperature in the tubes at a level of at least $300°F$, well above the dew point of the corrosive gas in the furnace, and deaerating and softening the water in the tubes will minimize these problems. The waste heat boiler plus the usual components of a conventional incinerator form the typical components of a waste heat recovery system.

In summary, waste heat recovery systems in the United States will probably increase in number with the depletion of fossil fuel supplies and corresponding fuel price rise. The success of these systems is primarily dependent upon long-term markets for the steam or hot water generated by the process.

9-4.3 CPU-400

The CPU-400 is a system being developed by the Combustion Power Company. The system is designed to recover energy from the combustible portion of solid wastes in the form of electric power through the use of a gas turbine-powered electric generator. Also, the CPU-400 system will include recovery of salvable noncombustible portions of the waste. According to Combustion Power Company engineers, a CPU-400 system will be capable of generating about 5 percent of the electrical power needs of the community supplying the waste. According to manufacturer forecasts, the income from the sale of electric power and recycled materials will substantially reduce the cost of solid waste disposal.

The CPU-400 system is made up of fluid bed combustor/gas turbine modules. Each module is capable of handling 150 tons per day of solid waste. The modular design will allow plant sizes to vary from 150 to 1500 tons per day. The only existing facility (1974) is an 80 ton per day pilot plant located in Menlo Park, California.

In the CPU-400 system, the solid waste is dumped in an enclosed receiving area. Front-end loaders push the waste onto conveyors which carry it directly to shredders. The shredded waste is then air-classified. Shredded metal and pulverized glass are separated and conveyed to the materials recovery module for further separation. Light materials, mostly paper and plastics, are pneumatically conveyed to the shredded refuse storage container. This container can hold several hours production of solid waste and is designed to deliver a steady flow of materials to the fluid bed combustor.

The shredded waste is fed to the combustor by high-pressure air-lock feeders. Combustion occurs at 1500°F. This temperature is intended to preclude the formation of air pollutants such as unburned hydrocarbons, oxides of nitrogen, and acid gases. Particulate matter is removed from the hot combustion gases in a two-stage inertial separation system. The entire operation is enclosed. A 450 ton per day unit would occupy about 2 acres and would have a capital cost of $15,000 (1974) per ton per day capacity.

The critical and limiting parameters of the CPU-400 system are primarily the composition of the refuse and the availability of markets for the generated electricity and recycled materials.

Two characteristics of the refuse are important: the amount of salvable materials (primarily metals and glass), and the BTU content of the combustible materials. Salvable materials in the refuse are potential sources of income. The BTU content of mixed municipal wastes ranges from 3000 to 8000 BTU/lb. The higher the BTU content, the more electric power which can be generated.

Markets for the generated electricity and the recycled materials are also important. The economic feasibility of the CPU-400 is based on its ability to recover energy from the solid waste, with sales from recycled materials considered an additional income.

The planned process by-products are the generated electricity and salvage materials. The 80 ton per day pilot plant produces about 1000 kilowatts of electric power per day.

Although pilot plant operation has been successful, full-scale development of the CPU-400 is necessary. The anticipated increase in electrical energy demand may make the CPU-400 a viable long-term solid waste management system.

9-4.4 Pyrolysis

Pyrolysis is the decomposition of organic compounds through the application of heat in an oxygen-free environment. Organic matter is converted into gases, liquids, and inert char. Solid waste pyrolysis products occupy about 50 percent of the volume of the original matter, and can be converted into energy to sustain the process and/or produce excess power.

The application of pyrolysis to energy recovery from solid wastes is a relatively new development. No pyrolysis method has been operated on a full-scale basis; however, plans are proceeding (1974) for the construction of at least three proprietary pyrolysis systems for municipal waste disposal.

The existing pyrolysis methods can be divided into two categories: high-temperature and conventional. Conventional pyrolysis can be further broken down into gas-producing and liquid fuel-producing systems.

Existing high-temperature pyrolysis systems are little more than high-temperature incinerators. The basic difference is that in pyrolysis, the product gases are burned separately from the waste in an afterburner; in high-temperature incineration, product gases are combusted along with the waste. For this reason, pyrolysis generally is more efficient in the recovery of heat produced through gas combustion.

In a typical high-temperature "pyrolysis" plant, as-received refuse is charged into a reactor through an air seal. Heated air is introduced to the bottom of the reactor; as the refuse descends through the reactor, it is dried, partially volatized, oxidized, and finally melted. Temperatures in the reactor of about $3000°F$ produce a molten slag in the bottom of the reactor and pyrolyze the refuse charged into the top of the reactor. The molten material is collected in a quench tank to form a glasslike aggregate.

The gases produced in the reactor are removed and burned in a secondary combustion chamber. Here temperatures exceed $2000°F$. The hot gases from the afterburner may be utilized to produce power and/or sustain the process.

For the conventional pyrolysis processes designed to optimize gas production, the first step is the shredding of the as-received refuse. The shredded refuse is continuously fed into a reactor opposite to the direction of flow of hot gases. The temperatures for this method are between 1500 and 2000°F. This is sufficient to pyrolyze the refuse into gases and solid residue.

At the end of the process, the solid residue is collected. If front end separation has not been performed on the refuse, back end separation can be practiced on this residue. The product gases are collected and are used to produce steam and/ or run gas turbine generators.

Another conventional pyrolysis method is designed to produce liquid fuel oil. This method employs a flash pyrolysis process which is designed to produce 1 barrel of synthetic fuel oil per ton of as-received refuse. This fuel is a low-sulfur oil with a heat value equivalent to 75 percent of No. 6 oil on a volumetric basis. In this method, refuse preparation is very important. The process requires a finely shredded, high organic content refuse for optimum operation. The refuse goes through the following preparation: primary shredding, air-classification, drying, screening, recovery of glass and metal, and secondary shredding. This refuse is then rapidly heated to produce fuel oil, gas, and solid residue.

There are four basic critical parameters associated with the pyrolysis of solid waste, including heating rate, temperature, waste composition, and the marketability of by-products.

The desirable by-products of pyrolysis are gases, liquids, and solid residues. The amounts and composition of each are controlled by the pyrolyzing method employed and are functions of temperature, retention time, and waste composition. These by-products are a possible source of revenue for the operation. Pilot plant operations indicate that undesirable by-products from the pyrolysis of solid waste can be held to a minimum.

When pyrolysis is used as a high temperature volume reduction process, it has little advantage over high-temperature incineration. Its potential as a pollution-free, self-sustaining process may provide it with some advantages over incineration. Conventional pyrolysis systems do not maximize volume reduction, but do convert the waste into a valuable resource. As the supplies of fossil fuels diminish, the pyrolysis of solid waste may provide an important source of power.

9-4.5 Methane Production by Anaerobic Fermentation

Laboratory studies have illustrated the feasibility of natural gas (methane) production from the anaerobic fermentation of solid wastes and wastewater sludge. Basically, the system consists of the "front end" separation of nondegradables followed by anaerobic digestion, dewatering, and ultimate disposal of the resulting residue.

The anaerobic fermentation process not only can produce methane but also results in an overall refuse weight reduction of about 40 to 50 percent. Consequently, there would also be process benefits attributable to the disposal and transportation of a reduced solid waste volume. Additionally, savings would be realized from disposing of wastewater sludge in the fermentation process. The projected revenue that could be generated from this process is estimated to be $0.44 to $1.74 per ton (1974) of refuse processed. Obviously, a comprehensive evaluation of the fermentation process must await pilot and full-scale application.

9-5. EUROPEAN PRACTICE

A brief description of European solid waste management practices has been included in this text for several purposes: it furnishes valuable comparative data on disposal and resource recovery in Europe, it explodes some prevailing myths, and it points toward potential improvements in American practice. The bulk of waste disposal in Europe consists of landfill operations, with the exception of Switzerland where incineration is predominant. Composting has been successful in some areas such as the Netherlands where compost finds ready markets; some composting is supported by the needs for soil conditioner (principally in vineyards) in France, Switzerland, and Germany. Incineration is the primary disposal method in central continental urban areas. Most large, modern incinerators in Europe are equipped for heat recovery. Recovered heat is used to generate steam for space heating and for the production of electricity. Modalities of heat utilization vary with local conditions. Grinding is being used to an ever-increasing degree in composting and landfill operations. Pyrolysis is being tested in experimental plants (very small scale). In Rome, materials and energy recovery are being practiced through metals and paper recycling, composting, and production of animal feed from refuse. Municipal waste management agencies are operated on a very professional basis with respect to provision of collection and disposal services. However, in many European nations, more than half of all generated solid wastes either are not collected or receive inadequate disposal on land.

9-6. SELECTION OF SHORT-TERM AND LONG-TERM SOLID WASTE MANAGEMENT TECHNIQUES

A basic need exists in most U.S. cities for a rational approach to the selection of a community's future solid waste management system. In most areas of the nation solid waste is currently being disposed of in poorly maintained landfill operations or open dumps. Improvement of solid waste disposal and resource recovery practices is necessary not only to meet increasingly stringent federal and local environmental standards, but also to possibly decrease solid waste management costs through proper planning. The selection of future solid waste management systems in this country is often made by political decision-makers with very little rational evaluation.

9-6.1 Short-Term and Long-Term Solid Waste Management Systems

The previous sections of this book have described the state of the art of both short-term and long-term solid waste disposal and resource recovery systems. Short-term systems may be defined as traditional solid waste disposal methods which are known to be technically feasible and able to be put into operation in a relatively short time period (1 to 2 years). Short-term solid waste disposal systems have been analyzed previously in this text with regard to their methodology, operational parameters, limiting parameters, by-products, equipment and personnel requirements, and economics and include:

1. Landfilling—either alone or with front end separation.
2. Incineration—either alone or with front end or back end separation.
3. Composting—either alone or with front end or back end separation.

Obviously an important characteristic to be considered when selecting a short range solid waste system is its compatibility with long range methods and planning.

Long-term systems, on the other hand, may be defined as those which have originated as innovative disposal or resource recovery techniques, currently show promise, require more detailed study before implementation, and consequently necessitate longer start-up times. Long-range systems have been described in this text with regard to their technical feasibility, critical operational parameters, process requirements, economic constraints, and feasibility of full-scale implementation. Long-range systems considered include:

1. Refuse as a supplementary fuel.
2. Waste heat recovery.
3. CPU-400.
4. Pyrolysis.
5. Methane production by anaerobic fermentation.

Obviously, any municipal official or design engineer has a large number of processes to choose from when formulating an area's future solid waste management plan. It is very difficult for any municipality to thoroughly analyze the applicability of all of these methods to its particular solid waste problem from both a technical and an economic point of view. Therefore, the development of a systematic preliminary evaluation scheme that would quickly select two or three solid waste disposal or resource recovery techniques that possess a high potential for the specific municipality in question would be invaluable. These resulting two or three processes could then be analyzed in-depth to determine which is the optimum solid waste management technique. The selection of short-term and long-term solid waste management methods would therefore be logical and systematic in its approach.

9-6.2   Development of a Selection Procedure

Probably the most pragmatic procedure that could be utilized to evaluate available solid waste management techniques for their applicability to a specific locality would be a matrix procedure. The development of an appropriate matrix would allow a municipality to record its specific characteristics (following the completion of a routine survey) and quickly determine which method or methods of solid waste disposal or resource recovery are properly suited to alleviate its solid waste management problems. Such a matrix would parallel ones developed by the Department of the Interior to assist in environmental impact assessment. Optimally, both short-term and long-term selection procedures could be formulated to complement each other.

One important consideration that must be addressed when developing a selection procedure of any sort is the range of municipal characteristics that will be pertinent to the selection of any solid waste disposal or resource recovery systems. Municipal characteristics of prime importance would include: population density, land costs, climate, and power availability. For example, low population densities would favor landfilling over most other methods. High land costs would detract from systems requiring large land areas such as landfilling or composting. Poor climatic areas (high rainfall or low temperatures) would favor enclosed operations over landfilling. Areas with high power costs would not be amenable to power-consuming methods, such as high-temperature incineration or wet-pulping. Other municipal characteristics of lesser importance in the selection of solid waste disposal or resource recovery methods would include: topography, geology, hydrology, and socio-economic conditions. These parameters may not be amenable to preliminary selection procedures; however, they should be addressed in the subsequent detailed evaluation.

Obviously, a second major consideration in the preliminary evaluation selection would be those process parameters that determine short-term and long-term process applicability and desirability. Important system parameters would include: materials recovery potential, energy recovery potential, start-up time, economics, and stage of development. Processes possessing high materials recovery potential (composting with front end separation) would be favored over those which would accomplish a negligible amount of materials recovery (conventional sanitary landfill). Likewise, high energy recovery potential systems (pyrolysis) would be more desirable than processes which recover little or no energy (wet-pulping). Methods having relatively short start-up times (landfilling) would be conducive to short-term operations. Naturally, one of the most important process parameters would be economics of the system. The cost of the system per ton of refuse processed would contribute heavily to preliminary system selection. The stage of process development (laboratory, pilot, or full-scale) would determine whether a process could be applied on a short-term or long-term basis. Obviously, a process which had not progressed beyond the small pilot-plant scale should not be considered to possess short-term applicability. Other process characteristics such as haul distances, transfer station requirements, etc., should be evaluated in the in-depth analysis which would follow the preliminary selection step.

The compositional characteristics of the refuse in the generation area must also be considered during the preliminary selection procedure. For example, the higher the paper and plastic content of the solid waste, the higher is its BTU value, and consequently the more amenable it would be to utilization in an energy recovery system. Similarly, a larger recoverable material percentage would suggest compatibility with a materials recovery system. On the other hand, a high refuse moisture content would be a deterrent to high-temperature incineration and favor leachate recirculation landfilling.

Any short range process selected must also be capable of smoothly phasing in with long-term systems. For example, a short-term conventional incineration process would not phase in well with long-term use of refuse as a supplementary fuel. Conversely, a short-term conventional landfill would be very compatible with a long-range pyrolysis process.

It seems apparent that a basic need exists for the development of a standardized, universally accepted, preliminary solid waste management selection procedure that would alleviate unnecessary and costly guesswork from this critical decision-making stage. The Advanced Solid Waste Study Group at the University of Louisville has formulated a preliminary solid waste management selection matrix which is currently (1974) in an embryonic state of development. Hopefully, a refinement of this matrix model may, at a later date, contribute to the enhancement of solid waste management decision-making.

9-7. SUMMARY

This text has presented an in-depth description and critique of the wide variety of solid waste disposal and resource recovery processes currently available. Because of the rapidly expanding technology that has developed in the solid waste management field during the last several years, it is mandatory that the practicing professional keep pace with this new wealth of information. It seems likely that the focus of solid waste management will continue to turn from simple refuse disposal to materials and energy recovery in the foreseeable future. Therefore, the informed solid waste management professional undoubtedly will be able to contribute immeasurably during the next decade to both the enhancement of environmental quality and the preservation of our natural resources.

Index

INDEX